A

The Haynes

# Small Engine Repair Manual

## 5.5 HP through 20 HP four-stroke engines

**by Alan Ahlstrand and John H Haynes**
Member of the Guild of Motoring Writers

**The Haynes Workshop Manual for small engine repair**

ABCDE
FGHIJ
KLMNO
PQRST

A
C.1

**Haynes Publishing Group**
Sparkford Nr Yeovil
Somerset BA22 7JJ England

**Haynes North America, Inc**
861 Lawrence Drive
Newbury Park
California 91320 USA

**Acknowledgements**

We are grateful for the cooperation of Briggs & Stratton Corporation and Tecumseh Products Company for assistance with technical information and certain illustrations. We also wish to thank the Power Equipment Center, Oxnard, California for their help. Technical authors who contributed to this book include Jay Storer and Robert Maddox.



**A book in the Haynes Automotive Repair Manual Series**

**Printed in the U.S.A.**



**ISBN 1 56392 298 3**

**Library of Congress Catalog Card Number 98-87691**



98-320

# Contents

## Chapter 6 Briggs & Stratton engines (continued)



## Chapter 7 Tecumseh/Craftsman engines



## Chapter 8 Kohler engines



## Chapter 9 Honda engines



## Chapter 10 Robin/Wisconsin Robin engines

# Introduction

There are literally millions of small engines in the garages, sheds and basements of homes all across America today. Some estimates are as high as five or six engines per household. They're mounted on lawn mowers, garden tillers, generators, air compressors, pumps, mini-bikes, go-carts and various other types of equipment and recreational vehicles . . . and many of them are badly neglected - in need of some type of maintenance or repair (often both). Since they're required to operate in hostile conditions (dust, heat, overloading and in many cases without proper lubrication), it's a tribute to the designers, as well as those who have a part in the manufacturing processes, that they perform as well and last as long as they do! However, you don't have to be guilty of neglecting the small engines in your possession, now that Haynes Manuals, the world's largest publisher of automotive repair manuals, has made available this second small engine repair manual covering the most popular and widely used 5.5 to 20 HP small engines from the leading manufacturers. This is a requested follow-up to our successful small engines manual for models up to 5 HP. Its proven approach, featuring easy-to-follow, step-by-step troubleshooting, maintenance and repair procedures, profusely illustrated with photographs taken in our own shop, has been refined over the years in our do-it-yourself automotive and motorcycle repair manuals.

The purpose of this manual is to help you maintain and repair small gas engines. It can do so in several ways. It can help you decide what work must be done, even if you choose to have it done by a repair shop, it provides information and procedures for tune-ups and routine maintenance and it offers diagnostic and repair procedures to follow when trouble occurs.

It's hoped you'll use the manual to tackle the work yourself. For many jobs, doing it yourself may be quicker than arranging an appointment to get the machinery into a shop and making the trips to drop it off and pick it up. More importantly, a lot of money can be saved by avoiding the expenses the shop must pass on to you to cover labor and overhead costs. An added benefit is the sense of satisfaction and accomplishment you feel after doing the job yourself. We also hope that as you gain experience and confidence working on small engines, you'll decide to move on to simple motorcycle, car or truck maintenance and repair jobs. When you do, Haynes can supply you with virtually all the service information you'll need.

# How to use this repair manual

The manual is divided into several chapters. Each chapter is sub-divided into well-defined sections, many of which consist of consecutively numbered Paragraphs (usually referred to as "Steps", since they're normally part of a maintenance or repair procedure). If the material is basically informative in nature, rather than a step-by-step procedure, the Paragraphs aren't numbered.

The first five chapters contain material that applies to all engines, regardless of manufacturer. The remaining chapters cover specific material related to the individual brand engines only. Since most people are initially exposed to practical mechanics working on small engines, comprehensive chapters covering tool selection and usage, safety and general shop practices have also been included. *Be sure to read through them before beginning any work.* All specifications are included in each brand's engine Chapter.

The term "**see illustration**" (in parentheses), is used in the text to indicate that a photo or drawing has been included to make the information easier to understand (the old cliché "a picture is worth a thousand words" is especially true when it comes to how-to procedures). Also, every attempt is made to position illustrations directly opposite the corresponding text to minimize confusion. The two types of illustrations used (photographs and line drawings) are referenced by a number preceding the caption. Illustration numbers in the first five Chapters, which are general in nature, denote chapter and numerical sequence within the chapter (i.e. 3.4 means Chapter 3, illustration number four in order). In the specific engine Chapters, there are many step-by-step procedures, and in these Chapters the illustration number represents the Section and Step (i.e. 6.24 represents Section 6, Step 24). This makes involved procedures easier to follow.

The terms "**Note**", "**Caution**" and "**Warning**" are used throughout the text with a specific purpose in mind - to attract the reader's attention. A "**Note**" simply provides information required to properly complete a procedure or information which will make the procedure easier to understand. A "**Caution**" outlines a special procedure or special steps which must be taken when completing the procedure where the Caution is found. Failure to pay attention to a Caution can result in damage to the component being repaired or the tools being used. A "**Warning**" is included where personal injury can result if the instructions aren't followed exactly as described.

**Note:** *Even though extreme care has been taken during the preparation of this manual, neither the publisher nor the authors can accept responsibility for any errors in, or omissions from, the information given.*

# Engine types and manufacturers included

The information in this repair manual is restricted to four-stroke, air-cooled gasoline engines rated from 5.5 through 20 horsepower, normally used to power riding lawn mowers, garden tillers, generators, air compressors, pumps and other types of commonly available equipment. The following manufacturers/engine types are included - for a complete list of engines, by model designation, refer to the chapter with the specific repair information for the particular manufacturer:

*Briggs & Stratton single and twin-cylinder engines, L-head and OHV*
*Tecumseh/Craftsman single-cylinder engines, L-head and OHV*
*Kohler single and twin-cylinder engines, L-head and OHV*
*Honda OHV single and twin-cylinder engines*
*Robin/Wisconsin Robin single-cylinder engines, L-head and OHV*

# How to identify an engine

To determine what repair information and specifications to use, and to purchase replacement parts, you'll have to be able to accurately identify the engine you're working on. Every engine, regardless of manufacturer, comes from the factory with a model number stamped or cast into it or a tag attached to it somewhere **(see illustration)**. The most common location is on the shroud used to direct the cooling air around the cylinder (look for the recoil starter - it's normally attached to the shroud as well). On some engines, the model number may be stamped or cast into or attached to the main engine casting and may not be visible, especially if the engine is dirty. To identify an engine from a known manufacturer covered in this manual, refer to the chapter with the specific repair information for the particular manufacturer.

If you can't find a model number or tag, you can determine if the engine is a two or four stroke (which will help a dealer decide what engine model you're dealing with) using one or more of the following quick checks:

Look for a cap used to check the oil level and add oil to the engine - if the engine has a threaded or friction fit cap or plug that's obviously intended for adding oil to the

**The engine model/serial number is usually located on the cooling shroud (as shown here), but it may be located on the main engine casting**

Four-stroke engines will have an oil level check/fill plug like this one somewhere on the lower part of the engine

Two-stroke engines require oil to be mixed with the gas for lubrication

crankcase **(see illustration)**, it's a four-stroke (the cap may be marked "ENGINE OIL" or "OIL FILL" and may have an oil level dipstick attached to it as well).

Look for instructions to mix oil with the gas - if the engine requires oil in the gasoline **(see illustration)**, it's a two-stroke. **Note:** *This manual covers only four-stroke engines.*

Look for a muffler near the cylinder head - if the muffler (usually a canister-shaped device with several holes or slots in the end) is threaded into or bolted to the engine near one end **(see illustration)**, it's a four-stroke. Two-stroke engines have exhaust ports on the cylinder itself, near the center.

Use the recoil starter to feel for compression strokes - detach the wire from the spark plug and ground it on the engine, then slowly operate the recoil starter. If you can feel resistance from cylinder compression every revolution of the crankshaft, the engine is a two-stroke. If compression resistance is felt every other revolution, the engine is a four-stroke.

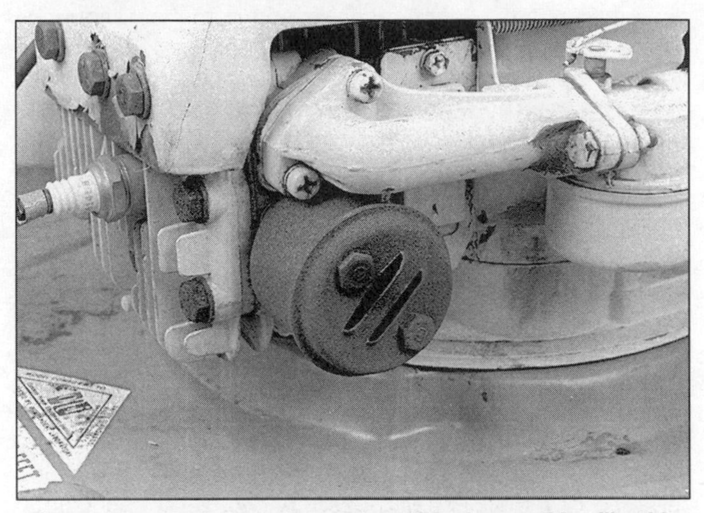
The muffler on a four-stroke engine will look something like this

# Buying parts

The best place (and sometimes the only place) to buy parts for any small engine is the dealer that sells and repairs the engine brand or the equipment the engine is mounted on. Some auto parts stores also stock small engine parts, but they normally carry only tune-up and maintenance items. Look in the yellow pages of your telephone directory under "Small engines" and "Lawn and garden equipment" for a list of dealers in your area.

Always purchase and install name-brand parts. Most manufacturers market new, complete replacement engines and also what is termed a "short block". A short block is a brand new engine sub-assembly that includes the main crankcase casting, piston, rings and connecting rod, valves and related components, cylinder and camshaft. If you purchase one, you'll have to bolt on the external parts, such as the cylinder head, magneto, carburetor, fuel tank and recoil starter/cooling shroud. A short block typically costs about half as much as a complete new engine and approximately twice as much as a new crankshaft. If you have an engine that's worn out, severely damaged or that requires more work than you're willing to invest, a short block - or an entire new engine - may be the best alternative to an overhaul or major repairs.

Be sure to have the engine model and serial number available when buying parts and, if possible, take the old parts with you to the dealer. Then you can compare the new with the old to make sure you're getting the right ones. Keep in mind that parts may have to be ordered, so as soon as you realize you're going to need something, see if it's in stock and allow extra time for completing the repair if parts must be ordered.

You may occasionally be able to purchase used parts in usable condition and save some money in the process. A reputable dealer normally won't sell substandard parts, so don't hesitate to inquire about used components.

## Notes

# 1 Setting up shop

## Finding a place to work

Before considering what tools to collect, or how to use them, a safe, clean, well-lit place to work should be located. If anything more than routine maintenance is going to be done, some sort of special work area is essential. It doesn't have to be particularly large, but it should be clean, organized and equipped especially for doing mechanic work. It's understood, and appreciated, that many home mechanics don't have a good workshop or garage available and end up servicing or repairing an engine out of doors; however, an overhaul or major repairs should be completed in a sheltered area with a roof (the main reason is to prevent parts from collecting dirt, which is abrasive and will cause wear if it finds its way into an engine).

### The workshop building

The size, shape and location of a shop building is usually dictated by circumstances rather than personal choice. Ideally, every do-it-yourselfer would have a spacious, clean, well-lit building specially designed and equipped for working on everything from small engines on lawn and garden equipment to cars and other vehicles. In reality, however, most of us must be content with a corner of the garage or basement or a small shed in the backyard.

As mentioned above, anything beyond minor maintenance and adjustments in nice weather should be done indoors. The best readily-available building would be a typical one or two car garage, preferably one that's detached from the house. A garage provides ample work and storage space and room for a large workbench. With that in mind, it must be pointed out that even the most extensive job possible on the typical small engine could - if necessary - be done in a small shed or corner of a garage. The bottom line is you'll have to make do with whatever facilities you have and adapt your workshop and methods of work to it.

Whatever the limitations of your own proposed or existing workshop area are, spend some time considering its potential and drawbacks - even a well-established workshop will benefit from occasional reorganization. Most do-it-yourselfers find that lack of space causes problems; this can be overcome to a great extent by carefully planning the locations of benches and storage facilities. The rest of this Section will cover some of the options available when setting up or reorganizing a workshop. Perhaps the best approach when designing a shop is to look at how others do it. Try approaching a local repair shop owner and asking to see his shop; note how work areas, storage and lighting are arranged, then try to scale it down to fit your own shop space, finances and needs.

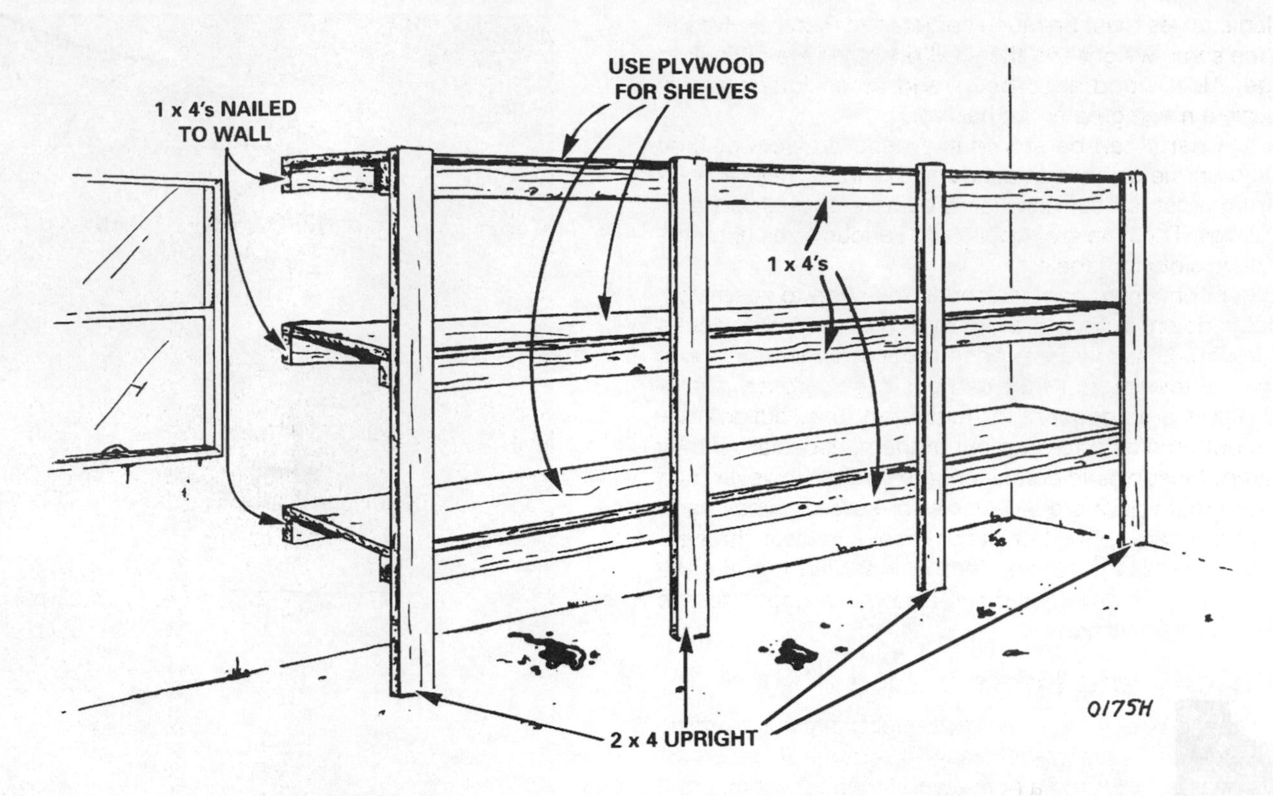


**1.1 Since they're relatively inexpensive and can be designed to fit available space, homemade wooden shelves may be the best choice of shop storage - however, keep in mind the obvious fire hazard they will become**

## General building requirements

A solid concrete floor is probably the best surface for any shop area used for mechanic work. The floor should be as even as possible and must also be dry. Although not absolutely necessary, it can be improved by applying a coat of paint or sealer formulated for concrete surfaces. This will make oil spills and dirt easier to remove and help cut down on dust - always a problem with concrete. A wood floor is less desirable and may sag or be damaged by the weight of equipment and machinery. It can be reinforced by laying sheets of thick plywood or chipboard over the existing surface. A dirt floor should be avoided at all costs, since it'll produce abrasive dust, which will be impossible to keep away from internal engine components. Dirt floors are also as bad as gravel or grass when it comes to swallowing up tiny dropped parts such as ball bearings and small springs.

Walls and ceilings should be as light as possible. It's a good idea to clean them and apply a couple of coats of white paint. The paint will minimize dust and reflect light inside the workshop. On the subject of light, the more natural light there is the better. Artificial light will also be needed, but you'll need a surprising amount of it to equal ordinary daylight.

A normal doorway is just wide enough to allow all but the biggest pieces of machinery and equipment through, but not wide enough to allow it through easily. If possible, a full-size garage door (overhead or hinged at each side) should allow access into the shop, especially when dealing with larger yard tractors. Steps (even one of them) can be difficult to negotiate - make a ramp out of wood to allow easier entry if the step can't be removed.

Make sure the building is adequately ventilated, particularly during the winter. This is essential to prevent condensation problems and is also a vital safety consideration where solvents, gasoline and other volatile liquids are being stored and used. You should be able to open one or more windows for ventilation. In addition, opening vents in the walls are desirable.

## Storage and shelving

All the parts from a small engine can occupy more space than you realize when its been completely disassembled - some sort of organized storage is needed to avoid losing them. In addition, storage space for hardware, lubricants, solvent, rags, tools and equipment will also be required.

If space and finances allow, install metal shelf units along the walls. Arrange the shelves so they're widely spaced near the bottom to take large or heavy items. Metal shelf units are expensive, but they make the best use of available space. An added advantage is the shelf positions are not fixed and can be changed if necessary.

A cheaper (but more labor intensive) solution is to build shelves out of wood **(see illustration 1.1)**. Remember that wooden shelves must be much heftier than metal shelves to carry the same weight and the shelf positions are difficult to change. Also, wood absorbs oil and other liquids and is obviously a much greater fire hazard.

Small parts can be stored in plastic drawers or bins mounted on metal racks attached to the wall. They're available from most lumber and home centers as well as hardware stores. The bins are available in various sizes and normally have slots for labels.

Other containers can be used in the shop to keep storage costs down, but try to avoid round tubs, which waste a lot of space. Glass jars are often recommended as cheap storage containers, but they can easily get broken. Cardboard boxes are adequate for temporary use, but eventually the bottoms tend to drop out of them, especially if they get damp. Most plastic containers are useful, however, and large ice cream pails are invaluable for keeping small parts together during a rebuild or major repairs (collect the type that has a cover that snaps into place). Old metal cake pans, bread pans and muffin tins also make good storage containers for small parts.

## Electricity and lights

Of all the useful shop facilities, electricity is by far the most essential. It's relatively easy to arrange if the workshop is near to or part of a house and it can be difficult and expensive if it isn't. It must be stressed that safety is the number one consideration when dealing with electricity; unless you have a very good working knowledge of electrical installations, any work required to provide power and lights in the shop should be done by an electrician.

You'll have to consider the total electrical requirements of the shop, making allowances for possible later additions of lights and equipment. Don't substitute extension cords for legal and safe permanent wiring. If the wiring isn't adequate or is substandard, have it upgraded.

Careful consideration should be given to lights for the workshop (two 150-watt incandescent bulbs or two 48-inch long, 40-watt fluorescent tubes suspended approximately 48-inches above the workbench would be a minimum). As a general rule, fluorescent lights are probably the best choice for even, shadow-free lighting. The position of the lights is important; for example, don't position a fixture directly above the area where the engine (or equipment it's mounted on) will be located during work - this will cause shadows even with fluorescent lights. Attach the light or lights slightly to the rear of or to each side of the workbench or work area to provide even lighting. A portable "trouble-light" is very helpful for use when overhead lights are inadequate. Note that if solvents, gasoline or other flammable liquids are present, which is usually the case in a mechanic's shop, special fittings should be used to minimize the risk of fire. Also, don't use fluorescent lights above machine tools (like a drill press). The flicker produced by alternating current is especially pronounced with this type of light and can make a rotating chuck appear stationary at certain speeds - a very dangerous situation.

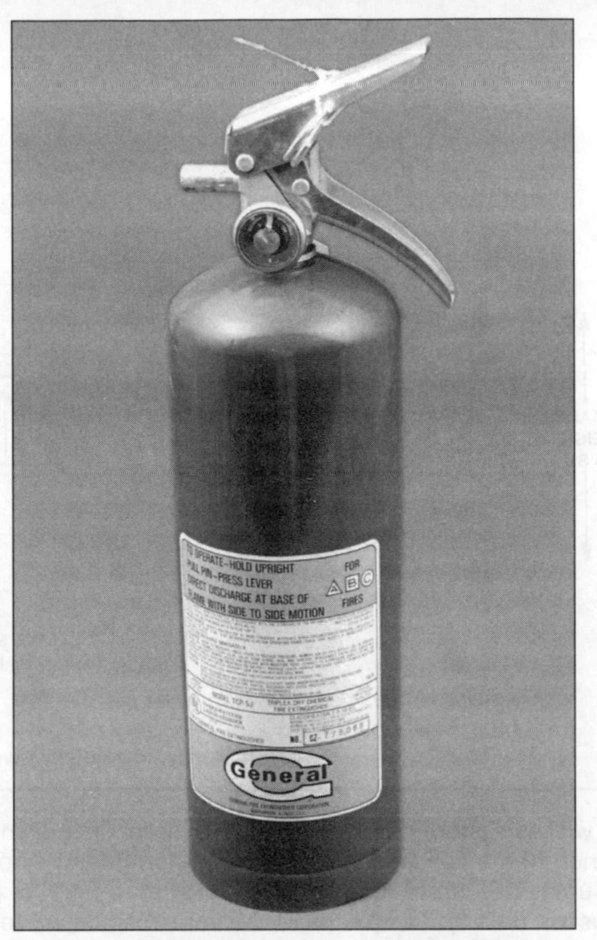


**1.2 Have a fire extinguisher designed for use on flammable liquid fires handy and know how to use it!**

# Tools and equipment needed

## Fire extinguisher

Since the use, maintenance and repair of any gasoline engine requires fuel to be handled and stored, buy a good-quality fire extinguisher before doing any maintenance or repair procedures **(see illustration 1.2)**. Make sure it's rated for flammable liquid fires, familiarize yourself with its use and be sure to have it checked/recharged at regular intervals. Refer to Chapter 2 for safety-related information - warnings about the hazards of gasoline and other flammable liquids are included there.

## Workbench

A workbench is essential - it provides a place to lay out parts and tools during repair procedures, which means they'll stay clean longer, and it's a lot more comfortable than working on a floor or the driveway. This very important piece of shop equipment should be as large and sturdy as space and finances will allow. Although many types of benches are commercially available, they're usually quite expensive and don't necessarily fit into the available space

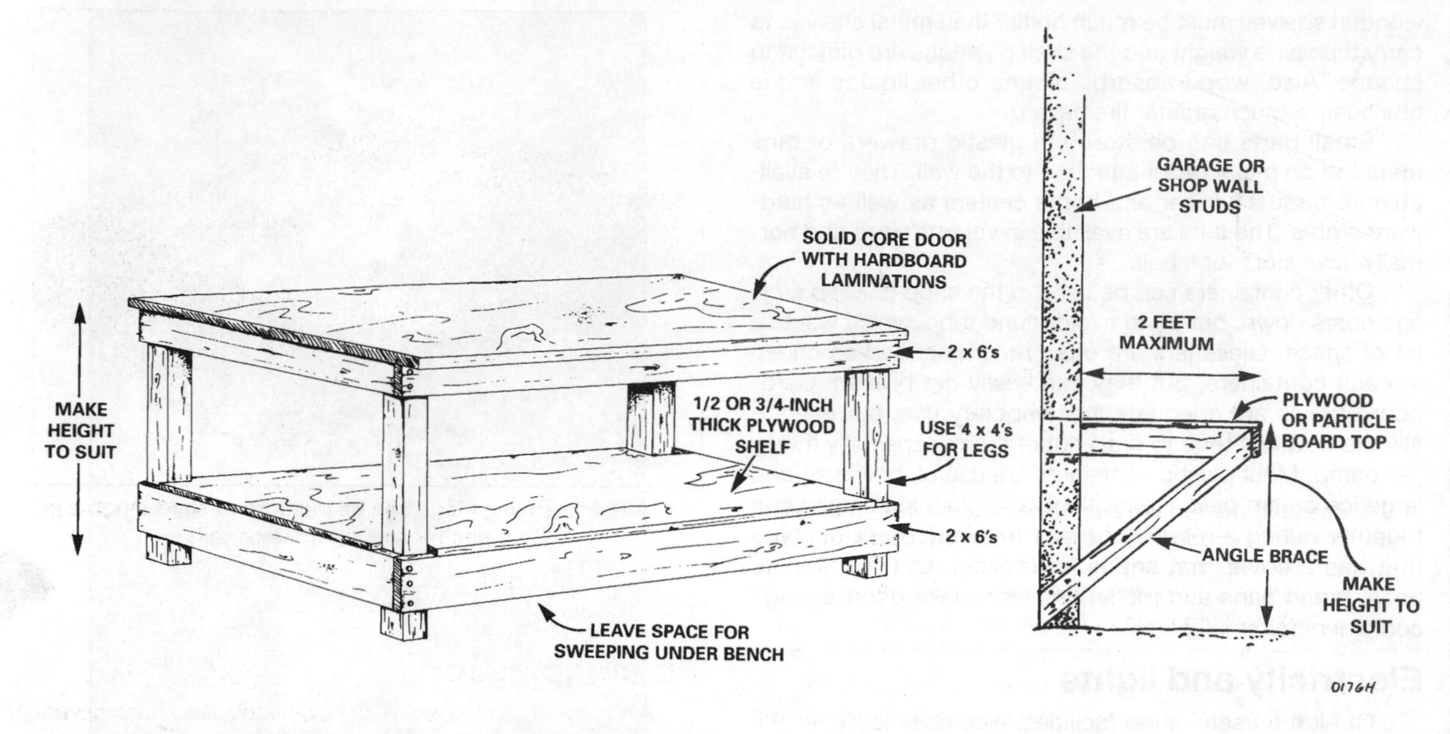


**1.3 A sturdy, inexpensive workbench can be constructed from 2 x 6's**

as well as custom-built ones will. An excellent free-standing bench frame can be fabricated from slotted angle-iron or Douglas fir lumber (use 2 x 6's rather than 2 x 4's) **(see illustration 1.3)**. The pieces of the frame can be cut to any required size and bolted together. A 30 or 36 by 80-inch wood, solid-core door with hardboard surfaces, available at any lumber or home center, makes a nice bench top and can be turned over to expose the fresh side if it gets damaged or worn out.

If you're setting up shop in a garage, a sturdy bench can be assembled very quickly by attaching the bench top frame pieces to the wall with angled braces, effectively using the wall studs as part of the framework. Regardless of the type of frame you decide to use for the workbench, be sure to position the bench top at a comfortable working height and make sure everything is level. Shelves installed below the bench will make it more rigid and provide useful storage space.

**1.4 A bench vise is one of the most useful pieces of equipment you can have in the shop**

One of the most useful pieces of equipment - and one that's usually associated with the workbench - is a vise. Size isn't necessarily the most important factor to consider when shopping for one; the quality of materials used and workmanship is. Good vises are very expensive, but as with anything else, you get what you pay for. Buy the best quality vise you can afford and make sure the jaws will open at least four inches. Purchase a set of soft jaws to fit the vise as well (they're used to grip engine parts that could be damaged by the hardened vise jaws) **(see illustrations 1.4 and 1.5)**.

## Engine stands

Many small engine manufacturers also distribute a special fixture to hold engines during disassembly and reassembly. Equipment of this type is undoubtedly very useful, but outside the scope of most home workshops. In practice, most do-it-yourselfers will have to make do with a selection of wood blocks that can be used to prop the engine up on a bench. They can be arranged as required so the engine is supported in almost any position. An engine stand can also be fabricated from short lengths of 2 x 4 lumber and lag bolts, screws or nails **(see illustration 1.6)**. When using wood blocks or a homemade engine stand, it's a good idea to have a helper available to assist in steadying

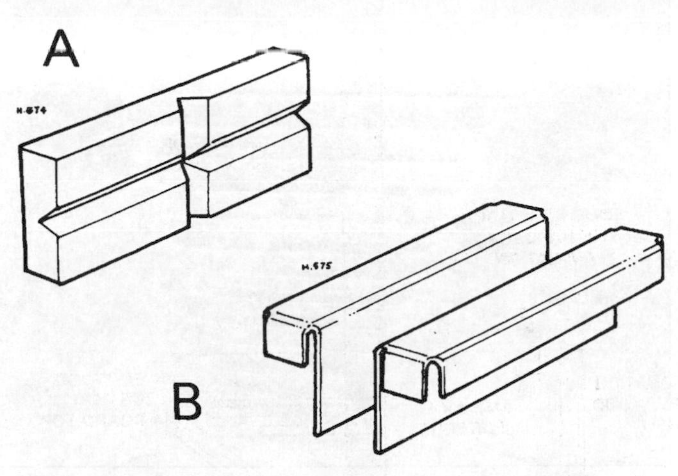


1.5 Some jobs will require engine parts to be held in the vise jaws - to avoid damage to the parts from the hardened vise jaws, use commercially available fiberglass or plastic "soft jaws" (A) or fabricated inserts from 1/8-inch thick aluminum to fit over the jaws (B)

1.6 A handy engine stand can be made from short lengths of 2 x 4 lumber and lag bolts or nails

the engine while fasteners are loosened or tightened. In some situations on smaller engines, the engine can be clamped in a vise, but be very careful not to damage the crankcase or cylinder castings. This is not suitable for larger engines. If you have access to welding equipment, you can fabricate a stand or mount for the engine you are working on from pieces of plate and angle-iron (for more information, see the Haynes *Welding Manual*).

Adjustable workbenches, like the Black & Decker Workmate, can also be very useful for holding a smaller engine while it's being worked on **(see illustration 1.7)**. You probably won't want to buy one just for working on one small engine, but if you already have one, it can easily be adapted for use as a holding fixture for all but the heaviest of the engines covered by this manual.

## Air compressor

Although it isn't absolutely necessary, an air compressor can make many tasks in the shop much easier and enable you to do a better job. (How else can you easily remove debris from the engine's cooling fins, dry off parts after cleaning them with solvent or blow out all the tiny passages in a carburetor?) If you can afford one, you'll wonder how you ever got along without it. In addition to supplying compressed air for cleaning parts, a compressor - if it's large enough - can also be used to power air tools, which are now widely available and quite inexpensive and can take much of the drudgery out of mechanical repair jobs **(see illustration 1.8)**. For example, an impact wrench (and special impact sockets) can be invaluable when it comes time to remove the large nut that holds a lawn mower blade

1.7 A Black & Decker Workmate comes in very handy for holding an engine while working on it - the quick-release clamping feature makes it easy to change the engine's position quickly

1.8 Although it's not absolutely necessary, an air compressor can make many jobs easier and produce better results, especially when air-powered tools are available to use with it

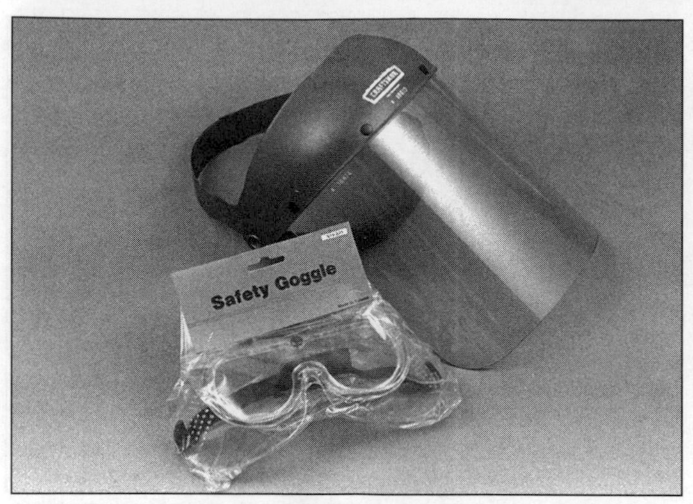

1.9 One of the most important items you'll need in the shop is a face shield/safety goggles - fortunately, it'll also be one of the least expensive

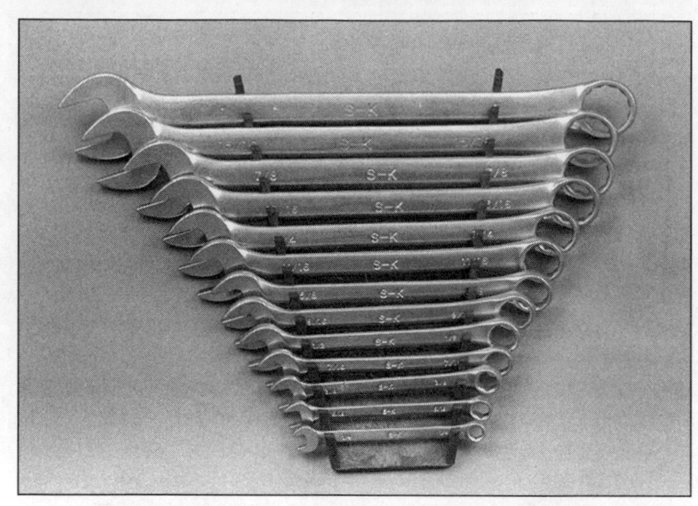
1.10 Combination wrenches - buy a set with sizes from 1/4 to 7/8-inch or 6 to 19 mm

or the magneto flywheel to the end of the crankshaft. On the down side, the cost involved, the need for maintenance on the equipment and additional electrical requirements must be considered before equipping your shop with compressed air.

## Hand tools

A selection of good mechanic's tools is a basic requirement for anyone who plans to maintain and repair small gasoline engines. For someone who has few tools, if any, the initial investment might seem high, but when compared to the spiraling costs of routine maintenance and repairs, it's a wise one; besides, most of the tools can also be used for other types of work. Keep in mind that this chapter simply lists the tools needed for doing the work - Chapter 2 explains in greater detail what to look for when shopping for tools and how to use them correctly.

To help the reader decide which tools are needed to perform the tasks detailed in this manual, two tool lists have been compiled: *Routine maintenance and minor repair* and *Repair and overhaul*. A separate section related to special factory tools is also included, but only the most serious do-it-yourselfers will be interested in reading about, purchasing and using them. Illustrations of most of the tools on each list are also included.

The newcomer to mechanic work should start off with the *Routine maintenance and minor repair* tool kit, which is adequate for simple jobs.
Then, as confidence and experience increase, you can tackle more difficult tasks, buying additional tools as they're needed. Eventually the basic kit will be built into the *Repair and overhaul* tool set. Over a period of time, the experienced do-it-yourselfer will assemble a set of tools complete enough for most repair and overhaul procedures and may begin adding special factory tools when it's felt the expense is justified by the frequency of use or the savings realized by not taking the equipment in to a shop for repair.

## Routine maintenance and minor repair tools

The tools on this list should be considered the minimum required for doing routine maintenance, servicing and minor repair work **(see illustrations 1.9 through 1.22)**. Incidentally, if you have a choice, it's a good idea to buy combination wrenches (box-end and open-end combined in one wrench); while more expensive than open-end ones, they offer the advantages of both types. Also included is a complete set of sockets which, though expensive, are invaluable because of their versatility (many types of interchangeable accessories are available). We recommend 3/8-inch drive over 1/2-inch drive for general small engine maintenance and repair, although a 1/4-inch drive set would also be useful (especially for ignition and carburetor work). Buy 6-point sockets, if possible, and be careful not to purchase sockets with extra thick walls - they can be difficult to use when access to fasteners is restricted.

*Safety goggles/face shield*
*Combination wrench set (1/4 to 7/8-inch or 6 to 19 mm)*
*Adjustable wrench - 10-inch*
*Socket set (6-point)*
*Reversible ratchet*
*Extension - 6-inch*
*Universal joint*
*Spark plug socket (with rubber insert)*
*Spark plug gap adjusting tool*
*Feeler gauge set*
*Standard screwdriver (5/16-inch x 6-inch)*
*Standard screwdriver (3/8-inch x 10-inch)*
*Phillips screwdriver (no. 2 x 6-inch)*
*Combination (slip-joint) pliers - 6-inch*
*Oil can*
*Fine emery cloth*
*Wire brush*

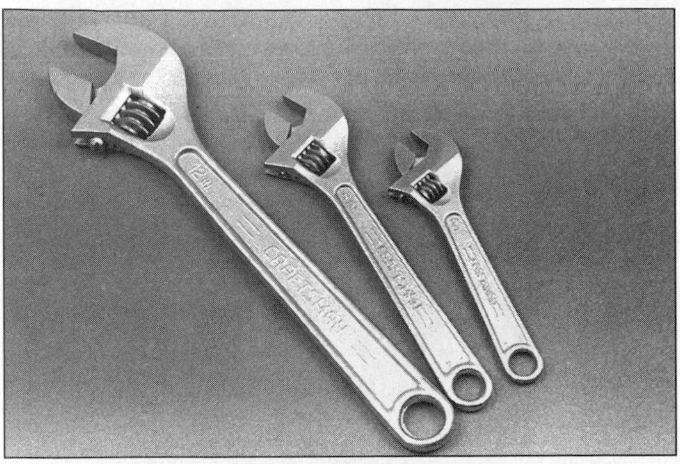

1.11 Adjustable wrenches are very handy - just be sure to use them correctly or you can damage fasteners by rounding off the hex head

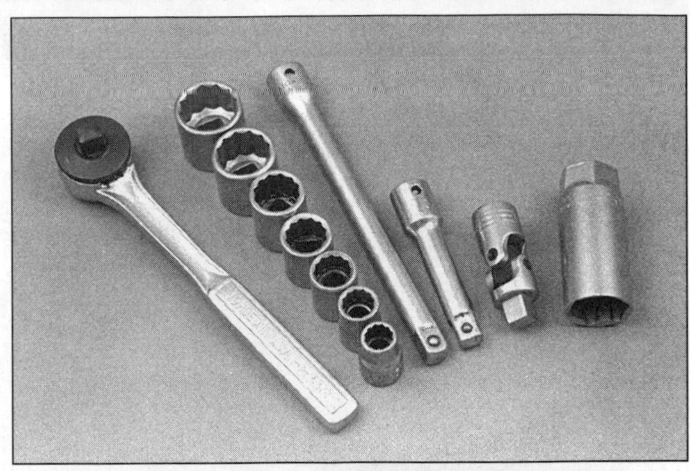

1.12 A 3/8-inch drive socket set with interchangeable accessories will probably be used more often than any other tool(s) (left-to-right; ratchet, sockets, extensions, U-joint, spark plug socket) - don't buy a cheap socket set!

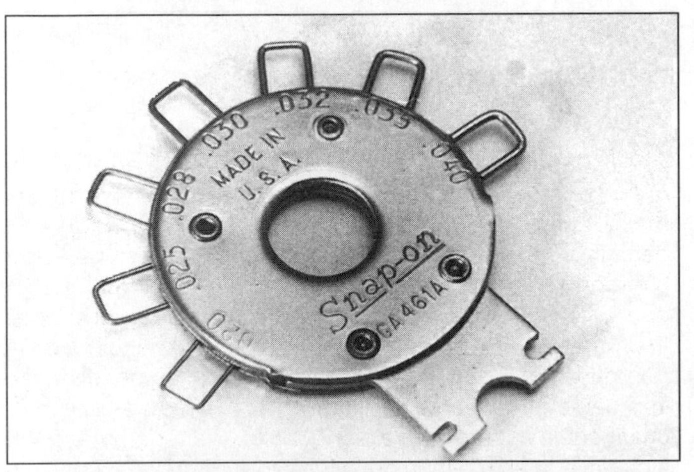


1.13 A spark plug adjusting tool will have several wire gauges for measuring the electrode gap and a device used for bending the side electrode to change the gap - make sure the one you buy has the correct size wire to check the spark plug gap on your engine

1.14 Feeler gauge sets have several blades of different thickness - if you need it to adjust ignition points, make sure the blades are as narrow as possible and check them to verify the required thickness is included

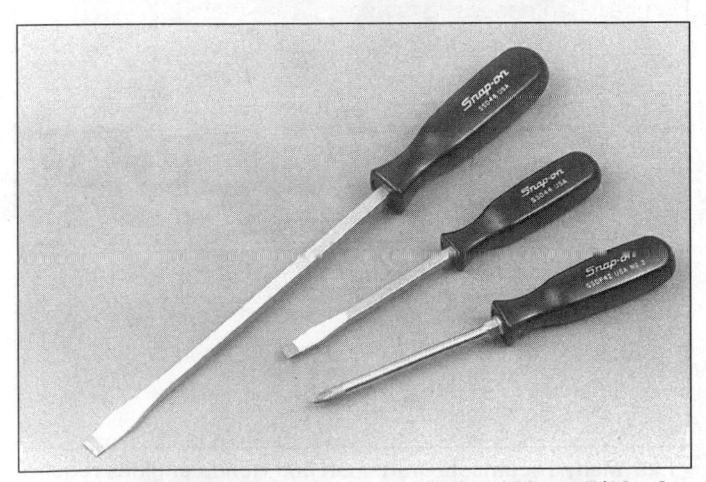


1.15 The routine maintenance tool kit should have 5/16 x 6-inch and 3/8 x 10-inch standard screwdrivers, as well as a no. 2 x 6-inch Phillips

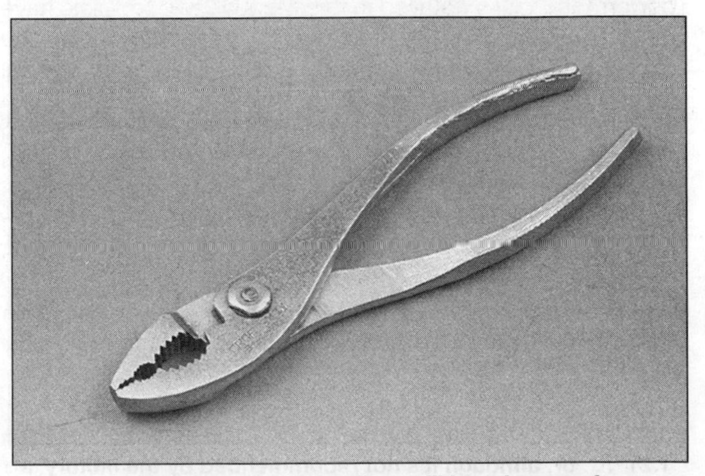

1.16 Common slip-joint pliers will be adequate for almost any job you end up doing

1.17 A shallow pan (for draining oil/cleaning parts with solvent), a wire brush and a medium size funnel should be part of the routine maintenance tool kit

1.18 To remove the starter clutch used on some Briggs & Stratton engines, a special tool (which is turned with a wrench), will be needed

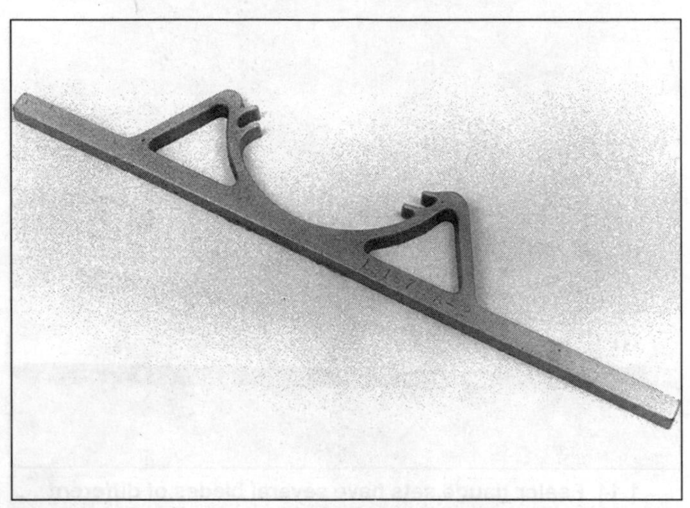

1.19 Briggs & Stratton also sells a special flywheel holder for use when loosening the nut or starter clutch

1.20 The flywheel on a Briggs & Stratton engine can be removed with a puller (shown here). . .

1.21 . . . or, although it's not recommended by the factory, a knock-off tool, which fits on the end of the crankshaft (Tecumseh/Craftsman flywheels can also be removed with one of these tools)

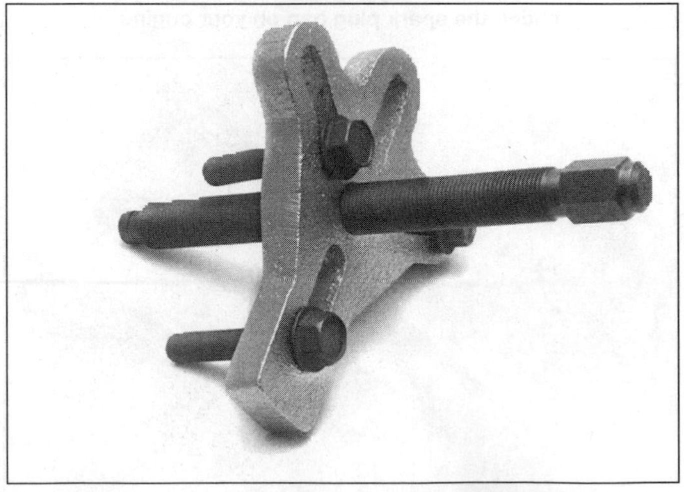

1.22 Many Tecumseh/Craftsman and Honda engines require a puller like the one shown here for flywheel removal

*Funnel (medium size)*
*Drain pan*
*Starter clutch wrench**
*Flywheel holder**
*Flywheel puller or knock-off tool**

* Although these tools are normally available exclusively through distributors/dealers (so technically they're "special factory tools"), they are included in this list because certain tune-up and minor repair procedures can't be done without them (specifically ignition point and flywheel key replacement on most Briggs & Stratton, Tecumseh and Craftsman engines). The factory tools may also be available at hardware and lawn and garden stores and occasionally you'll come across imported copies of the factory tools - examine them carefully before buying them.

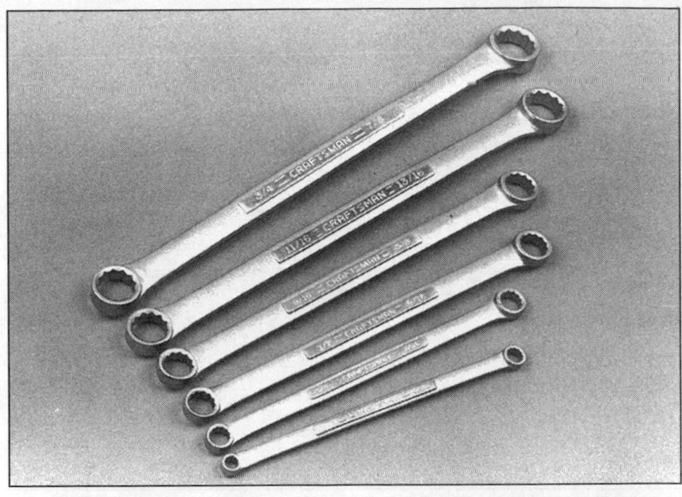

**1.23 A set of box-end wrenches will complement the combination wrenches in the routine maintenance tool kit**

## Repair and overhaul tools

These tools are essential if you intend to perform major repairs or overhauls and are intended to supplement those in the *Routine maintenance and minor repair* tool kit **(see illustrations 1.23 through 1.49)**.

The tools in this list include many which aren't used regularly, are expensive to buy, or which need to be used in accordance with their manufacturer's instructions. Unless these tools will be used frequently, it's not very economical to purchase many of them. A consideration would be to split the cost and use between yourself and a friend or neighbor.

*Box-end wrenches*
*Torque wrench (same size drive as sockets)*
*Ball-peen hammer - 12 oz (any steel hammer will do)*
*Soft-face hammer (plastic/rubber)*
*Standard screwdriver (1/4-inch x 6-inch)*
*Standard screwdriver (stubby - 5/16-inch)*
*Phillips screwdriver (no. 3 x 8-inch)*
*Phillips screwdriver (stubby - no. 2)*

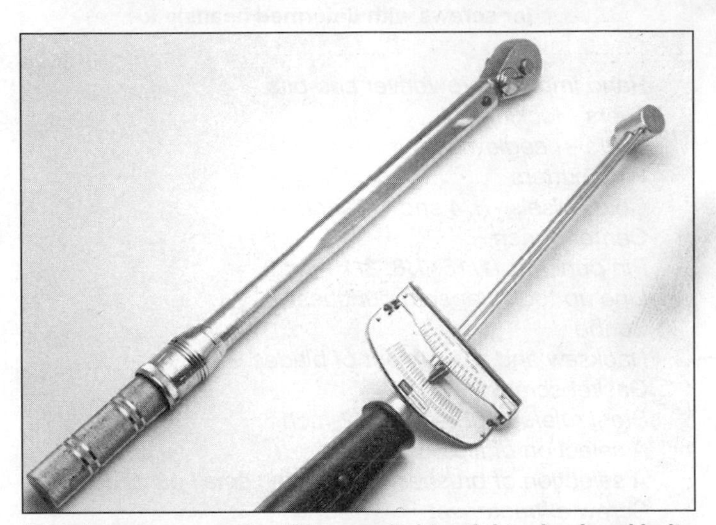

**1.24 A torque wrench will be needed for tightening head bolts and flywheel nuts (two types are available: click type - left; beam type - right)**

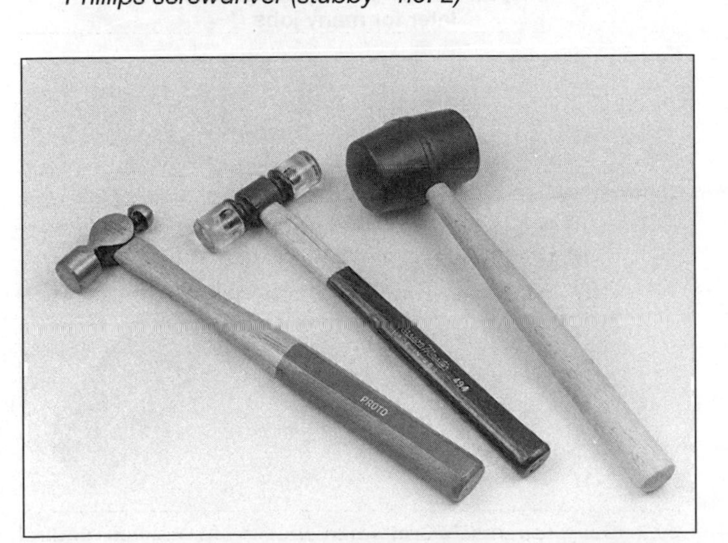


**1.25 A ball-peen hammer, soft-faced hammer and rubber mallet (left-to-right) will be needed for various tasks (any steel hammer can be used in place of the ball-peen hammer)**

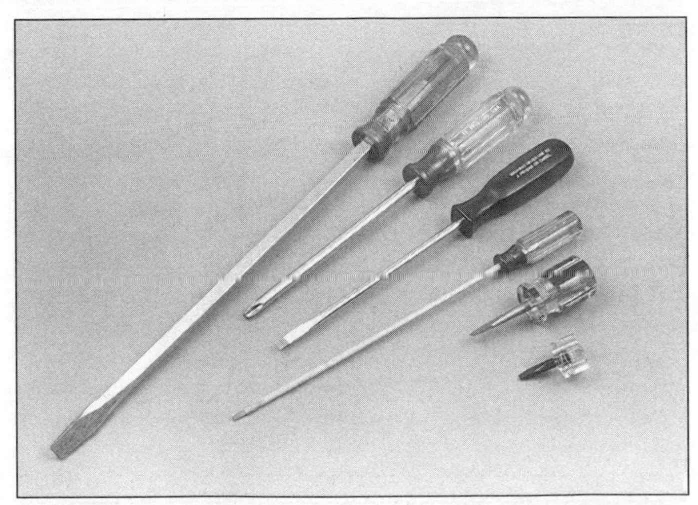

**1.26 Screwdrivers come in many different sizes and lengths**

1.27 A hand impact screwdriver (used with a hammer) and bits can be very helpful for removing stubborn, stuck screws (or screws with deformed heads)

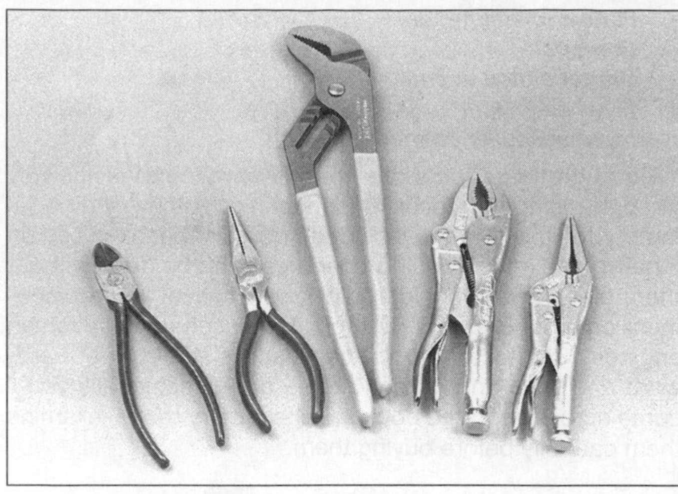

1.28 As you can afford them, arc-joint, needle-nose, Vise-grip and wire cutting pliers should be added to your tool collection

*Hand impact screwdriver and bits*
*Pliers - locking*
*Pliers - needle-nose*
*Wire cutters*
*Cold chisels - 1/4 and 1/2-inch*
*Center punch*
*Pin punches (1/16, 1/8, 3/16-inch)*
*Line up tools (tapered punches)*
*Scribe*
*Hacksaw and assortment of blades*
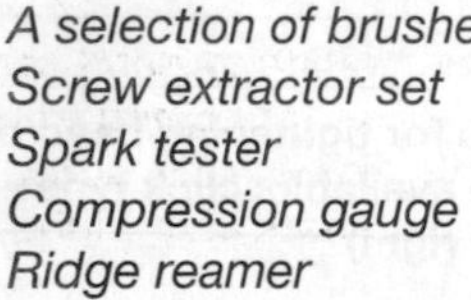
*Gasket scraper*
*Steel rule/straightedge - 12-inch*
*A selection of files*
*A selection of brushes for cleaning small passages*
*Screw extractor set*
*Spark tester*
*Compression gauge*
*Ridge reamer*

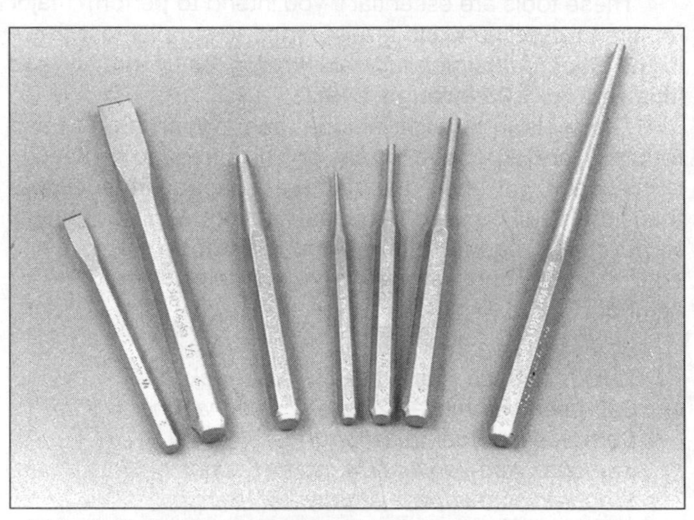

1.29 Cold chisels, center punches, pin punches and line-up punches (left-to-right) will be needed sooner or later for many jobs

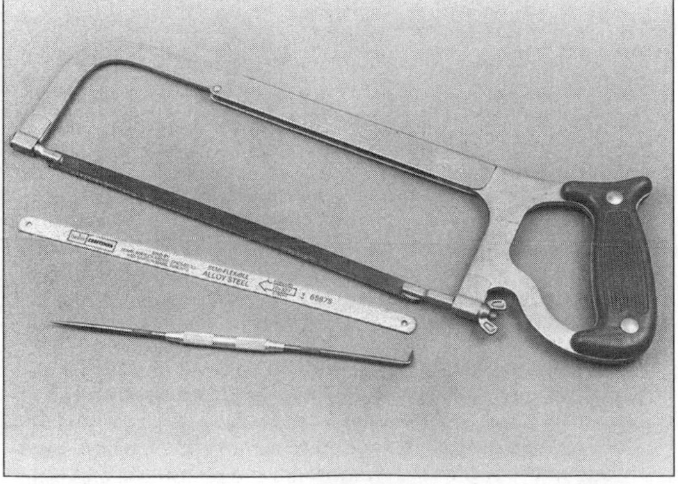

1.30 A scribe is used for making lines on metal parts and a hacksaw and blades will be needed for dealing with fasteners that won't unscrew

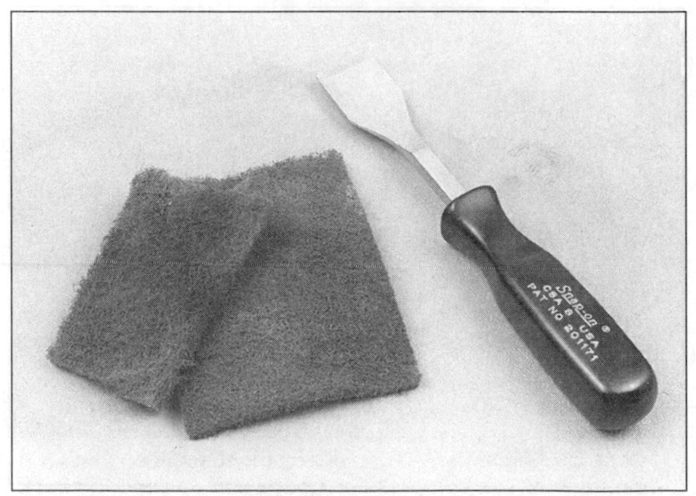

1.31 A gasket scraper is used for removing old gaskets from engine parts after disassembly - 3M "scrubbies" can be used to rough up the gasket surfaces prior to reassembly

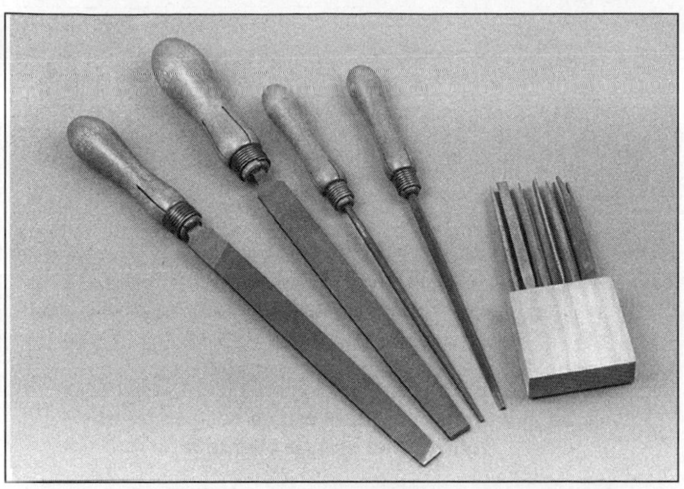

1.32 Files must be used with handles and should be stored so they don't contact each other

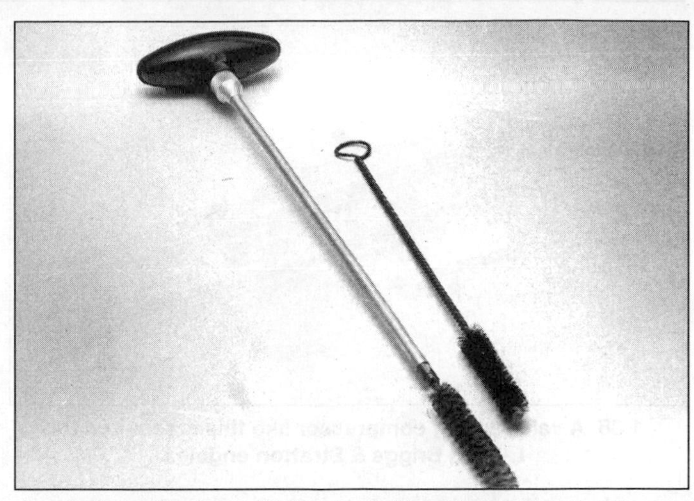

1.33 A selection of nylon/metal brushes is needed for cleaning passages in engines and carburetor parts

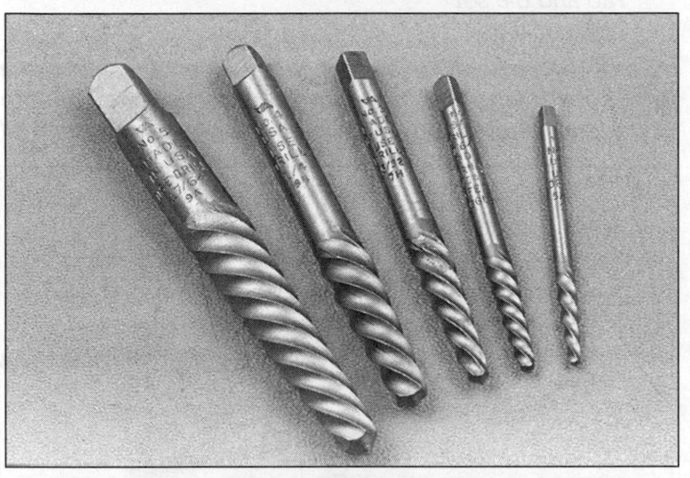

1.34 Screw extractors are used to remove broken screws and bolts from engine parts

1.35 A spark tester (for checking the ignition system) can be purchased at an auto parts store (left) or fabricated from a block of wood, a large alligator clip, some nails, screws and wire and the cap end of an old spark plug (right)

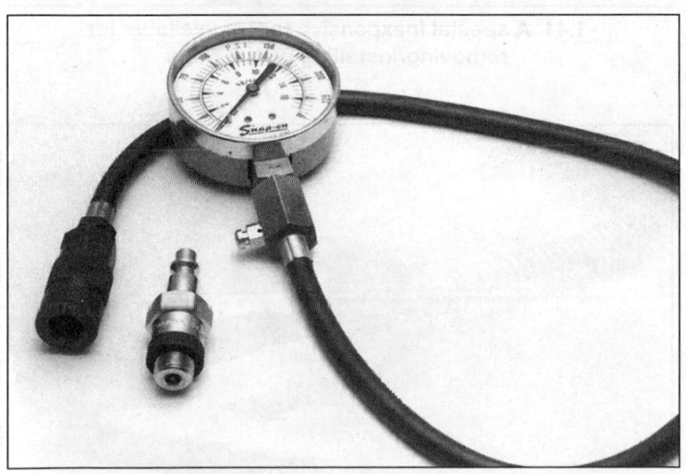

1.36 Although not required by most small engine manufacturers, a compression gauge can be used to check the condition of the piston rings and valves (two types are commonly available: The screw-in type - shown here - and the type that's held in by hand pressure)

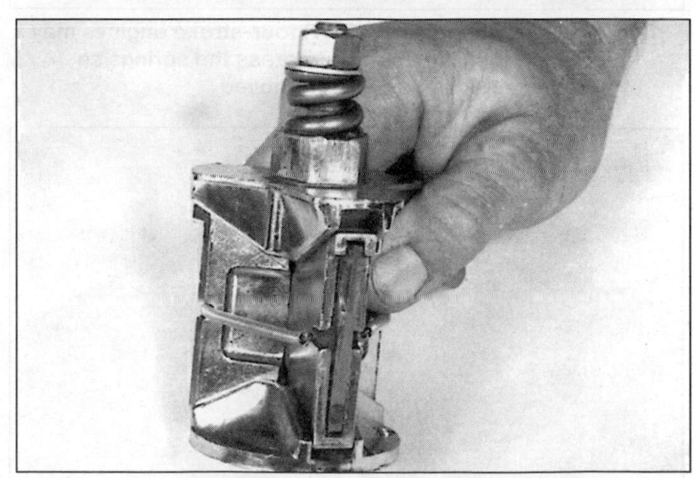

1.37 A ridge reamer is needed to remove the carbon/wear ridge at the top of the cylinder so the piston will slip out

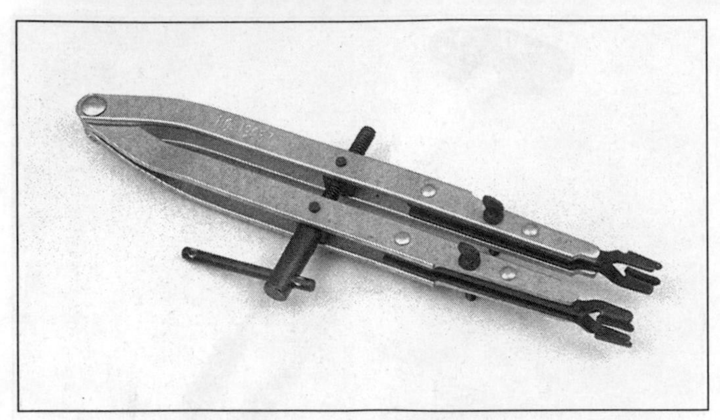

1.38 A valve spring compressor like this is required for L-head Briggs & Stratton engines

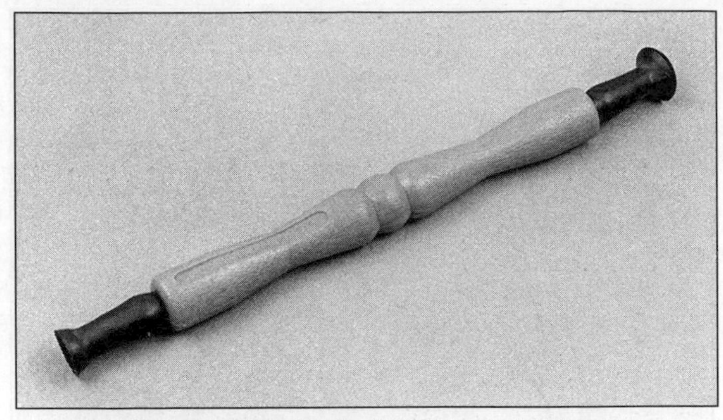

1.39 A valve lapping tool will be needed for any four-stroke engine overhaul

*Valve spring compressor*
*Valve lapping tool*
*Piston ring removal and installation tool*
*Piston ring compressor*
*Cylinder hone*
*Telescoping gauges*
*Micrometer(s) and/or dial/Vernier calipers*
*Dial indicator*
*Tap and die set*
*Torx socket(s)***

***Some Tecumseh/Craftsman two-stroke engines require a Torx socket (size E6) to remove the connecting rod cap bolts **(see illustration 1.50)**. If you're overhauling one of*

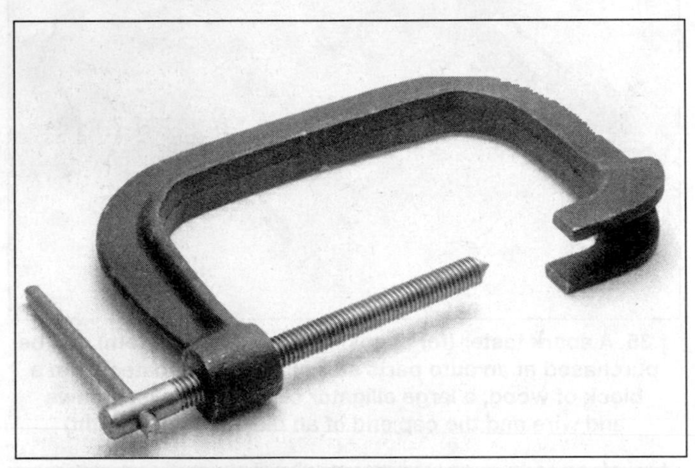

1.40 Some overhead valve (OHV) four-stroke engines may require a tool like this to compress the springs so valves can be removed

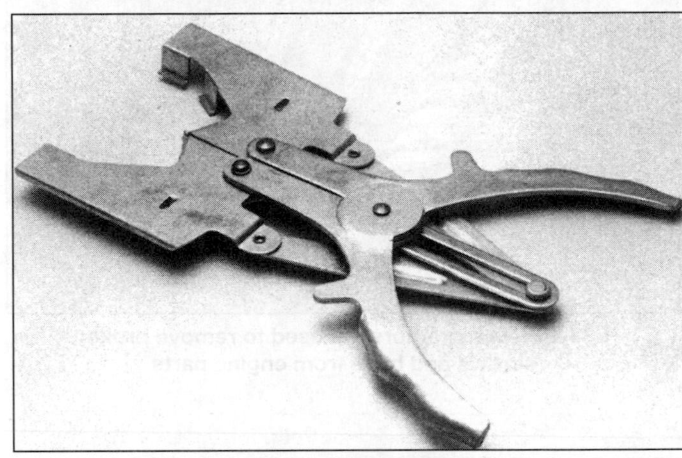

1.41 A special inexpensive tool is available for removing/installing piston rings

1.42 Piston ring compressors come in many sizes - be sure to buy one that will work on your engine

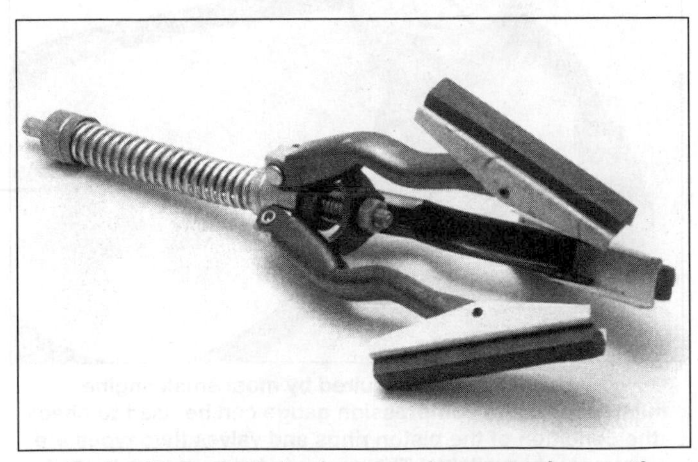

1.43 A cylinder surfacing hone can be use to clean up the bore so new rings will seat, but it won't resize the cylinder

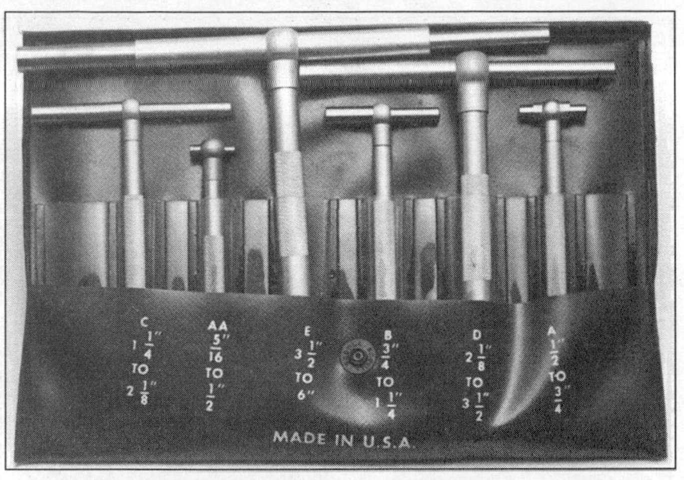


**1.44 Telescoping gauges are used with micrometers or calipers to determine the inside diameter of holes (like the cylinder bore) to see how much wear has occurred**

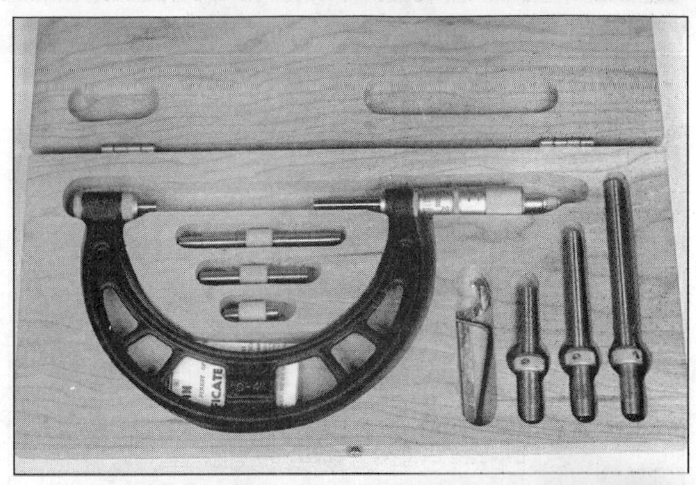

**1.45 Micrometers are needed for precision measurements to check for wear - they're available in two styles: The Mandrel type, shown here, which has one frame and interchangeable mandrels which allow for measurements from 0 to 4-inches, and . . .**

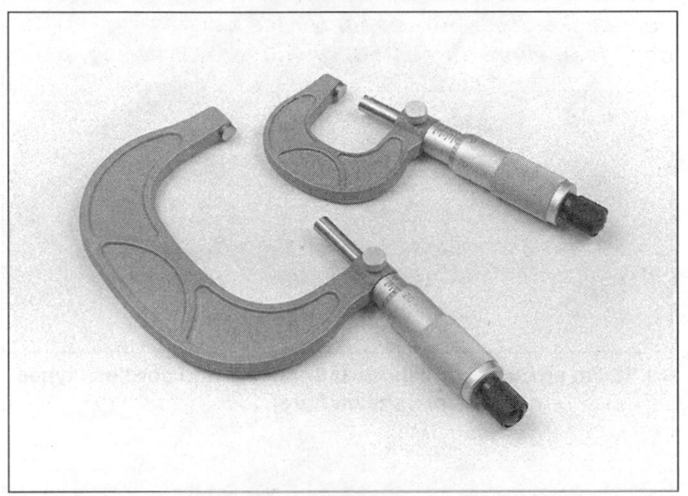

**1.46 . . . individual fixed-mandrel micrometers that are capable of making measurements in one inch increments (0 to 1, 1 to 2, 2 to 3, etc..)**

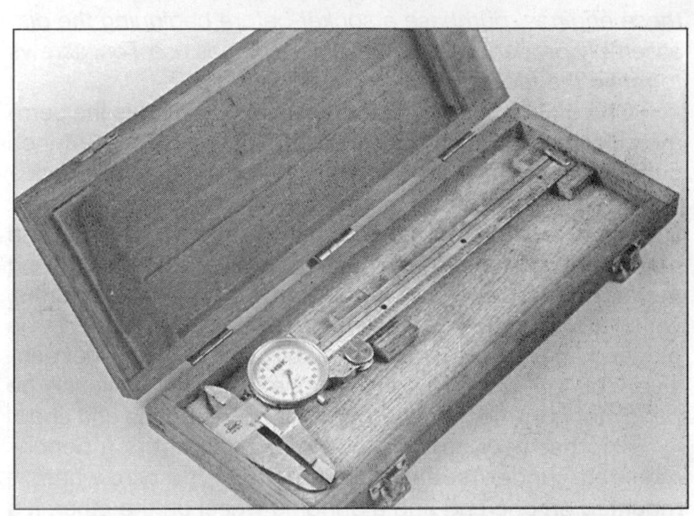

**1.47 Vernier or dial calipers (shown here) can be used in place of micrometers for most checks and can also be used for depth measurements**

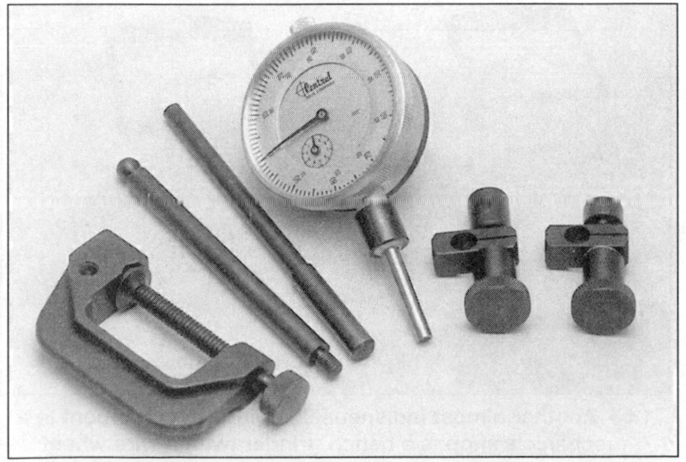

**1.48 A dial indicator can be used for end-play checks on crankshafts and camshafts**

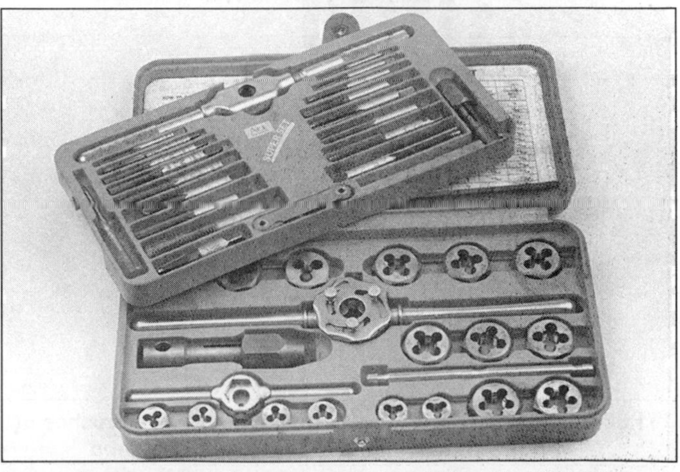

**1.49 A tap-and-die set is very handy for cleaning and restoring threads**

1.50 Some Tecumseh/Craftsman two-stroke engines require a no. 6 Torx socket for removal of the connecting rod bolts during an engine overhaul

*these engines, purchase a socket before beginning the disassembly procedure. Some Kohler engines use Torx screws to retain the fuel pump.*

One of the most indispensable tools around is the common electric drill **(see illustration 1.51)**. One with a 3/8-inch capacity chuck should be sufficient for most repair work - it'll be large enough to power a cylinder-surfacing hone. Collect several different wire brushes to use in the drill and make sure you have a complete set of sharp bits (for drilling metal, not wood) **(see illustration 1.52)**. Cordless drills, which are extremely versatile because they don't have to be plugged in, are now widely available and relatively inexpensive. You may want to consider one, since it'll obviously be handy for non-mechanical jobs around the house and shop.

Another very useful piece of equipment is a bench-mounted grinder **(see illustration 1.53)**. If a wire wheel is mounted on one end and a grinding wheel on the other, it's

1.51 An electrical drill (both 115-volt AC and cordless types are shown here), . . .

1.52 . . . a set of good-quality drill bits and wire brushes of various sizes will find many uses in the shop

1.53 Another almost indispensable piece of equipment in a mechanic's shop is a bench grinder (with a wire wheel mounted on one arbor) - make sure it's securely bolted down and never use it with the tool rests or eye shields removed!

very handy for cleaning up fasteners, sharpening tools and removing rust from parts. Make sure the grinder is fastened securely to the bench or stand, always wear eye protection when using it and never grind aluminum parts on the grinding wheel.

### Buying tools

For the do-it-yourselfer just starting to get involved in small engine maintenance and repair, there are a number of options available when purchasing tools. If maintenance and minor repair is the extent of the work to be done, the purchase of individual tools is satisfactory. If, on the other hand, extensive work is planned, it would be a good idea to purchase a modest tool set. A set can usually be bought at a substantial savings over the individual tool prices (and they often come with a tool box). As additional tools are needed, add-on sets, individual tools and a larger box can be purchased to expand the tool selection. Building a tool set gradually allows the cost to be spread over a longer period of time and gives the mechanic the freedom to choose only tools that will actually be used.

Tool stores and small engine distributors or dealers will often be the only source of some of the overhaul and special factory tools needed, but regardless of where tools are bought, try to avoid cheap ones (especially when buying screwdrivers, wrenches and sockets) because they won't last very long. The expense involved in replacing cheap tools will eventually be greater than the initial cost of quality tools. Read Chapter 2 for an in-depth, detailed look at choosing and using tools.

### Storage and care of tools

Good tools are expensive, so it makes sense to treat them with respect. Keep them clean and in usable condition and store them properly. Always wipe off dirt, grease and metal chips before putting them away. Never leave tools lying around in the work area.

Some tools, such as screwdrivers, pliers, wrenches and sockets, can be hung on a panel mounted on the garage or workshop wall, while others should be kept in a tool box or tray. Measuring instruments, gauges, cutting tools, etc. must be carefully stored where they can't be damaged by weather or impact from other tools.

When tools are used with care and stored properly, they'll last a very long time. However, even with the best of care, tools will wear out if used frequently. When a tool is damaged or worn out, replace it; subsequent jobs will be safer and more enjoyable if you do.

## Special factory tools

Each small engine manufacturer provides certain special tools to distributors and dealers for use when overhauling or doing major repairs on their engines. The distributors and dealers often stock some of the tools for sale to do-it-yourselfers and independent repair shops. A good example would be tools like the starter clutch wrench, flywheel holder and flywheel puller(s) supplied by Briggs & Stratton, which are needed for relatively simple procedures such as ignition point and flywheel key replacement (they're required to get the flywheel off for access to the ignition parts). If the special tools aren't used, the repair either can't be done properly or the engine could be damaged by using substitute tools. Fortunately, the tools mentioned are not very expensive or hard to find. For instance, depending on the carburetor used on your engine, a template tool may be needed to set the float height, but these tools are quite inexpensive.

Other special tools, like bushing drivers, bushing reamers, valve seat and guide service tools, cylinder sizing hones, main bearing repair sets, etc. are prohibitively expensive and not usually stocked for sale by dealers. If repairs requiring such tools are encountered, take the engine or components to a dealer with the necessary tools and pay to have the work done, then reassemble the engine yourself. If you do overhaul work on automotive engines, you may already have some of these specialized tools. In addition, these tools are found in almost every automotive machine shop, which are found in many more towns than a full-service small-engine shop.

## Notes

# 2 General shop practices

## Safety first!

Like it or not, a workshop can be a dangerous place. Electricity, especially if it's misused or taken for granted, is potentially harmful in an environment that's often damp. Hand and power tools, if misused, present opportunities for accidents and stored gasoline, solvents, lubricants and chemicals are a very real fire risk.

There's no way to make a shop totally safe (as long as people and potentially hazardous equipment/materials are involved) - the topic of safety really must focus on minimizing the risk of accidents by following safe shop practices (primary safety) and using the correct clothing and equipment to minimize injury in the event of an accident (secondary safety). The subject of safety is large and could easily fill a chapter on its own. To keep the subject within reasonable bounds - and because few people will bother to read an entire chapter on safety - its been confined here, initially, to a set of rules (Additional notes appear in the text and captions, where necessary, throughout the manual.)

The rest of this section covers some of the more important and relevant safety topics, but isn't intended to be definitive. Read through it, even if you've done mechanic work for years without scraping a knuckle. It should be emphasized that the most important piece of safety equipment of all is the human brain - try to get into the habit of thinking about what you're doing, and what could go wrong. A little common sense and foresight can prevent the majority of workshop accidents.

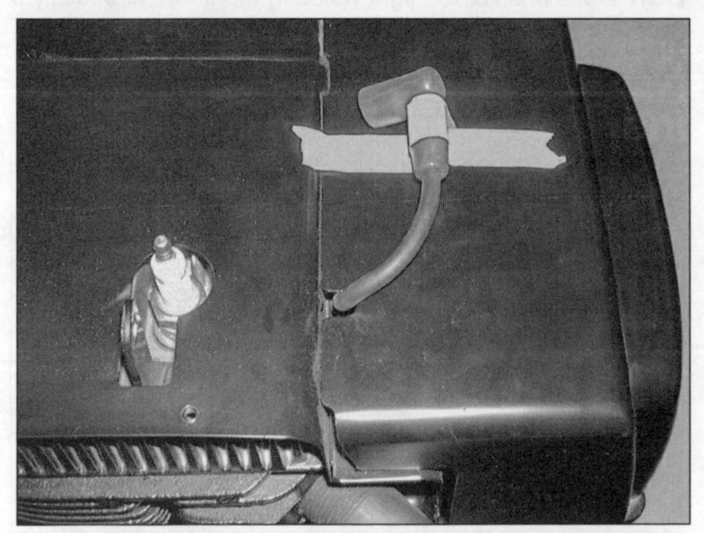

**2.1 Before doing any checks or maintenance of a small engine that requires you to turn the blade attached to the crankshaft, detach the wire from the spark plug and position it out of the way!**

## Safety rules

Professional mechanics are trained in safe working procedures. Regardless of how eager you are to start working on an engine or piece of equipment, take the time to read through the following list. As mentioned above, lack of attention, no matter how brief, can result in an accident. So can failure to follow certain simple safety precautions. The possibility of an accident will always exist, and the following points aren't intended to be a comprehensive list of all dangers; they are intended, however, to make you aware of the risks involved in mechanic work and encourage a safety-conscious approach to everything you do.

**DON'T** start the engine before checking to see if the drive is in Neutral (where applicable).

**DON'T** turn the blade attached to the engine unless the spark plug wire has been detached from the plug **(see illustration 2.1)** and positioned out of the way!

**DON'T** use gasoline for cleaning parts - ever!

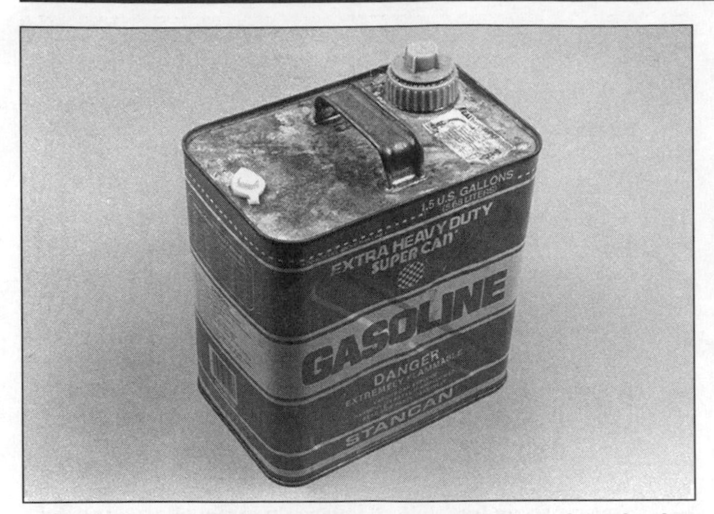


2.2 Store and transport gas in an approved metal or plastic container only - never use a glass bottle!

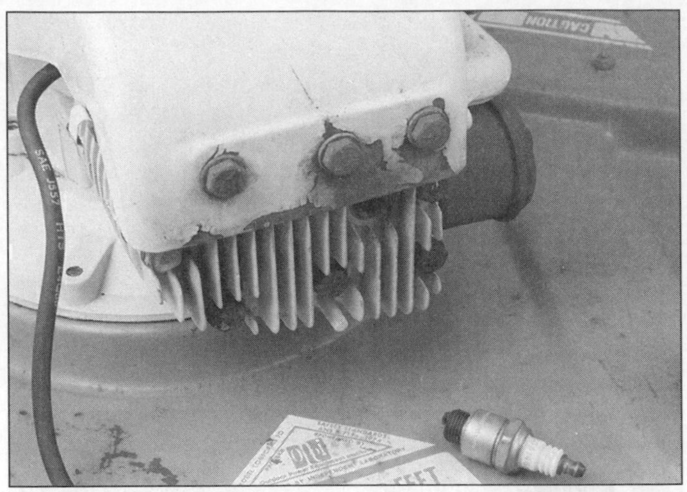

2.3 Don't try to clear a flooded engine by removing the spark plug and cranking the engine - the gasoline vapors coming out of the plug hole could be ignited

**DON'T** store gasoline in glass containers - use an approved metal or high-impact plastic gasoline container only **(see illustration 2.2)**!

**DON'T** store, pour or spill gasoline near an open flame or devices such as a stove, furnace, or water heater which utilizes a pilot light or devices that can create a spark.

**DON'T** smoke when filling the fuel tank!

**DON'T** fill the fuel tank while the engine is running. Allow the engine to cool for at least five minutes before refueling.

**DON'T** refuel equipment indoors where there's poor ventilation. Outdoor refueling is preferred.

**DON'T** operate the engine if a gasoline odor is present.

**DON'T** operate the engine if gasoline has been spilled. Move the machine away from the spill and don't start the engine until the gas has evaporated.

**DON'T** crank an engine with the spark plug removed **(see illustration 2.3)**. If the engine is flooded, open the throttle all the way and operate the starter until the engine starts.

**DON'T** attempt to drain the engine oil until you're sure its cooled so it won't burn you.

DON'T touch any part of the engine or muffler **(see illustration 2.4)** until its cooled down enough to avoid burns.

**DON'T** siphon toxic liquids, such as gasoline, by mouth or allow them to remain on your skin.

**DON'T** allow spilled oil or grease to remain on the floor - wipe it up before someone slips on it.

**DON'T** use loose fitting wrenches **(see illustration 2.5)** or other tools that may slip and cause injury.

**DON'T** push on wrenches when loosening or tightening nuts or bolts. Always try to pull the wrench towards you **(see illustration 2.6)**. If the situation calls for pushing the wrench away, push with an open hand to avoid scraped knuckles if the wrench slips.

**DON'T** use unshielded light bulbs in the shop, especially if gasoline is being used. Use an approved "trouble-light" only **(see illustration 2.7)**.

2.4 The cooling fins and mufflers can get extremely hot!

2.5 Wrenches that don't fit snugly on the fasteners can result in skinned knuckles, cuts and bruises

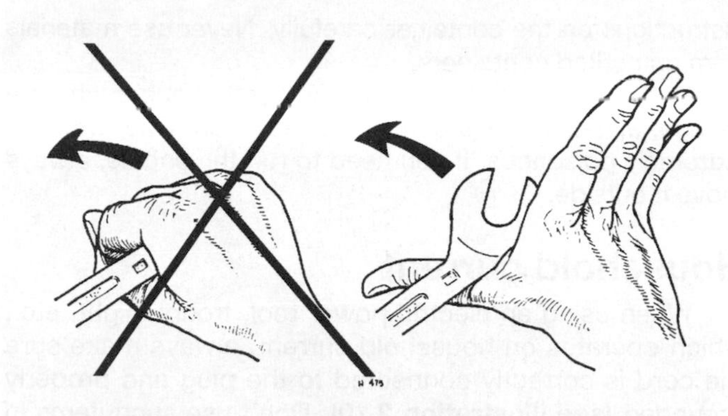

2.6 **Always pull on a wrench when loosening a fastener - if you can't pull on it, push with your hand open as shown here**

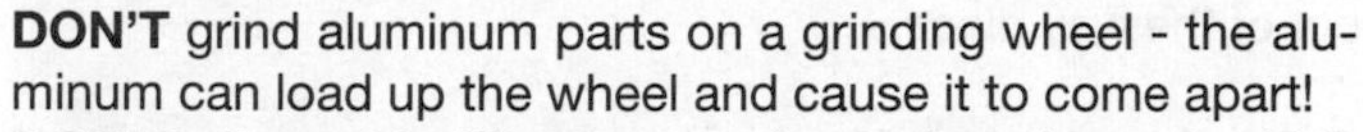

**DON'T** grind aluminum parts on a grinding wheel - the aluminum can load up the wheel and cause it to come apart!
**DON'T** attempt to lift a heavy piece of equipment which may be beyond your capability - get someone to help you.
**DON'T** rush or take unsafe shortcuts to finish a job.
**DON'T** allow children on or around equipment when you're working on it.
**DON'T** run an engine in an enclosed area. The exhaust contains carbon monoxide, an odorless, colorless, deadly poisonous gas.
**DON'T** operate an engine with a build-up of grass, leaves, dirt or other combustible material in the muffler area.
**DON'T** use equipment on any forested, brush-covered, or grass-covered unimproved land unless the engine has a spark arrester installed on the muffler.
**DON'T** run an engine with the air cleaner or cover (directly over the carburetor air intake) removed **(see illustration 2.8).**
**DON'T** store lubricants and chemicals near a heater or other sources of heat or sparks.
**DO** wear eye protection when using power tools such as a drill, bench grinder, etc. **(see illustration 2.9).**
**DO** keep loose clothing and long hair well out of the way of moving parts.

2.8 **Do not run the engine with the air cleaner removed**

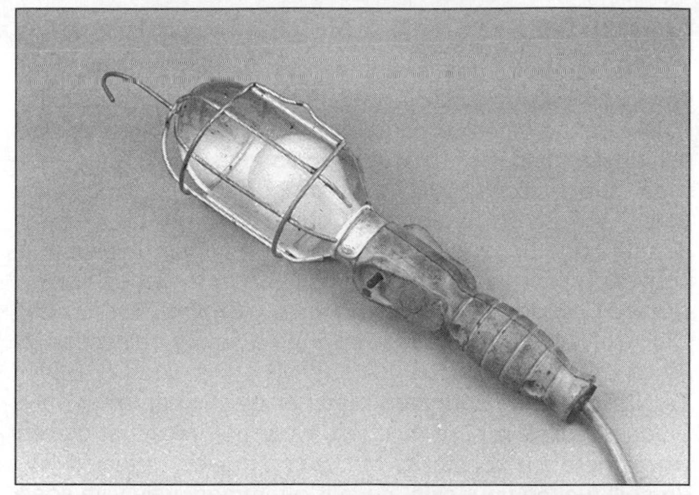

2.7 **Never use an unshielded light bulb in the shop - special "trouble-lights" designed to prevent broken bulbs and the accompanying safety hazards are commonly available**

**DO** wear steel-toe safety shoes when working on equipment on a bench. If heavy parts are dropped or fall, they won't crush your toes.
**DO** get someone to check on you periodically when working alone.
**DO** carry out work in a logical sequence and make sure everything is correctly assembled and tightened.
**DO** keep lubricants, chemicals and other fluids tightly capped and out of the reach of children and pets.

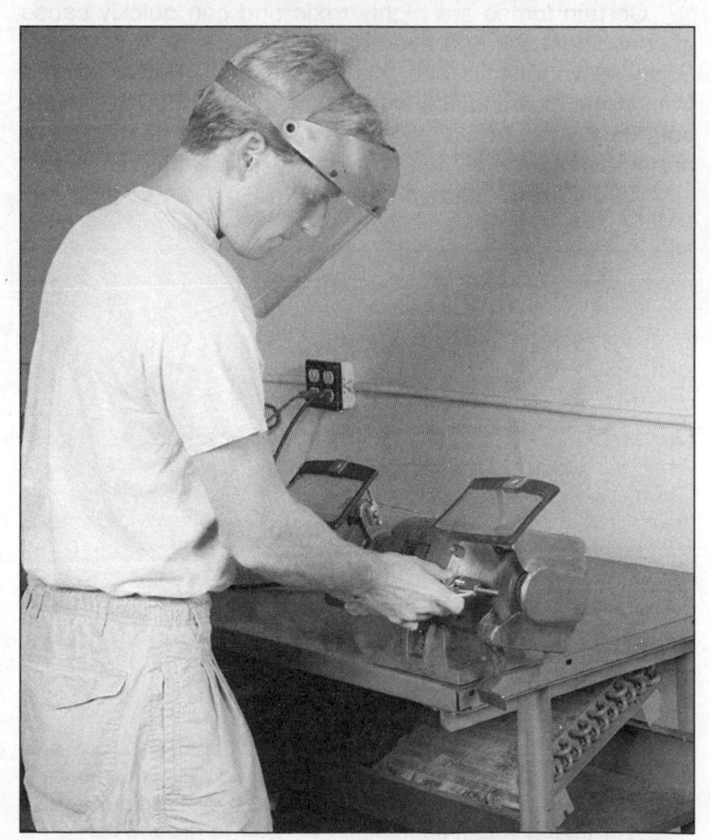

2.9 **Always wear eye protection when using power tools!**

## Gasoline

Remember - gasoline is extremely flammable! Never smoke or have any kind of open flames or unshielded light bulbs around when working on an engine in the shop. The risk doesn't end there however - a spark caused by an electrical short-circuit, by two metal surfaces striking each other, or even static electricity built up in your body under certain conditions, can ignite gasoline vapors, which in a confined space are highly explosive. As mentioned above, DO NOT, under any circumstances, USE GASOLINE FOR CLEANING PARTS; use an approved safety solvent only! Also, DO NOT STORE GASOLINE IN A GLASS CONTAINER - use an approved metal or plastic container only! Since gasoline is a carcinogen, wear fuel-resistant gloves when there's a possibility of being exposed to fuel. If you do spill fuel on your skin, rinse it off immediately with soap and water. Mop up any spills immediately and don't store fuel-soaked rags where they could ignite.

## Fire

Always have a fire extinguisher suitable for use on fuel and electrical fires handy in the garage or workshop. Never try to extinguish a gasoline or electrical fire with water! Have the fire department phone number posted near the telephone!

## Fumes

Certain fumes are highly toxic and can quickly cause unconsciousness and even death if inhaled to any extent. Gasoline vapor falls into this category, as well as vapors from some cleaning solvents. Draining and pouring of such volatile fluids should be done in a well-ventilated area, preferably outdoors.

When using cleaning fluids and solvents, read the instructions on the container carefully. Never use materials from unmarked containers.

Don't run the engine in an enclosed space such as a garage; exhaust fumes contain carbon monoxide, which is extremely poisonous. If you need to run the engine, always move it outside.

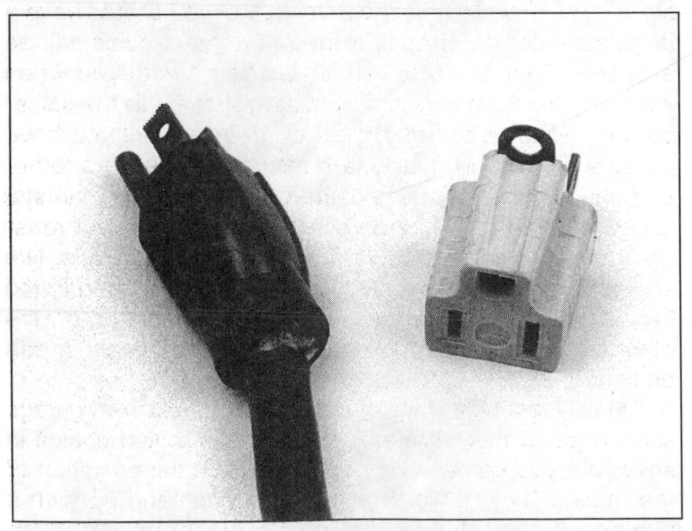

**2.10 Check the plugs on power tools and extension cords to make sure they're securely attached, with no burned or frayed wires, and use an adapter to ground the plug at the outlet if necessary**

## Household current

When using an electric power tool, trouble-light, etc., which operates on household current, always make sure the cord is correctly connected to the plug and properly grounded **(see illustration 2.10)**. Don't use such items in damp conditions and, again, don't create a spark or apply excessive heat in the vicinity of fuel or fuel vapor. Never string extension cords together to supply electricity to an out of the way place.

## Spark plug voltage

A severe electric shock can result from touching certain parts of the ignition system (such as the spark plug wire) when the engine is running or being cranked, particularly if components are damp or the insulation is defective. If an electronic ignition system is involved, the secondary system voltage is much higher and could prove fatal.

## Keep it clean

Get in the habit of taking a regular look around the shop, checking for potential dangers. The work area should always be kept clean and neat - all debris should be swept up and disposed of as soon as possible. Don't leave tools lying around on the floor.

Be very careful with oily rags. If they're left in a pile, it's not uncommon for spontaneous combustion to occur, so dispose of them properly in a covered metal container.

Check all equipment and tools for security and safety hazards (like frayed cords). Make necessary repairs as soon as a problem is noticed - don't wait for a shelf unit to collapse before fixing it.

## Accidents and emergencies

These range from minor cuts and skinned knuckles to serious injuries requiring immediate medical attention. The former are inevitable, while the latter are, hopefully, avoidable or at least uncommon. Think about what you would do in the event of an accident. Get some first aid training and have an adequate first aid kit somewhere within easy reach.

Think about what you would do if you were badly hurt and incapacitated. Is there someone nearby who could be summoned quickly? If possible, never work alone just in case something goes wrong.

If you had to cope with someone else's accident, would you know what to do? Dealing with accidents is a large and complex subject, and it's easy to make matters worse if you have no idea how to respond. Rather than attempt to deal with this subject in a superficial manner, buy a good First Aid book and read it carefully.

## Environmental safety

At the time this manual was being written, several state and Federal regulations governing the storage and disposal of oil and other lubricants, gasoline and solvents - petroleum-based substances in general - were pending (contact the appropriate government agency or your local auto parts store for the latest information). Be absolutely certain that all materials are properly stored, handled and disposed of. Never pour used or leftover oil or solvents down the drain or dump them on the ground. Also, don't allow volatile liquids to evaporate - keep them in sealed containers.

The old oil drained from the engine cannot be reused in its present state and should be recycled. Check with your local refuse disposal company, disposal facility or environmental agency to see if they will accept the oil for recycling. Don't pour used oil into drains or onto the ground. After the oil has cooled, it can be drained into a container (plastic jugs, bottles, milk cartons, etc.) for transport to one of these disposal sites.

# How to buy and use tools

Chances are you already own some of the tools in the lists included in Chapter 1. Many of them are the same ones needed for home maintenance and simple car repairs. This chapter will cover the types of tools to buy, assuming you'll need more, and how to use them properly so the repairs you tackle will be enjoyable and successful.

It's easy to fall into the trap of thinking you should only purchase individual, high-quality tools, gradually expanding your tool set as needs change and finances allow. This is good advice on the subject, and is normally suggested in how-to books and magazine articles, but it's difficult to follow through on. For starters, a glance through any mechanic's tool collection will reveal a very mixed assortment of tools. You'll usually find top-quality, lifetime guaranteed items alongside cheap tools purchased on the spur of the moment from many sources.

There seems to be a law governing the contents of tool boxes that dictates any expensive, well-made and indispensable tool will get lost or "disappear" very quickly, but during your short ownership of it, it'll never break, slip or damage fasteners. Conversely, a cheap, ill-fitting and poorly made tool will be with you for life, even when you thought you had thrown it away. It'll never quite fit properly and will probably drive you crazy. Although this is a broad generalization, it does happen in the real world. There are some very methodical and organized people out there who unfailingly clean and check each tool after use, before hanging it back up on the pegboard hook or placing it in its special drawer in the tool box. While there are few who practice this disciplined treatment of tools, there's no denying it's the correct approach and should be encouraged.

There are also those to whom the idea of using the correct tool is completely foreign and who will cheerfully tackle the most complex overhaul procedures with only a set of cheap open-end wrenches of the wrong type, a single screwdriver with a worn tip, a large hammer and an adjustable wrench. This approach is undeniably wrong and should be avoided - but while it often results in damaged fasteners and components, people often get away with it.

It's a good idea to strive for a compromise between these two extremes and, like most mechanics, cultivate a vision of the ideal workshop that's tempered by economic realities. This will inevitably lead to a mixed assortment of tools and seems to end up as the controlling factor in most workshops.

In this chapter we'll also try to give you some kind of idea when top-quality tools are essential and where cheaper ones will be adequate. As a general rule, if tools will be used often, purchase good-quality ones - if they'll be used infrequently, lower quality ones will usually suffice. If you're unsure about how much use a tool will get, the following approach may help. For example, if you need a set of combination wrenches but aren't sure which sizes you'll end up using most, buy a cheap or medium-priced set (make sure the jaws fit the fastener sizes marked on them). After some use over a period of time, carefully examine each tool in the set to assess its condition. If all the tools fit well and are undamaged, don't bother buying a better set. If one or two are worn, replace them with high-quality items - this way you'll end up with top-quality tools where they're needed most and the cheaper ones are sufficient for occasional use. On rare occasions you may conclude the whole set is poor quality. If so, buy a better set, if necessary, and remember never to buy that brand again.

The best place to buy hand tools is an auto parts store, tool store or the tool department at your nearest Sears store. You may not find cheap tools, but you should have a large selection to choose from and expert advice will be available. Take the tool lists in Chapter 1 with you when shopping for tools and explain what you want to the salesperson. Sources to steer clear of, at least until you have experience judging quality, are mail order suppliers (other than Sears or those selling name-brands) and flea markets. Some of them offer good value for the money, but most carry cheap, imported tools of dubious quality. Tools, like any other consumer product, are often counterfeited in the Far East. The resulting tools can be acceptable or, on the other hand, they might be unusable. Unfortunately, it can be hard to judge by looking at them.

Finally, consider buying secondhand tools from garage sales or used tool outlets. You may have limited choice in sizes, but you can usually determine from the condition of the tools if they're worth buying. You can end up with a number of unwanted or duplicate tools, but it's a cheap way of putting a basic tool kit together, and you can always sell off any surplus tools later.

## Buying wrenches and sockets

Wrenches of varying quality are available and cost is usually a good indication of quality - the more they cost, the better they are. In the case of wrenches, it's important to buy high-quality tools. Your wrenches will be some of the most often used tools in the shop, so buy the best you can afford.

Buy a set with the sizes outlined in Chapter 1. The size stamped on the wrench **(see illustration 2.11)** indicates the distance across the nut or bolt head (or the distance between the wrench jaws), in inches, not the diameter of the threads on the fastener. For example, a 1/4-inch bolt will almost always have a 7/16-inch hex-head - the size of the wrench required to loosen or tighten it. In the case of metric tools, the number is in millimeters. At the risk of confusing the issue, it should be mentioned the relationship between thread diameter and hex size doesn't always hold true; in some applications, an unusually small hex may be used, either for reasons of limited space around the fastener or to discourage over-tightening. Conversely, in some areas, fasteners with a disproportionately large hex-head may be encountered.

Wrenches tend to look similar, so it can be difficult to judge how well they're made just by looking at them. As with most other purchases, there are bargains to be had, just as there are overpriced tools with well-known brand names. On the other hand, you may buy what looks like a good set of wrenches only to find they fit badly or are made from poor-quality steel.

With a little experience, though, it's possible to judge the quality of a tool by looking at it. Often, you may have come across the brand name before and have a good idea of the quality. Close examination of the tool can often reveal some hints as to its quality. Prestige tools are usually polished and chrome-plated over their entire surface, with the working faces ground to size. The polished finish is largely cosmetic, but it does make them easy to keep clean. Ground jaws normally indicate the tool will fit well on fasteners.

A side-by-side comparison of a high-quality wrench with a cheap equivalent is an eye opener. The better tool will be made from a good-quality material, often a forged/chrome-vanadium steel alloy **(see illustration 2.12)**. This, together with careful design, allows the tool to be kept as small and compact as possible. If, by comparison, the cheap tool is thicker and heavier, especially around the jaws, it's usually because the extra material is needed to compensate for its lower quality.

If the tool fits properly, this is not necessarily bad - it is, after all, cheaper - but in situations where it's necessary to work in a confined area, the cheaper tool may be too bulky to fit.

## Open-end wrenches

The open-end wrench is the most common type, due mainly to its general versatility. It normally consists of two open jaws connected by a flat handle section. The jaws usually vary by one size, with an occasional overlap of sizes between consecutive wrenches in a set. This allows one wrench to be used to hold a bolt head while a similar-size nut is removed. A typical fractional-size wrench set might have the following jaw sizes: 1/4 x 5/16, 3/8 x 7/16, 1/2 x 9/16, 9/16 x 5/8 and so on.

Typically, the jaw end is set at an angle to the handle, a feature which makes them very useful in confined spaces; by turning the nut or bolt as far as the obstruction allows, then turning the wrench over so the jaw faces in the other direction, it's possible to move the fastener a fraction of a turn at a time **(see illustration 2.13)**. The handle length is generally determined by the size of the jaw and is calculated to allow a nut or bolt to be tightened sufficiently by hand with minimal risk of breakage or thread damage (though this doesn't apply to soft materials like brass or aluminum).

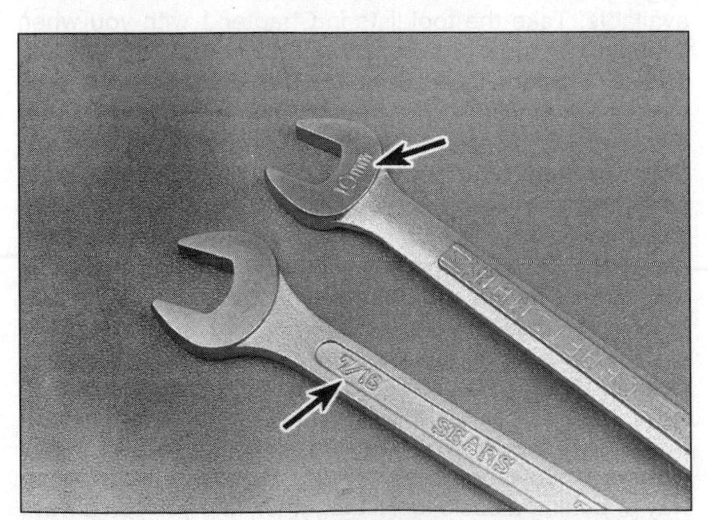


**2.11 Wrench sizes are clearly stamped on the ends or handle**

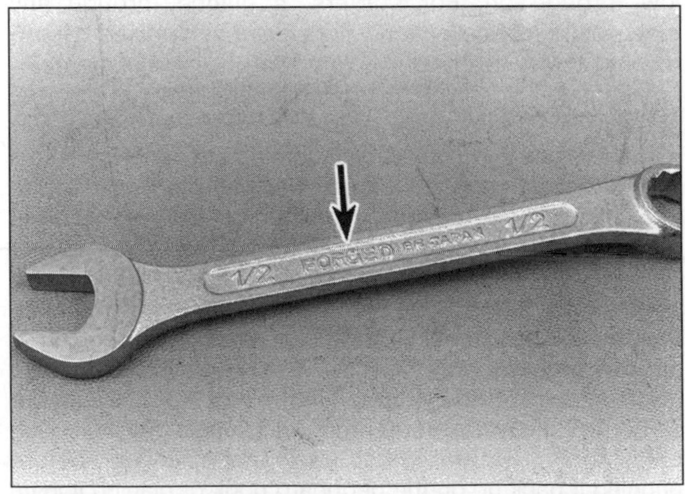


**2.12 Look for the words "chrome vanadium" or "forged" when trying to determine wrench quality**

Common open-end wrenches are usually sold in sets and it's rarely worth buying them individually unless it's to replace a lost or broken tool from a set.

Single tools invariably cost more, so check the sizes you're most likely to need regularly and buy the best set of wrenches you can afford in that range of sizes. If money is limited, remember that you'll use open-end wrenches more than any other type - it's a good idea to buy a good set and cut corners elsewhere.

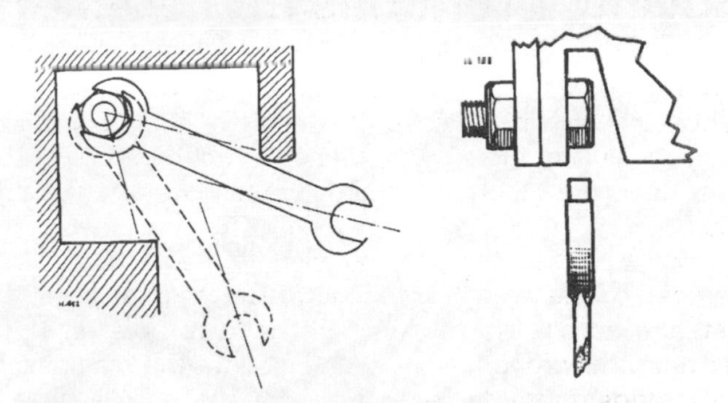

**2.13 Open-end wrenches are the most versatile for general uses**

## Box-end wrenches

A box-end wrench consists of a ring-shaped end with a 6-point (hex) or 12-point (double hex) opening **(see illustration 2.14)**. This allows the tool to fit on the fastener hex at 15 (12-point) or 30-degree (6-point) intervals. Normally, each tool has two ends of different sizes, allowing an overlapping range of sizes in a set, as described for open-end wrenches.

Although available as flat tools, the handle is usually offset at each end to allow it to clear obstructions near the fastener, which is normally an advantage. In addition to normal length wrenches, it's also possible to buy long handle types to allow more leverage (very useful when trying to loosen rusted or seized nuts). It is, however, easy to shear off fasteners if not careful, and sometimes the extra length impairs access.

As with open-end wrenches, box-ends are available in varying quality, again often indicated by finish and the amount of metal around the ends. While the same criteria should be applied when selecting a set of box-end wrenches, if your budget is limited, go for better quality open-end wrenches and a slightly cheaper set of box-ends.

## Combination wrenches

These wrenches combine a box-end and open-end of the same size in one tool and offer many of the advantages of both. Like the others, they're widely available in sets and as such are probably a better choice than box-ends only. They're generally compact, short-handled tools and are well suited for small engine repairs, where access is often restricted.

## Adjustable wrenches

These tools come in a wide variety of shapes and sizes with various types of adjustment mechanisms. The principle is the same in each case - a single tool that can handle fasteners of various sizes. Adjustable wrenches are not as good as single-size tools and it's easy to damage fasteners with them. However, they can be an invaluable addition to any tool kit - if they're used with discretion. **Note:** *If you attach the wrench to the fastener with the movable jaw pointing in the direction of wrench rotation* **(see illustration 2.15)**, *an adjustable wrench will be less likely to slip and damage the fastener head.*

The most common adjustable wrench is the open-end type with a set of parallel jaws that can be set to fit the head of a fastener. Most are controlled by a threaded spindle, though there are various cam and spring-loaded versions available. Don't buy large tools of this type; you'll rarely be able to find enough clearance to use them. The sizes specified in Chapter 1 are best suited to small engine repair work.

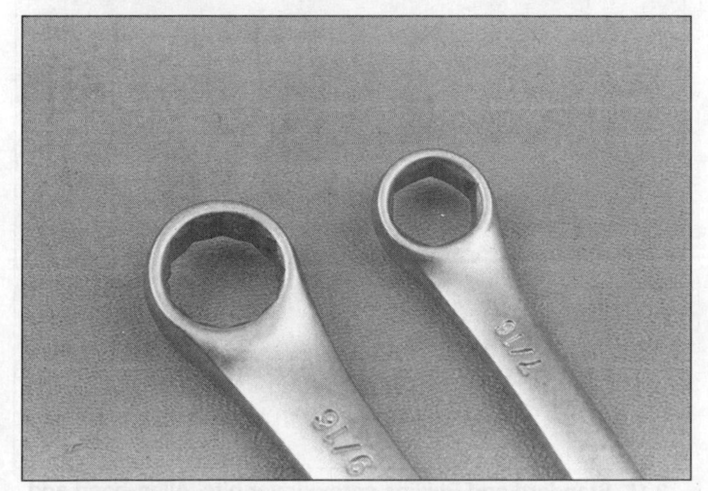

**2.14 Box-end wrenches are available in both 6 and 12-point openings - if you have a choice, buy 6-point wrenches**

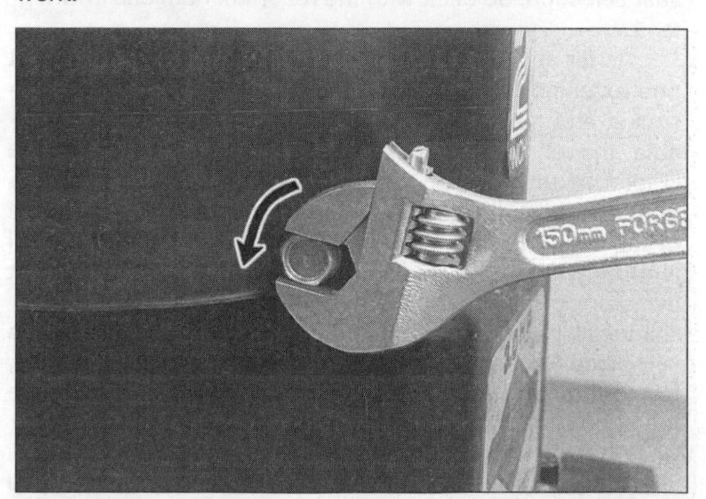


**2.15 When using an adjustable wrench, the movable jaw should point in the direction the wrench is being turned (arrow) so the wrench doesn't distort and slip off the fastener head**

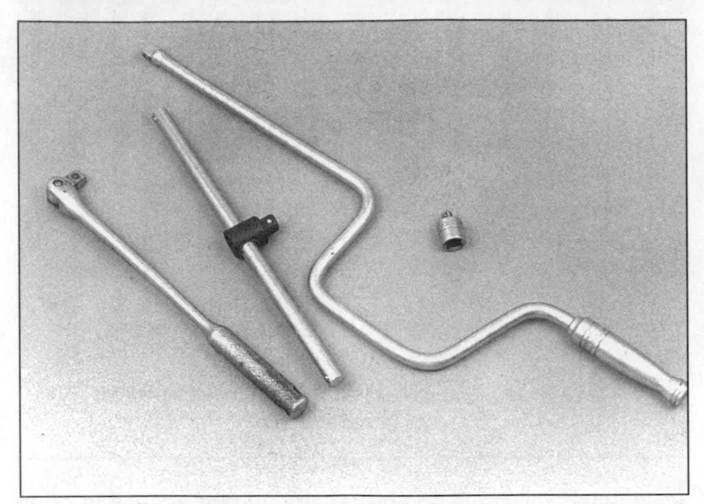

2.16 Many accessories are available in each drive size for use with sockets (left to right: Breaker bar, sliding T-handle, speed handle and 3/8 to 1/4-inch drive adapter)

2.17 Deep sockets are handy for loosening/tightening recessed bolts and nuts threaded onto long bolts or studs

## Socket sets

A refined version of the box-end wrench, interchangeable sockets consist of a forged steel alloy cylinder with a hex or double hex formed inside one end. The other end is formed into the square drive recess that engages over the corresponding square end of various socket drive tools.

Sockets are available in 1/4, 3/8, 1/2 and 3/4-inch drive sizes. Of these, a 3/8-inch drive set is most useful for small engine repairs, although 1/4-inch drive sockets and accessories may occasionally be needed.

The most economical way to buy sockets is in a set. As always, quality will govern the cost of the tools. Once again, the "buy the best" approach is usually advised when selecting sockets. While this is a good idea, since the end result is a set of quality tools that should last a lifetime, the cost is so high it's difficult to justify the expense for home use. Go shopping for a socket set and you'll be confronted with a vast selection, so stick with the recommendations in Chapter 1.

As far as accessories go, you'll need a ratchet, at least one extension (buy a three or six inch size), a spark plug socket and maybe a T-handle or breaker bar. Other desirable, though less essential items, are a speeder handle, a U-joint, extensions of various other lengths and adapters from one drive size to another **(see illustration 2.16)**. Some of the sets you find may combine drive sizes; they're well worth having if you find the right set at a good price, but avoid being dazzled by the number of pieces.
Above all, be sure to completely ignore any label that reads "86-piece Socket Set"; this refers to the number of pieces, not to the number of sockets (and in some cases even the metal box and plastic insert are counted in the total!).

Apart from well-known and respected brand names, you'll have to take a chance on the quality of the set you buy. If you know someone who has a set that has held up well, try to find the same brand, if possible. Take a few nuts and bolts with you and check the fit in some of the sockets. Check the operation of the ratchet. Good ones operate smoothly and crisply in small steps; cheap ones are coarse and stiff - a good basis for guessing the quality of the rest of the pieces.

One of the best things about a socket set is the built-in provision for expansion. Once you have a basic set, you can purchase extra sockets when needed and replace worn or damaged tools. There are special deep sockets for reaching recessed fasteners or to allow the socket to fit over a projecting bolt or stud **(see illustration 2.17)**. You can also buy screwdriver, Allen and Torx bits to fit various drive tools (they can be very handy in some applications) **(see illustration 2.18)**. Most socket sets include a special deep socket for spark plugs. They have rubber inserts to protect the spark plug porcelain insulator and hold the plug in the socket to avoid burned fingers.

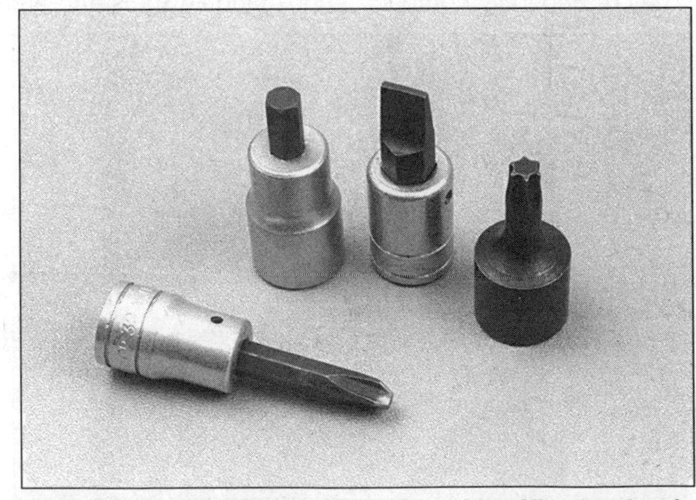

2.18 Standard and Phillips screwdriver bits, Allen-head and Torx drivers are available for use with ratchets and other socket drive tools

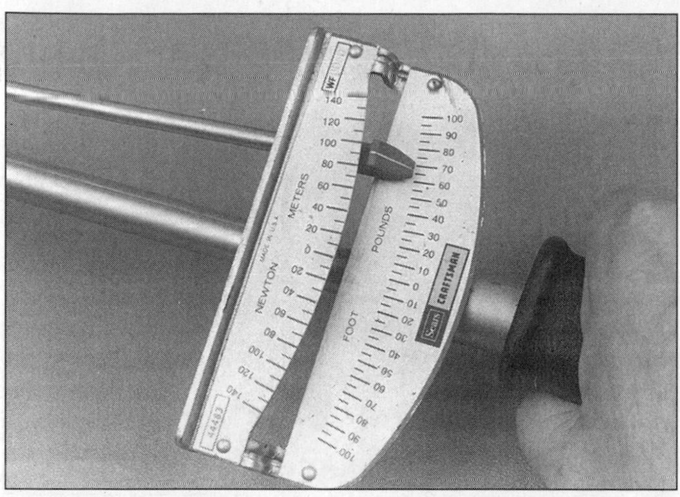


**2.19 A simple, inexpensive, deflecting beam torque wrench will be adequate for small engine repairs - the torque figure is read off the scale near the handle**

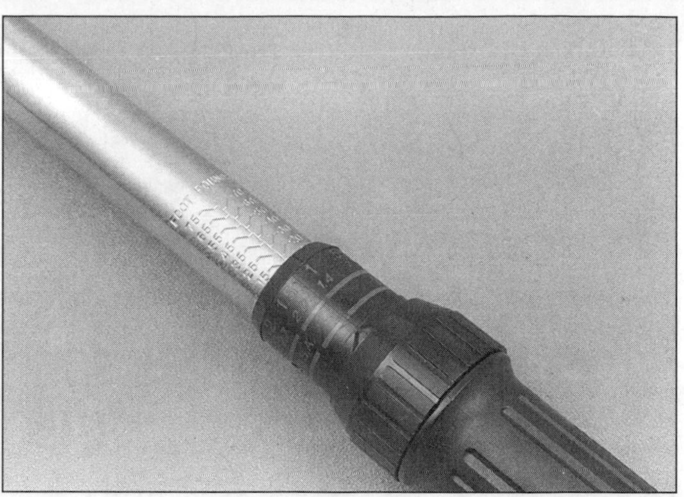

**2.20 "Click" type torque wrenches can be set to "give" at a pre-set torque, which makes them very accurate and easy to use**

## Torque wrenches

Torque wrenches compliment the socket set, since they require the use of a socket so a fastener can be tightened accurately to a specified torque figure. To attempt an engine overhaul without a torque wrench is to invite oil leaks, distortion of the cylinder head, damaged or stripped threads or worse.

The cheapest type of torque wrench consists of a long handle designed to bend as pressure increases. A long pointer is fixed to the drive end and reads off a scale near the handle as the fastener is tightened **(see illustration 2.19)**. This type of torque wrench is simple and usually accurate enough for most jobs. Another version is the pre-set or "click" type. The torque figure required is dialed in on a scale before use **(see illustration 2.20)**. The tool gives a positive indication, usually a loud click and/or a sudden movement of the handle, when the desired torque is reached. Needless to say, the pre-set type is far more expensive than the beam type - you alone can decide which type you need. For occasional use, go for the cheaper beam type.

Torque wrenches are available in a variety of drive sizes and torque ranges for particular applications. For small engine use, the range required is lower than for cars or trucks; 0 to 75 ft-lbs should be adequate. However, if you anticipate doing car repairs in the future, you may want to take that into consideration when buying a torque wrench - try to settle on one that'll be usable for both.

Actually, for almost all small-engine repair, an inch-pound torque wrench is more useable. Except for the flywheel bolt/nuts and the cylinder head bolts/nuts, torque values are rarely over 25 ft-lbs. Most cylinder head fasteners are torqued in the 25 to 35 ft-lb range, and flywheel fasteners are usually not much higher. The exception would be some engines that have only a single nut retaining the flywheel, which may require 100 ft-lbs or more. It may make more sense to own an inch-pound torque wrench (which will also be useful in working on today's automobiles), and rent the larger torque wrench for the few occasions when you need to torque a flywheel nut.

## Impact drivers

The impact driver belongs with the screwdrivers, but it's mentioned here since it can also be used with sockets (impact drivers normally are 3/8-inch square drive). An impact driver works by converting a hammer blow on the end of its handle into a sharp twisting movement. While this is a great way to jar a seized fastener loose, the loads imposed on the socket are excessive. Use sockets only with discretion and expect to have to replace damaged ones occasionally.

# Using wrenches and sockets

In the last section we looked at some of the various types of wrenches available, with a few suggestions about building up a tool collection without bankrupting yourself. Here we're more concerned with using the tools in actual work. Although you may feel it's self-explanatory, it's worth some thought. After all, when did you last see instructions for use supplied with a set of wrenches?

## Which wrench?

Before you start tearing an engine apart, figure out the best tool for the job; in this instance the best wrench for a hex-head fastener. Sit down with a few nuts and bolts and look at how various tools fit the bolt heads.

A good rule of thumb is to choose a tool that contacts

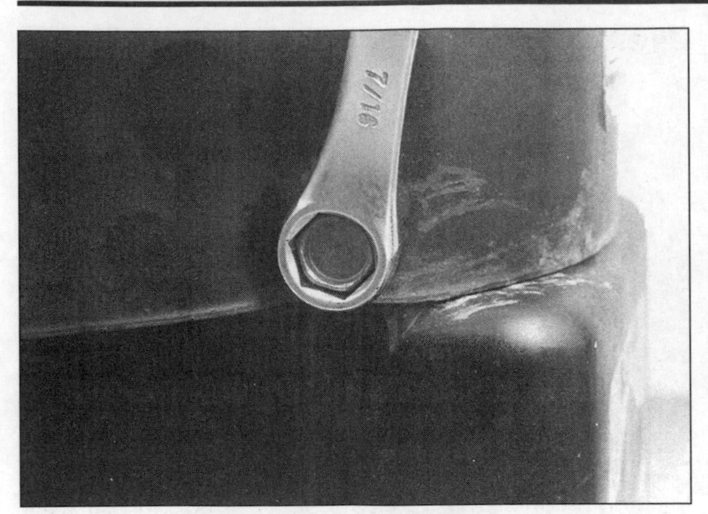

**2.21 A 6-point box-end wrench or socket contacts the nut or bolt head entirely, which spreads out the force and tends to prevent rounded-off corners**

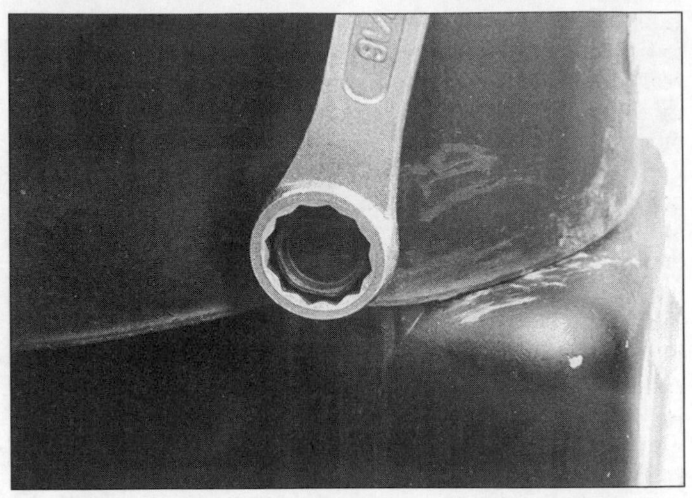

**2.22 A 12-point box-end wrench or socket only contacts the nut or bolt head near the corners and concentrates the force at specific points, which leads to rounded-off fasteners and a frustrated mechanic**

the largest area of the hex-head. This distributes the load as evenly as possible and lessens the risk of damage. The shape most closely resembling the bolt head or nut is another hex, so a 6-point socket or box-end wrench is usually the best choice **(see illustration 2.21)**. Many sockets and box-end wrenches have double hex (12-point) openings. If you slip a 12-point box-end wrench over a nut, look at how and where the two are in contact. The corners of the nut engage in every other point of the wrench. When the wrench is turned, pressure is applied evenly on each of the six corners **(see illustration 2.22)**. This is fine unless the fastener head was previously rounded off or is made of extremely poor quality (soft) material. If so, the corners will be damaged and the wrench will slip. If you encounter a damaged bolt head or nut, always use a 6-point wrench or socket if possible. If you don't have one in the right size, choose a 12-point wrench or socket that fits securely and proceed carefully.

If you slip an open-end wrench over a hex-head fastener, you'll see the tool is in contact on two faces only **(see illustration 2.23)**. This is acceptable provided the tool and fastener are both in good condition. The need for a snug fit between the wrench and nut or bolt explains the recommendation to buy good-quality open-end wrenches. If the wrench jaws, the bolt head or both are damaged, the wrench will probably slip, rounding off and distorting the head. In some applications, an open-end wrench is the only possible choice due to limited access, but always check the fit of the wrench on the fastener before attempting to loosen it; if it's hard to get at with a wrench, think how hard it will be to remove after the head is damaged.

The last choice is an adjustable wrench or self-locking plier/wrench (Vise-Grips). Use these tools only when all else has failed. In some cases, a self-locking wrench may be able to grip a damaged head that no wrench could deal with, but be careful not to make matters worse by damaging it further.

Bearing in mind the remarks about the correct choice of tool in the first place, there are several things worth noting about the actual use of the tool. First, make sure the wrench head is clean and undamaged. If the fastener is rusted or coated with paint, the wrench won't fit correctly. Clean off the head and, if it's rusted, apply some penetrating oil. Let it soak in for a while before attempting removal.

It may seem obvious, but take a close look at the fastener to be removed before using a wrench. On many mass-produced machines, one end of a fastener may be fixed or captive, which speeds up initial assembly and usually makes removal easier. If a nut is installed on a stud or a bolt threads into a captive nut or tapped hole, you may have only one fastener to deal with. If, on the other hand, you have a separate nut and bolt, you'll have to hold the bolt head while the nut is removed. In some areas this can be difficult, particularly where engine mount bolts are

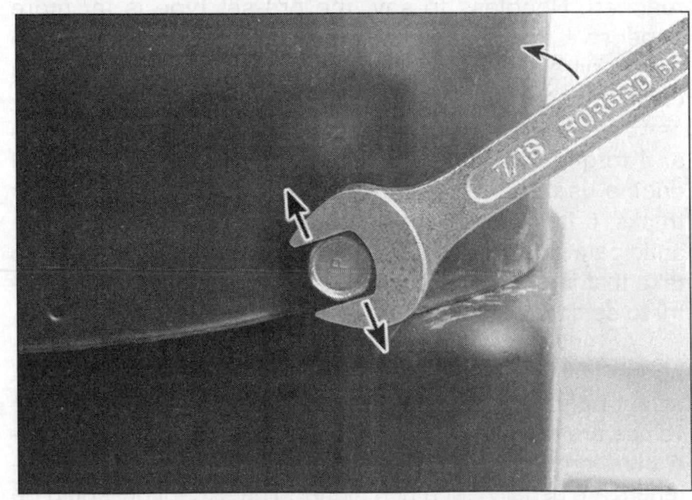


**2.23 Open-end wrench jaws tend to spread apart when loosening tight fasteners and can quickly damage a nut or bolt**

involved. In this type of situation you may need an assistant to hold the bolt head with a wrench, while you remove the nut from the other side. If this isn't possible, you'll have to try to position a box-end wrench so it wedges against some other component to prevent it from turning.

Be on the lookout for left-hand threads. They aren't common, but are sometimes used on the ends of rotating shafts to make sure the nut doesn't come loose during engine operation. If you can see the shaft end, the thread type can be checked visually. If you're unsure, place your thumbnail in the threads and see which way you have to turn your hand so your nail "unscrews" from the shaft. If you have to turn your hand counterclockwise, it's a conventional right-hand thread.

Beware of the upside-down fastener syndrome. If you're loosening an oil drain plug on the under side of a mower deck, for example, it's easy to get confused about which way to turn it. What seems like counterclockwise to you can easily be clockwise (from the plug's point of view). Even after years of experience, this can still catch you once in a while.

In most cases, a fastener can be removed simply by placing the wrench on the nut or bolt head and turning it. Occasionally, though, the condition or location of the fastener may make things more difficult. Make sure the wrench is square on the head. You may need to reposition the tool or try another type to obtain a snug fit. Make sure the engine you're working on is secure and can't move when you turn the wrench. If necessary, get someone to help steady it for you. Position yourself so you can get maximum leverage on the wrench.

If possible, locate the wrench so you can pull the end towards you. If you have to push on the tool, remember that it may slip, or the fastener may move suddenly. For this reason, don't curl your fingers around the handle or you may crush or bruise them when the fastener moves; keep your hand flat, pushing on the wrench with the heel of your thumb. If the tool digs into your hand, place a rag between it and your hand or wear a heavy glove.

If the fastener doesn't move with normal hand pressure, stop and try to figure out why before the fastener or wrench is damaged or you hurt yourself. Stuck fasteners may require penetrating oil, heat or an impact driver or air tool.

Using sockets to remove hex-head fasteners is less likely to result in damage than if a wrench is used. Make sure the socket fits snugly over the fastener head, then attach an extension, if needed, and the ratchet or breaker bar. Theoretically, a ratchet shouldn't be used for loosening a fastener or for final tightening because the ratchet mechanism may be overloaded and could slip. In some instances, the location of the fastener may mean you have no choice but to use a ratchet, in which case you'll have to be extra careful.

Never use extensions where they aren't needed. Whether or not an extension is used, always support the drive end of the breaker bar with one hand while turning it with the other. Once the fastener is loose, the ratchet can be used to speed up removal.

## Pliers

Generally speaking, three types of pliers are needed when doing mechanic work: Slip-joint, arc-joint or "Channel-lock" and Vise-Grips. Although somewhat limited in use, needle-nose pliers and side (wire) cutters should be included if your budget will allow it.

Slip-joint pliers have two open positions; a figure eight-shaped, elongated slot in one handle slips back-and-forth on a pivot pin on the other handle to change them. Good quality pliers have jaws made of tempered steel and there's usually a wire-cutter at the base of the jaws. The primary uses of slip-joint pliers are for holding objects, bending and cutting throttle wires and crimping and bending metal parts, not loosening nuts and bolts.

Arc-joint or "Channel-lock" pliers have parallel jaws that can be opened to various widths by engaging different tongues and grooves, or channels, near the pivot pin. Since the tool expands to fit many size objects, it has countless uses for small engine and equipment maintenance. Channel-lock pliers come in various sizes. The medium size is adequate for general work; small and large sizes are nice to have as your budget permits. You'll use all three sizes frequently.

Vise-Grips (a brand name) come in various sizes; the medium size with curved jaws is best for all-around work. However, buy a large and small one if possible, since they're often used in pairs. Although this tool falls somewhere between an adjustable wrench, a pair of pliers and a portable vise, it can be invaluable for loosening and tightening fasteners - it's the only pliers that should be used for this purpose.

The jaw opening is set by turning a knurled knob at the end of one handle. The jaws are placed over the head of the fastener and the handles are squeezed together, locking the tool onto the fastener **(see illustration 2.24)**. The design of the tool allows extreme pressure to be applied at the jaws and a variety of jaw designs enable the tool to grip

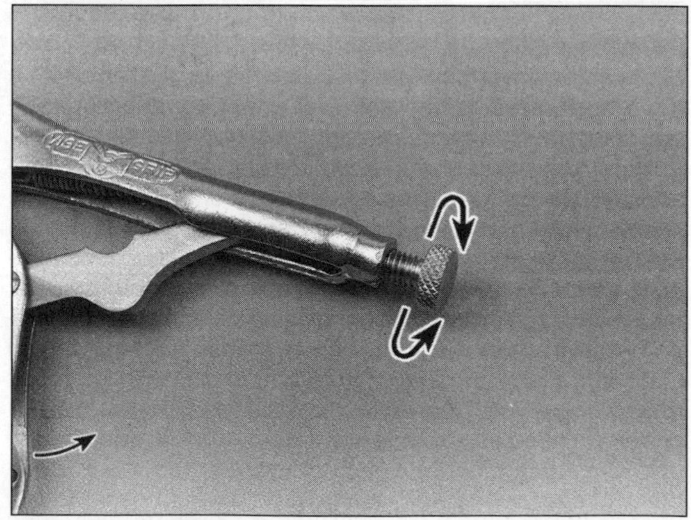

**2.24 Vise-Grips are adjusted with the knurled bolt, then the handles are closed to lock the jaws on the part**

2.25 As a last resort, you can use locking pliers to loosen a rusted or rounded-off nut or bolt

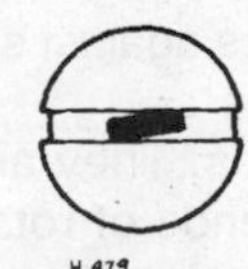



Misuse of a screwdriver - the blade shown is both too narrow and too thin and will probably slip or break off.

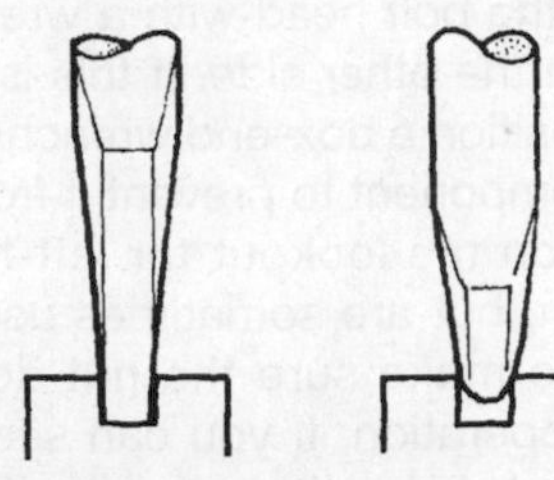

The left - hand example shows a snug - fitting tip. The right - hand drawing shows a damaged tip which will twist out of the slot when pressure is applied.

2.26 Standard screwdrivers - wrong size (left), correct fit in screw slot (center) and worn tip (right)

firmly even on damaged heads **(see illustration 2.25).** Vise-Grips are great for removing fasteners rounded off by badly-fitting wrenches.

As the name suggests, needle-nose pliers have long, thin jaws designed for reaching into holes and other restricted areas. Most needle-nose, or long-nose, pliers also have wire cutters at the base of the jaws.

Look for these qualities when buying pliers: Smooth operating handles and jaws, jaws that match up and grip evenly when the handles are closed, a nice finish and the word "forged" somewhere on the tool.

## Screwdrivers

Screwdrivers come in innumerable shapes and sizes to fit the various screw head designs in common use. Regardless of the tool quality, the screwdrivers in most tool boxes rarely fit the intended screw heads very well. This is attributable not only to general wear, but also to misuse. Screwdrivers make very tempting and convenient pry bars, chisels and punches, uses which in turn make them very bad screwdrivers.

A screwdriver consists of a steel blade or shank with a drive tip formed at one end. The most common tips are standard (also called straight slot and flat-blade) and Phillips. The other end has a handle attached to it. Traditionally, handles were made from wood and secured to the shank, which had raised tangs to prevent it from turning in the handle. Most screwdrivers now come with plastic handles, which are generally more durable than wood.

The design and size of handles and blades vary considerably. Some handles are specially shaped to fit the human hand and provide a better grip. The shank may be either round or square and some have a hex-shaped bolster under the handle to accept a wrench to provide more leverage when trying to turn a stubborn screw. The shank diameter, tip size and overall length vary too. As a general rule, it's a good idea to use the longest screwdriver possible, which allows the greatest possible leverage.

If access is restricted, a number of special screwdrivers are designed to fit into confined spaces. The "stubby" screwdriver has a specially shortened handle and blade. There are also offset screwdrivers and special screwdriver bits that attach to a ratchet or extension.

## Standard screwdrivers

These are used to remove and install conventional slotted screws and are available in a wide range of sizes denoting the width of the tip and the length of the shank (for example: A 3/8 x 10-inch screwdriver is 3/8-inch wide at the tip and the shank is 10-inches long). You should have a variety of screwdrivers so screws of various sizes can be dealt with without damaging them. The blade end must be the same width and thickness as the screw slot to work properly, without slipping. When selecting standard screwdrivers, choose good-quality tools, preferably with chrome moly, forged steel shanks. The tip of the shank should be ground to a parallel, flat profile (hollow ground) and not to a taper or wedge shape, which will tend to twist out of the slot when pressure is applied **(see illustration 2.26).**

All screwdrivers wear in use, but standard types can be reground to shape a number of times. When reshaping a tip, start by grinding the very end flat at right angles to the shank. Make sure the tip fits snugly in the slot of a screw of the appropriate size and keep the sides of the tip parallel. Remove only a small amount of metal at a time to avoid overheating the tip and destroying the temper of the steel.

## Phillips screwdrivers

Some engines have Phillips screws that are installed during initial assembly with air tools and are next to impossible to remove later without ruining the heads, particularly if the wrong size screwdriver is used. Be sure to use only Phillips type screwdrivers on them; other cross-head patterns are available, but they won't work on Phillips screws.

The only way to ensure the tools you buy will fit properly, is to take a couple of screws with you to make

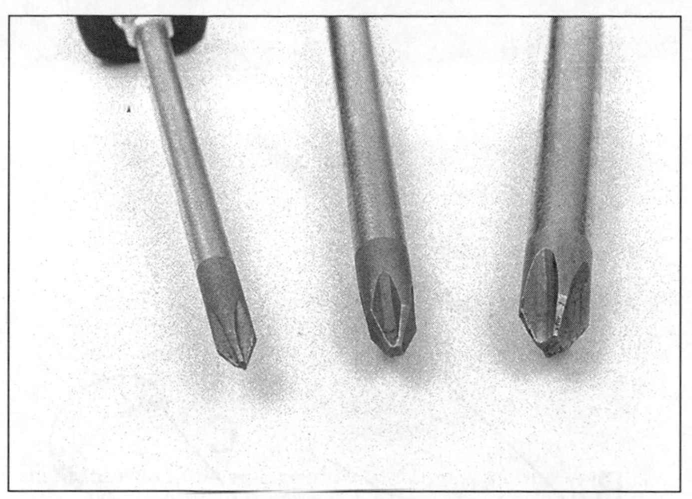

2.27 The tip size on a Phillips screwdriver is indicated by a number from 1 to 4, with 1 being the smallest (left - no. 1; center - no. 2; right - no. 3)

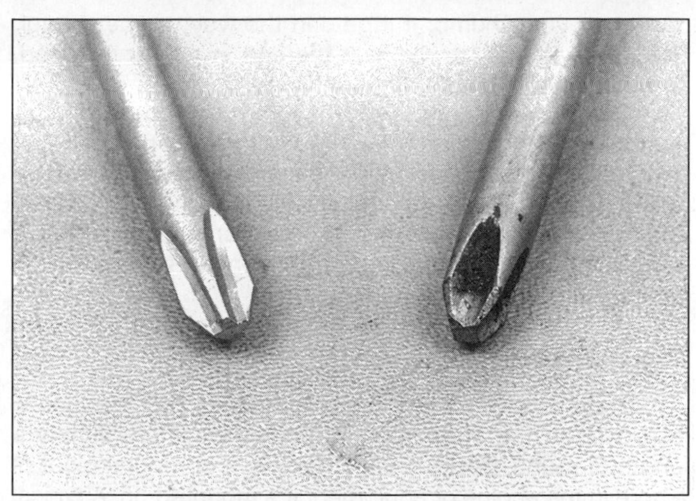

2.28 New (left) and worn (right) Phillips screwdriver tips

sure the fit between the screwdriver and fastener is snug. If the fit is good, you should be able to angle the blade down almost vertically without the screw slipping off the tip. Use only screwdrivers that fit exactly - anything else is guaranteed to chew out the screw head instantly.

The idea behind all cross-head screw designs is to make the screw and screwdriver blade self-aligning. Provided you aim the blade at the center of the screw head, it'll engage correctly, unlike conventional slotted screws, which need careful alignment. This makes the screws suitable for machine installation on an assembly line (which explains why they're usually so tight and difficult to remove). The drawback with these screws is the driving tangs on the screwdriver tip are very small and must fit very precisely in the screw head. If this isn't the case, the huge loads imposed on the small flats of the screw recess simply tear the metal away, at which point the screw ceases to be removable by normal methods. The problem is made worse by the normally-soft material chosen for screws.

To deal with these screws on a regular basis, you'll need high-quality screwdrivers with various size tips so you'll be sure to have the right one when you need it. Phillips screwdrivers are sized by the tip number and length of the shank (for example: A number 2 x 6-inch Phillips screwdriver has a number 2 tip - to fit screws of only that size recess - and the shank is 6-inches long). Tip sizes 1, 2 and 3 should be adequate for small engine repair work **(see illustration 2.27)**. If the tips get worn or damaged, buy new screwdrivers so the tools don't destroy the screws they're used on **(see illustration 2.28)**.

Here's a tip that may come in handy when using Phillips screwdrivers - if the screw is extremely tight and the tip tends to back out of the recess rather than turn the screw, apply a small amount of valve lapping compound to the screwdriver tip so it will grip better.

## Hammers

You'll need at least one ball-pein hammer, although almost any steel hammer will work in most cases. A ball-pein hammer has a head with a conventional cylindrical face at one end and a rounded ball end at the other and is a general-purpose tool found in almost any type of shop. It has a shorter neck than a claw hammer and the face is tempered for striking punches and chisels. A fairly large hammer is preferred over a small one. Although it's possible to find small ones, you won't need them very often and it's much easier to control the blows from a heavier head. As a general rule, a single 12 or 16-ounce hammer will work for most jobs, though occasionally larger or smaller ones may be useful.

A soft-face hammer is used where a steel hammer could cause damage to the component or other tools being used. A steel hammer head might crack an aluminum part, but a rubber or plastic hammer can be used with more confidence. Soft-face hammers are available with interchangeable heads (usually one made of rubber and another made of relatively hard plastic). When the heads are worn out, new ones can be installed. If finances are really limited, you can get by without a soft-face hammer by placing a small hardwood block between the component and a steel hammer head to prevent damage.

Hammers should be used with common sense; the head should strike the desired object squarely and with the right amount of force. For many jobs, little effort is needed - simply allow the weight of the head to do the work, using the length of the swing to control the amount of force applied. With practice, a hammer can be used with surprising finesse, but it'll take a while to achieve. Initial mistakes include striking the object at an angle, in which case the hammer head may glance off to one side, or hitting the edge of the object. Either one can result in damage to the part or to your thumb, if it gets in the way, so be careful. Hold the hammer handle near the end, not near the head, and grip it firmly but not too tightly.

Check the condition of your hammers on a regular basis. The danger of a loose head coming off is self-evident, but check the head for chips and cracks too. If damage is noted, buy a new hammer - the head may chip in use and the resulting fragments can be extremely dangerous. It goes without saying that eye protection is essential whenever a hammer is used.

## Punches and chisels

These tools are used along with a hammer for various purposes in the shop. Drift punches are often simply a length of round steel bar used to drive a component out of a bore in the engine or equipment it's mounted on. A typical use would be for removing or installing a bearing or bushing. A drift of the same diameter as the bearing outer race is placed against the bearing and tapped with a hammer to drive it in or out of the bore. Most manufacturers offer special drifts for the various bearings in a particular engine. While they're useful to a busy dealer service department, they are prohibitively expensive for the do-it-yourselfer who may only need to use them once. In such cases, it's better to improvise. For bearing removal and installation it's usually possible to use a socket of the appropriate diameter to tap the bearing in or out; an unorthodox use for a socket, but it works.

Smaller-diameter drift punches can be purchased or fabricated from steel bar stock. In some cases, you'll need to drive out items like corroded engine mounting bolts. Here, it's essential to avoid damaging the threaded end of the bolt, so the drift must be a softer material than the bolt. Brass or copper is the usual choice for such jobs; the drift may be damaged in use, but the thread will be protected.

Punches are available in various shapes and sizes and a set of assorted types will be very useful. One of the most basic is the center punch, a small cylindrical punch with the end ground to a point. It'll be needed whenever a hole is drilled. The center of the hole is located first and the punch is used to make a small indentation at the intended point. The indentation acts as a guide for the drill bit so the hole ends up in the right place. Without a punch mark the drill bit will wander and you'll find it impossible to drill with any real accuracy. You can also buy automatic center punches. They're spring-loaded and are pressed against the surface to be marked, without the need to use a hammer.

Pin punches are intended for removing items like roll pins (semi-hard, hollow pins that fit tightly in their holes). Pin punches have other uses, however. You may occasionally have to remove rivets or bolts by cutting off the heads and driving out the shanks with a pin punch. They're also very handy for aligning holes in components while bolts or screws are inserted.

Of the various sizes and types of metal-cutting chisels available, a simple cold chisel is essential in any mechanic's workshop. One about 6-inches long with a 1/2-inch wide

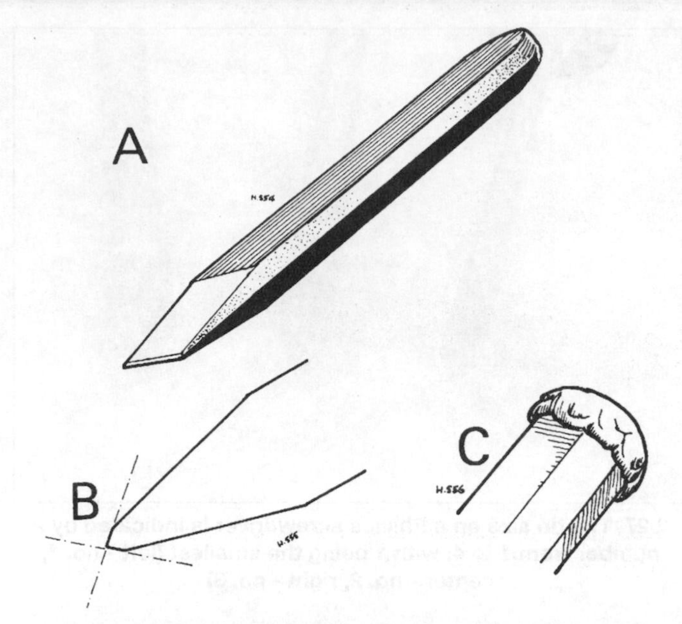


**2.29 A typical general purpose cold chisel (A) - note the angle of the cutting edge (B), which should be checked and resharpened on a regular basis; the mushroomed head (C) is dangerous and should be filed to restore it to its original shape**

blade should be adequate. The cutting edge is ground to about 80-degrees **(see illustration 2.29),** while the rest of the tip is ground to a shallower angle away from the edge. The primary use of the cold chisel is rough metal cutting - this can be anything from sheet metal work (uncommon on small engines) to cutting off the heads of seized or rusted bolts or splitting nuts. A cold chisel can also be useful for turning out screws or bolts with messed-up heads.

All of the tools described in this section should be good quality items. They're not particularly expensive, so it's not really worth trying to save money on them. More significantly, there's a risk that with cheap tools, fragments may break off in use - a potentially-dangerous situation.

Even with good-quality tools, the heads and working ends will inevitably get worn or damaged, so it's a good idea to maintain all such tools on a regular basis. Using a file or bench grinder, remove all burrs and mushroomed edges from around the head. This is an important task because the build-up of material around the head can fly off when it's struck with a hammer and is potentially dangerous. Make sure the tool retains its original profile at the working end, again, filing or grinding off all burrs. In the case of cold chisels, the cutting edge will usually have to be reground quite often because the material in the tool isn't usually much harder than materials typically being cut. Make sure the edge is reasonably sharp, but don't make the tip angle greater than it was originally; it'll just wear down faster if you do.

The techniques for using these tools vary according to the job to be done and are best learned by experience. The one common denominator is the fact they're all normally

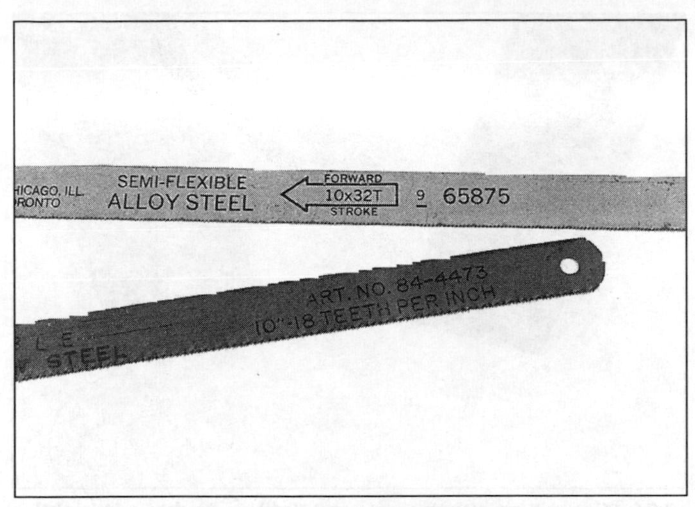


**2.30 Hacksaw blades are marked with the number of teeth per inch (TPI) - use a relatively coarse blade for aluminum and a fine blade for steel**

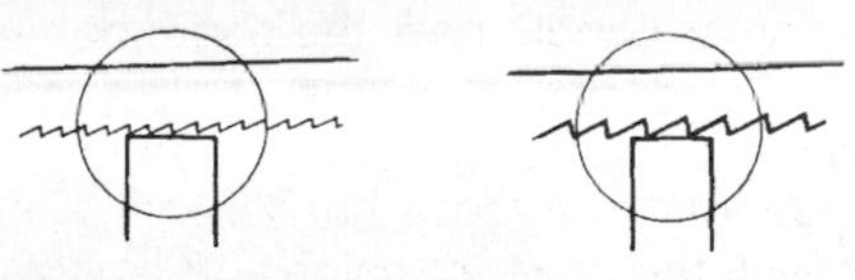

**When cutting thin materials, check that at least three teeth are in contact with the workpiece at any time. Too coarse a blade will result in a poor cut and may break the blade. If you do not have the correct blade, cut at a shallow angle to the material.**

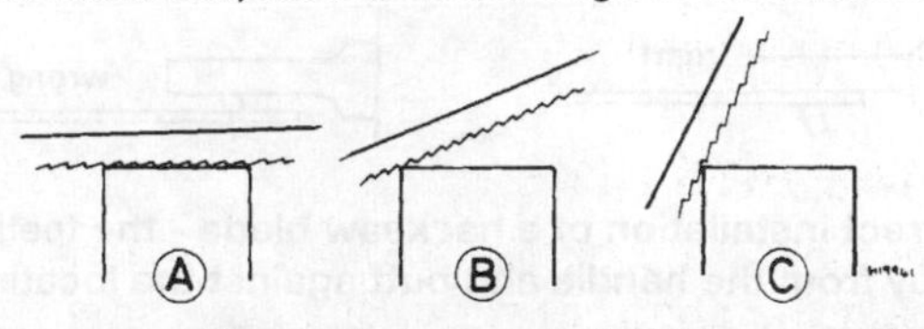


**The correct cutting angle is important. If it is too shallow (A) the blade will wander. The angle shown at (B) is correct when starting the cut, and may be reduced slightly once under way. In (C) the angle is too steep and the blade will be inclined to jump out of the cut.**

**2.31 Correct procedure for use of a hacksaw**

struck with a hammer. It follows that eye protection should be worn. Always make sure the working end of the tool is in contact with the part being punched or cut. If it isn't, the tool will bounce off the surface and damage may result.

## Hacksaws

A hacksaw consists of a handle and frame supporting a flexible steel blade under tension. Blades are available in various lengths and most hacksaws can be adjusted to accommodate the different sizes. The most common blade length is 10-inches.

Most hacksaw frames are adequate and since they're simple tools, there's not much difference between brands. Try to pick one that's rigid when assembled and allows the blade to be changed or repositioned easily.

The type of blade to use, indicated by the number of teeth per inch (TPI) **(see illustration 2.30)** is determined by the material being cut. The rule of thumb is to make sure at least three teeth are in contact with the metal being cut at any one time **(see illustration 2.31)**. In practice, this means a fine blade for cutting thin sheet materials, while a coarser blade can be used for faster cutting through thicker items such as bolts or bar stock. It's worth noting that when cutting thin materials it's helpful to angle the saw so the blade cuts at a shallow angle. This way more teeth are in contact and there's less chance of the blade binding and breaking or teeth being broken off. This approach can also be used when a fine enough blade isn't available; the shallower the angle, the more teeth are contacting the workpiece.

When buying blades, choose a well-known brand. Cheap, unbranded blades may be perfectly acceptable, but you can't tell by looking at them. Poor quality blades will be insufficiently hardened on the teeth edge and will dull quickly. Most reputable brands will be marked "Flexible High Speed Steel" or something similar, giving some indication of the material they're made of **(see illustration 2.32)**. It is possible to buy "unbreakable" blades (only the teeth are hardened, leaving the rest of the blade less brittle).

In some situations, a full-size hacksaw is too big to allow access to a frozen nut or bolt. Sometimes this can be overcome by turning the blade 90-degrees - most saws allow this to be done. Occasionally you may have to position the saw around an obstacle and then install the blade on the other side of it. Where space is really restricted, you may have to use a handle that clamps onto a saw blade at one end. This allows access when a hacksaw frame would not work at all and has another advantage in that you can make use of broken off hacksaw blades instead of throwing them away. Note that because only one end of the blade is supported, and it's not held under tension, it's difficult to control and less efficient when cutting.

Before using a hacksaw, make sure the blade is suitable for the material being cut and installed correctly in the

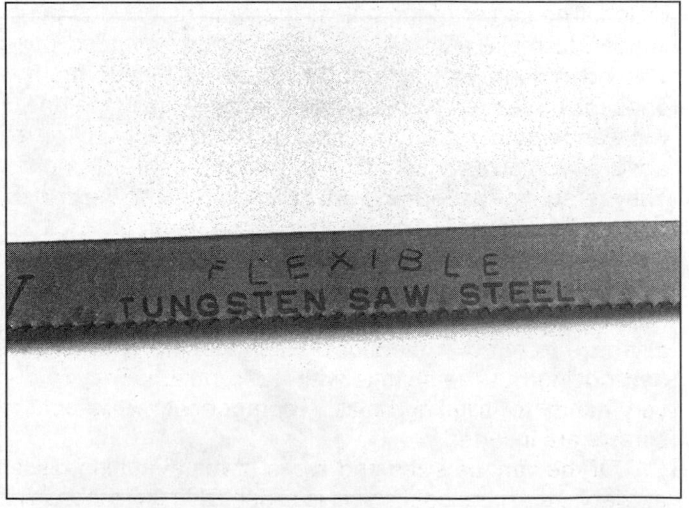


**2.32 Good quality hacksaw blades will be marked like this**

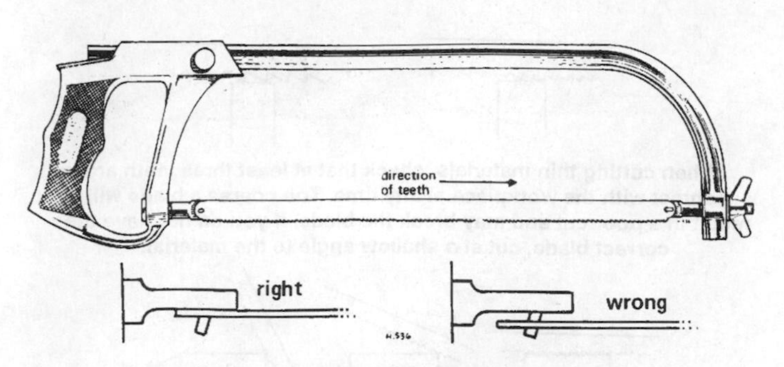


2.33 **Correct installation of a hacksaw blade - the teeth must point away from the handle and butt against the locating lugs**

frame **(see illustration 2.33)**. Whatever it is you're cutting must be securely supported so it can't move around. The saw cuts on the forward stroke, so the teeth must point away from the handle. This might seem obvious, but it's easy to install the blade backwards by mistake and ruin the teeth on the first few strokes. Make sure the blade is tensioned adequately or it'll distort and chatter in the cut and may break. Wear safety glasses and be careful not to cut yourself on the saw blade or the sharp edge of the cut.

## Files

Files come in a wide variety of sizes and types for specific jobs, but all of them are used for the same basic function of removing small amounts of metal in a controlled fashion. Files are used by mechanics mainly for deburring, marking parts, removing rust, filing the heads off rivets, restoring threads and fabricating small parts.

File shapes commonly available include flat, half-round, round, square and triangular. Each shape comes in a range of sizes (lengths) and cuts ranging from rough to smooth. The file face is covered with rows of diagonal ridges which form the cutting teeth. They may be aligned in one direction only (single cut) or in two directions to form a diamond-shaped pattern (double-cut) **(see illustration 2.34)**. The spacing of the teeth determines the file coarseness, again, ranging from rough to smooth in five basic grades: Rough, coarse, bastard, second-cut and smooth.

**2.34 Files will be either single-cut (left) or double-cut (right) - generally speaking, use a single-cut file to produce a very smooth surface; use a double-cut file to remove large amounts of material quickly**

You'll want to build up a set of files by purchasing tools of the required shape and cut as they're needed. A good starting point would be flat, half-round, round and triangular files (at least one each - bastard or second-cut types). In addition, you'll have to buy one or more file handles (files are usually sold without handles, which are purchased separately and pushed over the tapered tang of the file when in use) **(see illustration 2.35)**. You may need to buy more than one size handle to fit the various files in your tool box, but don't attempt to get by without them. A file tang is fairly sharp and you almost certainly will end up stabbing yourself in the palm of the hand if you use a file without a handle and it catches in the workpiece during use. Adjustable handles are also available for use with files of various sizes, eliminating the need for several handles **(see illustration 2.36)**.

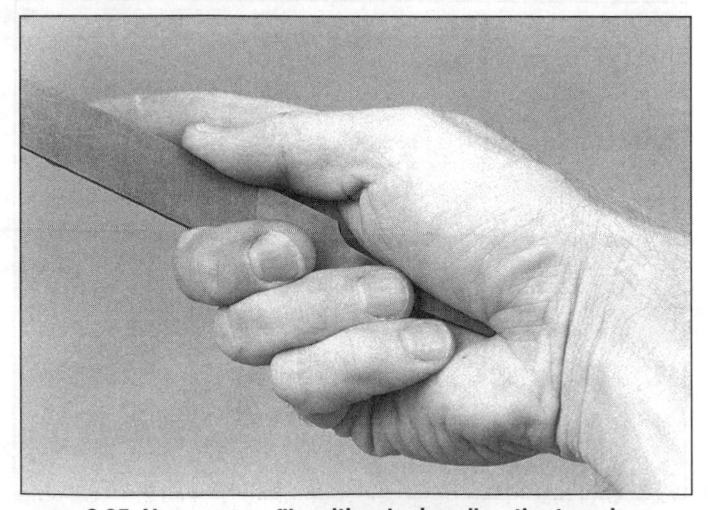

**2.35 Never use a file without a handle - the tang is sharp and could puncture your hand**

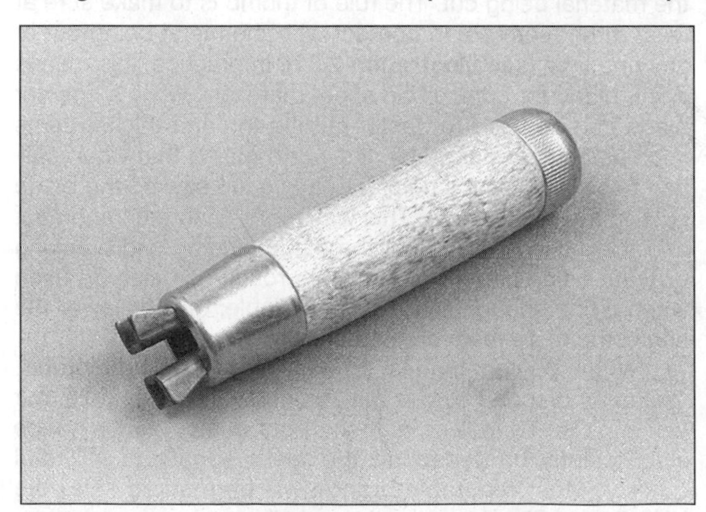

**2.36 Adjustable handles that will work with many different size files are also available**

Exceptions to the need for a handle are fine Swiss-pattern files, which have a rounded handle instead of a tang. These small files are usually sold in sets with a number of different shapes. Originally intended for very fine work, they can be very handy for use in inaccessible areas. Swiss files are normally the best choice if piston ring ends require filing to obtain the correct end gap.

The correct procedure for using files is fairly easy to master. As with a hacksaw, the work should be clamped securely in a vise, if needed, to prevent it from moving around while being worked on. Hold the file by the handle, using your free hand at the file end to guide it and keep it flat in relation to the surface being filed. Use smooth cutting strokes and be careful not to rock the file as it passes over the surface. Also, don't slide it diagonally across the surface or the teeth will make grooves in the workpiece. Don't drag a file back across the workpiece at the end of the stroke - lift it slightly and pull it back to prevent damage to the teeth.

Files don't require maintenance in the usual sense, but they should be kept clean and free of metal filings. Steel is a reasonably easy material to work with, but softer metals like aluminum tend to clog the file teeth very quickly, which will result in scratches in the workpiece. This can be avoided by rubbing the file face with chalk before using it. General cleaning is done with a file card or a fine wire brush. If they're kept clean, files will last a long time - when they do eventually dull, they must be replaced; there is no satisfactory way of sharpening a worn file.

# Twist drills and drilling equipment

Drills are often needed to remove rusted or broken off fasteners, enlarge holes and fabricate small parts.

Drilling operations are done with twist drills, either in a hand drill or a drill press. Twist drills (or drill bits, as they're often called) consist of a round shank with spiral flutes formed into the upper two-thirds to clear the waste produced while drilling, keep the drill centered in the hole and finish the sides of the hole.

The lower portion of the shank is left plain and used to hold the drill in the chuck. In this section, we'll cover only parallel shank drill bits **(see illustration 2.37)**. There is another type of bit with the plain end formed into a special taper designed to fit directly into a corresponding socket in a heavy-duty drill press. These drills are known as Morse Taper drills and are used primarily in machine shops.

At the cutting end of the drill, two edges are ground to form a conical point. They're generally angled at about 60-degrees from the drill axis, but they can be reground to other angles for specific applications. For general use, the standard angle is correct - this is how drill bits are sold.

When buying bits, purchase a good-quality set (sizes 1/16 to 3/8-inch). Make sure they're marked "High Speed Steel" or "HSS", which indicates they're hard enough to

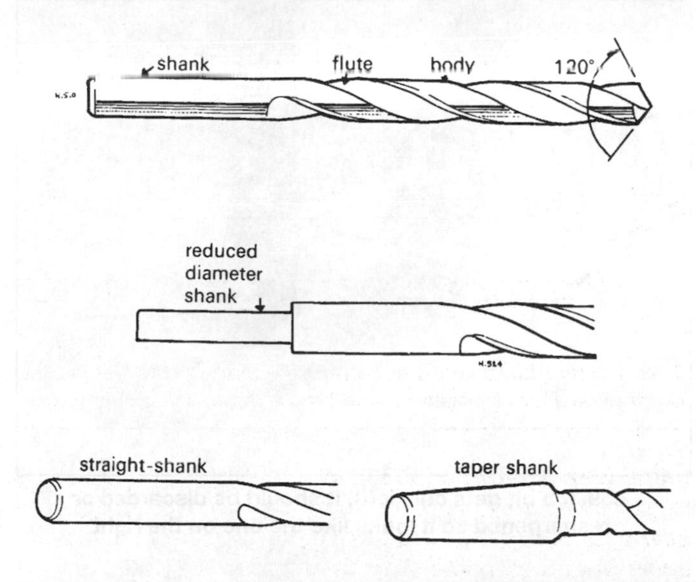


**2.37 A typical drill bit (top), a reduced shank bit (center) and a tapered shank bit (bottom right)**

withstand continual use in metal; many cheaper, unmarked bits are suitable only for use in wood or other soft materials. Buying a set ensures the right size bit will be available when it's needed.

## Twist drill sizes

Twist drills are available in a vast array of sizes, most of which you'll never need. There are three basic drill sizing systems: Fractional, number and letter **(see illustration 2.38)** (we won't get involved with the fourth system, which is metric sizes).

Fractional sizes start at 1/64-inch and increase in increments of 1/64-inch. Number drills range in descending order from 80 (0.0135-inch), the smallest, to 1 (0.2280-inch), the largest. Letter sizes start with A (0.234-inch), the smallest, and go through Z (0.413-inch), the largest.

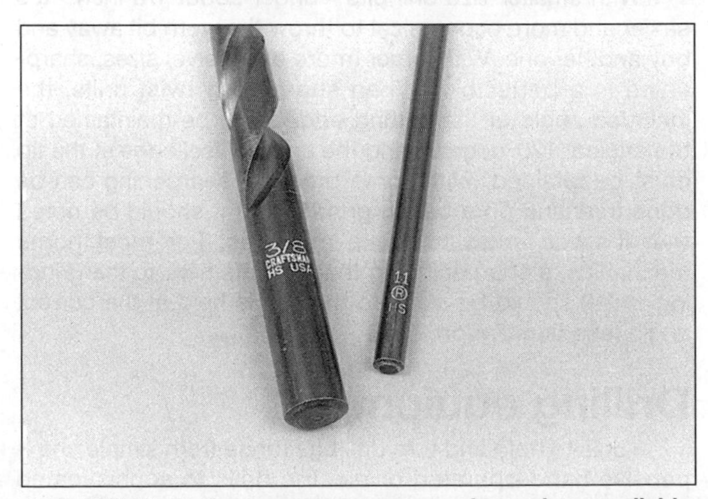

**2.38 Drill bits in the range most commonly used are available in fractional sizes (left) and number sizes (right) so almost any size hole can be drilled**

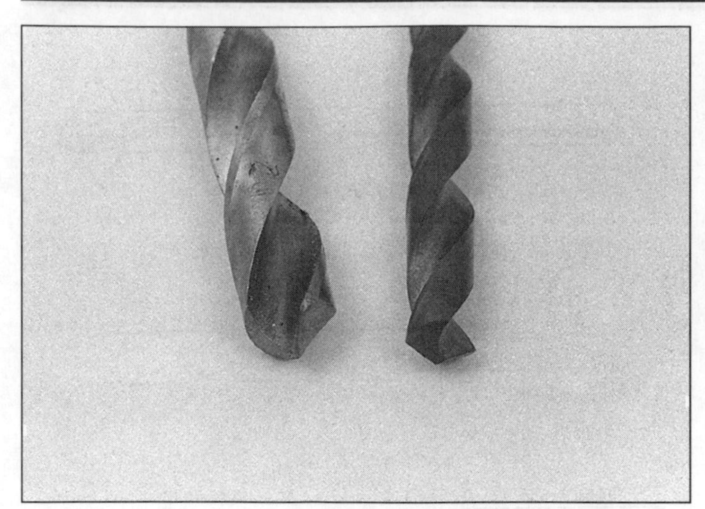

**2.39 If a bit gets dull (left), it should be discarded or resharpened so it looks like the one on the right**

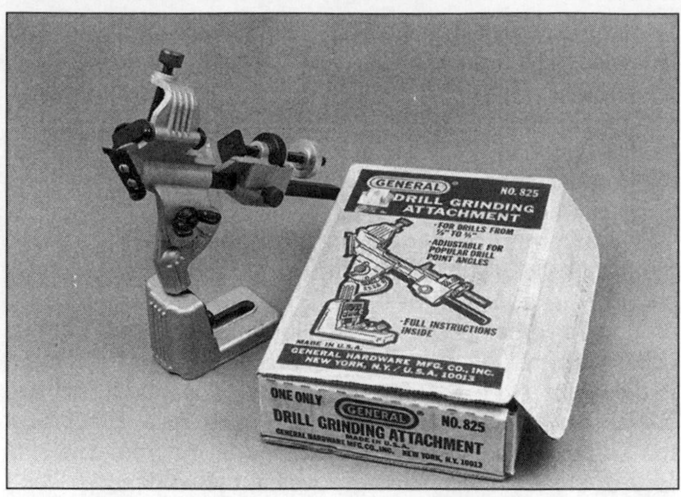


**2.40 Inexpensive drill bit sharpening jigs for use with a bench grinder are widely available - even if you use it infrequently to resharpen drill bits, it'll pay for itself quickly**

This bewildering range of sizes means it's possible to drill an accurate hole of almost any size within reason. In practice, you'll be limited by the size of chuck on your drill (normally 3/8 or 1/2-inch). In addition, very few stores stock all sizes, so you'll have to shop around for the nearest available size to the one you need.

## Sharpening twist drills

Like any tool with a cutting edge, twist drills will eventually get dull **(see illustration 2.39)**. How often they'll need sharpening depends to some extent on whether they're used correctly. A dull twist drill will be obvious in use. A good indication of the condition of the cutting edges is to watch the waste emerging from the hole being drilled. If the tip is in good condition, two even spirals of waste metal will be produced; if this fails to happen or the tip gets hot, it's safe to assume that sharpening is required.

With smaller size drill bits - under about 1/8-inch - it's easier and more economical to throw the worn bit away and buy another one. With larger (more expensive) sizes, sharpening is a better bet. When sharpening twist drills, the included angle of the cutting edge must be maintained at the original 120-degrees and the small chisel edge at the tip must be retained. With some practice, sharpening can be done freehand on a bench grinder, but it should be noted that it's very easy to make mistakes. For most home mechanics, a sharpening jig that mounts next to the grinding wheel should be used so the drill is held at the correct angle **(see illustration 2.40)**.

## Drilling equipment

Tools to hold and turn drill bits range from simple, inexpensive hand-operated or electric drills to sophisticated and expensive drill presses. Ideally, all drilling should be done on a drill press with the workpiece clamped solidly in a vise. These machines are expensive and take up a lot of bench or floor space, so they're out of the question for many do-it-yourselfers. An additional problem is the fact that many of the drilling jobs you end up doing will be on the engine itself or the equipment it's mounted on, in which case the tool has to be taken to the work.

The best tool for the home shop is an electric drill with a 3/8-inch chuck. As mentioned in Chapter 1, both cordless and AC drills (that run off household current) are available. If you're purchasing one for the first time, look for a well-known, reputable brand name and variable-speed model with reversing capability as minimum requirements. A 1/4-inch chuck, single-speed drill will work, but it's worth paying a little more for the larger, variable-speed type.

All drills require a key to lock the bit in the chuck. When removing or installing a bit, make sure the cord is unplugged to avoid accidents. Initially, tighten the chuck by hand, checking to see if the bit is centered correctly. This is especially important when using small drill bits that can get caught between the jaws. Once the chuck is hand tight, use the key to tighten it securely - remember to remove the key afterwards! Small bits can be tightened in the chuck off-center if you're not careful. Before drilling, spin the drill while watching the bit to make sure it is not wobbling.

## Drilling and finishing holes

### Preparation for drilling

If possible, make sure the part you intend to drill in is securely clamped in a vise. If it's impossible to get the work to a vise, make sure it's stable and secure. Drill bits often catch during drilling - this can be dangerous, particularly if the work suddenly starts spinning on the end of the drill. Obviously, there's not much chance of a complete engine or piece of equipment doing this, but you should make sure it's supported securely.

Start by locating the center of the hole you're drilling. Use a center punch to make an indentation for the drill bit so it won't wander. If you're drilling out a broken-off bolt, be sure to position the punch in the exact center of the bolt **(see illustration 2.61)**.
If you're drilling a large hole (above 1/4-inch), you may want to make a pilot hole first. As the name suggests, it will guide the larger drill bit and minimize bit wandering. Before actually drilling a hole, make sure the area immediately behind the bit is clear of anything you don't want drilled.

## Drilling

When drilling steel, especially with smaller bits, no lubrication is needed. If a large bit is involved, oil can be used to ensure a clean cut and prevent overheating of the drill tip. When drilling aluminum, which tends to cling to the cutting edges and clog the drill bit flutes, use kerosene as a lubricant.

Wear safety goggles or a face shield and assume a comfortable, stable stance so you can control the pressure on the drill easily. Position the drill tip in the punch mark and make sure, if you're drilling by hand, the bit is perpendicular to the surface of the workpiece. Start drilling without applying much pressure until you're sure the hole is positioned correctly. If the hole starts off center, it can be very difficult to correct. You can try angling the bit slightly so the hole center moves in the opposite direction, but this must be done before the flutes of the bit have entered the hole. It's at the starting point that a variable-speed drill is invaluable; the low speed allows fine adjustments to be made before it's too late. Continue drilling until the desired hole depth is reached or until the drill tip emerges from the other side of the workpiece.

Cutting speed and pressure are important - as a general rule, the larger the diameter of the drill bit, the slower the drilling speed should be. With a single-speed drill, there's little that can be done to control it, but two-speed or variable speed drills can be controlled. If the drilling speed is too high, the cutting edges of the bit will tend to overheat and dull. Pressure should be varied during drilling. Start with light pressure until the drill tip has located properly in the work. Gradually increase pressure so the bit cuts evenly. If the tip is sharp and the pressure correct, two distinct spirals of metal will emerge from the bit flutes. If the pressure is too light, the bit won't cut properly, while excessive pressure will overheat the tip.

Decrease pressure as the bit breaks through the workpiece. If this isn't done, the bit may jam in the hole; if you're using a hand-held drill, it could be jerked out of your hands, especially when using larger-size bits.

Once a pilot hole has been made, install the larger bit in the chuck and enlarge the hole. The second bit will follow the pilot hole - there's no need to attempt to guide it (if you do, the bit may break off). It is important, however, to hold the drill at the correct angle.

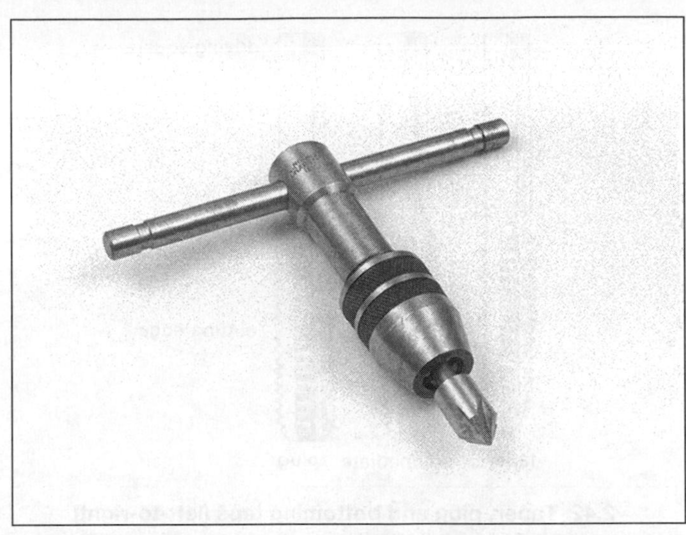

**2.41 Use a large bit or a countersink mounted in a tap wrench to remove burrs from a hole after drilling or enlarging it**

After the hole has been drilled to the correct size, remove the burrs left around the edges of the hole. This can be done with a small round file, or by chamfering the opening with a larger bit or a countersink **(see illustration 2.41)**. Use a drill bit that's several sizes larger than the hole and simply twist it around each opening by hand until any rough edges are removed.

## Enlarging and reshaping holes

The biggest practical size for bits used in a hand drill is about 1/2-inch. This is partly determined by the capacity of the chuck (although it's possible to buy larger drills with stepped shanks). The real limit is the difficulty of controlling large bits by hand; drills over 1/2-inch tend to be too much to handle in anything other than a drill press. If you have to make a larger hole, or if a shape other than round is involved, different techniques are required.

If a hole simply must be enlarged slightly, a round file is probably the best tool to use. If the hole must be very large, a hole saw will be needed, but they can only be used in sheet metal and other thin materials.

Large or irregular-shaped holes can also be made in relatively thin materials by drilling a series of small holes very close together. In this case the desired hole size and shape must be marked with a scribe. The next step depends on the size bit to be used; the idea is to drill a series of almost touching holes just inside the outline of the large hole. Center punch each hole location, then drill them out. A cold chisel is used to knock out the waste material at the center of the hole, which can then be filed to size. This is a time consuming process, but it's the only practical approach for the home shop. Success is dependent on accuracy when marking the hole shape and using the center punch.

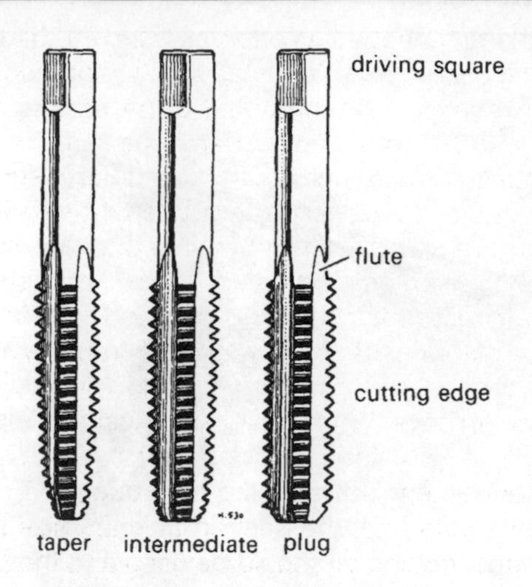


2.42 Taper, plug and bottoming taps (left-to-right)

2.43 You'll run into many situations where a tap is needed to clean up or restore threads when working on small engines

# Taps and dies

## Taps

Taps, which are available in inch and metric sizes, are used to cut internal threads and clean or restore damaged threads. A tap consists of a fluted shank with a drive square at one end. It's threaded along part of its length - the cutting edges are formed where the flutes intersect the threads **(see illustration 2.42)**. Taps are made from hardened steel so they'll cut threads in materials softer than what they're made of.

Taps come in three different types: taper, plug and bottoming. The only real difference is the length of the chamfer on the cutting end of the tap. Taper taps are chamfered for the first 6 or 8 threads, which makes them easy to start but prevents them from cutting threads close to the bottom of a hole. Plug taps are chamfered up about 3 to 5 threads, which makes them a good all around tap because they're relatively easy to start and will cut nearly to the bottom of a hole. Bottoming taps, as the name implies, have a very short chamfer (1-1/2 to 3 threads) and will cut as close to the bottom of a blind hole as practical. However, to do this, the threads should be started with a plug or taper tap.

Although cheap tap and die sets are available, the quality is usually very low and they can actually do more harm than good when used on threaded holes in aluminum engines. The alternative is to buy high-quality taps if and when you need them, even though they aren't cheap, especially if you need to buy two or more thread pitches in a given size. Despite this, it's the best option - you'll probably only need taps on rare occasions, so a full set isn't absolutely necessary.

Taps are normally used by hand (they can be used in machine tools, but not when doing engine repairs). The square drive end of the tap is held in a tap wrench (an adjustable T-handle). For smaller sizes, a T-handled chuck can be used **(see illustration 2.43)**. The tapping process starts by drilling a hole of the correct diameter. For each tap size, there's a corresponding twist drill that will produce a hole of the correct size.

This is important; too large a hole will leave the finished thread with the tops missing, producing a weak and unreliable grip. Conversely, too small a hole will place excessive loads on the hard and brittle shank of the tap, which can break it off in the hole. Removing a broken-off tap from a hole is no fun!

The correct tap drill size is normally marked on the tap itself or the container it comes in **(see illustration 2.44)**.

## Dies

Dies are used to cut, clean or restore external threads. Most dies are made from a hex-shaped or cylindrical piece of hardened steel with a threaded hole in the center. The threaded hole is overlapped by three or four cutouts, which equate to the flutes on taps and allow waste to escape during the threading process. Dies are held in a T-handle holder (called a die stock) **(see illustration 2.45)**. Some dies are split at one point, allowing them to be adjusted slightly (opened and closed) for fine control of thread clearances.

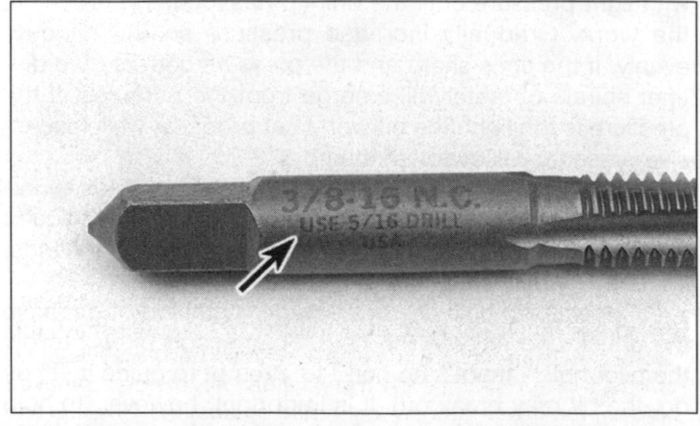


2.44 If you need to drill and tap a hole, the drill bit size to use for a given bolt (tap) size is marked on the tap

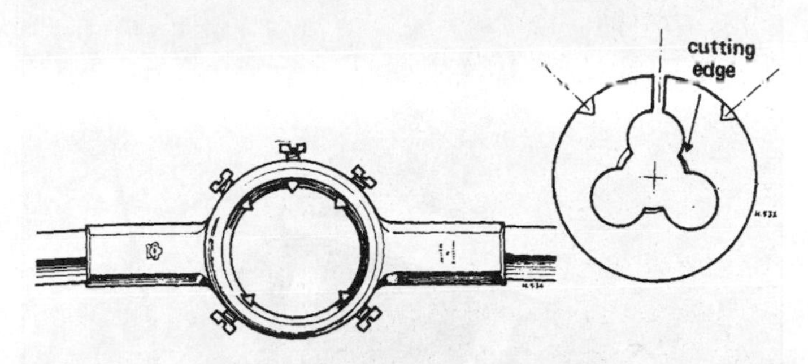


**2.45 A die (right) is used for cutting external threads (this one is a split-type/adjustable die) and is held in a tool called a die stock (left)**

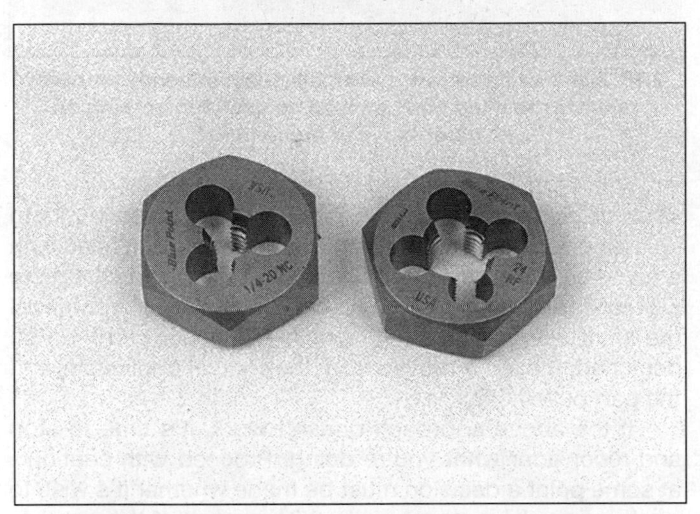

**2.46 Hex-shaped dies are especially handy for mechanic work because they can be turned with a wrench**

**2.47 In some cases, a special puller (available from the engine manufacturer) will be needed for removing the flywheel - in this example (Briggs & Stratton engine shown), the puller body is slipped over the end of the crankshaft, the bolts are threaded into the flywheel holes (they may have to cut their own threads if the flywheel has never been removed before), the lower nuts are tightened against the flywheel and the upper nuts are tightened in 1/4-turn increments until the flywheel pops off the shaft taper**

Dies aren't needed as often as taps, for the simple reason it's normally easier to install a new bolt than to salvage one. However, it's often helpful to be able to extend the threads of a bolt or clean up damaged threads with a die. Hex-shaped dies are particularly useful for mechanic work, since they can be turned with a wrench **(see illustration 2.46)** and are usually less expensive than adjustable ones. When working on small engines, you will probably use dies mostly for restoring threads on studs that are part of an engine or component and not as easily replaced as a bolt.

The procedure for cutting threads with a die is similar to the one for taps. When using an adjustable die, the initial cut is made with the die open as far as possible. The adjustment screw is then used to reduce the diameter of successive cuts until the desired finished size is reached. As with taps, cutting oil should be used and the die must be backed off every few turns to clear waste from the cutouts.

# Pullers

During every engine overhaul, and many simple repairs, you'll often need some type of puller. The most common pullers are required for removal of the magneto flywheel from the end of the crankshaft. Other less common tasks that'll require some sort of puller include the removal of bushings and bearings. Common to all of these jobs is the need to exert pressure on the part being removed while avoiding damage to the surrounding area or components. The best way to do this is with a puller specially designed for the job.

As mentioned in Chapter 1, it's a good idea to have procedures that require a puller (other than flywheel removal) done by a dealer or repair shop, then you won't have to invest in a tool that won't be used very often.

## Special pullers

A good example of a special puller is the one needed for removal of the magneto flywheel. On most engines, the flywheel fits over the tapered end of the crankshaft, where it's secured by a large nut and located by a Woodruff key. Even after the nut has been removed, you have to exert lots of pressure to draw the flywheel off the shaft. This is because the nut draws the tapered faces on the shaft and the inside of the flywheel hub together very tightly during assembly.

The method normally used to remove the flywheel requires a specially designed puller that fits over the crankshaft end and has bolts that thread into holes in the flywheel hub. After the bolts are threaded into place (they often will have to cut their own threads in the holes, but they're designed to do so), the lower nuts are tightened down against the flywheel and the upper nuts are tightened in 1/4-turn increments until the flywheel pops loose **(see illustration 2.47)**.

Flywheel "knock-off" tools are also available for Briggs & Stratton and Tecumseh engines. **Note:** *Briggs and Stratton does not endorse or recommend the use of flywheel knock-off tools on their engines, while Tecumseh does. A knock-off tool should not be used on any engine that has a ball bearing on the flywheel side of the crankshaft.* The tool is slipped over or threaded onto the end of the crankshaft until it contacts the flywheel, then it's backed off one or two turns. Moderate pressure is applied to the flywheel with a large screwdriver and the end of the knock-off tool is struck with a hammer - the blow from the hammer will usually pop the flywheel loose. **Note:** *On some engines with a single nut retaining the flywheel, a roll pin or setscrew may also be securing the nut. Check first before attempting to remove the nut.*

Using a puller means the pressure on the flywheel is applied where it does the most good - at the center of the hub, rather than at the edge, where it would be more likely to distort the flywheel than release the hub from the shaft taper. Read Chapter 1 to find out about specific pullers for the engines covered in this manual. On most engines covered by this manual, a standard automotive puller, one that can be used on either two-bolt or three-bolt patterns, is suitable.

## General-purpose pullers

You're likely to need some sort of general-purpose puller at some point in most overhauls, often where parts are seized or corroded, or where bushings or bearings must be removed. Universal two and three-jaw pullers are widely available in numerous designs and sizes.

These tools generally have jaws attached to a large center boss, which has a threaded hole to accept the puller bolt. The ends of the jaws have hooks which locate on and grip the part to be pulled off **(see illustration 2.48)**. Normally, the jaws can be reversed to allow the tool to be used on internal bushings and bearings as well. The jaws are hooked over the part being removed and the puller bolt is positioned against the end of the shaft. As the bolt is tightened, the component is drawn off the shaft.

It's possible to adapt pullers by making special jaws for specific jobs, but it'll take extra time and may not be successful. If you decide to try this approach, remember that the force should be concentrated as close to the center of the component as possible to avoid damaging it.

When using a puller, it should be assembled and a careful check should be made to ensure it doesn't snag on anything and the loads on the part to be removed are distributed evenly. If you're dealing with a part held on a shaft by a nut, loosen the nut, but don't remove it entirely. It will help prevent distortion of the shaft end under pressure from the puller bolt and will also stop the part from flying off the shaft when it comes loose.

Pullers of this type should be tightened gradually until moderate pressure is applied to the part being removed. **Caution:** *The puller bolt should never be tightened excessively or damage will occur!* Once it's under pressure, try to jar the component loose by striking the puller bolt head with a hammer. If this doesn't work, tighten the bolt a little more and repeat the procedure. The component should come off the shaft with a distinctive "pop". The puller can then be detached, the nut removed from the shaft (if applicable) and the part pulled off.

**2.48 A two or three-jaw puller will come in handy for many tasks in the shop and can also be used for working on other types of equipment**

If the above approach doesn't work, it's time to stop and reconsider what you're doing. Proceed with caution - at some point a decision must be made whether it's wise to continue applying pressure in this manner. If the component is unusually tight, something will probably break before it comes off. If you find yourself in this situation, try applying penetrating oil around the joint and leaving it overnight, with the puller in place and tightened securely. In some cases, the taper will separate and the problem will resolve itself by the next morning.

If you have the necessary equipment, are skilled in its use and take the necessary safety precautions, you can try heating the component with a propane or gas welding torch. This can be a good way to release a stubborn part, but isn't recommended unless you have experience doing it. **Caution:** *This approach should be used with extreme caution on a flywheel - the heat can easily demagnetize it or cause damage to the coil windings.* The heat should be applied to the hub area of the component to be removed, keeping the flame moving to avoid uneven heating and the risk of distortion. Keep pressure applied with the puller and make sure you're able to deal with the resulting hot component and the puller jaws if it does come free (wear gloves to protect your hands). Be very careful to keep the flame away from aluminum parts.

If all rational methods fail, don't be afraid to give up an attempt to remove something; it's cheaper than repairing a badly damaged engine. Either buy or borrow the correct tool or take the engine to a dealer and ask him to remove the part for you.

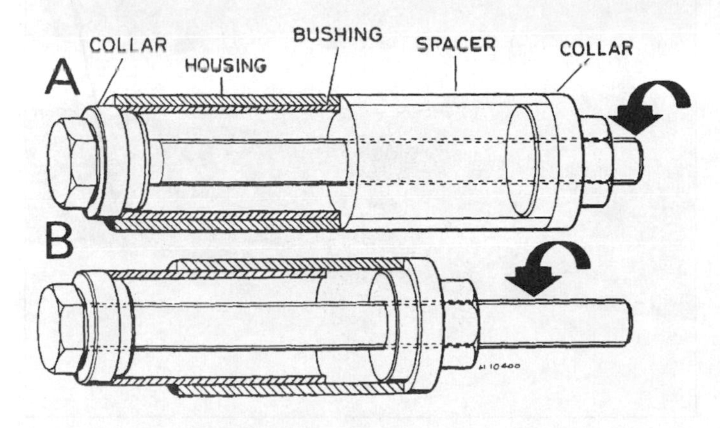


**2.49 Typical drawbolt uses - in A, the nut is tightened to pull the collar and bushing into the large spacer; in B, the spacer is left out and the drawbolt is repositioned to install the new bushing**

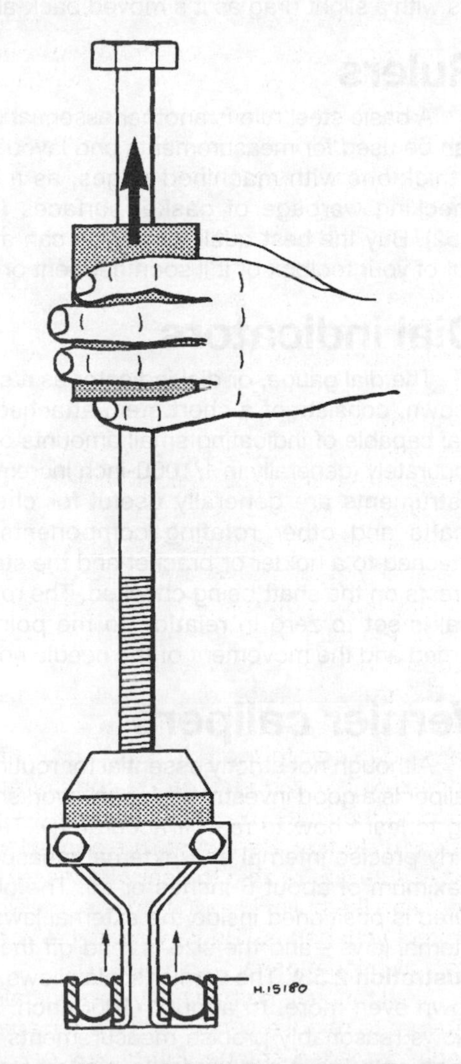

**2.50 A slide hammer with special attachments can be used for removing bearings and bushings from blind holes**

# Drawbolts

A simple drawbolt extractor is easy to make and invaluable in many situations. There are no standard, commercially-available tools of this type; you simply make a tool to suit a particular application. You can use a drawbolt to pull out stubborn piston pins and to remove bearings and bushings.

To make a drawbolt extractor, you'll need an assortment of threaded rods in various sizes (available at hardware stores), along with nuts to fit them. In addition, you'll need assorted washers, spacers and pieces of pipe. Don't forget to improvise where you can. A socket set can provide several sizes of spacers for short parts like bushings. For things like piston pins you'll usually need a longer piece of pipe.

Some typical drawbolt uses are shown in **illustration 2.49** - they also reveal the order of assembly of the various pieces. The same arrangement, minus the spacer, can usually be used to install a new bushing or piston pin. Using the tool is quite simple - the main thing to watch out for is to make sure you get the bushing or pin square in the bore when it's installed. Lubricate the part being pressed into place, if appropriate.

# Pullers for use in blind holes

You may encounter bushings or bearings installed in blind holes in almost any engine; there are special pullers designed to deal with them as well. In the case of engine bearings, it's sometimes possible to remove them without a puller if you heat the engine or component evenly (in an oven) and tap it face down on a clean wooden surface to dislodge the bearing. If you use this method, be careful not to burn yourself when handling the heated components - wear heavy gloves! If a puller is needed, a slide-hammer with interchangeable tips is your best bet. They range from universal two or three jaw puller arrangements to special bearing pullers. Bearing pullers are hardened steel tubes with a flange around the bottom edge. The tube is split at several places, which allows a wedge to expand the tool once it's in place. The tool fits inside the bearing inner race and is tightened so the flange or lip is locked under the edge of the race.

A slide-hammer consists of a steel shaft with a stop at the upper end. The shaft carries a sliding weight, which is moved along the shaft until it strikes the stop. This allows the tool holding the bearing to drive it out of the bore **(see illustration 2.50)**. A bearing puller set is an expensive and infrequently-used piece of equipment - to avoid the expense of buying one, as mentioned in Chapter 1, take the engine to a dealer and have the bearings/bushings replaced.

2.51 Feeler gauges are usually marked with inch and metric equivalents

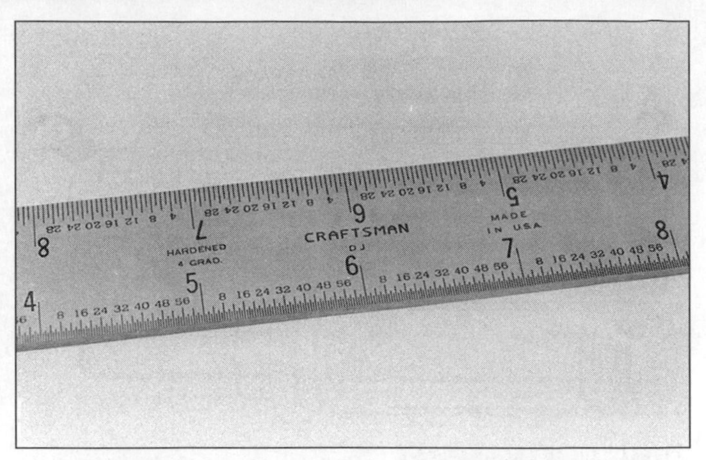


2.52 A steel rule will come in handy in any shop - if it's a good-quality tool, it can be used for checking engine components (like the cylinder head) for distortion

# Precision measurements

During any overhaul or major repair job, you'll need certain precision measuring devices to determine the amount of wear that has occurred and whether or not a component can be reused in the rebuilt engine. In addition, some of the more basic tools, like feeler gauges, are needed for routine service and tune-up work. In this section we'll look at the most commonly needed tools, starting with those that are considered essential, and working up to more specialized and expensive items. Some of them, such as vernier calipers and micrometers, are specialized pieces of equipment, but - unless you have someone else do the measuring for you - no substitutes are available.

## Feeler gauges

These are essential for work on almost any engine. You'll need feeler gauges to check/set the valve or (if equipped) ignition point clearances.

Feeler gauges normally come in sets. In smaller sets, different feeler gauges must be combined to make up thicknesses not included separately. Larger sets have a wider range of sizes to avoid this problem. Feeler gauges are available in both inch and metric sizes; they're usually marked both ways **(see illustration 2.51)**. Blade-type feeler gauges are thin steel strips and are the best choice for most purposes. There are also wire-type feeler gauges, which may be preferable in some circumstances.

You'll need feeler gauges whenever you have to make an accurate measurement of a small gap (typical applications are checking valve clearances on four-stroke engines and endplay in crankshafts or camshafts). You can also use feeler gauges when checking for distortion of gasket surfaces. The cover or casting is placed, gasket surface down, on a flat plate and any gap (which indicates distortion) can be measured directly with feeler gauges.

To measure a gap with feeler gauges, slide progressively thicker blades into the gap until you find the size that fits with a slight drag as it's moved back-and-forth.

## Rulers

A basic steel rule is another essential workshop item. It can be used for measurements and layout work and, if it is a thick one with machined edges, as a straightedge for checking warpage of gasket surfaces **(see illustration 2.52)**. Buy the best quality tool you can afford and keep it out of your toolbox or it'll soon get bent or damaged.

## Dial indicators

The dial gauge, or dial indicator as it's more commonly known, consists of a short stem attached to a clock-type dial capable of indicating small amounts of movement very accurately (generally in 1/1000-inch increments). These test instruments are generally useful for checking runout in shafts and other rotating components. The gauge is attached to a holder or bracket and the stem positioned so it rests on the shaft being checked. The rotating face of the dial is set to zero in relation to the pointer, the shaft is turned and the movement of the needle noted.

## Vernier calipers

Although not strictly essential for routine work, a vernier caliper is a good investment for any workshop (if you're willing to learn how to read it accurately). The tool allows for fairly precise internal and external measurements, up to a maximum of about 6-inches or so. The object to be measured is positioned inside the external jaws - or outside the internal jaws - and the size is read off the main scale **(see illustration 2.53)**. The vernier scale allows it to be narrowed down even more, to about 1/1000-inch. A vernier caliper allows reasonably precise measurements of a wide variety of objects, so it's a versatile piece of equipment. Even though it lacks the absolute accuracy of a micrometer, it's much cheaper to buy and can be used in more ways.

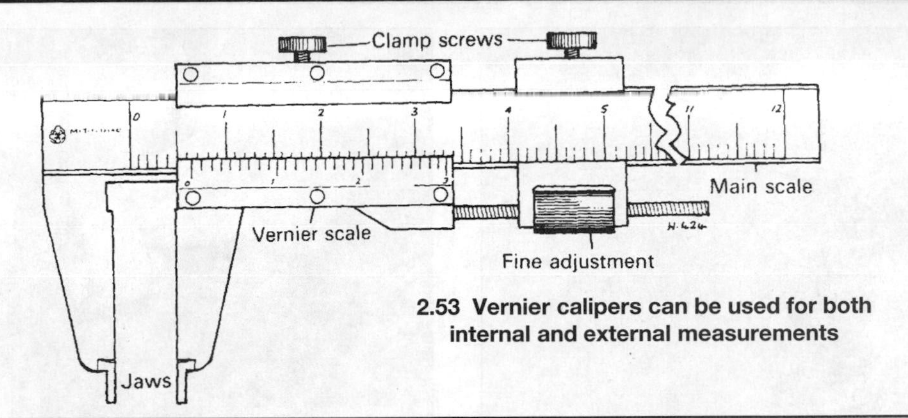


2.53 Vernier calipers can be used for both internal and external measurements

## Micrometers

A micrometer is the most accurate measuring tool likely to be needed in a home shop and you could successfully argue that the cost of the tool, weighed against occasional use, makes it an unaffordable luxury. This is particularly true since individual micrometers are limited to measurements in 1-inch increments and you would need a set of two or three micrometers to be completely prepared for any measuring job.

The basic outside micrometer consists of a U-shaped metal frame covering a 1-inch size range. At one end is a precision ground stop, called the anvil, while at the other end is an adjustable stop called the spindle **(see illustration 2.54)**. The spindle is moved in-or-out of the frame on a very precise, fine thread by a calibrated thimble, usually incorporating a ratchet to prevent damage to the threads. In use, the spindle is turned very carefully until the object to be measured is gripped very lightly between the anvil and spindle. A calibrated line on the fixed sleeve below the spindle indicates the rough size (base figure), while a more accurate measurement (down to 1/1000-inch) is calculated by adding an additional number indicated on the thimble scale to the base figure.

Micrometers are available in a wide range of sizes, starting with 0-1 inch (for small engine repair, anything over 2-3 inch is useless). There are also versions called inside micrometers, designed for making internal measurements such as cylinder bore sizes. Small hole gauges and telescoping gauges can be used along with outside micrometers to avoid the need for them **(see illustrations 2.55 and 2.56)**.

All micrometers are precision instruments and easily damaged if misused or stored with other tools. They also require regular checks and calibration to maintain their accuracy. Given the fragile nature and high cost of these tools, as a general rule you should try to get by without them until you know for sure there's a definite need for them.

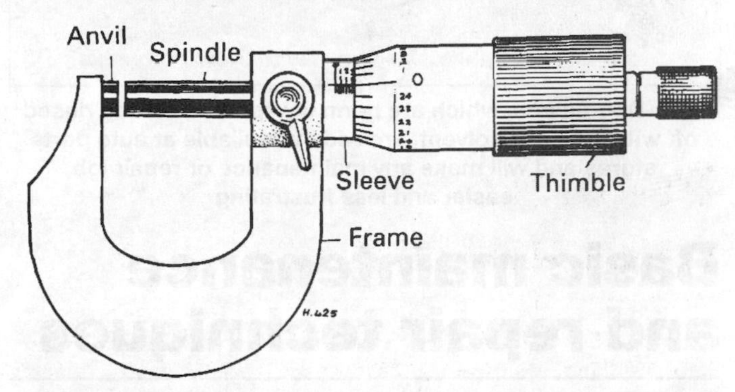


**2.54 Micrometers, though expensive, are very accurate and almost indispensable when checking engine parts for wear**

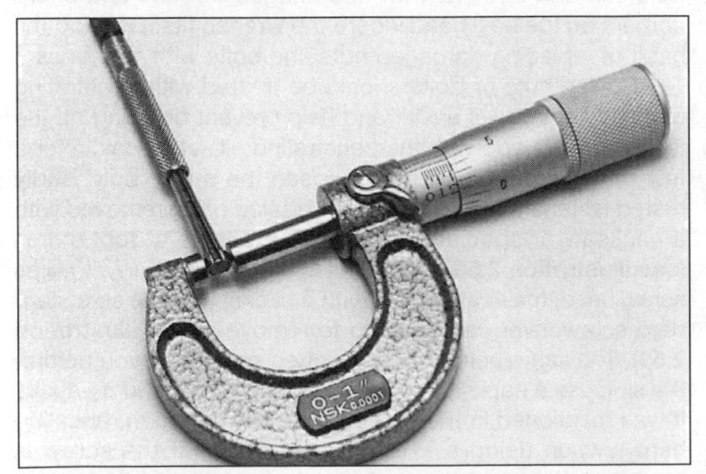


**2.55 When used in conjunction with a small hole gauge. . .**

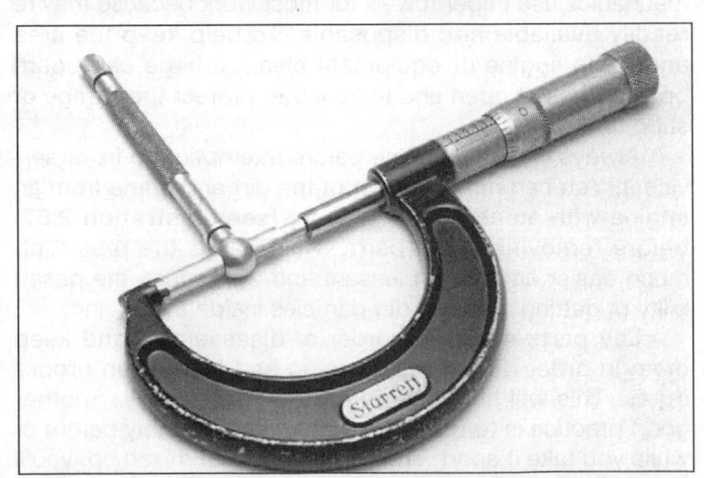


**2.56 . . . or a telescoping gauge, a micrometer can also make internal measurements - valve guide and cylinder bore checks are two typical examples**

2.57 Degreasers, which are normally sprayed on and rinsed off with water or solvent, are widely available at auto parts stores and will make any maintenance or repair job easier and less frustrating

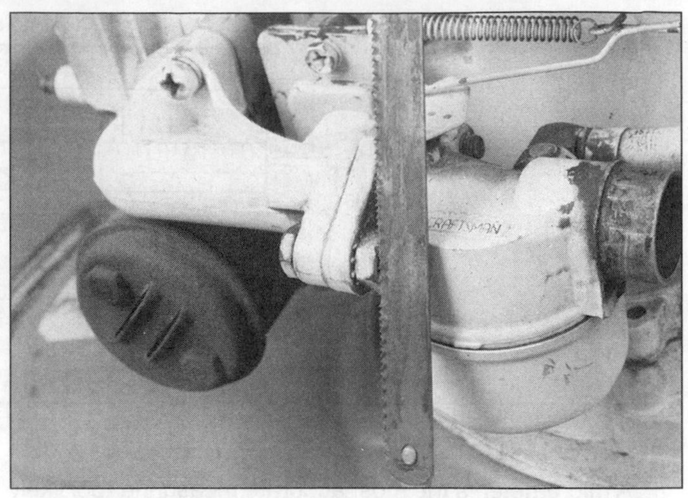

2.58 To remove a frozen or rounded-off nut, use a hacksaw to saw off one side, then open the nut with a chisel or turn it with a Vise-Grips

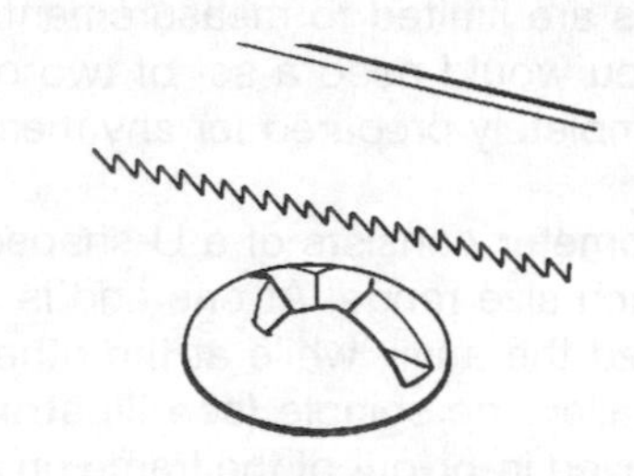

2.59 A hacksaw can be used to make a slot in a damaged Phillips screw head so it can be removed with a standard screwdriver

# Basic maintenance and repair techniques

## General repair hints

Although it was mentioned in the *Environmental safety* section, it's worth repeating here - sometimes waste oil, drained from the engine during normal maintenance or repairs, presents a disposal problem. To avoid pouring oil on the ground or into the sewage system, pour it into large containers, seal them with caps and take them to an authorized disposal site or service station. Plastic jugs are ideal for this purpose. **Note:** *Do not contaminate the oil with any other fluids - service stations will not accept it if you do!*

Keep a supply of old newspapers and clean rags available. Old towels are excellent for mopping up spills. Many mechanics use paper towels for most work because they're readily available and disposable. To help keep the area under the engine or equipment clean, a large cardboard box can be cut open and flattened to protect the garage or shop floor.

Always clean an engine before attempting to fix or service it. You can remove most of the dirt and grime from an engine with an aerosol degreaser **(see illustration 2.57)** before removing many parts. This makes the repair job much easier and more pleasant and will reduce the possibility of getting abrasive dirt particles inside the engine.

Lay parts out in the order of disassembly and keep them in order during the cleaning and inspection procedures. This will help ensure correct reassembly. Another good practice is to draw a sketch of an assembly before or while you take it apart. Then if the parts get mixed up, you'll have a guide to follow when putting them together again.

When working on an engine, look for conditions that may cause future trouble. Check for unusual wear and damage. You may be able to prevent future problems by making an adjustment or repairing a part before it fails.

## Fasteners

Fasteners, basically, are nuts, bolts and screws used to hold two or more parts together. There are a few things to keep in mind when working with fasteners. Many of them require a locking device of some type (either a lock washer, locknut, locking tab or thread cement). All fasteners should be clean and straight with undamaged threads and sharp corners on the hex-head where the wrench fits. Develop the habit of replacing damaged nuts and bolts with new ones.

Rusted nuts or bolts should be treated with penetrating oil to make removal easier and help prevent breaking off the fastener. After applying the penetrating oil, let it soak in for a few minutes before trying to loosen the nut or bolt. Badly rusted fasteners may have to be chiseled off or removed with a hacksaw or special nut breaker, available at tool stores **(see illustration 2.58)**. If you mess up the recess in a Phillips screw head, make a slot in it with a hacksaw blade so a standard screwdriver can be used to remove it **(see illustration 2.59)**. The same holds true for slotted screws - if you deform the slot, use a hacksaw to enlarge or deepen it and try again. It was mentioned in the tool section, but it's worth repeating here - when using a Phillips screwdriver, if the screw is extremely tight and the tip tends to back out of the recess rather than turn the screw, apply a small amount of valve lapping compound to the screwdriver tip so it'll grip better.

Flat washers and lock washers, when removed from an assembly, should always be replaced in their original locations. Discard damaged washers and replace them with new ones. Always use a flat washer between a lock washer and any soft metal surface (such as aluminum), thin sheet metal or plastic. Special locknuts can only be used once or twice before they lose their locking ability and must be replaced.

If a bolt or stud breaks off in an assembly, it can be drilled out and removed with a special tool called an E-Z out. Broken fastener removal and thread repairs are covered later in this chapter. If you don't have the tools or don't want to do it yourself, most small engine dealers and repair shops - as well as automotive machine shops - can perform these tasks.

## Tightening sequences and procedures

When threaded fasteners are tightened, they're often tightened to a specific torque value (torque is basically a twisting force). Over-tightening the fastener can weaken it and cause it to break, while under-tightening can cause it to eventually come loose from engine vibration. Important fasteners, depending on the material they're made of, the diameter of the thread and, in the case of bolts, the material they're threaded into, have specific torque values. Be sure to follow the torque recommendations closely. For fasteners not requiring a specific torque, use common sense when tightening them.

Fasteners laid out in a pattern (such as cylinder head bolts) must be loosened and tightened in a sequence to avoid warping the component. Initially, the bolts should go in finger-tight only. Next, they should be tightened 1/2-turn each, in a criss-cross or diagonal pattern. After each one has been tightened 1/2-turn, return to the first one and tighten each of them 1/4-turn at a time until each fastener has been tightened to the proper torque. To loosen the fasteners the procedure can be reversed.

## Disassembly sequence

Engine disassembly should be done slowly and deliberately to make sure the parts go back together properly during reassembly. Always keep track of the sequence parts are removed in. Note special characteristics or marks on parts that can be installed more than one way. It's a good idea to lay the disassembled parts out on a clean surface in the order they were removed. As mentioned before, it may also be helpful to make sketches or take instant photos of components before removal.

When removing fasteners from a component, keep track of their locations. Sometimes threading a bolt back in a part, or putting the washers and nut back on a stud, can prevent mixups later. If nuts and bolts can't be returned to their original locations, they should be kept in a compartmented box or a series of small boxes. A cupcake or muffin tin is ideal for this purpose, since each cavity can hold the bolts and nuts from a particular area or sub-assembly. A pan of this type is especially helpful when working on components with very small parts (such as the carburetor and valve train). The cavities can be marked with a felt-tip pen or tape to identify the contents.

## Gasket sealing surfaces

Gaskets are used to seal the mating surfaces between components and keep lubricants, fuel, vacuum or pressure contained in an assembly.

Gaskets are often coated with a liquid or paste-type gasket sealant before assembly. Age, heat and pressure can sometimes cause the two parts to stick together so tightly they're very difficult to separate. In most cases, the part can be loosened by striking it with a soft-face hammer near the joint. A regular hammer can be used if a block of wood is placed between the hammer and part. **Caution:** *Do not hammer on cast parts or parts that could be easily damaged.* With any particularly stubborn part, always recheck to make sure all fasteners have been removed.

Avoid using a screwdriver or bar to pry components apart, as they can easily mar the gasket sealing surfaces of the parts (which must remain smooth). If prying is absolutely necessary, use a piece of wood, but keep in mind that extra clean-up will be necessary if the wood splinters.

After the parts are separated, the old gasket must be carefully scraped off and the engine surfaces cleaned. Stubborn gasket material can be soaked with gasket removal solvent (available in aerosol cans) to soften it so it can be easily removed. Gasket scrapers are widely available - just be careful not to gouge the sealing surfaces if you use one. Some gaskets can be removed with a wire brush, but regardless of the method used, the mating surfaces must be left clean and smooth. If the gasket surface is scratched or gouged, then a gasket sealant thick enough to fill scratches should be used during reassembly of the components. For most applications, non-drying (or semi-drying) gasket sealant is best.

# How to remove broken bolts and repair stripped threads

## Removing broken-off bolts

If a bolt breaks off in the hole, a drill, drill bit and E-Z out (screw extractor) will be required to remove it. First, select the E-Z out needed for the job (based on the bolt size), then select the drill bit required to make the hole for the E-Z out. Follow the procedure shown in the accompanying photos **(see illustrations 2.60 through 2.64)**.

2.60 Before attempting to remove a broken-off bolt, apply penetrating oil and let it soak in for awhile

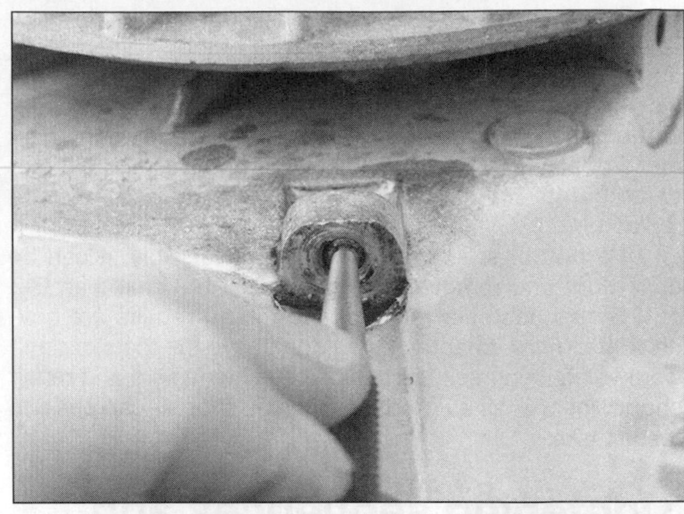

2.61 Use a center punch to make an indentation as close to the center of the bolt as you can

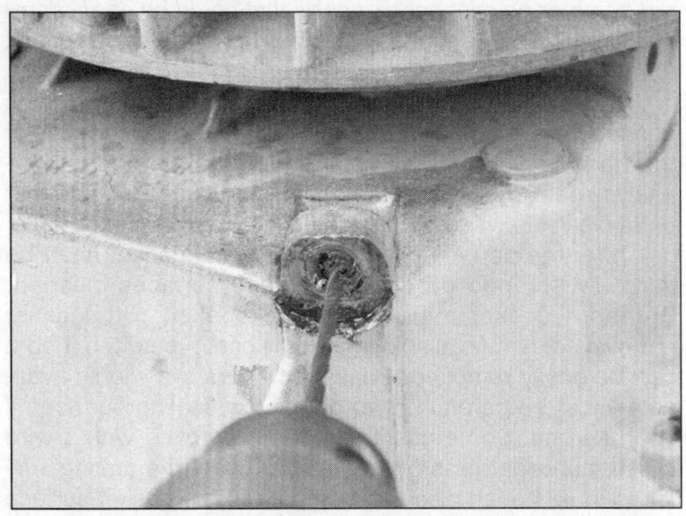

2.62 Carefully drill a hole in the bolt (hold the drill so the bit is parallel to the bolt - the bit should be about two-thirds the diameter of the bolt and the hole should be as deep as possible without going through the bolt

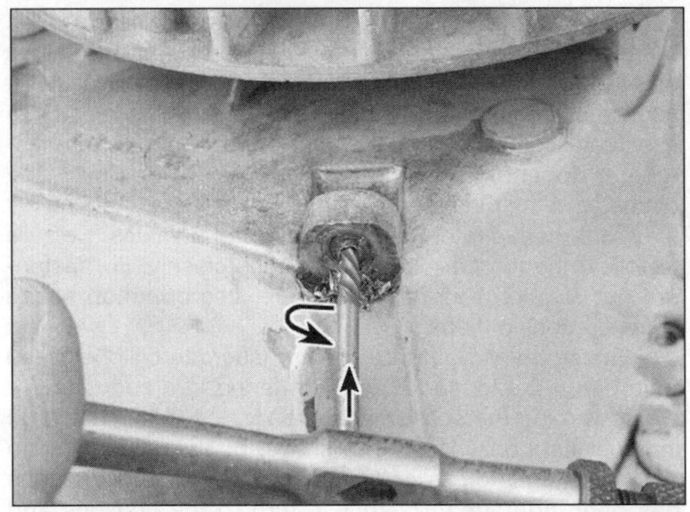

2.63 Push the E-Z out into the hole and turn it with a tap handle or adjustable wrench - keep pressure on the extractor so it doesn't turn in the hole instead of moving the bolt

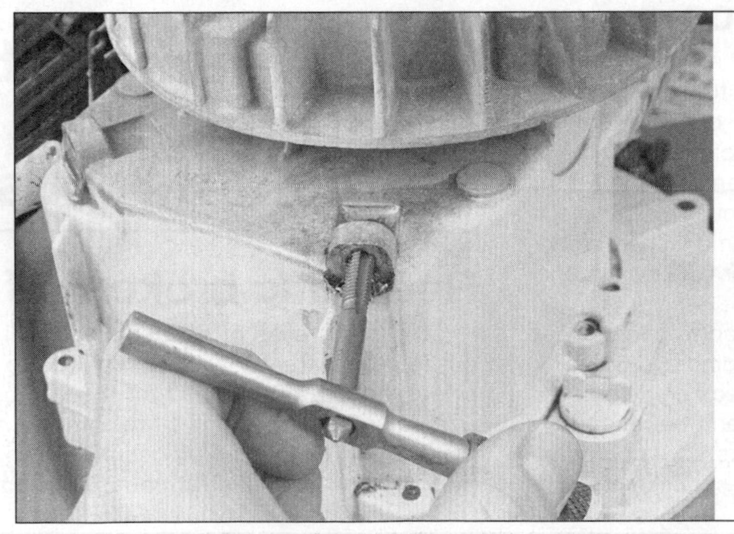

2.64 If you have one of the right size, run a tap into the hole after the bolt is out to clean up the threads and remove any rust or corrosion

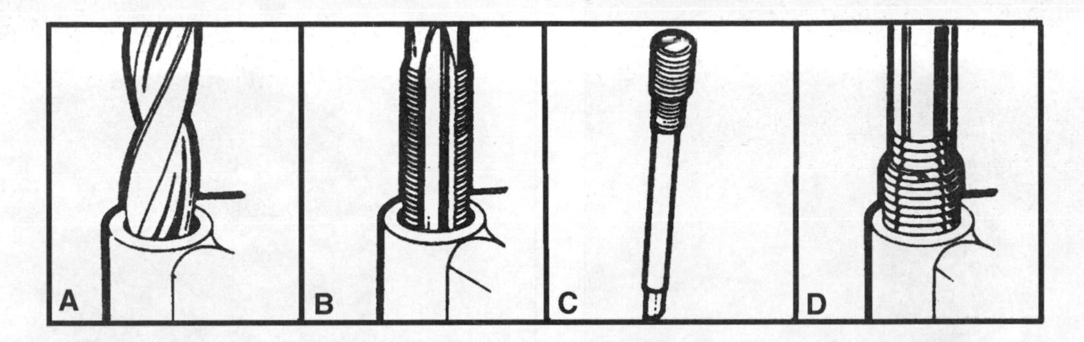


**2.65 Installing a thread insert**

*A Drill out the hole to remove the old threads, (this isn't required with all insert brands)*
*B Use a tap to cut new threads in the hole (the tap is included with some thread repair kits)*
*C Attach the insert to the installation tool (included with the kit)*
*D Screw the insert into the newly-threaded hole (when it's flush with the top of the hole, break off the drive tang and remove the tool)*

## Repairing stripped threads

If the thread isn't totally ruined, a tap or die can be used to clean it up so the fastener can be reused. If the nut or bolt is a standard size, or an extremely important fastener like a head bolt or flywheel nut, don't worry about salvaging it - buy and install a new one. Remember, if a bolt is stripped, the threads in the bolt hole may also be damaged. **Note:** *Always use thread cement on the threads of a restored nut or bolt when it's reinstalled and tighten it carefully to avoid further damage.*

If the thread in a bolt hole is completely stripped or seriously damaged, retapping may not work. In such cases a thread insert will be needed. The most commonly available inserts require drilling out the hole and cutting an oversize thread with a special tap. The resulting new thread is then reduced to the original size by installing a stainless steel wire insert in the threaded hole **(see illustration 2.65)**. This allows the original bolt or stud to be reinstalled **(see illustration 2.66)**. Heli-Coil thread inserts are the most common and sets with the required tap, several inserts, an installation tool (mandrel) for common thread sizes and comprehensive instructions are available.

Thread inserts look like springs prior to installation and have a small drive tang at the lower end. The thread insert is attached to a special mandrel and the tang engages in a slot in the end of the tool. The insert is screwed into the hole until the upper end is flush with or slightly below the surface. Once in position, the drive tang is broken off, leaving the insert locked in place. If the drive tang doesn't break off when backing out the mandrel, use a pin punch or needle-nose pliers to snap it off.

This type of repair is a good way to reclaim badly worn or stripped threads in aluminum parts (very common in small engines). The new thread formed by the stainless steel insert is permanent and more durable than the original. In some cases, the original hole doesn't have to be enlarged prior to retapping, making the repair quick and simple to carry out. The only drawback for the home shop is the high cost of purchasing a range of taps and inserts. The taps needed for thread inserts are not standard sizes - they're made specially for use with each insert size. The inserts aren't expensive individually, but stocking up on them in each size that could be required on a small engine is probably too expensive for most home shops.

STANDARD SCREW FITS IN . . .
HELI-COIL INSERT IN . . .
HELI-COIL TAPPED HOLE.

**2.66 The thread insert allows the original fastener to be used in the repaired hole**

# Small engine lubricants and chemicals

A number of lubricants and chemicals are required for small engine maintenance and repair. They include a wide variety of products ranging from cleaning solvents and degreasers to lubricants and penetrating oil. **Caution 1:** *Always follow the directions and heed the precautions and warnings printed on the containers of lubricants and chemicals designed for shop use.*

2.67 Contact point/spark plug cleaner

**Caution 2:** *When using any cleaners or degreasers, you should wear protective gloves to prevent toxic materials being absorbed by your skin. Disposable latex gloves can be purchased in a box of twenty or more relatively inexpensively.*

**Caution 3:** *When using any type of spray cleaner, or a solvent that you are working vigorously in with a brush, always wear eye protection. Even a tiny amount of chemical spattered in your eye can cause damage.*

**Ignition point/spark plug cleaner (see illustration 2.67)** is a solvent used to clean oily film and dirt off points and oil deposits off spark plugs. It's oil free and leaves no residue. It can also be used to remove gum and varnish from carburetor jets and other orifices.

**Carburetor cleaner** is similar to contact point/spark plug cleaner but it has a much stronger solvent and may leave a slight oily reside. It isn't normally needed on small engine carburetors, but if deposits are heavy it will work faster and better than solvent.

**Silicone-based lubricants** are used to protect rubber parts such as hoses and grommets.

Multi-purpose grease **(see illustration 2.68)** is an all purpose lubricant used wherever grease is more practical than a liquid lubricant such as oil. Some multi-purpose grease is colored white and specially formulated to be more resistant to water than ordinary grease.

**Motor oil**, of course, is the lubricant specially formulated for use in an engine. It normally contains a wide variety of additives to prevent corrosion and reduce foaming and wear. Motor oil comes in various weights (viscosity ratings) of from 0 to 50. The recommended weight of the oil depends on the seasonal temperature and the demands on the engine. Light oil is used in cold climates and under light load conditions; heavy oil is used in hot climates and where high loads are encountered. Multi-viscosity oils are designed to have characteristics of both light and heavy oils and are available in a number of weights from 0W-30 to 20W-50. Be sure to follow the engine manufacturer's recommendations.

2.68 Multi-purpose grease

**Anti-seize** is a liquid or paste material usually sold in a small can with a brush applicator in the top of the can. Anti-seize is used where dissimilar metals are joined, like a steel bolt threaded into an aluminum casting, or on high-heat applications such as exhaust fasteners. As the name implies, the compound prevents fasteners from seizing up due to rust, heat or electrolytic action caused when ferrous and non-ferrous parts are joined.

**Gas additives** perform several functions, depending on their chemical makeup. They usually contain solvents that help dissolve gum and varnish that build up on carburetor and intake parts. They also serve to break down carbon deposits that form on the inside surfaces of the combustion chamber. Some types contain upper cylinder lubricants for valves and piston rings. For small engine use, the most common additive is a gas stabilizer used when equipment is stored for long periods.

**Degreasers** are heavy-duty solvents used to remove grease and grime that may accumulate on engines and machinery. They can be sprayed or brushed on and, depending on the type, are rinsed off with either water or solvent.

**Solvents** are used alone or in combination with degreasers to clean parts during repairs and overhauls. The home mechanic should use only solvents that are non-flammable and that don't produce irritating fumes.

**Gasket sealants (see illustration 2.69)** may be used in conjunction with gaskets, to improve their sealing capabilities, or alone, to seal metal-to-metal joints. Many gasket sealants can withstand extreme heat, some are impervious to gasoline and lubricants, while others are capable of filling and sealing cavities. Depending on the intended use, gasket sealants either dry hard or stay relatively soft and pliable. They're usually applied by hand, with a brush, or are sprayed on the gasket sealing surfaces.

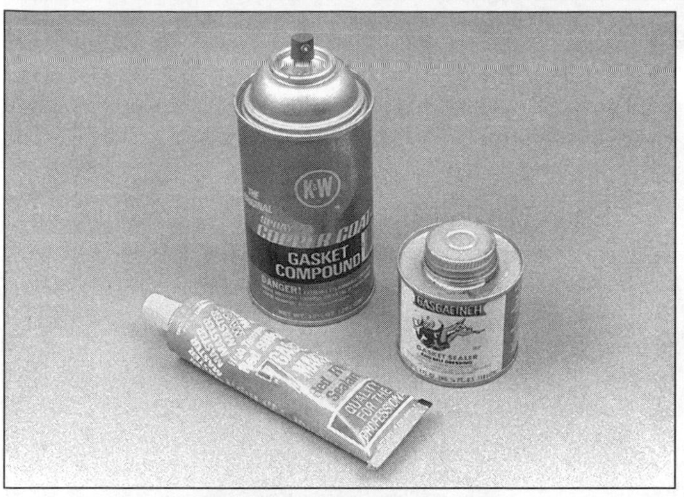


2.69 Sealant needed for small engine repair work

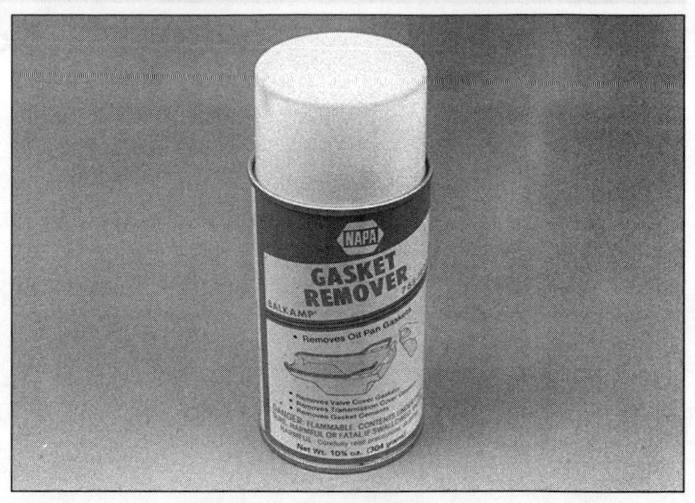


2.70 Aerosol gasket remover

**Gasket/sealant removal solvents (see illustration 2.70)** are available at auto parts stores and are helpful when removing gaskets that are baked on or stuck to engine components. They're usually powerful chemicals and should be used with care.

**Thread cement (see illustration 2.71)** is an adhesive compound that prevents threaded fasteners from loosening because of vibration. It's available in a variety of types for different applications.

**Moisture dispersants (see illustration 2.72)** are usually sprays that can be used to dry out ignition system components and wire connections. Some types also are very good solvents and lubricants for cables and other components.

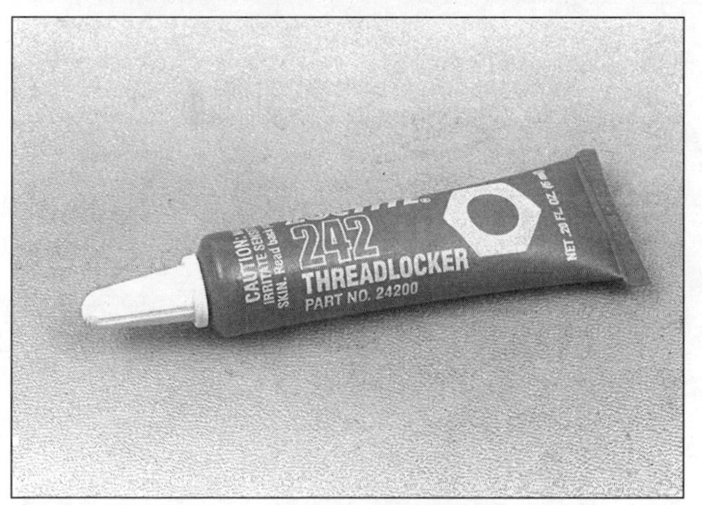


2.71 Thread cement

2.72 WD-40 is a good moisture dispersant, solvent and lubricant

## Notes

# 3 Troubleshooting

## How an engine works

All small, air-cooled, gasoline engines are internal-combustion engines much like the ones used in cars, trucks and motorcycles. The term "internal-combustion" is used because energy for turning the crankshaft is developed inside the engine.

This happens when the fuel/air mixture is burned inside a confined space called a combustion chamber (or cylinder). Because of the heat produced, the mixture expands, which then forces the piston to move. The piston is connected to the crankshaft, which changes linear motion into rotary motion. The crankshaft may be oriented vertically or horizontally, depending on the engine application. The crankshaft is situated at a right angle to the cylinder bore.

To supply power - motion - to the crankshaft, a series of events must occur. This series of events is called a combustion cycle. The events in the cycle are:

*a) Intake of the fuel/air mixture into the cylinder*
*b) Compression of the fuel/air mixture*
*c) Ignition/expansion of the fuel/air mixture*
*d) Expulsion of the burned gases*

The movement of the piston in one direction, either toward the crankshaft or away from it, is called a stroke. Some small engines complete a cycle during one revolution of the crankshaft (two strokes of the piston). Other types require two revolutions of the crankshaft (four strokes of the piston). In this manual, the shortened terms "two-stroke" and "four-stroke" are used in place of the technically correct terms "two-stroke cycle" and "four-stroke cycle." The major differences between four-stroke and two-stroke engines are:

*The number of power strokes per crankshaft revolution*
*The method of getting the fuel/air mixture into the combustion chamber and the burned gases out*
*The number of moving parts in the engine*
*The method used to lubricate the internal engine components*

## Four-stroke engines

As mentioned above, a four-stroke engine completes one combustion cycle during two revolutions of the crankshaft, or four strokes of the piston. This is due to the design of the engine and the way the fuel/air mixture is introduced into the cylinder. A camshaft, which is driven off the crankshaft, opens valves that allow the fuel/air mixture in and exhaust gases out of the engine. The valves are closed by spring pressure. The four strokes of the piston have been labeled to describe what happens during each one **(see illustration 3.1)**:

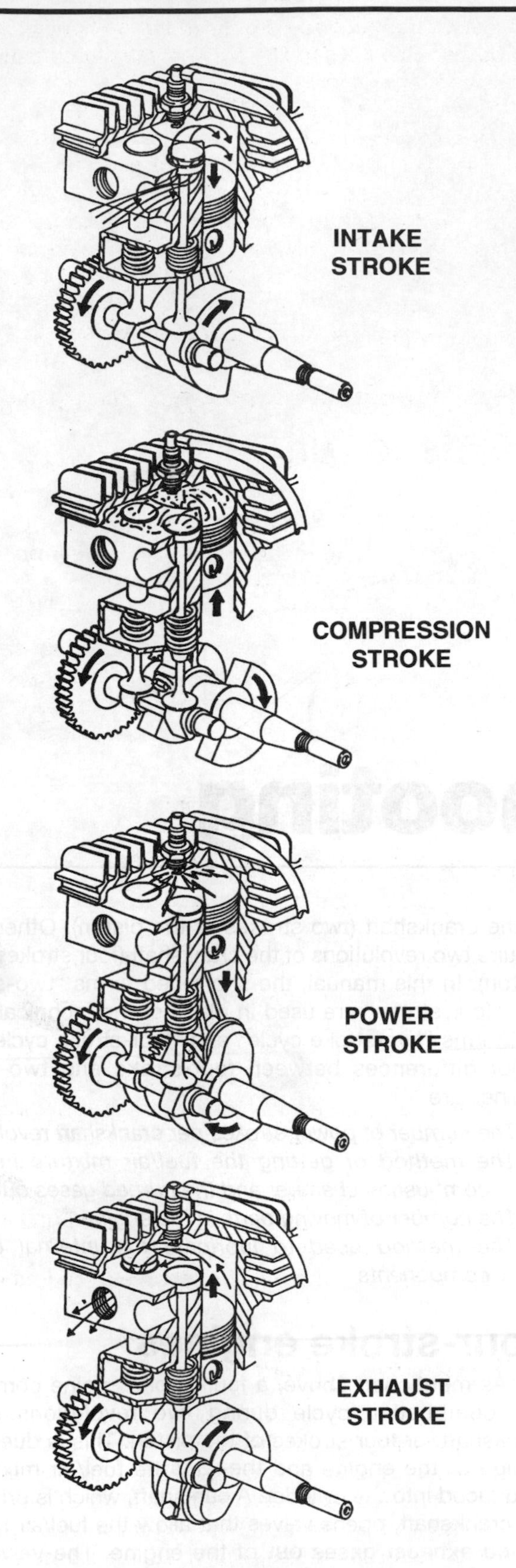


**3.1 The four-stroke engine combustion cycle**

*Illustration courtesy of and with permission of Briggs & Stratton Corp.*

**INTAKE STROKE** - During the intake stroke, the piston moves down in the cylinder with the intake valve open. The fuel/air mixture is forced into the cylinder through the open intake valve. Near the end of the piston's movement, the intake valve closes, sealing off the combustion chamber.

**COMPRESSION STROKE** - The piston changes direction at the end of the intake stroke and begins to move up in the cylinder, which compresses and heats the fuel/air mixture.

**POWER STROKE** - As the piston reaches the end of the compression stroke, the ignition system fires the spark plug, which ignites the compressed fuel/air mixture in the combustion chamber. The piston changes direction again and moves down in the cylinder with great force produced by the burning/expanding gases. The connecting rod transfers the movement of the piston to the crankshaft, which changes the linear motion to rotary (turning) motion, which can be used to do work.

**EXHAUST STROKE** - When the piston reaches the end of the power stroke, it changes direction again and the exhaust valve opens. As the piston moves up in the cylinder, the exhaust gases are pushed out through the exhaust valve. When the piston reaches the end of the exhaust stroke, the cycle starts all over again.

## Two-stroke engines

Two stroke engines have a slightly different design that doesn't require valves. Instead, openings called ports are used to route the fuel/air mixture to the cylinder. Also, the crankcase is used as a temporary storage area as the fuel mixture is on its way to the cylinder, so it must be leakproof. The piston serves to open and close off the ports to seal off the combustion chamber.

A two-stroke engine is designed to complete the same cycle described for a four-stroke engine, but it does it during one revolution of the crankshaft. The INTAKE and COMPRESSION strokes actually occur simultaneously, during one stroke of the piston, and so do the POWER and EXHAUST STROKES **(see illustration 3.2)**.

As the piston moves down in the cylinder, pushed by the burning/expanding gases, the fresh fuel/air charge in the crankcase is being pressurized slightly.

Once the piston uncovers the exhaust port, the spent gases begin to exit the cylinder. Next, the piston uncovers the transfer ports and the fresh fuel/air charge begins to flow out of the crankcase, through the transfer ports and into the cylinder, helping to push out the exhaust gases. At the bottom of the stroke, the piston changes direction and begins to move up in the cylinder, sealing off the transfer and exhaust ports and compressing the fuel/air mixture in the cylinder. At the same time, the intake port (which may be sealed off by a one-way valve called a reed valve or by the piston skirt) opens and more fuel/air mixture is forced into the crankcase; this is because a slight vacuum is created as the piston moves up in the cylinder. At the top of the piston's stroke, the spark plug fires, which ignites the compressed fuel/air mixture in the combustion chamber. The piston changes direction again and moves down in the

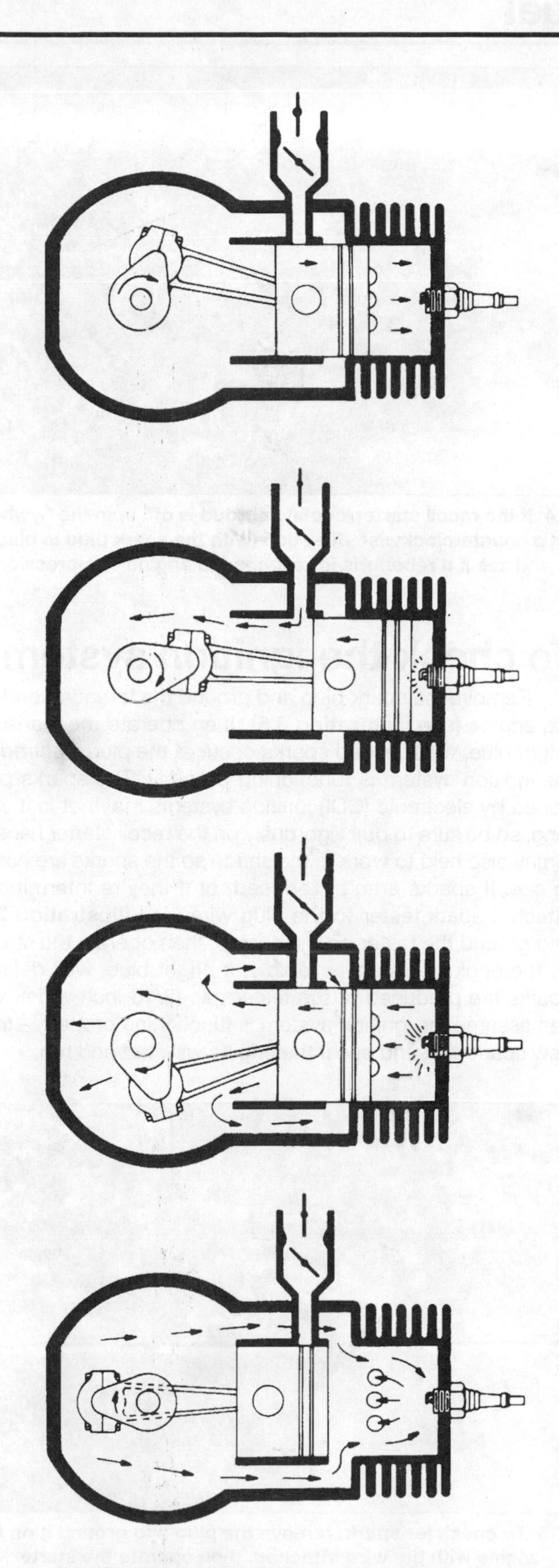

3.2 The two-stroke engine combustion cycle

*Illustration courtesy of Tecumseh Products Co.*

cylinder with great force produced by the burning/expanding gases. This overlapping cycle is repeated continuously as the engine is running.

Since a two-stroke engine uses the crankcase for storing a reserve charge of fuel/air mixture, the crankcase can't be used as an oil reservoir for lubricating the engine. Instead, lubrication is supplied by a specific quantity of oil that's mixed with the gas and circulated through the engine as it runs. Never put gasoline in a two-stroke engine without mixing oil with it - the engine will overheat because of improper lubrication. It won't run very long before the piston and bearings will overheat, score and seize!

## The three things an engine needs to run

Any engine, whether it's a two or four-stroke, must have three essential elements to run:

a) *Fuel and air mixed together in the correct proportions*
b) *Compression of the fuel/air mixture*
c) *Ignition at the right time*

If any one of these three elements is missing, the engine simply will not run. Troubleshooting is the process of determining which one is missing and why. Once you figure this out, repairs are done to restore the missing element.

# Introduction to troubleshooting

Possible causes for various engine problems and recommendations for correcting them are covered in detail in this chapter. **Note:** *What appears to be an engine malfunction may be a problem in the power equipment rather than the engine.*

You won't find any absolutely foolproof troubleshooting procedures for small engines that won't start and run properly. Some symptoms can be so obscure that it takes a professional mechanic to spot the problem. However, in general, malfunctioning engines have symptoms that are relatively easy to identify. By thoroughly checking the problem in an orderly manner, as outlined in this chapter, you usually will be able to find the problem and save a trip to the repair shop.

## Where to start

When the cause of a malfunction isn't obvious, check the compression, ignition and carburetor, in that order (the three elements required for an engine to run). These checks must be done in a systematic manner. It's the quickest and surest way to find the problem and may reveal possible causes of future problems, which can be corrected at the same time. The procedure is basically the same for all engines.

3.3 As a general rule, if the engine's compression will blow your thumb off the spark plug hole, it should be adequate for the engine to run

3.4 If the recoil starter/cooling shroud is off, spin the flywheel in a counterclockwise direction (with the spark plug in place) and see if it rebounds in response to engine compression

# To check the compression

Remove the spark plug and ground the plug wire on the engine, then seal off the plug hole with your thumb **(see illustration 3.3)**. Operate the starter - if the compression pressure blows your thumb off the hole, the compression is adequate for the engine to run; be careful not to touch the plug wire as this is done - you'll get quite a jolt if you do! Another way to check the compression with the spark plug in place is to remove the cooling shroud/recoil starter mechanism and spin the flywheel in reverse (counterclockwise) **(see illustration 3.4)**. The flywheel should rebound (change directions) sharply; if it does, the compression is adequate for the engine to run.

If the compression is low, it may be due to:

*a) Loose spark plug*
*b) Loose cylinder head bolts*
*c) Blown head gasket*
*d) Damaged valves/valve seats (four-stroke engine only)*
*e) Insufficient valve tappet clearance (four-stroke engine only)*
*f) Warped cylinder head*
*g) Bent valve stem(s) (four-stroke engine only)*
*h) Worn cylinder bore and/or piston rings*
*i) Broken connecting rod or piston*

**Note:** *There are some Kohler, Honda and other engines that have a feature called Automatic Compression Release (ACR), which uses a mechanism to hold the exhaust valve open more during starting, to reduce compression at that time and makes the engine easier to start in hot or cold weather. Once the engine is running, normal compression is restored. On these engines it may not be possible to take a standard compression test, since the engine must be at 600 rpm or more to achieve normal compression.*

# To check the ignition system

Remove the spark plug and ground the threaded end on the engine **(see illustration 3.5)**, then operate the starter. If bright blue, well-defined sparks occur at the plug electrodes, the ignition system is functioning properly. The sparks produced by electronic (CDI) ignition systems may not last very long, so be sure to pull vigorously on the recoil starter handle. It may also help to work in the shade so the sparks are easier to see. If sparks aren't produced, or if they're intermittent, attach a spark tester to the plug wire **(see illustration 3.6)** and ground the tester on the engine, then operate the starter so the engine turns over rapidly. If bright blue, well-defined sparks are produced at the tester gap (3/16-inch wide), you can assume the ignition system is functioning properly - try a new spark plug and see if the engine will start and run.

3.5 To check for spark, remove the plug and ground it on the engine with the wire attached, then operate the starter - if sparks occur at the plug, the ignition system is OK; if no sparks occur, the plug may be bad, so don't automatically condemn the ignition system

3.6 A spark tester with the gap set at about 3/16-inch is the best tool to use for checking the ignition system -if it will produce a spark strong enough to jump the tester gap, it's adequate for the engine to run

If sparks do not occur, it may be due to:

*a) A sheared flywheel key*
*b) Incorrect ignition point gap**
*c) Dirty or burned ignition points**
*d) Coil failure/electronic ignition failure*
*e) Incorrect armature air gap*
*f) Worn bearings and/or crankshaft on flywheel side only*
*g) Ignition point plunger stuck or worn**
*h) Shorted ground wire (if equipped)*
*i) Shorted stop switch (if equipped)*
*j) Condenser failure**
*k) Malfunctioning starter interlock system*
*l) Defective spark plug wire*
*m) Malfunctioning oil level switch*

* *Engines with ignition points only*

If the engine runs but misses during operation, a quick check to determine if the ignition system is or is not at fault can be made by attaching a spark tester between the plug wire and the spark plug. An ignition system misfire will be apparent when watching the tester (it should spark continually).

3.8 If the engine has good compression and spark, try priming it by carefully pouring gasoline directly into the plug hole with a small funnel - if the engine starts, you have a fuel delivery problem to find and fix

3.7 The choke (arrow) must be closed when starting the engine cold - if it isn't closing, free it up so it will (typical choke shown, air cleaner removed)

# To check the carburetor

Before proceeding, be sure the fuel tank is filled with fresh, clean gasoline, the fuel shut-off valve is open (if equipped) and fuel flows freely through the fuel line. Check/adjust the mixture adjusting screw (if equipped) to make sure it's open and make sure the choke closes completely **(see illustration 3.7)**.

If the engine won't start, remove and check the spark plug - if it's wet or fouled, it may be due to:

*a) Stuck choke*
*b) Overly-rich fuel mixture*
*c) Water in fuel*
*d) Inlet needle valve stuck open (not all carburetors)*
*e) Clogged muffler*
*f) Plugged crankcase breather*
*g) Too much oil in crankcase*
*h) Worn piston rings*

If the spark plug is dry, check for:

*a) Leaking carburetor mounting gaskets*
*b) Gummy or dirty carburetor internal parts*
*c) Inlet needle valve stuck shut (not all carburetors)*
*d) Damaged or deteriorated rubber diaphragms in carburetor (not all carburetors)*
*e) Plugged fuel line or filter (not all engines)*
*f) Failed fuel pump (not all engines)*

A simple check to determine if fuel is getting to the combustion chamber through the carburetor is to remove the spark plug and pour a small amount of gasoline (about one teaspoonful) into the engine through the spark plug hole (this is called "priming" the engine) **(see illustration 3.8)**. Reinstall the plug. If the engine fires a few times and then stops, look for the same conditions described above.

If your engine is equipped with a fuel pump and you suspect the carburetor is not getting fuel, disconnect the fuel line from the carburetor. Aim the fuel line into a suitable container and operate the starter with the spark plug wire disconnected from the spark plug and grounded on the engine **(see illustration 3.9)**. If you don't see fuel coming out, the pump may be at fault.

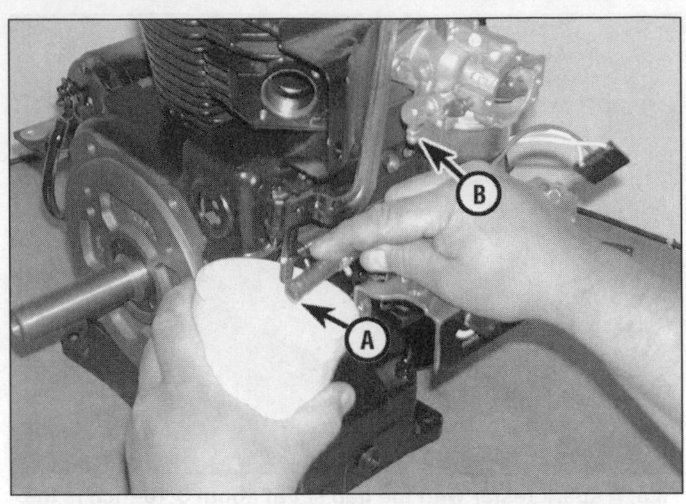


**3.9 To check the fuel pump, disconnect the fuel line (A) from the fitting at the carburetor (B) - aim the hose into a container and rotate the engine (with the spark plug wire disconnected and grounded) - fuel should come out of the hose**

# Troubleshooting a four-stroke engine

Most four-stroke engine problems will fall into one or more of the following categories:

a) *Won't start*
b) *Hard to start/kicks back when starting*
c) *Stops suddenly*
d) *Lack of power/erratic operation*
e) *Excessive vibration*
f) *Noise*
g) *Engine smokes*
h) *Overheating*
i) *Excessive oil use*

## Won't start

**Note:** *When an engine won't start, it's usually because the controls are improperly set, the safety interlock devices are interrupting the ignition system or the ignition system is malfunctioning, although carburetor problems and low compression can also prevent an engine from starting (see the information under the heading Where to start? to rule out lack of compression as a cause).*

### Electrical system checks

#### Charging system

1 On models so equipped, check the battery with a voltmeter and hydrometer (see Chapter 4). If the battery is consistently discharged, check the charging system.

2 Charging systems generally exhibit only two problems, either the system is overcharging the battery and evaporating the electrolyte, or the system is undercharging and the battery keeps going down. All batteries tend to discharge slowly with disuse over a long period of time. If your equipment is used only infrequently, the charging system may not be at fault when the battery goes dead. If, with regular usage, the battery is consistently discharging enough to be incapable of starting the engine, make a few checks of the charging system.

3 If yours is a model with an alternator for lighting only (no battery or electric starter), disconnect the wires to the light(s), but do not allow then to ground out on the equipment. Hook a digital voltmeter positive lead to the power wire to the light (the other wire will be a ground), and ground the negative lead from the meter. Start the engine and run it at around 3200 rpm. If the meter reads less than 15 volts, the stator should be checked. With the engine Off, use an ohmmeter to check the two leads of the stator. There should be some resistance, which varies according to the manufacturer. No resistance or infinite resistance indicate the stator should be replaced. Refer to the engine Chapter for your model for stator replacement procedure.

4 On models with battery charging systems, run the engine at 3000 to 3200 rpm and check the DC battery voltage at the battery terminals. If the reading is 12.5 volts or higher, the charging system is working.

5 If the charging system is suspect, start the engine (3000 rpm) and disconnect the positive cable from the battery. Check the voltage from the battery cable to ground. On models with a 1.25-amp charging system, the voltage should be 11.5 volts or more, on models with 3-amp charging systems, voltage should be 28 volts or more. This indicates that the stator is OK.

6 On models with a regulator, disconnect the two stator electrical connectors from the regulator and attach an AC voltmeter **(see illustration 3.10)**. On most small engine regulator/rectifiers, there are usually three wires, two stator leads and a positive lead connected to the battery. Though the position of the leads may vary with manufacturer, only one is the lead to the battery, the other two are the stator leads to be tested. With the engine running at 3200 to 3600 rpm, voltage should be 23 to 28 volts or higher, indicating the stator is OK, and the regulator is probably at fault. To check further, stop the engine and measure the resistance between the two stator leads. On most charging systems, resistance should be small, but zero or infinite resistance means the stator should be replaced. **Note:** *On some models, there is only one stator wire. Connect the positive lead of the AC voltmeter to this wire, and the other test lead to an engine ground and test. If the voltage is low, replace the stator.*

7 The electrical systems on some engines have a fuse in their control panel. If the fuse is blown, it could prevent

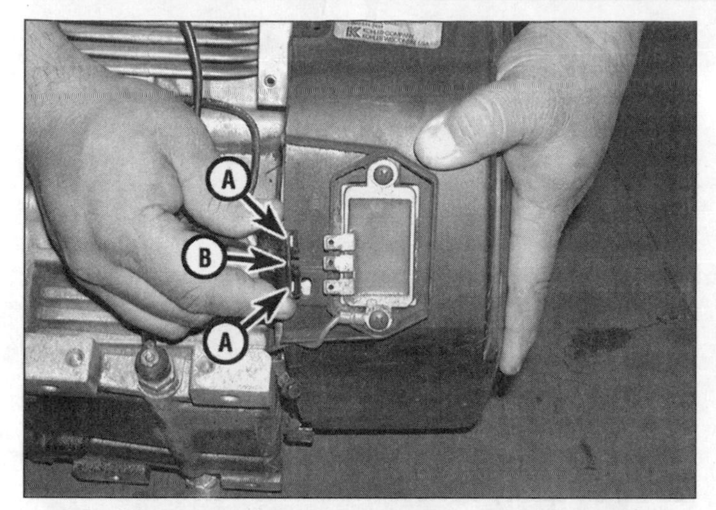


3.10 Disconnect the two stator connectors (A) from the regulator and measure running AC voltage between them - with engine stopped measure resistance between them to test the stator - on some models, all three wires are in one connector - (B) leads to the battery

3.11 Typical starter-mounted solenoid (A), battery cable terminal (B) and ignition switch terminal (C)

starting. Check the back of the control panel and pull out the fuse for examination. If it's blown replace it and try starting again. If the fuse blows again, check for a short to ground somewhere in the wiring and repair it. Other systems use a one-way diode in the wiring from the regulator to the battery. This diode prevents the battery from being drained if the charging system is faulty, but if the diode fails, the battery can discharge. To test the diode, hook a test light to the battery side of the diode and to a good ground. If the test light goes on, the diode must be replaced. Further testing of the diode is performed by checking both sides of the diode with a continuity tester. A good diode should indicate continuity with the test leads connected one way and no continuity when the leads are reversed. **Warning:** *Never replace a fuse with a higher-rated fuse. It may cure the immediate problem, but may cause the wiring to overheat, possibly causing a fire.*

## Starting system

8 If the battery is fully charged, and the engine won't start, check the starter operation **Caution:** *Do not attempt to jump-start your equipment with another battery if the onboard battery is weak. The increased current can damage your starting system. Charge the onboard battery and start the engine without using a jumper battery.*

9 If the battery is fully charged and the starter doesn't operate when the key is turned to Start, first test the ignition switch. On the back of the switch use a voltmeter to check for battery voltage at the B terminal ( in all switch positions). Some engines have only a pushbutton for starter operation. Such pushbuttons are momentary ON switches with two wires in back. Disconnect the two wires and attach a continuity tester to the switch terminals. There should be no continuity when the switch isn't pressed, and continuity when it is pressed. If there is no continuity when the switch is pressed, replace it.

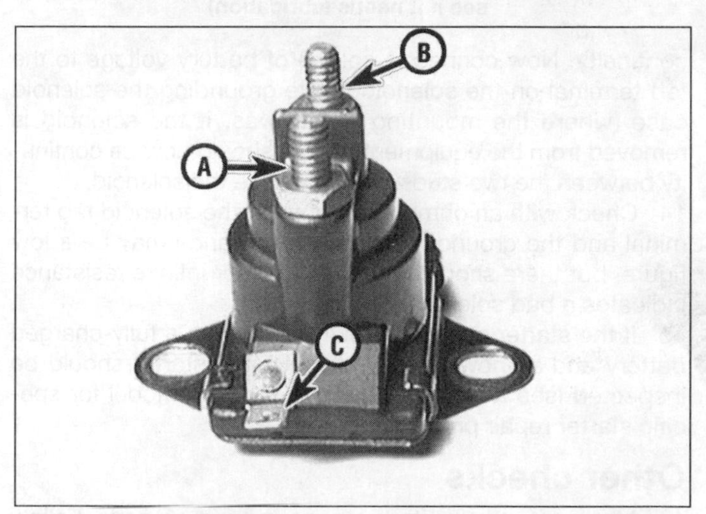


3.12 Typical remote starter solenoid - battery-to-solenoid terminal (A), solenoid-to-starter terminal (B) and ignition switch terminal (C)

10 When the key is turned to Start, check that there is battery voltage at the S terminal. If not, replace the ignition switch. If there is voltage there, test the starter solenoid. On most models the solenoid is a separate component connected by cables to the starter, while on some models the solenoid is mounted directly on top of the starter **(see illustration 3.11)**.

11 The solenoid usually has two large stud-type posts for the cables, a tab connector for the wire from the ignition switch, and the solenoid grounds through its mounting bolts **(see illustration 3.12)**. On some models with optional ammeter, another wire will be connected to the stud where the battery positive cable connects.

12 Check that there is battery voltage at the battery stud on the solenoid. If not, there is a problem with the battery or the cable from the battery.

13 Disconnect the negative battery cable, then remove the wire and cables from the solenoid. Connect an ohmmeter to the two stud terminals on the solenoid. There should be no

3.13 Make sure the control cable (arrow) is securely clamped to the engine so it doesn't slip when the control lever is moved (if the clamp is loose, tighten it and check the cable to see if it needs lubrication)

3.14 If the spark plug wires terminal fits loosely on the spark plug, crimp the terminal with pliers so it fits snugly on the plug tip - on models with a rubber boot, pull the terminal out of the boot, crimp it and slide it back into the boot

continuity. Now connect a source of battery voltage to the tab terminal on the solenoid, while grounding the solenoid case (where the mounting screw was, if the solenoid is removed from the equipment). There should now be continuity between the two studs. If not, replace the solenoid.

14 Check with an ohmmeter between the solenoid tab terminal and the grounding mount. Resistance may be a low figure, but there should be some. Zero or infinite resistance indicates a bad solenoid.

15 If the starter operates slowly, even with a fully-charged battery and a known-good solenoid, the starter should be inspected (see the engine Chapter for your model for specific starter repair procedures).

## Other checks

1 Make sure the controls are positioned properly. Follow the control cable from the lever to the carburetor. The throttle should be all the way open and the choke should operate when the lever is set on START. As you move the control lever from START to FAST to STOP, the cable should be clamped so the throttle operates properly **(see illustration 3.13)**. The cable may be slipping in the clamp just enough to cause the throttle to malfunction and you might not see the slight movement.

2 Move the throttle to the open or START position with your fingers. You may have to move the control lever with your other hand to open the throttle.

3 If the engine now starts, let it run for several minutes, then pull the control lever back to STOP. If the engine slows down but doesn't stop, loosen the cable clamp with a screwdriver and pull the cable toward the control lever very slightly until the engine stops, then retighten the clamp. Start the engine again and run through the control positions. The engine should start, run slowly, run fast and stop when the lever is positioned next to the appropriate label on the control.

4 On some engines, there are other possible controls that could prevent starting.

5 Many models have a low-oil-level switch or solenoid, which is designed to prevent starting if the oil level is too low in the crankcase. Test the switch with a test light or continuity tester. Use an ohmmeter to check for continuity between the two terminals on the sensor. If the switch is good, there should be no continuity when the engine oil level is normal, and continuity when the oil level is low. If the switch is faulty, it could prevent the engine from starting.

6 Some engines also have a fuel-shut solenoid on the carburetor, which is designed to prevent fuel from draining through the carburetor (into the engine) when the equipment isn't running. This feature is found on equipment with battery starting systems. To check the solenoid, remove the solenoid and test it by apply battery voltage to one terminal and ground to the other (if there is only one terminal, apply voltage to that terminal and ground the body of the solenoid). With voltage supplied, the solenoid plunger should retract, and extend once the voltage is removed **(see the carburetor section of your engine's Chapter)**. A faulty solenoid can prevent engine starting.

7 Make sure the gas tank is at least half full of fresh fuel and that fuel is reaching the carburetor (the line between the tank and carburetor could be plugged, kinked or detached or the filter, if equipped, could be plugged).

8 Check the plug wire to make sure it's securely attached to the spark plug.

The terminal on the end of the spark plug wire can come loose and get corroded. Crimp the loop with a pair of pliers **(see illustration 3.14)** and remove corrosion with sandpaper, a wire brush or a round file.

9 Check the spark plug to make sure it's tight.

10 Check the spark plug ground strap to make sure it's not malfunctioning. Some engines have a metal strap that's used to ground the spark plug wire to stop the engine. If the engine has one, make sure it's not touching the terminal.

11 If the engine has a fuel priming device, make sure it's working properly.

It should be pushed four or five times when the engine is

**3.15 If the gas tank cap vent is plugged, the engine will starve for fuel until you remove the cap (which relieves the vacuum) - the engine will then start and run for a while before it stalls again**

**3.16 If the air filter is dirty or clogged with debris, remove and clean it or install a new one - foam types and pleated paper types are the most common**

**3.17 One way to check to make sure gas is getting to the carburetor is to remove the float bowl (if used) - it's usually held in place with a nut on the bottom, which is hard to get at and easily damaged**

cold to fill the carburetor. If the engine is hot, don't operate the primer - it may flood the carburetor. Instead, pull the starter handle several times with the control lever in the STOP position. This will help clear excess fuel out of the engine. Put the control lever on START and start the engine normally.

12 Make sure the grass catcher is properly installed. Some lawnmowers have a safety switch for the grass catcher where it attaches to the mower housing. This device prevents the engine from starting until the grass catcher is properly installed. If the grass catcher isn't being used, make sure the chute is properly attached to the mower deck.

13 See if the gas tank cap vent is clogged. If the gas tank vent is clogged, a vacuum will eventually form in the tank and prevent fuel from reaching the carburetor (the engine will act like it's out of gas).

14 Remove the cap and check the gasket in it. Sometimes the space between the gasket and cap gets clogged with debris, shutting off the air supply to the tank. Check the vent hole(s) in the cap to make sure they're open **(see illustration 3.15)**. If you're not sure if the vent is open or not, leave the cap off and try to start the engine. If the engine runs, the cap vent is the problem. Either clean the vent or install a new cap. DO NOT run the engine without a cap on the gas tank. Some engines are equipped with a mechanical or crankcase-pressure-operated fuel pump (see the Section *If the spark plug is dry. . .*

15 Check the air filter to see if it's clean and make sure the gasket between the filter and carburetor is in good shape **(see illustration 3.16)**.

16 Check to see if the plug is firing (see To check the ignition system in the section headed *Where to start?*). You can clean spark plugs, but a new one should be installed - they're not expensive and you can usually be sure the new plug is good. Every once in a while, a new plug will turn out to be faulty. If you install a new plug and the engine won't fire even though everything else seems to be okay, try another new plug or a spark tester.

17 Make sure the plug wire is in good condition. Bend the wire by hand and look for cracks in the insulation. Also look for burned or melted insulation. If damage is noted, you may have to replace the ignition coil, since the wire usually is permanently attached to it.

18 If equipped, check the blade for looseness - it must be tight on the shaft or adapter.

19 If the equipment has a chain, drivebelt or belts, check to see if they're loose. A loose chain or belt, like a loose blade, can cause a backlash effect, which will work against engine cranking effort.

20 You may be trying to start the engine when it's under load. Make sure the equipment is disengaged when the engine is started, or, if engaged, doesn't create an unusual starting load.

21 Remove the float bowl (not all carburetors have one) and see if the carburetor is dirty or gummed up **(see illustration 3.17)**. **Caution:** *Shut off the fuel valve or remove the tank before detaching the float bowl - if you don't, gas will run out until the tank is empty!*

**3.18 Craftsman engines with an "Auto-prime" carburetor often won't run because this fuel inlet gets clogged with dirt or sludge - clean it with a fine piece of wire and the engine should run fine**

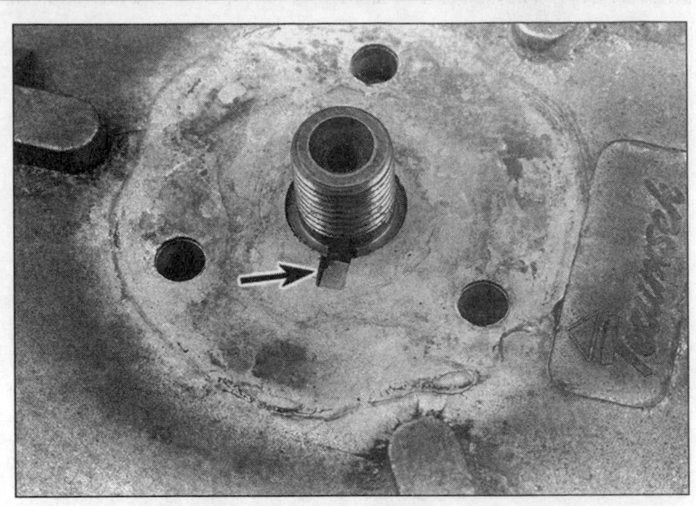

**3.19 If the flywheel key (arrow) is sheared, the ignition timing will be off and the engine wont' run**

If there's dirt or sludge in the float bowl, the carburetor passages may be clogged. Remove the carburetor and clean it thoroughly. Some carburetors have a small spring-loaded valve on the bottom of the float bowl that's used to drain out sediment and water. Push up on it with a small screwdriver and let gas run out until it looks clean (you'll see little droplets if water comes out). **Note:** *Many Tecumseh/Craftsman engines are equipped with an "Auto-prime" float-type carburetor that has a fuel inlet which gets clogged easily by gasoline residue, particularly if the engine isn't run for a long time (suspect this in lawn mowers that are stored all winter without draining the gasoline out of the carburetor or adding a gas stabilizer to the tank).*

22 Drain the fuel out of the tank or detach the tank from the engine and set it aside. Remove the float bowl and clean it out, then clear the fuel inlet with a very fine wire **(see illustration 3.18)**. Reinstall the float bowl and tank, add gasoline and try to start the engine.

# Hard to start/kicks back when starting

1 These problems are quite common and are usually caused by belts or chains that aren't disengaged when the engine is cranked or by a loose blade. If equipped, check the blade for looseness - it must be tight on the shaft or adapter.

2 If the equipment has a chain, drivebelt or belts, check to see they're loose. A loose chain or belt, like a loose blade, can cause a backlash effect, which will work against engine cranking effort.

3 You may be trying to start the engine when it's under load. See if the equipment is disengaged when the engine is started, or, if engaged, doesn't create an unusual starting load.

4 The recoil starter may not be operating properly.

5 If your engine is a model with automatic compression release feature (designed to reduce compression below 600 rpm for easier starting), the compression-release mechanism may not be operating (see the engine Chapter for your model).

6 The cylinder head may be coated with carbon deposits. Heavy deposits built up over a long period of time can raise the compression and make staring difficult. Refer to Chapter 4 for decarbonizing procedure.

# Stops suddenly

1 Check the gas tank to make sure it hasn't run dry.

2 See if the gas tank cap vent is clogged. If the gas tank vent is clogged, a vacuum will eventually form in the tank and prevent fuel from reaching the carburetor (the engine may start but then act like it's out of gas).

3 Remove the cap and check the gasket in it. Sometimes the space between the gasket and cap gets clogged with debris, shutting off the air supply to the tank. Check the vent hole(s) in the cap to make sure they're open **(refer to illustration 3.15)**. If you're not sure if the vent is open or not, leave the cap off and try to start the engine. If the engine runs, the cap vent is the problem. Either clean the vent or install a new cap. DO NOT run the engine without a cap on the gas tank.

4 Sometimes when everything is running perfectly, the engine will stop suddenly and you won't be able to get it started again. If the problem isn't a dry gas tank, this problem can usually be traced to one of two things:

*a) A sheared flywheel key*
*b) A defective condenser (engines with ignition points only)*

5 The flywheel key will shear if the equipment blade strikes an immovable object (this is to protect the crankshaft and other expensive engine components). It can happen without any indication to the equipment operator. The flywheel key is important for proper ignition timing - if it's damaged, the engine won't run.

**3.20 Even though you can see oil in the engine after the plug is removed, it must be at the very top of the hole or the crankcase will be dangerously low! - some engines have a dipstick with the safe level clearly marked on it**

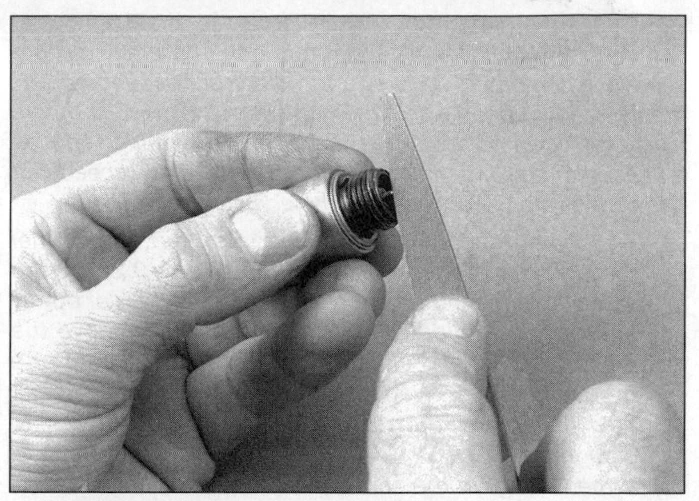

**3.21 If the plug is covered with deposits, clean it with a wire brush, then file the electrode tips until they have sharp edges (this makes it easier for the ignition system to fire the plug when the engine is running)**

6 If the engine is equipped with ignition points, the condenser (located under the flywheel) can stop working at any time without warning. Even a brand new condenser can go bad, so don't be fooled into thinking the problem is something else if the engine has new, or nearly new, ignition parts. Instructions for servicing to the condenser are included in Chapter 4 (under ignition point replacement).

7 You can also check the flywheel key at the same time you check the condenser. It's a rectangular piece of soft metal that fits into a slot in the flywheel and crankshaft taper to index the flywheel to the crankshaft **(see illustration 3.19)**. As mentioned above, the metal is soft for a reason - when the blade strikes something hard, the key shears and releases the force on the crankshaft. This prevents damage to the crankshaft, piston, valves, gears and other major (expensive) parts of the engine. Replacement of the key is covered in each engine chapter (flywheel removal and installation).

8 Check the oil in the crankcase. When you remove the oil check/fill cap, the engine may appear to be full but still need oil. Add oil with a funnel until it runs out the hole **(see illustration 3.20)**. If the engine has a dipstick, the safe level will be clearly indicated. **Note:** *If your engine has a low-oil-level shutoff system and you are operating the equipment on a steep slope, the engine may be shutting off because it reads the oil level as low. Start and run the engine on level ground to see if this is the problem.*

Start the engine. If it's noisy, the lack of lubrication has probably damaged the engine. It may have to be overhauled or replaced.

9 Check the fuel lines and filter (if used) to make sure they're clear.

10 See if the muffler/exhaust pipe is clogged. A plugged exhaust system can stop the engine and damage it. So can a defective muffler. Remove the muffler and see if the engine will start and run.

11 Check the plug wire to make sure it's securely attached to the spark plug.

The terminal on the end of the spark plug wire can come loose and get corroded. Crimp the loop with pliers **(refer to illustration 3.14)** and remove corrosion with sandpaper, a wire brush or a round file.

12 Check the cable and linkage between the handle controls, throttle and governor to see if they're binding or if anything has come loose. Lubricate the cable if necessary.

# Lack of power/erratic operation

1 These symptoms are usually caused by problems in the ignition system (especially on engines with ignition points) or the carburetor (particularly the idle adjustment).

2 Check the gas in the tank to make sure it's fresh and doesn't have any water in it. Drain the tank and refill it with new gas.

3 Check the flywheel key (see the section headed *Stops suddenly*).

4 Check the engine compression (see the section headed *Where to start?*).

5 Check the plug wire to make sure it's securely attached to the spark plug. The terminal on the end of the spark plug wire can come loose and get corroded. Crimp the loop with pliers **(refer to illustration 3.14)** and remove corrosion with sandpaper, a wire brush or a round file.

6 Make sure the control lever is free and working properly (not stuck on any of the settings).

7 Make sure the air filter is clean **(refer to illustration 3.16)**.

8 Remove the spark plug and check the gap and the base of the plug for dirt. Clean the electrodes with a wire brush and use a fine file to square the side electrode tip so any worn edges are sharp **(see illustration 3.21)**. **Note:** *As a general rule, the spark plug gap can be set at 0.025-inch.*

9 Check the carburetor mixture screw to see if it's out of adjustment. **Note:** *Some carburetors don't have any mixture screws, while others have one for either high-speed adjustments or low speed adjustments, but not both. Still others have one screw to adjust the fuel/air mixture at high speeds and another screw that controls the mixture at low speeds - if two screws are used, they must be adjusted separately.*

a) *The mixture screw is used to control the flow of fuel through the carburetor* **(see illustration 3.22)**. *If the screw is damaged or incorrectly adjusted, loss of power and erratic engine operation will result. Remove the mixture screw and check the tip - if it looks bent or a groove has been worn in the tapered portion, install a new one* **(see illustration 3.23)**. *Do not attempt to straighten it.*

b) *If the O-ring on the screw is damaged or deteriorated, replace it before attempting to adjust the mixture.*

c) *If the screw isn't bent or worn, reinstall it and turn it in until it stops - tighten it with your fingers only, don't force it. Back it out about 1-1/4 turns (counterclockwise).* **Note:** *The actual factory-recommended number of turns out is different for each carburetor type, but the figure given here is in the ballpark for most engines (see the engine Chapter for your model for specific carburetor screw adjustment).*

d) *Start the engine and turn the screw clockwise until the engine starts to slow down. This means the fuel mixture is too lean (not enough gas). Slowly turn the screw out (counterclockwise) until the engine begins to run smoothly. Keep going very slowly until the engine just begins to run rough again (rich condition). Finally, turn the screw in again (clockwise) to a point about half-way between lean operation and rich operation. This is the perfect setting.*

e) *If an idle (low-speed) mixture screw is used, adjust it in the same manner with the engine idling. After the low-speed mixture has been set, recheck the high-speed adjustment (if applicable), it may be affected by the idle adjustment. Some carburetors also have an idle speed adjusting screw that's used to open or close the throttle valve slightly to change the idle speed only, not the fuel/air mixture. Turning it will cause the engine to speed up or slow down.*

9 Check to see if the engine is flooded. Running equipment on hills and slopes can cause flooding. Flooding can also occur when the engine is cranked with the spark plug wire disconnected and when the mixture is too rich (when the carburetor mixture screw is out of adjustment).

10 Set the control lever to the STOP position. Pull the starter rope or crank the engine over several times. The closed throttle produces a high vacuum and opens the choke, which cleans excess fuel out of the engine. Start the engine.

11 If the engine continues to flood, adjust the mixture screws as outlined above. If this doesn't work, there is another alternative, but you must be very careful when doing it (there is a possibility of gasoline igniting in the carburetor).

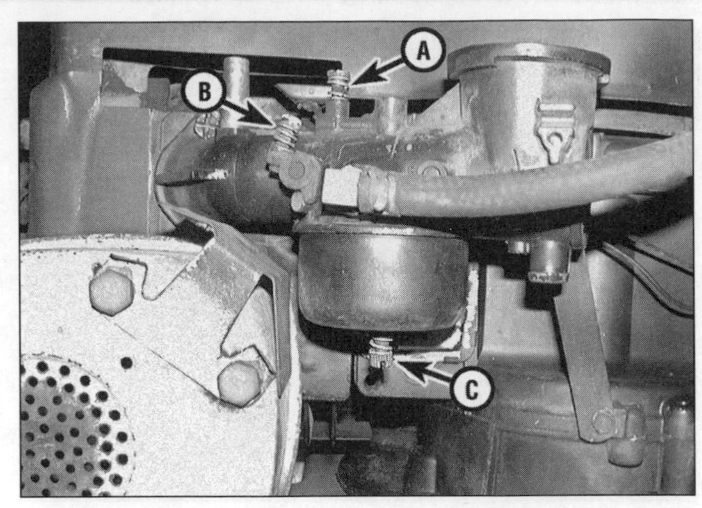


**3.22 Some carburetors have several adjusting screws:**

A *Idle mixture screw*
B *Idle speed screw*
C *High-speed mixture screw (not on all models)*

12 Remove the air filter and hold the choke/throttle valves open with a small screwdriver **(see illustration 3.24)**. Keep your face and hands away from the carburetor and crank the engine several times. The engine should start. When it starts, remove the screwdriver and reinstall the air filter.

13 When the engine is running, adjust the mixture screws as described above.

# Excessive vibration

1 Engine vibration can be caused by a bent crankshaft, which results from hitting something with the blade. Many small engine repair shops can straighten a crankshaft without disassembling the engine. If the power take-off end of the crankshaft wobbles as it's turned by hand, remove the engine from the equipment and take it to a repair shop for crankshaft straightening.

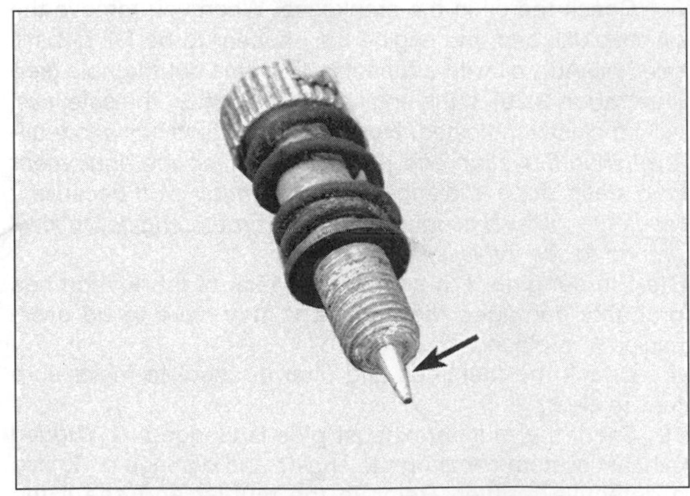

**3.23 If the mixture screw tip is bent or worn, the correct adjustment will be difficult to maintain**

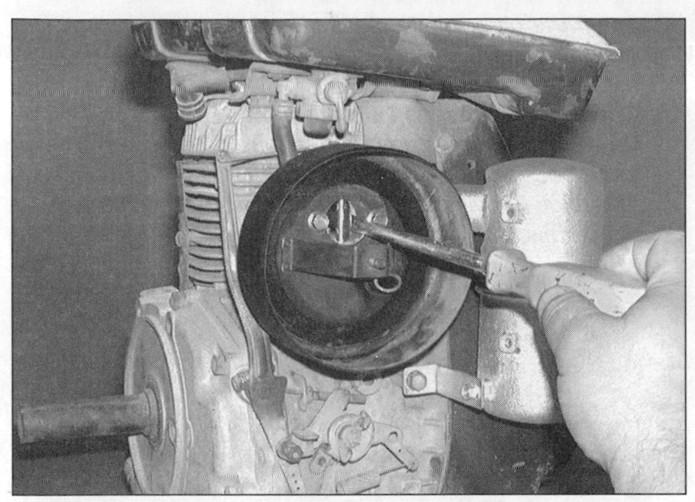

3.24 If the engine is flooded (the spark plug is wet with gasoline), hold the choke and throttle valves - if possible - wide open with a screwdriver while cranking the engine to clear the excess fuel

3.25 If the mounting bolts (arrows) are loose, the engine will vibrate and damage will result

2 Look for an out-of-balance blade. Mower blades can get twisted or badly nicked and may have poor lift. Sometimes sharpening the blade unevenly can throw it off. Resharpen the blade or install a new one.
3 Check to see if the blade is tight. Tighten the blade mounting bolt or bolts. Turn them counterclockwise.
4 Check the engine mounting bolts to see if they're tight. These bolts hold the engine to the equipment **(see illustration 3.25)**. You may have to hold the bolts on top while you tighten the nuts under the deck.
5 Check the deck for damage. If it's broken, cracked or badly rusted, the engine may get out of alignment and vibrate. If the deck is damaged, it must be repaired or replaced. Repairs usually involve welding (a job for a professional). If the deck must be replaced, get several estimates; a new mower may be cheaper.
6 Check the carburetor adjustment (see *Lack of power/erratic operation* above.) A rough-running engine can cause vibration.

# Noise

1 Excessive or unusual noise is almost always the result of a muffler that's rusted out, deformed or missing completely. A new muffler will remedy the problem.
2 If the noise isn't caused by the exhaust system, but is in the engine itself, check the oil level immediately!
3 If you notice the engine is suddenly running quieter, check the muffler or exhaust system and remove matted grass clippings, dirt and other debris.

# Engine smokes

1 This may be caused by a broken piston or ring or a damaged cylinder, although other causes are more likely.
2 Check the carburetor to see if it's adjusted properly (see *Lack of power/erratic operation* above). Start the engine and allow it to idle (if possible), then open it up. If black smoke comes out of the exhaust, the mixture is probably too rich. Adjust the fuel mixture (if possible) (follow the procedure in Chapter 4 or your engine Chapter).
3 If the smoke is blue and smells like burned oil, the trouble is probably with the rings, piston or cylinder bore (more than likely all three). The blue smoke may be accompanied by excessive internal engine noises - the engine may make a dull, knocking sound. It will have to be overhauled.

# Overheating

1 This is usually a minor problem, but don't ignore it. Prolonged overheating can cause serious engine damage.
2 Check the oil level immediately. Add clean, fresh oil as needed. Make sure the oil used is the proper viscosity - if it's too thin or contaminated with fuel (from flooding the engine), lubrication will be inadequate.
3 Make sure the cooling fins aren't clogged with debris. Clean them with a putty knife and paint brush.
4 Make sure all shrouds and blower housings are correctly installed. If they aren't, air can't circulate properly through the cooling fins.
5 Check the carburetor to see if it's adjusted properly (see *Lack of power/erratic operation* above).
6 See if the muffler is obstructed by dirt or debris.
7 Make sure the engine isn't overloaded. You may be running it too fast for too long (has the governor been tampered with?). Also, the engine may be overburdened with too much equipment. Stop the engine, let it cool and disengage the extra equipment.
8 Check the cylinder head for excessive carbon build-up (see Chapter 4 for instructions to remove the head and clean it).
9 Check the valve tappet clearances to see if they're too tight (see Chapter 4).
10 Make sure the correct spark plug is installed.

## Excessive oil use

1 This problem can have several causes:

a) *Check the oil level - if the engine is overfilled, excess oil will be blown out through the crankcase breather and get all over the engine.*

b) *Check the governor. The engine may be operating at speeds that are too high.*

c) *See if the oil level check/fill plug gasket is missing. The oil may be leaking out around the plug (it'll run down the engine if it is).*

d) *Check the crankcase breather assembly* **(see Chapter 4)***. Clean or replace it as required. Make sure the oil drain back hole is open.*

e) *The rings or cylinder bore may be worn or damaged. To check this you'll have to disassemble the engine.*

# 4 Tune-up and routine maintenance

## 1 Introduction

This chapter covers the checks and procedures necessary for the tune-up and routine maintenance of typical 5.5 to 20 HP gas engines. It includes a checklist of service procedures designed to keep the engine in proper running condition and prevent possible problems. Separate sections contain detailed instructions for doing the jobs on the checklist, as well as additional maintenance information designed to increase the engine's reliability.

The schedule below is a general guide for routine maintenance. Always refer to the manufacturer's recommendations for maintenance intervals for your specific engine. If your equipment is operated in especially dusty or dirty conditions, perform maintenance procedures more frequently on air cleaners, oil changes, and cleaning of fins and grass screens.

The sections detailing the maintenance and inspection procedures are written as step-by-step comprehensive guides to the actual performance of the work. References to additional information in other chapters is also included and shouldn't be overlooked.

The first step in this or any maintenance plan is to prepare yourself before the actual work begins. Read through the appropriate sections covering the procedures to be done before you begin. Gather up all necessary parts and tools. If it appears that you could have a problem during a particular job, don't hesitate to seek advice from a dealer, repair shop or experienced do-it-yourselfer.

Before attacking the engine with wrenches and screwdrivers, clean it with a degreaser to ensure that dirt doesn't contaminate the internal parts. This will also allow you to detect wear and damage that could otherwise easily go unnoticed.

## 2 Tune-up and maintenance checklist

### Every time the engine is refueled

Check the oil level
Clean the grass screen over the flywheel

### Every 25 hours of operation

Clean the air cleaner pre-filter (foam element over the paper filter) if equipped
Change the engine oil
Check the electrolyte level in the battery (if equipped)
Check for dirt in the engine cooling fins/shrouds and clean if necessary
Check control operation

### Every 50 hours of operation

Check and clean or replace air filter
Clean the terminals and cable ends on the battery (if equipped)
Inspect/adjust the flywheel brake, (if equipped)
Inspect/clean crankcase breather (if equipped)

### Every 100 hours of operation

Inspect and clean/adjust spark plug(s)
Inspect and clean/replace paper air filter element
Check the coil and ignition wire(s)

### Every 300 hours of operation

Drain and clean the fuel tank and replace the fuel filter*
Clean the carburetor float bowl*
Inspect/replace ignition points (if equipped)*
Check the recoil starter*
Check the governor and linkage*
Check engine compression
Check the muffler*
Check the engine mount bolts/nuts*
Check/service valves and adjust the valve clearance
Adjust the carburetor
Decarbonize the cylinder head

**Perform annually on seasonal-use equipment, regardless of hours.*

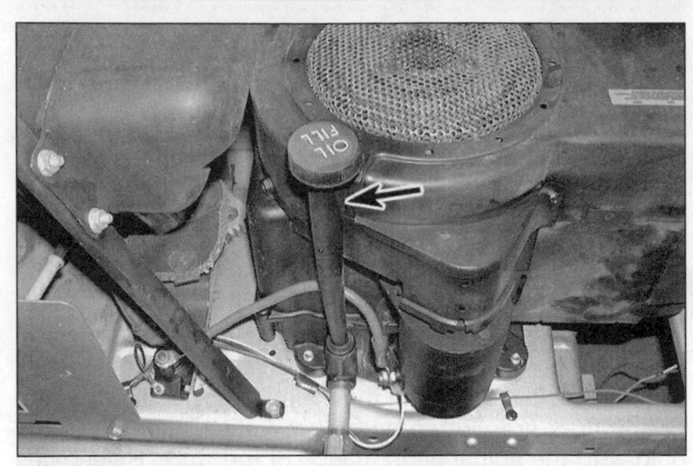


**4.1 The engine oil check/fill cap or plug should be clearly marked - clean it off before removing it - dipstick-type shown**

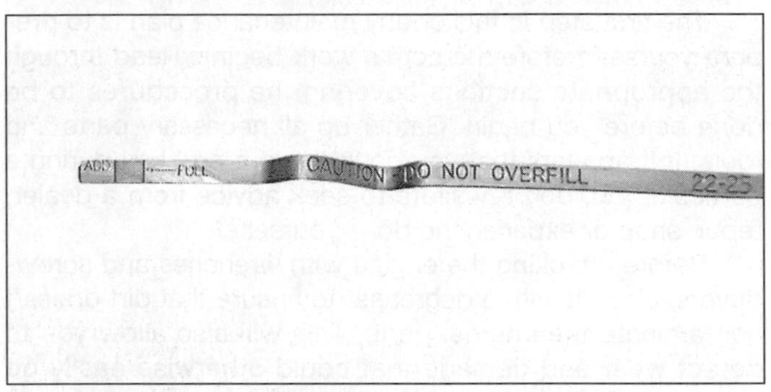


**4.3 The oil level should be at or near the full mark**

# 3 Tune and maintenance procedures

## 1 Check the oil level

Each time you refill the gas tank, or every two or three hours of engine operation, check the crankcase oil level and add more oil as needed. Some manufacturers may recommend more or less frequent oil checks - follow the instructions in your owner's manual if they differ from the information here.

1 Locate the cap used to check the oil level and add oil to the engine - it may be a threaded or friction fit cap, a plug, or an automotive type dipstick. The cap may be marked "Engine oil" or "Oil fill" **(see illustration 4.1)**.

2 Clean the plug and the area around it to prevent dirt from falling into the engine when the plug is removed.

3 Make sure the engine is level, then remove the oil check/fill cap or plug.

4 If the cap or plug doesn't have a dipstick attached to it, the oil level should be at the top of the opening **(see illustration 4.2)** or even with a mark or the top of a slot that indicates the FULL level.

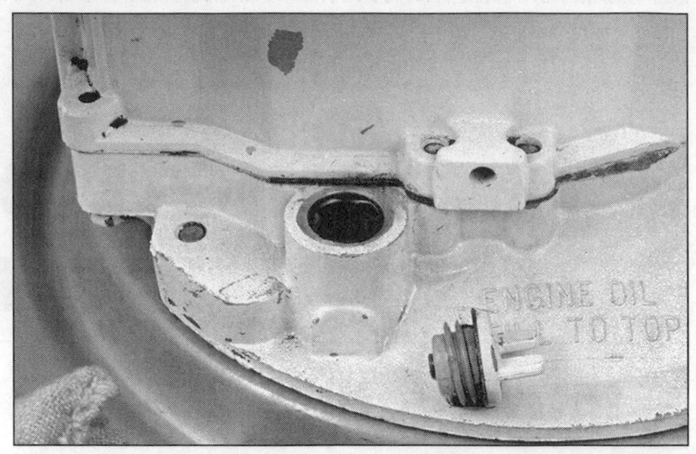


**4.2 The oil level on engines that don't have a dipstick should be even with the top of the check/fill plug opening as shown here**

5 If the cap or plug has a dipstick, wipe the oil off, then reinsert it into the engine and pull it out again. Follow the instructions on the dipstick - sometimes the plug must be threaded back in to get an accurate reading.

6 Note the oil level on the dipstick. It should be between the marks on the dipstick (usually ADD and FULL), not above the upper mark or below the lower mark **(see illustration 4.3)**.

7 Add oil to bring it up to the correct level. If it's time to change the oil, don't add any now - change the oil instead (see Section 4).

## 2 Inspect the grass screen

1 The grass screen over the flywheel is an important part of the cooling system on many air-cooled engines, since it is the entry point for air being drawn in by the fins on the flywheel and circulated through the shrouds. Some screens are metal-mesh type held in place by a large circular clip, while other grass screens are stamped metal or molded plastic, usually either snapped in place with tabs or retained by screws **(see illustration 4.4)**.

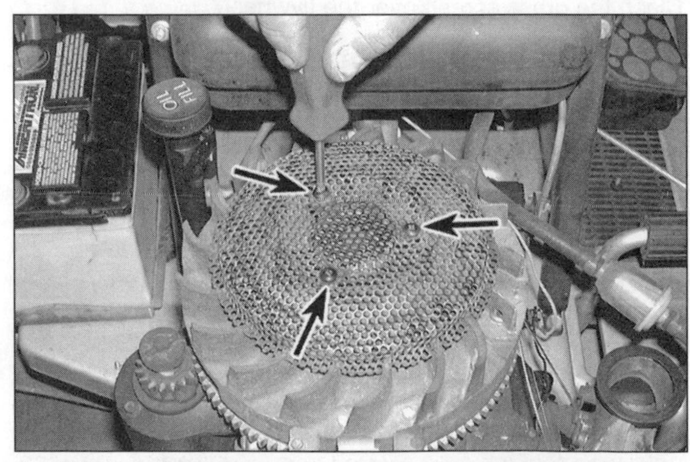

**4.4 Remove the grass screen to clean it, and clean behind it - screens are usually retained by clips or screws (arrows)**

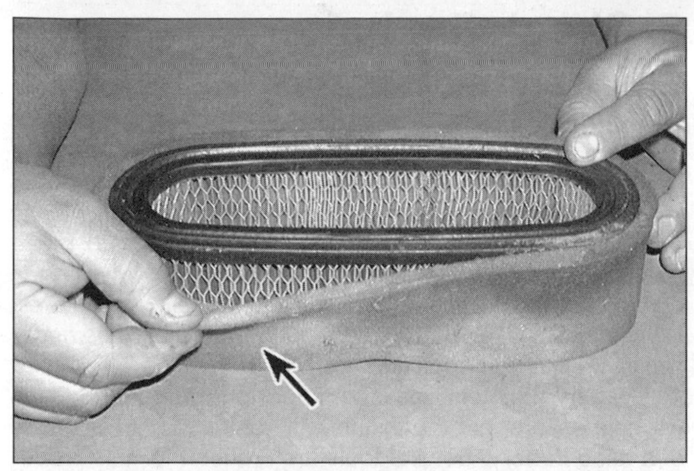

4.5 Most foam pre-filters (arrow) can be washed in soapy water and reused, although some of them are disposable and should be replaced with a new one

2 Remove the screen and clean the flywheel side with a brush, or soapy water if necessary. Do not clean the screen in place with compressed air, or dirt could be blown into the cooling fins of the engine.

## 3 Service the air cleaner pre-filter

Most small engine air cleaners are either foam or pleated-paper types that should be checked/cleaned frequently to ensure proper engine operation. Some engines have a pleated-paper filter that's covered with a foam type pre-filter. If the filter isn't serviced regularly, dirt will get into the engine or it'll build up on the filter and cause an excessively rich fuel mixture - either condition will shorten the engine's life.

1 The foam pre-filter is usually wrapped around the outside of the main paper filter **(see illustration 4.5)**.

2 Remove the air cleaner cover (if necessary) and pull off the foam element.

3 Wash the foam with warm, soapy water, rinse, then allow it to dry. **Note:** *Do not wring out the element, the foam could start to deteriorate.* Re-oil it with clean engine or light oil, blot out the excess oil, and reinstall the foam over the air cleaner.

## 4 Change the engine oil

Oil is the lifeblood of an engine; check and change it often to ensure maximum performance and the longest engine life possible. If the equipment is operated in dusty conditions, change the oil more frequently than you normally would. **Note:** *Most small engine manufacturers recommend 30-weight oil - check your owner's manual for exact recommendations. It is not recommended to use multi-viscosity oil in temperatures above 40-degrees F unless it's the only type available, as increased oil consumption may result.*

1 Start the engine and allow it to warm up (warm oil will

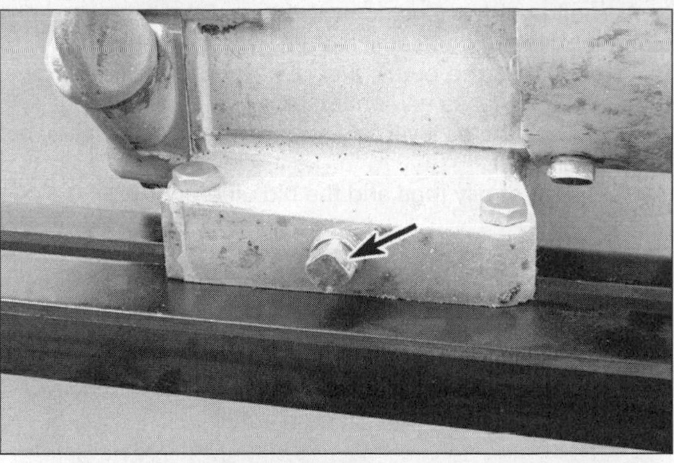

4.6 On some engines, the oil drain plug(s) is on the side of the crankcase . . .

4.7 . . . while on others it's at the bottom - don't work underneath a mower deck unless the spark plug wire is disconnected!

drain easier and more contaminants will be removed with it).

2 Stop the engine - never attempt to drain the oil with the engine running!

3 Disconnect the spark plug wire from the spark plug and position it out of the way.

4 Locate the drain plug. Some are located on the outside edge of the bottom of the engine **(see illustration 4.6)**, while others (particularly on engines used on rotary mowers) are on the bottom of the engine **(see illustration 4.7)**. **Note:** *Some engines don't have a drain plug - the oil is drained out through the filler hole by tilting the engine.*

5 Clean the plug and the area around it, then remove it from the engine and allow the oil to drain into a container. Don't rush this part of the procedure - let the oil drain until the engine is completely empty. Tip the engine so oil runs toward the opening if necessary.

6 Remove the oil check/fill plug also.

7 Clean the drain plug and reinstall it in the engine. If a gasket is used, be sure it's in place and undamaged. Tighten the plug securely.

8 Refill the crankcase with new, clean oil. Use a funnel to avoid spills, but be sure to wipe it out before pouring oil into

it. Add oil until the level is at the top of the opening, then clean the plug and reinstall it. Wipe up any spilled oil.
9 Reconnect the spark plug wire and start the engine, then check for leaks and shut it off.
10 Recheck the oil level and add more oil if necessary, but don't over- fill it.
11 Dispose of oily rags and the old oil properly.

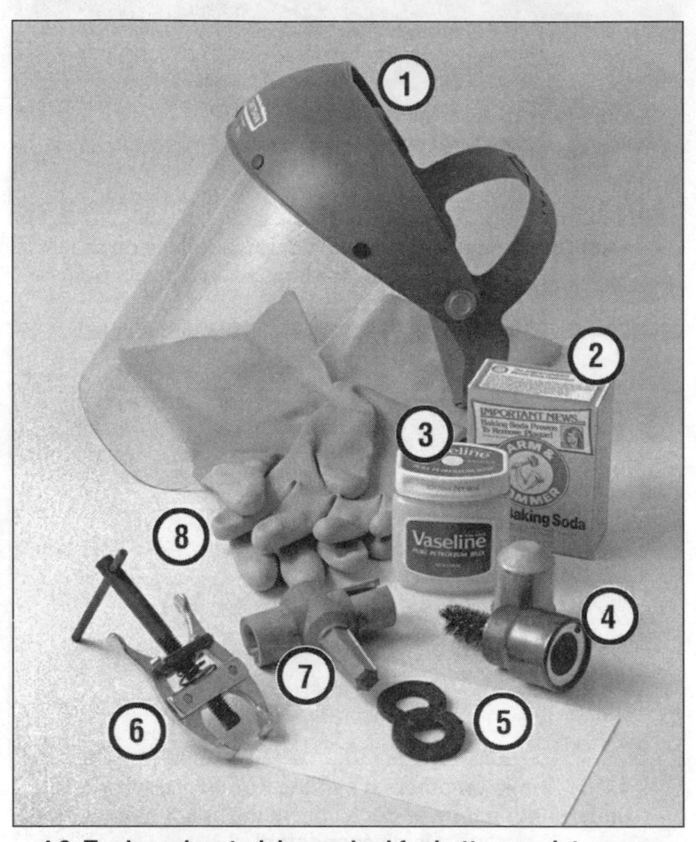


**4.8 Tools and materials required for battery maintenance**

1 ***Face shield/safety goggles*** - *When removing corrosion with a brush, the acidic particles can easily fly up into your eyes*
2 ***Baking soda*** - *A solution of baking soda and water can be used to neutralize corrosion*
3 ***Petroleum jelly*** - *A layer of this on the battery posts will help prevent corrosion*
4 ***Battery post/cable cleaner*** - *This wire brush cleaning tool will remove all traces of corrosion from the battery posts and cable clamps*
5 ***Treated felt washers*** - *Placing one of these on each post, directly under the cable clamps, will help prevent corrosion*
6 ***Puller*** - *Sometimes the cable clamps are very difficult to pull off the posts, even after the nut/bolt has been completely loosened. This tool pulls the clamp straight up and off the post without damage.*
7 ***Battery post/cable cleaner*** - *Here is another cleaning tool which is a slightly different version of number 4 above, but it does the same thing*
8 ***Rubber gloves*** - *Another safety item to consider when servicing the battery; remember that's acid inside the battery!*

# 5 Check the battery

## Check and maintenance

**Warning:** *Certain precautions must be followed when checking and servicing the battery. Hydrogen gas, which is highly flammable, is always present in the battery cells, so keep lighted tobacco and all other flames and sparks away from it. The electrolyte inside the battery is actually dilute sulfuric acid, which will cause injury if splashed on your skin or in your eyes. It will also ruin clothes and painted surfaces. When removing the battery cables, always detach the negative cable first and hook it up last!*

1 Battery maintenance is an important procedure which will help ensure that the engine will not fail to start because of a dead battery. Several tools are required for this procedure **(see illustration 4.8)**.

2 Before servicing the battery, always turn the engine and all lights off and disconnect the cable from the negative terminal of the battery.

3 Some equipment is equipped with a sealed (sometimes called maintenance free) battery. The cell caps cannot be removed, no electrolyte checks are required and water cannot be added to the cells. However, any battery that has removable caps is a type that requires regular maintenance, and the following procedures can be used.

4 Check the electrolyte level in each of the battery cells **(see illustration 4.9)**. It must be above the plates. There's usually a split-ring indicator in each cell to indicate the correct level. If the level is low, add distilled water only, then install the cell caps. **Caution:** *Overfilling the cells may cause electrolyte to spill over during periods of heavy charging, causing corrosion and damage to nearby components.*

5 If the positive terminal and cable clamp on your vehicle's battery is equipped with a rubber protector, make sure that it's not torn or damaged. It should completely cover the terminal.

6 The external condition of the battery should be

**4.9 On batteries that are not sealed, remove the cell caps to check the water level in the battery - if the level is low, add distilled water only**

**4.10 Regardless of the type of tool used on the battery posts, a clean, shiny surface should be the result**

**4.11 When cleaning the cable clamps, all corrosion must be removed (the inside of the clamp is tapered to match the taper on the post, so don't remove too much material)**

checked periodically. Look for damage such as a cracked case.

7 Check the tightness of the battery cable clamps to ensure good electrical connections and inspect the entire length of each cable, looking for cracked or abraded insulation and frayed conductors.

8 If corrosion (visible as white, fluffy deposits) is evident, remove the cables from the terminals, clean them with a battery brush or emery paper and reinstall them **(see illustrations 4.10 and 4.11)**. Corrosion can be kept to a minimum by installing specially treated washers available at auto parts stores or by applying a layer of petroleum jelly or grease to the terminals and cable clamps after they are assembled.

9 Make sure that the battery carrier is in good condition and that the hold-down clamp bolt is tight. If the battery is removed, make sure that no parts remain in the bottom of the carrier when it's reinstalled. When reinstalling the hold-down clamp, don't overtighten the bolt.

10 Corrosion on the carrier, battery case and surrounding areas can be removed with a solution of water and baking soda. Apply the mixture with a small brush, let it work, then rinse it off with plenty of clean water.

11 Any metal parts damaged by corrosion should be coated with a zinc-based primer, then painted.

## Charging

12 Remove all of the cell caps (if equipped) and cover the holes with a clean cloth to prevent spattering electrolyte. Disconnect the negative battery cable and hook the battery charger leads to the battery posts (positive to positive, negative to negative), then plug in the charger. Make sure it is set at 12-volts if it has a selector switch.

13 If you're using a charger with a rate higher than two amps, check the battery regularly during charging to make sure it doesn't overheat. If you're using a trickle charger, you can safely let the battery charge overnight after you've

**4.12 The condition of the battery's cells can be checked with a hydrometer (arrow) - sample the electrolyte in each cell of a charged battery - the scale on the hydrometer will tell you the condition of the cells**

checked it regularly for the first couple of hours.

14 If the battery has removable cell caps, measure the specific gravity with a hydrometer every hour during the last few hours of the charging cycle. Hydrometers are available inexpensively from auto parts stores - follow the instructions that come with the hydrometer **(see illustration 4.12)**. Consider the battery charged when there's no change in the specific gravity reading for two hours and the electrolyte in the cells is gassing (bubbling) freely. The specific gravity reading from each cell should be very close to the others. If not, the battery probably has a bad cell(s).

15 Some batteries with sealed tops have built-in hydrometers on the top that indicate the state of charge by the color displayed in the hydrometer window. Normally, a bright-colored hydrometer indicates a full charge and a dark hydrometer indicates the battery still needs charging. Check the battery manufacturer's instructions to be sure

4.13 Use a stiff paintbrush or compressed air (if available) to remove grass and dirt from the engine cooling fins

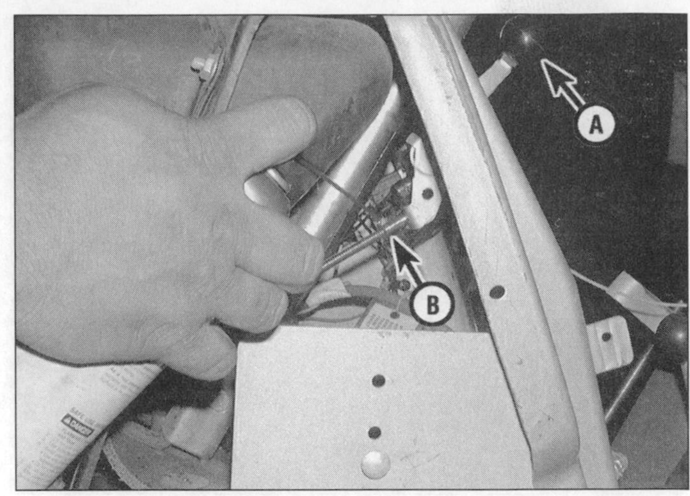

4.14 Check the operation of the control lever (A) and lubricate the cable (B) near the control unit

4.15 The cable must be securely attached at the engine or the controls won't work properly

4.16 A typical pleated-paper air cleaner - this one is held in place with plastic clips (arrows)

you know what the colors mean.

16 If the battery has a sealed top and no built-in hydrometer, you can hook up a voltmeter across the battery terminals to check the charge. A fully charged battery should read 12.6-volts or higher.

# 6 Clean the cooling fins and shroud

1 The shroud, air intake and the engine cooling fins must be clean so air can circulate properly to prevent overheating and prolong the engine's life.

1 Refer to Chapter 5 and remove the shroud from the engine.

2 Use a brush or compressed air to clean the grass screen **(see illustration 4.4)**.

3 Do the same for the fins on the cylinder and head **(see illustration 4.13)**.

# 7 Check/adjust the controls

1 The engine controls normally consist of a single lever that operates a cable connected to the governor linkage and/or choke valve on the carburetor. Some types of power equipment also have safety-related controls that shut down the engine if the operator releases his grip on the equipment. There are so many different control configurations in use on small engines that it would be impossible to cover the correct hook-up and adjustment of all of them, so the following information is general in nature.

2 Check the lever to make sure it operates smoothly and moves the cable. Lubricate the lever pivot and cable if necessary **(see illustration 4.14)**.

3 The cable must be clamped in a stationary position at the engine. Tighten the clamp if necessary **(see illustration 4.15)**.

4 When the lever is moved to the STOP position, the switch on the carburetor must operate and short out (ground) the ignition system.

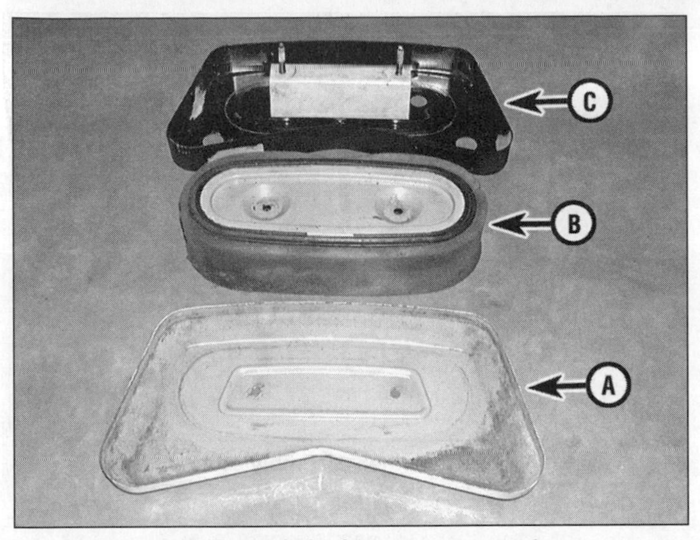


4.17 **Typical air cleaner components**

*A Cover*
*B Paper element with foam pre-filter*
*C Base assembly*

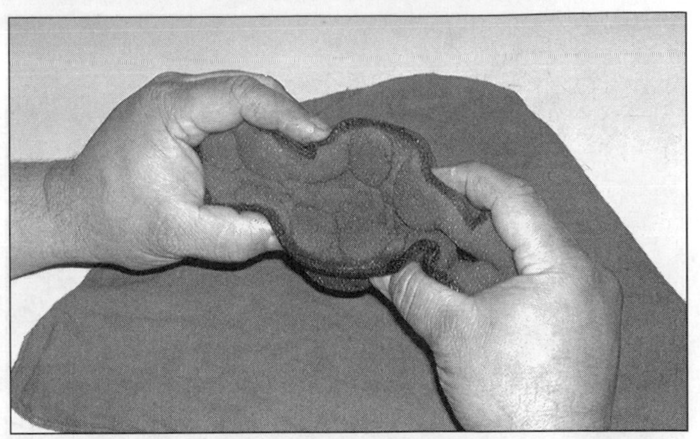

**4.18 After adding oil to the filter element, squeeze it several times to distribute the oil and drain the excess**

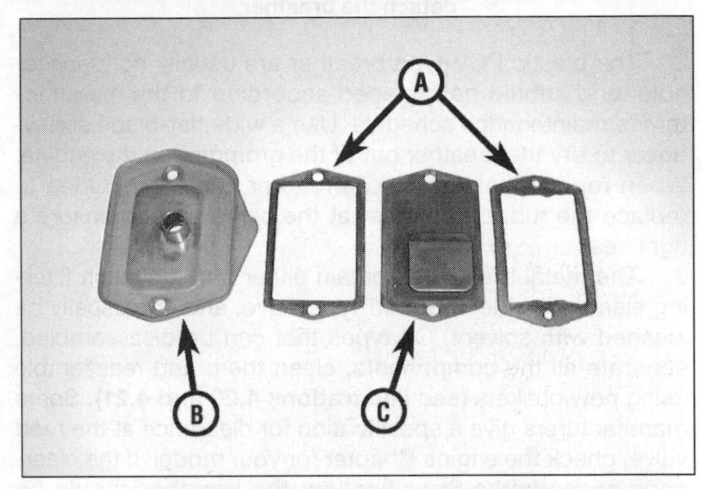


**4.19 Crankcase breather components**

*A Gasket* *B Cover* *C Filter*

# 8 Clean/replace air filter

1 Remove the filter from the engine. Some filters simply snap into place, while others are under a cover attached with screw(s) or wing nut(s) **(see illustrations 4.16 and 4.17)**.
2 If the filter is made of pleated-paper, tap it on a workbench to dislodge the dirt or blow it out from the inside with LOW PRESSURE compressed air. **Warning:** *Always wear eye protection when using compressed air!* If it's torn, bent, crushed, wet or damaged in any other way, install a new one. DO NOT wash a pleated-paper filter to clean it!
3 If the filter is foam, wash it in hot soapy water and wring it out, then let it dry thoroughly. Add about two teaspoons of engine oil to the filter and squeeze it several times to distribute the oil evenly **(see illustration 4.18)**. This is very important - the oil is what catches the dirt in the filter. If the filter is torn or falling apart, install a new one.
4 Remove any dirt from the air cleaner housing and check the gasket between it and the carburetor. If the gasket is deteriorated or missing, dirt will get past the filter into the engine.
5 Reinstall the filter.

# 9 Clean the battery terminals

Refer to Section 5 for battery testing and maintenance procedures.

# 10 Check the flywheel brake

1 Some kinds of equipment are equipped with a flywheel brake as a safety device that can stop the engine from turning within three seconds. Such a brake is usually connected to a device on the operator controls that shuts off the ignition and simultaneously initiates the flywheel brake when the operator's hand leaves the controls. The intention is to prevent injuries caused by "runaway" equipment. Just shutting off the ignition may leave the crankshaft (and the driven unit or blade) turning and still capable of causing injury.
2 Test the system periodically by letting go of the controls while the engine is operating. The engine should shut off immediately and the engine brake should apply to stop the flywheel.
3 Depending on the model of equipment, the condition of the brake can be checked by removing the fan/flywheel cover and inspecting the brake pad. If it is worn past its built-in wear indicator, replace the brake pad.

# 11 Inspect/clean the crankcase breather

1 The crankcase breather must be inspected or cleaned periodically. Each engine manufacturer has its own style breather, but there are three basic types: replaceable plastic type similar to an automotive PCV valve, a metal type that is unitized, and a metal type that can be disassembled for cleaning **(see illustration 4.19)**.

4.20 **Most breather assemblies also function as the cover for the valve springs and tappets - remove the bolts (arrows) and detach the breather**

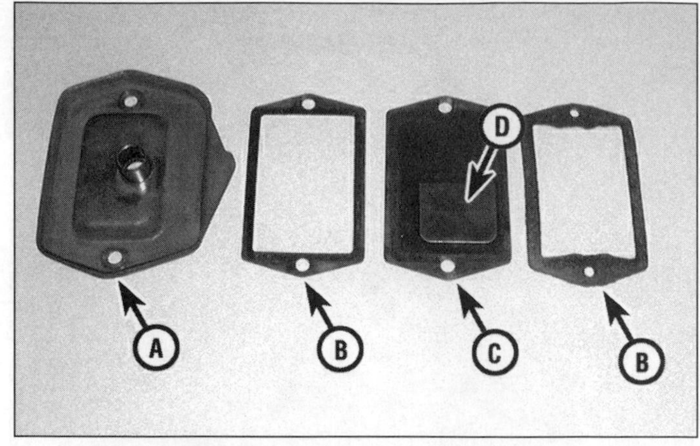


**4.21 Typical breather assembly components:**

*A Cover B Gasket C Plate D Reed*

2 The plastic PCV-type breather are usually not serviceable, and should be replaced according to the manufacturer's maintenance schedule. Use a wide flat-blade screwdriver to pry the breather out of the grommet on the engine. When replacing this type of breather, it's a good idea to replace the rubber grommet at the same time to ensure a tight seal.

3 The metal breathers contain either a metal mesh filtering element inside or a reed-type valve, and can usually be cleaned with solvent. On types that can be disassembled, separate all the components, clean them and reassemble using new gaskets **(see illustrations 4.20 and 4.21)**. Some manufacturers give a specification for clearance at the reed valve, check the engine Chapter for your model. If the clearance exceeds the Specification, the breather should be replaced.

# 12 Install a new spark plug

1 A defective or worn spark plug will increase fuel consumption, lead to formation of deposits in the cylinder head, cause hard starting, contribute to engine oil dilution (from contamination with raw gas) and cause the engine to misfire.

2 Detach the wire from the spark plug.

3 Remove the spark plug from the engine **(see illustration 4.22)**. Compare the spark plug to those shown in the spark plug condition chart on the inside of the back cover to get an indication of the general running condition of the engine.

4 If the plug is coated with deposits, it can be cleaned with a wire brush **(see illustration 4.23)** and sprayed with plug/point cleaner (available in aerosol cans).

5 If the deposits are thick or hard, use a knife blade or file to remove them, then resort to the wire brush and aerosol cleaner.

6 If the electrodes are slightly rounded off, use a small file to square them up **(see illustration 4.24)**. The sharp edges will make it easier for the spark to occur.

7 If the electrodes are worn smooth or the porcelain insulator is cracked, install a new spark plug - the cost is minimal. Make sure the new plug has the same length threads

**4.22 Use a spark plug socket to remove and install the plug - on some engines, a flex-socket is needed**

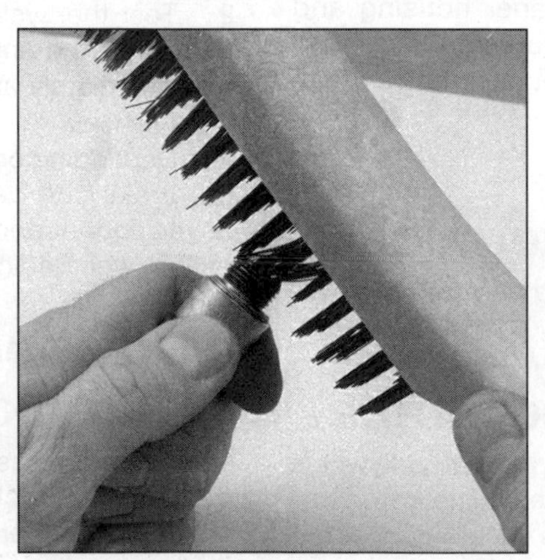

**4.23 Use a wire brush and aerosol cleaner to remove deposits from the plug tip**

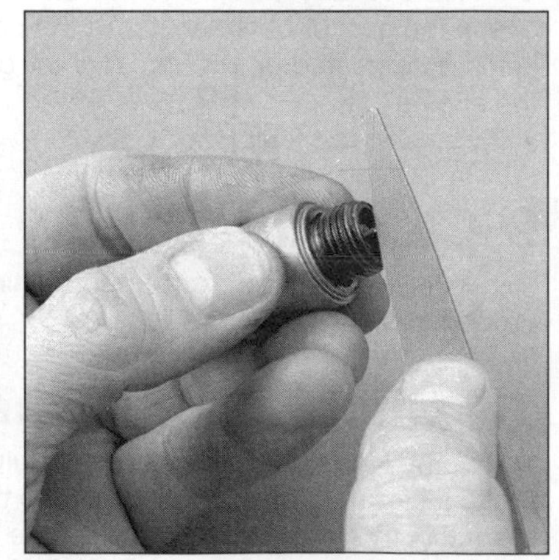

**4.24 The spark plug electrodes should be square and sharp - use a fine file to dress them**

**4.25 Spark plug manufacturers recommend using a wire-type gauge when checking the gap - if the wire doesn't slide between the electrodes with a slight drag, adjustment is required**

**4.26 To change the gap, bend the side electrode only, as indicated by the arrows, and be very careful not to crack or chip the porcelain insulator surrounding the center electrode**

and tip as the original.

8 Check the gap with a wire-type gauge **(see illustration 4.25)**. The correct gap is listed in the Specifications section of each individual engine Chapter.

9 If adjustment is required, bend the side electrode only with the special notched adjuster on the gap gauge **(see illustration 4.26)**.

10 Check the threaded hole in the cylinder head. If the threads are damaged or stripped out, a special insert can be installed to salvage the head (see Chapter 2).

11 Install the plug in the engine and tighten it finger-tight. A torque wrench should be used for final tightening of the spark plug to a specified torque value, but the torque figure isn't always readily available (it'll vary depending on the size of the plug, the type of seat and the material the head is made of). As a general rule, the plug should be tightened 1/2-to-3/4 turn after the gasket contacts the cylinder head. **Note:** *On engines with an aluminum cylinder head, apply a thin coating of anti-seize compound to the spark plug threads before installing the new plug. This will prevent the plug becoming stuck in the head due to the dissimilar metals of the plug and head.*

12 Reconnect the spark plug wire. If it's loose on the plug, crimp the wire terminal loop with a pair of pliers **(see illustration 4.27)**.

## 13 Check the coil and ignition wires

1 The ignition coil is usually mounted next to the flywheel, so the shroud will have to be removed to check the wires.

2 Check the spark plug wire for cracked and melted insulation and make sure it's securely attached to the ignition coil or module **(see illustration 4.28)**.

3 Make sure the terminal fits snugly on the spark plug end. Crimp it with a pair of pliers if necessary.

4 Check the primary (small) wires as well. Look for loose

**4.27 Use a pair of pliers to crimp the plug wire terminal so it fits snugly on the plug**

**4.28 Check the spark plug wire for cracked and melted insulation and make sure it's securely attached at the coil or module**

**4.29 Many engines have an easy-to-replace inline fuel filter (arrow)**

and corroded connections and abraded or melted insulation. Now is also a good time to check the engine stop switch. Make sure the switch is actuated when the control lever is moved to STOP. If it isn't, adjust the cable.

## 14 Clean the gas tank and fuel line

**Warning:** *Gasoline is extremely flammable and highly explosive under certain conditions - safety precautions must be followed when working on any part of the fuel system! Don't smoke or allow open flames or unshielded light bulbs in or near the work area. Don't do this procedure in a garage with a natural gas appliance (such as a water heater or clothes dryer).*
**Note:** *The gas tank collects dust, grass clippings, dirt, water and other debris under normal circumstances and must be cleaned so the contaminants don't find their way into the carburetor. The tank may be mounted separately or attached directly to the carburetor.*

1 Remove any covers or shrouds mounted over the gas tank, then remove the tank mounting screws (if used). **Note:** *If the tank is attached to the carburetor, removing the screws may free it, but there's usually not enough room to maneuver it out of position unless the carburetor is removed first.*

2 If a shut-off valve is installed, turn it off.

3 Detach the fuel line from the tank and plug the fitting with your finger so gas doesn't run out (this isn't necessary if a shut-off valve is used).

4 Lift the tank off the engine.

5 Drain the fuel out of the tank, through a strainer and into a gas can, then rinse the tank with solvent. If the tank has a strainer at the outlet fitting, make sure it's clean.

6 Loosen the hose clamps, if used, and detach the fuel line from the carburetor fitting.

7 Make sure the line is clean and unobstructed. If it's hardened, cracked or otherwise deteriorated, install a new line and new clamps. **Note:** *If a filter is installed in the line or the tank outlet fitting, clean it or install a new one* **(see illustration 4.29)**.

8 Proceed to section 4 and clean the float bowl (if equipped), then reinstall the tank.

**4.30 Some carburetors have a spring-loaded drain valve on the bottom of the float bowl (arrow) to get rid of sediment and water in the carburetor**

## 15 Clean the carburetor float bowl

1 Some carburetors have a float bowl (a reservoir for gasoline) that collects sediment and water which will clog the jets and cause the engine to run poorly or not at all. The float bowl should be drained/cleaned frequently.

2 Some carburetors have a drain plug or small spring-loaded valve on the bottom of the float bowl that's used to drain out sediment and water **(see illustration 4.30)**. Lay a rag under the carburetor, then push up on the valve with a small screwdriver or remove the plug and let gas run out until it looks clean (you'll see little droplets if there's water in the fuel.

3 On engines that don't have a drain valve, DO NOT remove the float bowl until the gas tank is drained, the fuel line is pinched off or the tank is removed, otherwise gas will run out when the float bowl is detached.

4 You'll have to remove a bolt or fitting to detach the float bowl **(see illustration 4.31)**. Use a flare-nut wrench, if you

**4.31 The float bowl is usually attached to the carburetor with a bolt or other fitting at the bottom (and it's usually hard to get at)**

4.32 Wipe out the float bowl and check the gasket before installing it

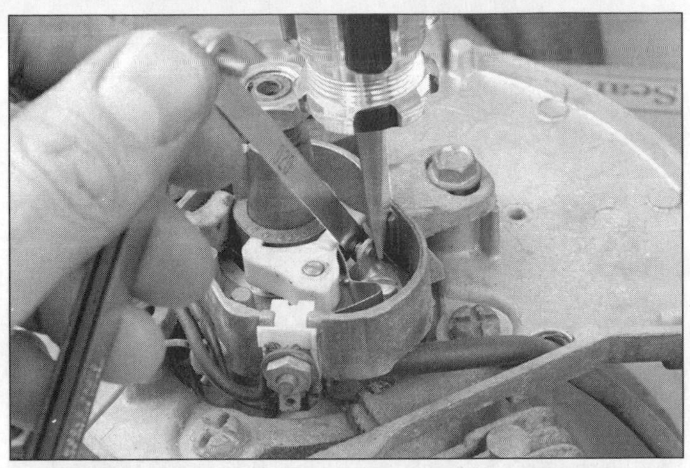

4.33 Insert a clean feeler gauge between the contact points, and move the fixed point with a screwdriver until the gap is the same as the gauge

have one, to avoid rounding off the bolt. On some engines you may have to remove the carburetor to get the bolt out so the float bowl will come off.

5 Dump the gas out of the float bowl and clean it with a rag **(see illustration 4.32)**.

6 Check the condition of the gasket - if it's deteriorated or deformed, install a new one.

7 Reinstall the float bowl and tighten the bolt securely. Make sure the fiber washer is in place on the bolt (if used).

8 Reinstall the tank, remove the fuel line clamp or add gas to the tank, then start the engine and make sure it runs okay.

## 16 Replace or clean/adjust the ignition points

1 The points, on engines so equipped, should be checked and replaced or cleaned and adjusted at least once a year. On most engines covered in this manual, the ignition points are mounted under the flywheel, so it must come off first (see appropriate engine chapter). **Warning:** *Be sure to remove the spark plug from the engine before working on the ignition system!*

2 Since the flywheel is off to get at the points, be sure to check for oil leakage past the crankshaft seal under the flywheel. If the seal is leaking, oil more than likely will eventually foul the points and you'll have to remove the flywheel for additional repairs. Seal replacement is covered in Chapter 5.

3 Because the flywheel must be removed to access most point sets, points/condenser sets are usually replaced, rather than cleaned. The point/condenser sets are inexpensive, and can be obtained at home improvement stores and small engine shops.

4 To check the point adjustment, remove the cover (if equipped) over the points. Rotate the flywheel until the maximum opening clearance is observed between the stationary and moving point. Measure the gap at this point with a clean feeler gauge and compare the results to the Specifications in your model engine Chapter **(see illustration 4.33)**. Replacement usually consists of removing the old points and condenser, installing the new components, then adjusting the point gap to the specified opening point (see the Chapter covering your engine).

## 17 Check the recoil starter (if equipped)

1 This is a simple check that can be done without removing anything from the engine. **Note:** *Disconnect the wire from the spark plug to prevent the engine from starting.*

2 Pull the starter rope out slowly.

3 If the starter is noisy, binding or rough, the return spring, pulley or rope may be jammed.

4 If the crankshaft doesn't turn as the rope is pulled out, the ratcheting drive mechanism isn't engaging.

5 After the rope is all the way out, check it for wear along its entire length.

6 Allow the rope to rewind, but don't release the handle so the rope flies back.

7 If the rope won't rewind, the pulley may be binding, the return spring may be broken, disengaged or insufficiently tensioned or the starter may be assembled incorrectly. See Chapter 5 for recoil starter removal, and the specific engine Chapter for your model engine for recoil starter servicing.

## 18 Check the governor and linkage

1 Two types of governors are in common use on air-cooled engines: the air-vane type and the mechanical (centrifugal) type. Routine checks of an air-vane governor require removal of the shroud (see Chapter 5). The mechanical governor is usually mounted inside the engine, but the linkage connected to the carburetor is visible on the outside of the engine. **Note:** *If the governor isn't hooked up or seems to be malfunctioning, it should be repaired and the*

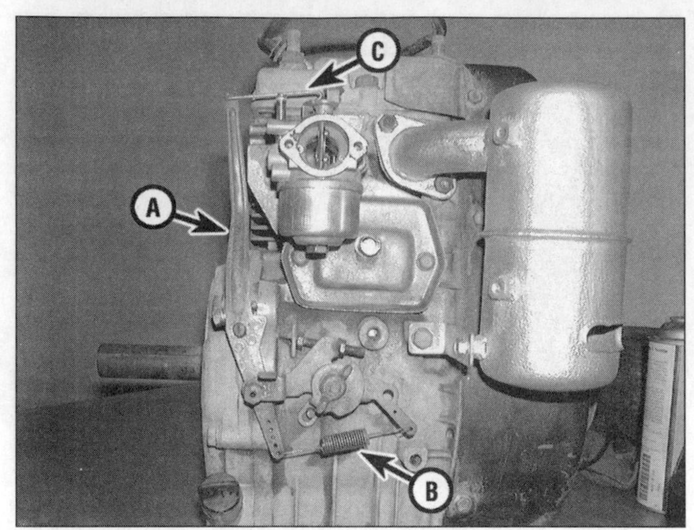

**4.34 Clean debris from the governor linkage and check for free operation of: (A), governor arm; (B), governor spring; and (C), governor linkage rods**

*engine operating speed adjusted by a dealer or repair shop with the necessary special tools.*

2 Clean grass clippings and other debris out of the governor linkage **(see illustration 4.34)**.

3 See if the linkage moves freely.

4 The throttle on the carburetor should be wide-open with the engine stopped. If it isn't, the linkage may be binding or hooked up incorrectly.

5 Look for worn links and holes and disconnected springs.

6 If the engine has an air-vane governor, the vane should move freely and operate the linkage. If the vane is bent or distorted, the governor may not operate correctly.

7 If the engine has a mechanical governor, make sure the lever is securely attached to the shaft where it exits the crankcase. **Warning:** *On most air-cooled engines, maximum speed is usually around 3500 rpm. If the governor is not working properly, engine overspeed could damage the engine.* For more specifics on governor adjustment, refer to the Chapter for your engine model.

# 19 Check the compression

1 Among other things, poor engine performance may be caused by leaking valves, incorrect valve tappet clearances, a leaking head gasket or worn piston(s), rings and/or cylinder(s). A compression check will help pinpoint these conditions.

2 The compression should be checked anytime the engine is hard to start or power loss is evident. If the engine is run with low compression, fuel and oil consumption will increase and engine wear will be accelerated.

## Simple check

3 Remove the spark plug and ground the plug wire on the engine, then seal off the plug hole with your thumb **(see illustration 3.3** in Chapter 3).

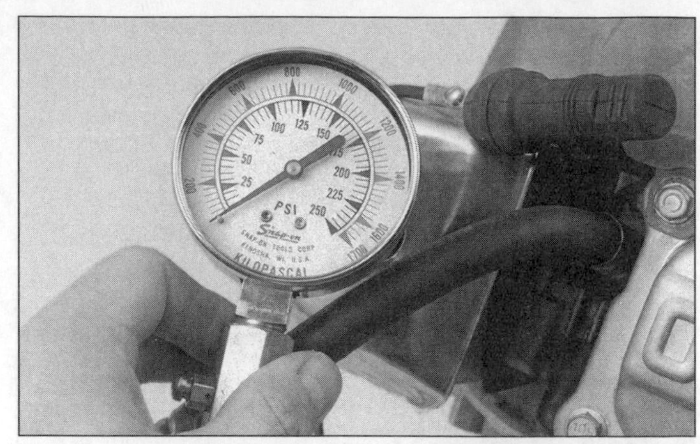
**4.35 A compression gauge with a threaded fitting is preferred over the type that requires hand pressure to maintain the seal**

4 Operate the starter - if the compression pressure blows your thumb off the hole, the compression is adequate for the engine to run. **Warning:** *Be careful not to touch the plug wire as this is done - you'll get quite a jolt if you do!*

5 Another way to check the compression with the spark plug in place is to remove the cooling shroud/recoil starter mechanism and spin the flywheel in reverse (counterclockwise) **(see illustration 3.4** in Chapter 3). It should return sharply; if it does, the compression is adequate for the engine to run.

## Using a gauge

6 The only tools required are a compression gauge and a spark plug wrench. Depending on the results of the initial test, a squirt-type oil can may also be needed. A compression gauge that screws into the spark plug hole is preferred over the type that requires hand pressure to maintain the seal at the plug hole.

7 Warm up the engine to normal operating temperature and remove any dirt around the spark plug with compressed air or a small brush, then remove the plug. Work carefully, don't strip the spark plug hole threads and don't burn your hands.

8 Ground the spark plug wire on the engine.

9 Install the compression gauge in the spark plug hole **(see illustration 4.35)**. Make sure the choke is open and hold or block the throttle wide open.

10 Crank the engine over a minimum of five to seven revolutions and note the initial movement of the compression gauge needle as well as the final gauge reading.

11 If the compression built up quickly and evenly, you can assume the engine upper end is in reasonably good mechanical condition. Worn or sticking piston rings and a worn cylinder will produce very little initial movement of the gauge needle, but compression will tend to build up gradually as the engine spins over. Valve and valve seat leakage, or head gasket leakage, is indicated by low initial compression which doesn't tend to build up.

12 To further confirm your findings, add about 1/2-ounce of engine oil to the cylinder by inserting the nozzle of a squirt-type oil can through the spark plug hole. The oil will

tend to seal the piston rings if they're leaking. Repeat the compression test.

13 If the compression increases significantly after the addition of oil, the piston rings and/or cylinder are definitely worn. If the compression doesn't increase, the pressure is leaking past the valves or the head gasket. Leakage past the valves may be caused by burned or cracked valve seats or faces, warped or bent valves or insufficient valve tappet clearance.

14 To summarize, if the compression is low, it may be due to:

*Loose spark plug*
*Loose cylinder head bolts*
*Blown head gasket*
*Damaged valves/valve seats*
*Insufficient valve tappet clearance*
*Warped cylinder head*
*Bent valve stem(s)*
*Worn cylinder bore and/or piston rings*
*Broken connecting rod or piston*

**Note:** *There are some Kohler, Honda and other engines that have a feature called Automatic Compression Release (ACR), which uses a mechanism to hold the exhaust valve open more during starting (to reduce compression at that time) and makes the engine easier to start. Once the engine is running, normal compression is restored. On these engines it may not be possible to take a standard compression test using a gauge, since the engine must be at 600 rpm or more to achieve normal compression.*

# 20 Check the muffler

1 Make sure the muffler isn't restricted (if it's bent, dented, rusted or falling apart, install a new one).

2 Check the mounting bolts to make sure they're tight. A loose muffler can damage the engine.

3 If the muffler threads directly into the engine, make sure it's tight.

# 21 Check the engine mount bolts/nuts

1 If the engine mounting bolts/nuts are loose, the engine will vibrate excessively and damage the equipment it's mounted on.

2 Disconnect the spark plug wire from the plug and ground it on the engine.

3 Tighten the mounting fasteners securely.

4 If the nuts/bolts are stripped, install new ones.

5 Reconnect the spark plug wire.

# 22 Check the valve tappet clearance

1 Correct valve tappet clearance is essential for efficient fuel use, easy starting, maximum power output, prevention of overheating and smooth engine operation. It also ensures the valves will last as long as possible.

2 When the valve is closed, clearance should exist between the end of the stem and the tappet. The clearance is very small - measured in thousandths of an inch - but it's very important. The recommended clearance is listed in the specifications in each engine Chapter. Note that intake and exhaust valves often require a different clearance specification. **Note:** *The engine must be cold when the clearance is checked.* A feeler gauge with a blade thickness equal to the valve clearance will be needed for this procedure.

3 On some engines, if the clearance is too small, the valves will have to be removed and the stem ends ground down carefully and lapped to provide more clearance (this is a major job, covered in Chapter 5). If the clearance is too great, new valves will have to be installed (again, a major repair procedure). On many other engines, the tappets are adjustable, and with two wrenches, the clearance can be set to Specifications. **Note:** *Some OHV engines have adjustable rocker arms for adjusting the valve clearance, while other OHV engines (Kohler for one) feature hydraulic lifters, so valve adjustment is not necessary*. Refer to the engine Chapter for your model for more specific valve adjustment procedure.

4 Disconnect the wire from the spark plug and ground it on the engine.

5 Remove the bolts and detach the tappet cover plate or the crankcase breather assembly **(see illustration 4.20)**. **Note:** *On some engines the crankcase breather is behind the carburetor, so the carburetor will have to be removed first.*

6 Turn the crankshaft by hand and watch the valves to see if they stick in the guide(s).

7 Turn the crankshaft until the intake valve is wide open, then turn it an additional 360-degrees (one complete turn). This will ensure the valves are completely closed for the clearance check **(see illustration 4.36)**.

8 Select a feeler gauge thickness equal to the specified valve clearance and slip it between the valve stem end and

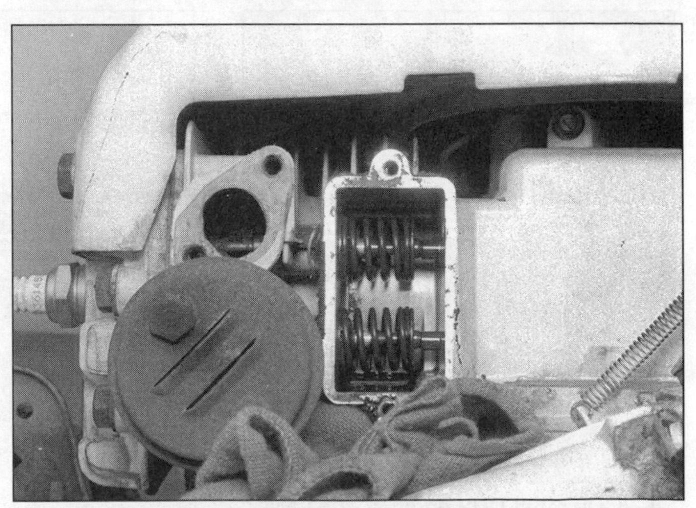

**4.36 Make sure the valves are completely closed when checking the clearance**

the tappet **(see illustration 4.37)**.

9 If the feeler gauge can be moved back-and-forth with a slight drag, the clearance is correct. If it's loose, the clearance is excessive; if it's tight (watch the valve to see if it's forced open slightly), the clearance is inadequate.

10 If the clearance is incorrect, refer to the appropriate chapter for valve service procedures.

11 Reinstall the crankcase breather or tappet cover plate.

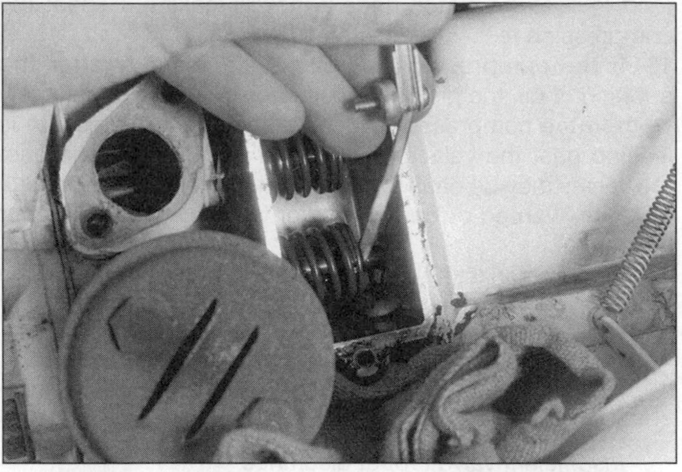

**4.37 If the clearance is correct, the feeler gauge will fit between the valve stem and tappet with a slight drag**

## OHV engines

12 Remove the bolts and detach the valve cover from the engine.

13 Remove the spark plug and place your thumb over the plug hole, then slowly turn the crankshaft with the starter or blade until you feel pressure building up in the cylinder. Use a flashlight to look into the spark plug hole and see if the piston is at the top of it's stroke. Continue to turn the crankshaft until it is. **Note:** *Some OHV engines have external timing marks to indicate TDC (Top Dead Center) for the piston on the compression stroke (see the engine Chapter for your model).*

14 Select a feeler gauge thickness equal to the specified valve clearance and slip it between the valve stem end and the rocker arm **(see illustration 4.38)**.

15 If the feeler gauge can be moved back-and-forth with a slight drag, the clearance is correct. If it's loose, the clearance is excessive; if it's tight (watch the valve to see if it's forced open slightly), the clearance is inadequate.

16 To adjust the clearance, loosen the rocker arm locknut and turn the pivot in or out as required (turn it out to increase the clearance; turn it in to decrease the clearance).

17 Hold the pivot with a wrench and tighten the locknut securely, then recheck the clearance.

# 23 Adjust the carburetor

1 Carburetor adjustments are done by turning the high and/or low speed mixture screws. Some carburetors don't have any mixture screws, while others have one for either high-speed adjustments or low speed adjustments, but not both. Still others have one screw to adjust the fuel/air mixture at high speeds and another screw that controls the mixture at low speeds - the low speed screw is usually the one closest to the engine end of the carburetor. If two screws are used, they must be adjusted separately.

2 The mixture screws control the flow of fuel through the carburetor circuit(s) **(see illustration 4.39)**. If the tip is damaged or the screw is incorrectly adjusted, loss of power and erratic engine operation will result.

3 Remove the mixture screw and check the tip - if it looks bent or a groove has been worn in the tapered portion, install a new one **(see illustration 4.40)**. Do not attempt to straighten it. If the O-ring on the screw is damaged or deteriorated, replace it before attempting to adjust the mixture.

4 If the screw isn't bent or worn, reinstall it and turn it in until it stops - tighten it with your fingers only, don't force it.

5 Back it out about 1-1/4 turns (counterclockwise). **Note:**

**4.38 On most OHV engines, the valve clearance is checked between the valve stem and rocker arm**

**4.39 Typical mixture adjusting screws**

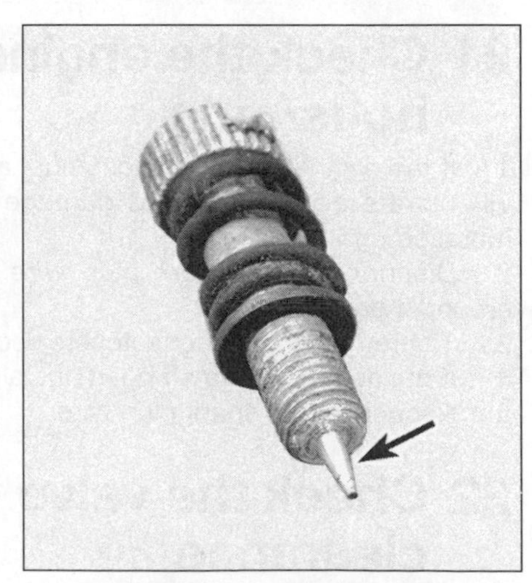

**4.40 Check the mixture screw tip to make sure it isn't damaged**

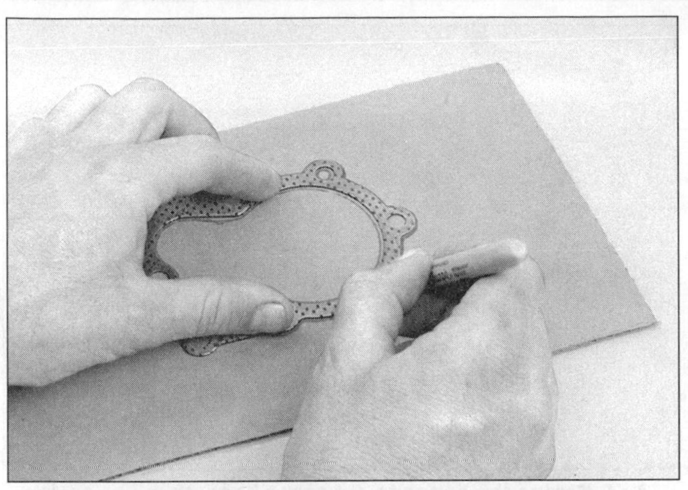
4.41 To avoid mixing up the head bolts (just in case they're different lengths), use the new gasket to transfer the hole pattern to a piece of cardboard, punch holes to accept the bolts and push each bolt through the matching hole in the cardboard as it's removed

4.42 Use a putty knife to remove the deposits from the piston and valves - don't nick or gouge the block or piston (if deposits are oily, the rings may be bad)

*The actual recommended number of turns out is different for each carburetor type (see the engine Chapter for your model), but the figure given here is in the ballpark for most engines.*

6 Start the engine and turn the screw clockwise until the engine starts to slow down. This means the fuel mixture is too lean (not enough gas).

7 Slowly turn the screw out (counterclockwise) until the engine begins to run smoothly. Keep going very slowly until the engine just begins to run rough again (rich condition). Also watch for black smoke from the exhaust.

8 Finally, turn the screw in again (clockwise) to a point about half-way between lean and rich operation. This is the perfect setting.

9 If an idle (low-speed) mixture screw is used, adjust it in the same manner with the engine idling. **Note:** *Honda engines are equipped with a pilot air screw, rather than a low-speed mixture screw. Turning the pilot screw has the same effect (it changes the fuel/air mixture), but it's reversed. When the pilot screw is turned in, it causes a richer mixture; conversely, when it's backed out, the mixture becomes leaner.* After the low-speed mixture has been set, recheck the high-speed adjustment - it may be affected by the idle adjustment.

10 Some carburetors also have an idle speed adjusting screw that's used to open or close the throttle valve slightly to change the idle speed only, not the fuel/air mixture. Turning it will cause the engine to speed up or slow down. You can tell the idle speed screw from the mixture screw(s) because it acts on the throttle linkage and doesn't screw into the carburetor body.

## 24 Decarbonize the cylinder head and piston

1 Now that leaded gasoline has been out of use for some years, carbon build-up in the cylinder head is not the problem it used to be. However, it's still a good idea to remove the head during a Spring tune-up to scrape out the carbon and other deposits. Before beginning this procedure, buy a new head gasket for your engine. **Note:** *This procedure does not apply to OHV engines.*

2 Begin by disconnecting the wire from the spark plug.

3 Next, remove the shroud and any covers that prevent direct access to the cylinder head and bolts. **Note:** *On many engines, some of the head bolts also are used to attach the shroud or carburetor mounting bracket to the engine. If you're working on one, loosen all of the head bolts in 1/4-turn increments, following a criss-cross pattern, until the shroud mounting bolts can be removed by hand.*

4 Using the new head gasket, outline the head bolt pattern on a piece of cardboard **(see illustration 4.41)**. Punch holes at the bolt locations.

5 If not already done, loosen the cylinder head bolts in 1/4-turn increments until they can be removed by hand. Follow a criss-cross pattern to avoid warping the head.

6 Store the bolts in the cardboard holder as they're removed - this will guarantee that they're reinstalled in their original locations, which is essential (different length bolts are used on some engines).

7 Detach the head from the engine. If it's stuck, tap it with a soft-face hammer to break the gasket seal - DO NOT pry it off with a screwdriver!

8 Remove and discard the gasket - use the new one when the cylinder head is reinstalled.

9 Turn the crankshaft until the piston is at the top of the cylinder, then use a scraper or putty knife and wire brush to remove all deposits from the top of the piston and the area around the valves **(see illustration 4.42)**. Be careful not to nick the gasket mating surface.

10 Turn the crankshaft to open each valve and check them for burned and cracked faces and seats **(see illustration**

4.43 Turn the crankshaft to open each valve and check the seats and faces (arrows) for cracks and other damage

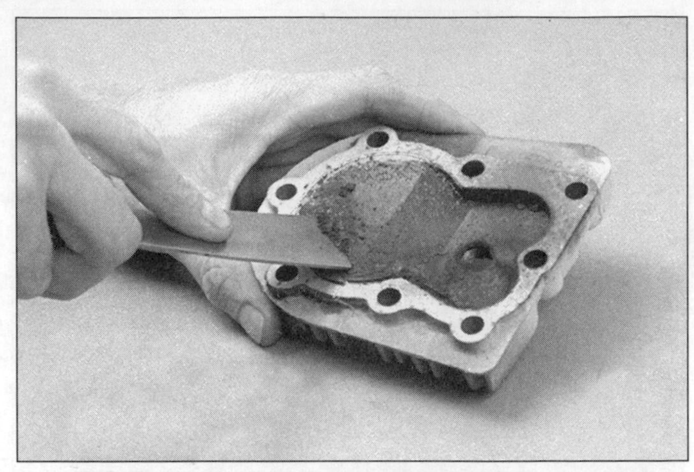
4.44 Scrape the deposits out of the head, then use a wire brush and solvent to finish the job

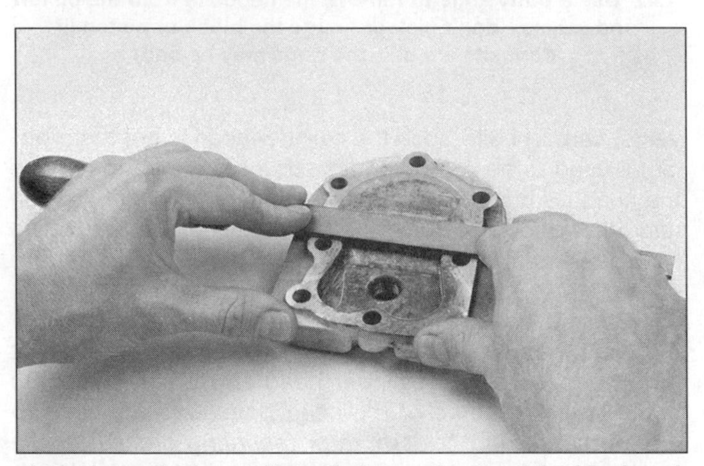
4.45 Use a single-cut file to flatten and restore the block and head mating surface - move the file sideways (arrow) and don't apply excessive pressure

4.46 Use a tap to clean and restore the head bolt holes in the block

**4.43).** If the valves are cracked, pitted or bent and the seats are in bad shape, major engine repairs are required.

11 Remove the deposits from the combustion chamber in the head **(see illustration 4.44)**.

12 The mating surfaces of the head and block must be perfectly clean when the head is reinstalled.

13 Use a gasket scraper or putty knife to remove all traces of carbon and old gasket material, then clean the mating surfaces with lacquer thinner or acetone. If there's oil on the mating surfaces when the head is installed, the gasket may not seal correctly and leaks could develop.

14 Check the block and head mating surfaces for nicks, deep scratches and other damage. If damage is slight, it can be removed with a file **(see illustration 4.45)**.

15 Use a tap of the correct size - if you have one - to chase the threads in the head bolt holes **(see illustration 4.46)**, then clean the holes with compressed air (if available) - make sure that nothing remains in the holes. **Warning:** *Wear eye protection when using compressed air!*

16 If you have the correct size die, mount each bolt in a vise and run the die down the threads to remove corrosion and restore the threads **(see illustration 4.47)**. Dirt, corrosion, sealant and damaged threads will affect torque readings.

4.47 Mount each head bolt in a vise and restore the threads with a die of the correct size

17 Reinstall the head using the new gasket. Do not use sealant on the gasket.

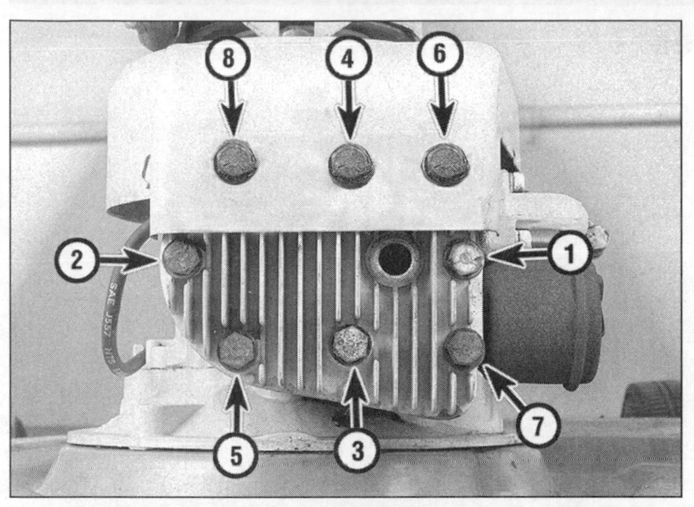


**4.48 When tightening the head bolts, follow a criss-cross pattern - never tighten them in order around the edge of the head**

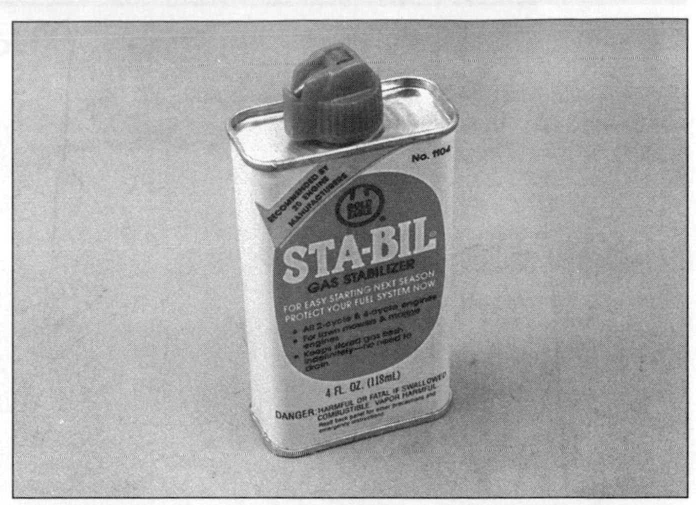


**4.49 The most important thing when preparing an engine for storage is to drain the fuel system or add a gas stabilizing chemical to the tank**

18 Once the head bolts are finger-tight, tighten them in 1/4-turn increments to the torque listed in the Specifications Section in the engine Chapter for your model. The engine Chapter will also show specific head-bolt tightening sequences for your model. The principle behind tightening sequences is to evenly clamp the head by tightening bolts in a criss-cross pattern **(see illustration 4.48)**. **Note:** *Don't forget to install the shroud first if some of the bolts are used to hold it in place!*

19 If you don't have a torque wrench, tighten the bolts evenly and securely with a socket and ratchet or breaker bar.

# 4 Preparing an engine for storage

Since power equipment is often designed for use only during certain times of the year (like lawn mowers and snow blowers, for example), small engines usually end up being stored for months at a time. As a result, needed repairs are neglected, corrosion takes place, gas left in the tank and carburetor gums up, moisture collects in ignition and fuel system components and the equipment is subjected to physical damage as it's moved to get at something stored behind it.

After a long dormant period, the equipment is hauled out, gas is added to the tank, the oil is checked (not always!) and the engine is fired up - but it won't start or it won't run very well. To avoid problems caused by seasonal storage, run down the following checklist of things to do and make sure the engine is properly prepared to survive a long period of non-use so it'll start and run well when you need it.

1 Operate the engine until it runs out of gas, then drain the float bowl (if equipped)

2 An alternative to running the engine out of gas is to add a gasoline stabilizer (available at auto parts stores and power equipment dealers) to the tank - follow the instructions on the container **(see illustration 4.49)**.

3 Wipe off all dust and remove debris from engine parts.

4 Service the air cleaner (see above in this chapter)

5 Remove the spark plug and squirt some clean engine oil into the spark plug hole, then operate the starter to distribute the oil in the cylinder.

6 Clean and regap the spark plug, then reinstall it (see Section 15 in this Chapter).

7 Store the equipment in a dry place and cover the engine with plastic - don't seal the plastic around the base of the engine or condensation may occur.

## Notes

# 5 Repair procedures common to all engines

## Contents



## 1 Engine removal and installation

**Note:** *Engine removal is usually done only if major repairs or an overhaul are required (or, obviously, if a new engine is being installed on the equipment). In most cases, minor repairs can be accomplished without removing the engine.*

1 Detach the spark plug wire and ground it on the engine block.

2 Disconnect the control cables from the engine. **Note:** *On many newer pieces of equipment, a flywheel brake cable and wire harness may have to be detached as well as the throttle cable.*

3 Drain and/or remove the fuel tank so gas doesn't run out if the equipment must be tipped to get at the blade or engine mount bolts.

4 Remove the blade and hub or drivebelt(s)/chain from the power take-off end of the crankshaft.

5 The bolt(s) holding a blade, pulley or sprocket or in

**1.5a Some pulleys are retained by Allen set screws, soak them overnight with penetrating oil before using an Allen wrench to remove them**

1.5b When a blade, pulley or sprocket is held by a bolt, then use a breaker bar and six-point socket to remove it - here the equipment is raised by a floor jack to access the pulley

1.6 Remove the bolts (arrows) holding the engine to the chassis - generator application shown)

1.7 Before beginning any major engine work, soak the engine with a degreasing spray cleaner, let it soak, then hose off with water

1.8 A large block of wood drilled with a crankshaft-sized hole makes a simple bench support for vertical crankshaft engines

place are usually very tight and often corroded. Apply penetrating oil and let it soak in for several minutes, then use a six-point socket and breaker bar for added leverage **(see illustrations 1.5a and 1.5b)**. If you're working on a mower, wear a leather glove so the blade doesn't cut your hand or wedge a block of wood between the mower deck and blade so it doesn't turn.

6 Remove the mounting nuts/bolts and detach the engine from the equipment **(see illustration 1.6)**.

7 If major repairs are planned, use a degreaser to clean the engine before disassembling it **(see illustration 1.7)**. Engine degreasing spray and brush-on cleaners are available at any auto parts store. After letting the degreaser soak, rinse thoroughly with a hose, especially of your engine has aluminum parts. Some chemical cleaners can harm aluminum if left to sit too long.

8 On vertical crankshaft engines, obtain a large block of wood an drill a hole through it large enough for the PTO end of the crankshaft to fit through **(see illustration 1.8)**. This will make a simple stand, so the engine sits upright on your bench during overhaul, without the shaft being damaged.

9 Installation is the reverse of removal.

# 2 Muffler removal and installation

**Note:** *Some mufflers screw into the engine, while others are attached with bolts. If the muffler is in good condition, it can be cleaned by tapping it with a soft-face hammer and dumping out the carbon that's dislodged.*

## Screw-in mufflers

1 To remove a screw-in type, first apply penetrating oil to the threads and let it sit for several minutes. You may have

2.2 Screw-in mufflers can be removed/installed with a pipe wrench - if the muffler has a built-in hex for an open-end wrench, be sure to take advantage of it (don't use a pipe wrench on the muffler body, only the pipe screwed into the engine)

to tip the equipment to do this.

2 Try to remove the muffler with a pipe wrench by turning it counterclockwise **(see illustration 2.2)**. Some mufflers have a lock ring that must be loosened with a punch and hammer before turning the pipe. Others have a built-in hex to accept an open-end wrench (usually there's not enough room to use a pipe wrench with this type).

3 If it breaks off, try to remove the part left in the engine (if there's enough left to grasp with the wrench). If it won't come out, use a large screw extractor or cold chisel to remove it. The muffler material is fairly soft, so don't damage the threads in the engine.

4 Screw in the new muffler, but don't overtighten it. If a lock ring is used, tighten it with a hammer and punch. **Note:** *In order to prevent problems removing the muffler in the future, apply a thin coat of anti-seize compound to the threads of the exhaust pipe where it enters the engine.*

## Bolt-on mufflers

5 Apply penetrating oil to the bolt(s) and let it soak in for several minutes.

6 Remove the bolt(s) and detach the muffler **(see illustration 2.6)**. Some mufflers also have a gasket.

7 If a bolt breaks off in the engine block (which they often do), you may be able to remove it with a screw extractor (read Chapter 2 before deciding to tackle this job). If it protrudes far enough, you may be able to grip it with locking pliers and unscrew it. Apply more penetrating oil before attempting this.

8 Install the new muffler (with a new gasket, if used) and tighten the bolt(s) securely. **Note:** *In order to prevent problems removing the muffler in the future, apply a thin coat of anti-seize compound to the threads of the exhaust bolts before installation. Use new bolts whenever possible.*

2.6 Larger engines generally have mufflers held on with bolted flanges (arrows indicate bolts)

# 3 Shroud/recoil starter removal and installation

1 Many larger air-cooled engines have optional onboard battery electrical systems with electric starters, but recoil starters are standard equipment on a number of basic single-cylinder engines. The recoil starter on most engines is an integral part of the shroud that's used to direct the cooling air around the cylinder and head. On some engines, the starter is attached to the shroud or engine with nuts or bolts and can be removed separately for repairs or replacement **(see illustration 3.1)**.

2 Detach any control cables/wire harnesses clamped to the shroud.

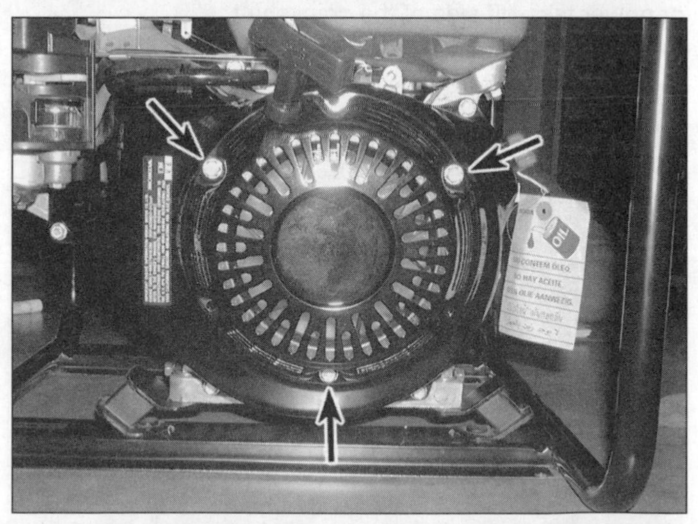

3.1 Most recoil starters are attached to the shroud with screws (arrows)

3.4 Three or four bolts or screws are usually used to hold the shroud to the engine -cylinder head bolts often are used to attach it at one end - here bolts (A) retain the blower housing and bolt (B) retains a side cover

3 If the gas tank is mounted on the shroud, remove it or detach the fuel line from the carburetor and plug it so fuel doesn't run out. **Note:** *Some engines are equipped with a shut-off valve on the tank - if your engine has one, turn it off and detach the line from the carburetor.*

4 Remove the nuts/bolts and lift the shroud off the engine **(see illustration 3.4).**

5 Before installing the shroud, clean it to remove grass clippings and other debris. Also, make sure the bolt threads in the engine are clean and in good condition.

# 4 Oil seal replacement

1 Two seals are normally used to keep oil inside the crankcase - one on the flywheel side and one on the drive or power take-off side of the crankshaft.

If an oil seal goes bad, oil will leak out of the engine and/or performance will suffer (particularly on four-stroke engines with ignition points that can get fouled by the oil).

Seals can usually be replaced without removing the crankshaft. If the seal on the power take-off end of the crankshaft is leaking, the blade or drive pulley/sprocket will have to be removed first. On some engines, the oil seal is installed in the gearcase cover, not in the block. The flywheel will have to be removed first if the seal under it is leaking (refer to the appropriate engine chapter for the flywheel removal procedure). Once the seal is exposed, proceed as follows:

2 Note how the seal is installed (what the side that faces out looks like and how far it's recessed into the bore), then remove it.

3 On most engines, the seal can be pried out with a screwdriver **(see illustrations 4.3a and 4.3b).** Be careful not to nick or otherwise damage the seal bore if this is done.

4 Clean the seal bore and the crankshaft. Remove any burrs that could damage the new seal from the crankshaft with a file or whetstone.

5 If necessary, wrap electrician's tape around the

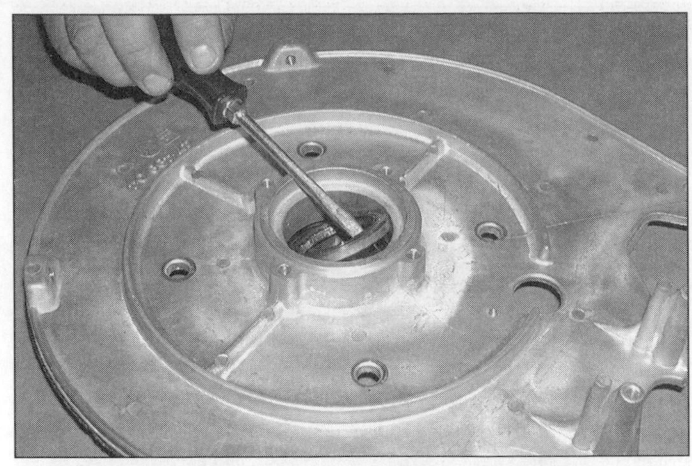

4.3a Carefully pry out the oil seal with a screwdriver or seal-removal tool - bearing plate seal shown

4.3b On the PTO end of the crankshaft, the seal can be pried out with the crankshaft in place

4.6 Apply multi-purpose grease to the outer edge and the lip(s) of the new seal before installing it

4.8 A socket or piece of pipe makes a handy seal installation tool

crankshaft to protect the new seal as it's installed. The keyway in the crankshaft is particularly apt to cut or otherwise damage the seal as it's slipped over it.

6 Apply a thin layer of multi-purpose grease to the outer edge of the new seal and lubricate the seal lip(s) with grease **(see illustration 4.6)**.

7 Place the seal squarely in the bore with the open side facing into the engine.

8 Carefully tap the seal into place with a large socket or section of pipe and a hammer until it's seated in the bore **(see illustration 4.8)**. The outer diameter of the socket or pipe should be the same size as the seal outer diameter.

9 If the seal consists of several pieces, install the retainer and lock ring and make sure the lock ring is seated in the groove.

# 5 Carburetor removal

**Warning:** *Gasoline is extremely flammable and highly explosive under certain conditions - safety precautions must be followed when working on the carburetor or gas tank! Don't smoke or allow open flames or unshielded light bulbs in or near the work area. Don't do this procedure in a garage with a natural gas appliance (such as a water heater or clothes dryer) and have a fire extinguisher handy.*

1 If equipped, turn the fuel valve off.

2 Remove the air cleaner assembly.

3 Disconnect the governor spring(s) **(see illustration 5.3)**. This is very important - most governor linkages have several holes for hooking things up and it can get very confusing. Don't rely on your memory or you may not be able to get everything hooked up correctly. Make a simple sketch to refer to later. Sometimes it's easier to disconnect the governor spring(s) after the carburetor is detached from the engine.

4 Disconnect the throttle cable and kill switch wire

5.3 Governor linkages are unique and somewhat complex, so make a sketch of how all the parts fit before disconnecting anything - (A) is the governor arm, (B) is the governor spring (typical)

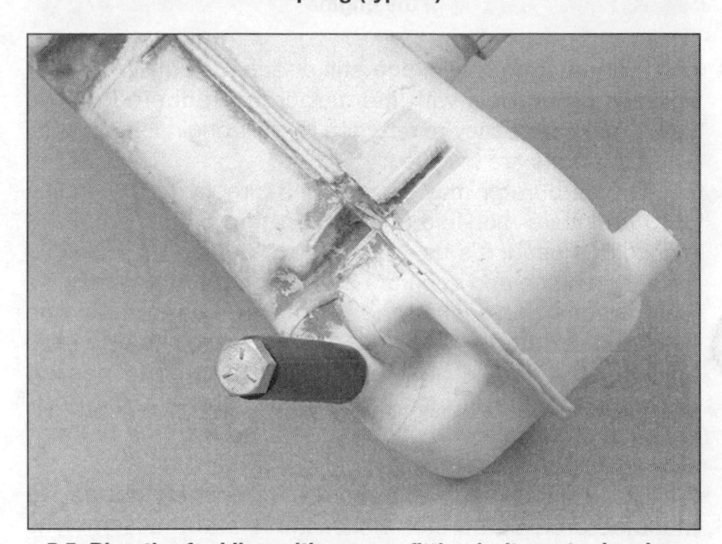

5.5 Plug the fuel line with a snug-fitting bolt or steel rod so gas doesn't run out

(and/or fuel-cut solenoid (if equipped) from the carburetor. This isn't necessary on all engines - try to determine if the cable/wire will interfere with the actual carburetor removal before disconnecting them (sometimes they're attached to the governor linkage and don't have to be removed). Note that after the mounting bolts are removed, the governor link will have to be manipulated out of the throttle lever as the carburetor is detached.

5 Detach the fuel line from the carburetor or gas tank fitting and plug it (if a shut-off valve isn't used) **(see illustration 5.5)**. Now is a good time to inspect the fuel line and install a new one if it's damaged or deteriorated. **Note:** *Some carburetors are mounted directly on the fuel tank and no fuel line is used.*

6 Remove the nuts/bolts and detach the carburetor (or fuel tank/carburetor assembly) from the engine, then dis-

5.7 Most carburetors are attached to a manifold, which is bolted to the engine - don't try to separate the carburetor from the manifold until after the assembly is detached from the engine

6.7 Soak all metal carburetor parts in a carburetor dip tank - available at most auto parts stores

connect any control linkage still attached to it. Watch for spacers on models with the carburetor mounted on the tank - make sure they're returned to their original location(s) when the bolts are installed.

7 The carburetor may be attached directly to the engine or to an intake manifold **(see illustration 5.7)**. If an intake manifold is used, it's usually easier to detach the manifold from the engine and separate the carburetor afterwards.

8 Remove the gasket and discard it - use a new one when the carburetor is reinstalled. Some engines also have an insulator and/or heat shield (and a second gasket) between the carburetor and engine. **Note:** *On Honda engines, the insulator must be reinstalled with the grooved side against the carburetor.*

9 Due to the many differences from manufacturer to manufacturer, carburetor disassembly and reassembly is covered in each engine chapter.

10 Installation is the reverse of removal.

# 6 Carburetor overhaul

**Warning:** *Gasoline is extremely flammable and highly explosive under certain conditions - safety precautions must be followed when working on the carburetor! Don't smoke or allow open flames or unshielded light bulbs in or near the work area. Don't do this procedure in a garage with a natural gas appliance (such as a water heater or clothes dryer) and have a fire extinguisher handy.*

1 Poor engine performance, hesitation, black smoke and little or no engine response to fuel/air mixture adjustments are all signs that major carburetor maintenance is required. Keep in mind that many so-called carburetor problems are really not carburetor problems at all, but mechanical problems in the engine or ignition system faults. Establish for certain the carburetor needs servicing before assuming an overhaul is necessary. For example, fuel starvation is often mistaken for a carburetor problem. Make sure the fuel filter (if used), the fuel line and the gas tank cap vent hole aren't plugged before blaming the carburetor for this relatively common malfunction.

2 Most carburetor problems are caused by dirt particles, varnish and other deposits which build up in and block the fuel and air passages. Also, in time, gaskets and O-rings shrink and cause fuel and air leaks which lead to poor performance.

3 When the carburetor is overhauled, it's generally disassembled completely - disassembly is covered in the appropriate engine chapter - and the metal components are soaked in carburetor cleaner (which dissolves gasoline deposits, varnish, dirt and sludge). The parts are then rinsed thoroughly with solvent and dried with compressed air. The fuel and air passages are also blown out with compressed air to force out any dirt that may have been loosened but not removed by the carburetor cleaner. **Warning:** *Always wear eye protection when using compressed air!* Once the cleaning process is complete, the carburetor is reassembled using new gaskets, O-rings, diaphragms and, generally, a new inlet needle and seat (not used in all carburetors).

4 Before taking the carburetor apart, make sure you have a rebuild kit (which will include all necessary gaskets and other parts), some carburetor cleaner, solvent, a supply of rags, some means of blowing out the carburetor passages and a clean place to work.

5 Some of the carburetor settings, such as the sizes of the jets and the internal passageways are predetermined by the manufacturer. Under normal circumstances, they won't have to be changed or modified and they should never be enlarged.

6 Before disassembling the carburetor, clean the outside

7.2 **Use a scraper or putty knife to remove old gaskets from the engine components - if the gasket is difficult to remove, use a gasket removal solvent**

with solvent and lay it on a clean sheet of paper or a shop towel.

7 After it's been completely disassembled, submerge the metal components in carburetor cleaner and allow them to soak for approximately 30 minutes **(see illustration 6.7)**. **Caution:** *Do not soak plastic or rubber parts in carburetor cleaner - they'll be damaged or dissolved. Also, don't allow carburetor cleaner to get on your skin and keep the carburetor cleaner away from fire or flame (open pilot light).*

8 After the carburetor has soaked long enough for the cleaner to loosen and dissolve the varnish and other deposits, rinse it thoroughly with solvent and blow it dry with compressed air. Also, blow out all the fuel and air passages in the carburetor body. Never clean the jets or passages with a piece of wire or drill bit - they could be enlarged, causing the fuel and air metering rates to be upset. **Warning:** *Always wear eye protection when using compressed air!*

9 Reassembly and carburetor adjustment is covered in the appropriate engine chapter.

# 7 Engine block cleaning

1 After the engine has been completely disassembled, clean the block as described here before conducting a thorough inspection to determine if it's reusable.

2 Using a gasket scraper, remove all traces of gasket material and old sealant from the block **(see illustration 7.2)**. Be very careful not to nick or gouge the gasket sealing surfaces.

3 Clean the block with solvent to remove dirt, sludge and oil, then dry it with compressed air (if available). Take your time and do a thorough job. **Warning:** *Always wear eye protection when using compressed air!*

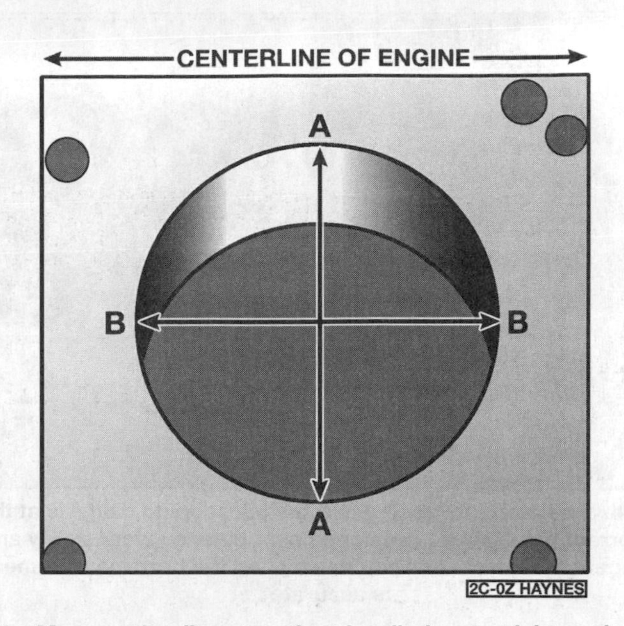


8.4a **Measure the diameter of each cylinder at a right angle to the engine centerline (A), and parallel to the engine centerline (B) - out-of-round is the difference between A and B; taper is the difference between A and B at the top of the cylinder and A and B at the bottom of the cylinder**

4 The threaded holes in the block must be clean to ensure accurate torque readings/prevent damaged threads during reassembly. Run the proper-size tap into each of the holes to remove rust, corrosion, thread cement or sludge and restore damaged threads. If possible, use compressed air to clear the holes of debris produced by this operation. Now is a good time to clean the threads on the head bolts and the connecting rod cap bolts as well. **Warning:** *Always wear eye protection when using compressed air!*

# 8 Engine block inspection

1 Before the block is inspected, it should be cleaned as described above. Double-check to make sure the carbon or wear ridge at the top of the cylinder has been completely removed. If you can still feel the ridge with your fingernail, remove the ridge with fine emery paper.

2 Visually check the block for cracks, rust and corrosion. Look for stripped threads in the threaded holes. It's also a good idea to have the block checked for hidden cracks by an automotive machine shop that has the special equipment to do this type of work. If defects are found, have the block repaired, if possible, or replaced.

3 Check the cylinder bore for scuffing and score marks.

4 Measure the diameter of the cylinder bore. This should be done at the top (just under the ridge area), center and bottom of the cylinder bore, parallel to the crankshaft **(see illustrations 8.4a, 8.4b and 8.4c)**.

8.4b The ability to "feel" when the telescoping gauge is at the correct point will be developed over time, so work slowly and repeat the check until you're satisfied the bore measurement is accurate . . .

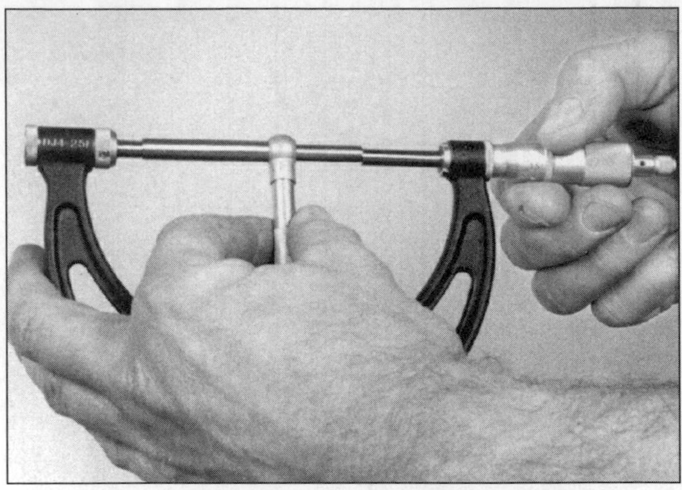

8.4c . . . then measure the telescoping gauge with a micrometer to determine the actual bore size in inches

5 Next, measure the cylinder diameter at the same three locations 90-degrees to the crankshaft. Compare the results to the Specifications in the Chapter for your model engine. If the cylinder is badly scuffed or scored, or if it's out-of-round or tapered beyond the limits given in the Specifications, have the engine block rebored and honed at a small engine dealer or an automotive machine shop. If a rebore is done, an oversize piston and rings will be required.

6 If the cylinder is in reasonably good condition and not worn to the outside of the limits, and if the piston-to-cylinder clearance can be maintained properly, then it doesn't have to be resized. Honing is all that's necessary (see the next section).

9.3 For the do-it-yourselfer, better results can usually be obtained with a "bottle brush" hone than a traditional spring-loaded hone

# 9 Cylinder honing

1 Prior to engine reassembly, the cylinder bore should be honed so the new piston rings will seat correctly and provide the best possible combustion chamber seal. **Note:** *This procedure applies to engines with an iron bore only - aluminum cylinder bores do not require honing for the rings to seat. Also, most small engine manufacturers provide chrome ring sets (for both aluminum and iron-bore engines) that don't require cylinder honing before installation. If you don't have the tools or don't want to tackle the honing operation, most automotive machine shops and small engine dealers will do it for a reasonable fee.*

2 Two types of cylinder hones are commonly available - the flex hone or "bottle brush" type and the more traditional surfacing hone with spring-loaded stones. Both will do the job, but for the less experienced mechanic the "bottle brush" hone will probably be easier to use. You'll also need plenty of light oil or honing oil, some rags and an electric drill motor. Proceed as follows:

3 Mount the hone in the drill, compress the stones and slip it into the cylinder **(see illustration 9.3)**. Be sure to wear safety goggles or a face shield!

4 Lubricate the cylinder with plenty of oil, turn on the drill and move the hone up-and-down at a pace that'll produce a fine crosshatch pattern on the cylinder walls. Ideally, the crosshatch lines should intersect at approximately a 60-degree angle **(see illustration 9.4)**. Be sure to use plenty of lubricant and don't take off any more material than absolutely necessary to produce the desired finish. **Note:** *Piston ring manufacturers may specify a smaller crosshatch angle than the traditional 60-degrees - read and follow any instructions included with the new rings.*

5 Don't withdraw the hone from the cylinder while it's running. Instead, shut off the drill and continue moving the hone up-and-down in the cylinder until it comes to a complete stop, then compress the stones and withdraw the

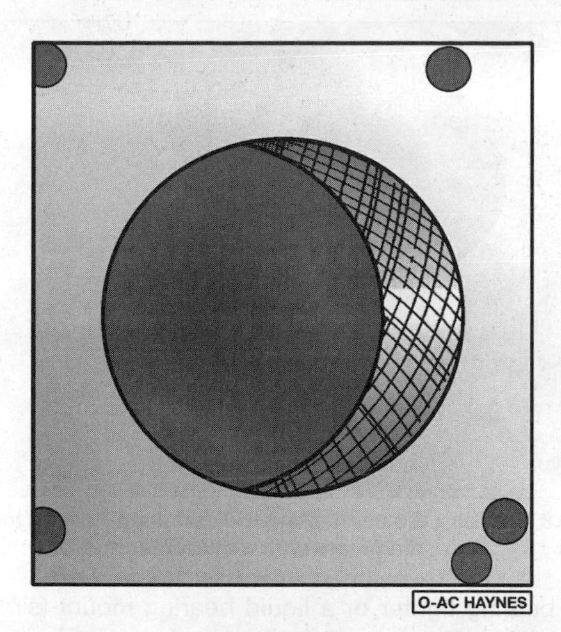


**9.4 The cylinder hone should leave a smooth, crosshatch pattern with the lines intersecting at approximately a 60-degree angle**

**9.7 After honing the cylinder, run a file around the top edge of the bore to remove the sharp edge so the rings don't catch when the piston is reinstalled - note the tape on the end of the file to prevent nicks in the cylinder wall**

hone. If you're using a "bottle brush" type hone, stop the drill, then turn the chuck in the normal direction of rotation while withdrawing the hone from the cylinder.

6 Wipe the oil out of the cylinder.

7 After honing is complete, chamfer the top edge of the cylinder bore with a small file so the rings won't catch when the piston is installed **(see illustration 9.7)**. Be very careful not to nick the cylinder wall with the end of the file!

8 The engine block must be washed again very thoroughly with warm, soapy water to remove all traces of abrasive grit produced during the honing operation. **Note:** *The bore can be considered clean when a white cloth - dampened with clean engine oil - used to wipe it down doesn't pick up any more honing residue, which will show up as gray areas on the cloth.*

9 After rinsing, dry the block and apply a coat of light oil to the cylinder to prevent the formation of rust. If the engine isn't going to be reassembled right away, store the block in a plastic trash bag to keep it clean and set it aside until reassembly.

# 10 Crankshaft and bearing inspection

## Crankshaft

1 After the crankshaft has been removed from the engine, it should be cleaned thoroughly with solvent and dried with compressed air (if available). **Warning:** *Wear eye protection when using compressed air!* If the crankshaft has oil passages drilled in it, clean them out with a wire or stiff plastic bristle brush, then flush them with solvent.

**10.3 Check the gear teeth for wear and damage . . .**

2 Check the connecting rod journal for uneven wear, score marks, pits, cracks and flat spots. If the rod journal is damaged or worn, check the connecting rod bearing surface as well. If the crankshaft rides in plain bearings, check the main bearing journals and the thrust faces in the same manner (the thrust faces contact the bearings to restrict the end play of the crankshaft).

3 Check the gear teeth for cracks, chips and excessive wear **(see illustration 10.3)**.

4 Check the threads on each end of the crankshaft - if they're worn or damaged, they may be salvageable with a die or thread file. Check the power take-off end to make sure it's not bent.

5 Check the crankshaft taper for rust and damage **(see**

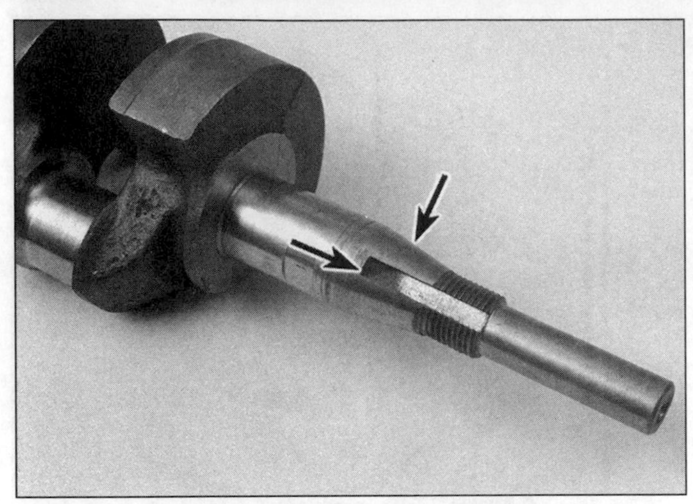

10.5 . . . and make sure the taper and keyway are in good shape

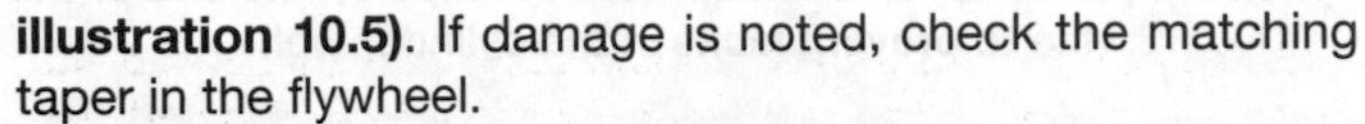

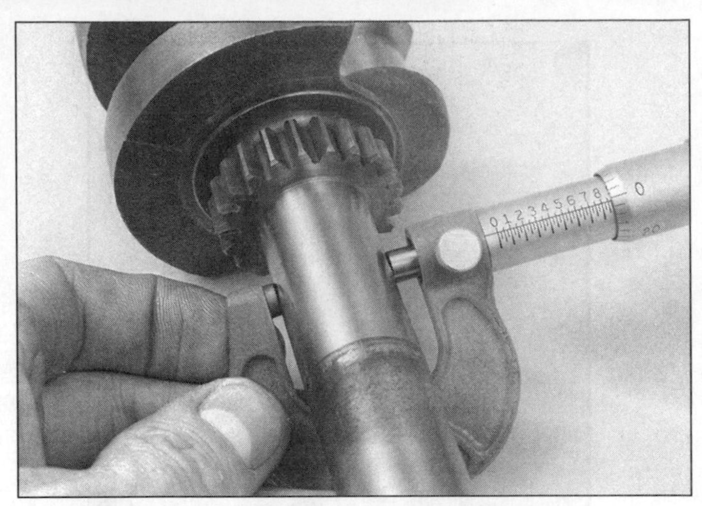

10.8 Measure the connecting rod and main bearing journal diameters with a micrometer

**illustration 10.5)**. If damage is noted, check the matching taper in the flywheel.

6 Inspect each keyway for deformation - if the one in the taper is worn or spread open, the ignition timing will be off. A new crankshaft will be needed.

7 Check the rest of the crankshaft for cracks and other damage.

8 Using a micrometer, measure the diameter of the main and connecting rod journals **(see illustration 10.8)**.

9 Compare the results to the Specifications in the engine Chapter for your model. By measuring the diameter at a number of points around each journal's circumference, you'll be able to determine whether or not the journal is out-of-round. Take the measurement at each end of the rod journal, near the crank throws, to determine if the journal is tapered. If the crankshaft journals are damaged, tapered, out-of-round or worn beyond the limits given in the Specifications, a new crankshaft will be required.

10 Check the oil seal journals at each end of the crankshaft for wear and damage. If the seal has worn a groove in the journal, or if the journal is nicked or scratched, the new seal may leak when the engine is reassembled.

## Bearings

11 The bearings shouldn't be removed from the crankcase unless they're defective or they have to come out with the crankshaft. **Note:** *On many engines, the bearings must be pressed in or out of the case at a machine shop.*

12 Clean the bearings with solvent and allow them to air dry. **Caution:** *Do not use compressed air to spin ball bearings - spinning a dry bearing will cause rapid wear and damage.*

13 Check ball or roller bearings for wear, damage and play in the bore. Rotate them by hand and feel for smooth operation with no axial or radial play. If the bearing is in the engine, make sure the outer race is securely fastened in the bore. If it's loose, the block may have to be peened to grip the bearing tighter or a liquid bearing mount (similar to thread-locking compound) may have to be used.

14 Check plain bearings for wear, score marks and grooves or deep scratches. Be sure to check the thrust faces (they keep the crankshaft from moving end-to-end too much) as well. The bearing face should be smooth with a satin finish, not brightly polished.

15 Use a telescoping gauge and a micrometer to measure the bearing inside diameter. Any bearing worn beyond the specified limits (see the engine Chapter for your model), must be replaced with a new one. If new plain bearings are needed, have them installed by a dealer so they can be reamed to size as well. **Note:** *On some engines, the crankshaft rides directly in the aluminum material used for the engine block. If the bearing surfaces are worn or damaged, a dealer or machine shop can bore out the holes and install bushings. The bushings are then precision-reamed to size.*

# 11 Camshaft and bearing inspection

1 After the camshaft has been removed from the engine, it should be cleaned thoroughly with solvent and dried with compressed air (if available). **Caution:** *Wear eye protection when using compressed air!* Visually inspect the camshaft for wear and/or damage to the gear teeth, lobe surfaces and bearing journals. If the cam lobes are worn or damaged, check the matching tappets as well.

2 Measure the camshaft lobe heights **(see illustration 11.2)** and compare the results to the Specifications in your engine Chapter. **Note:** *Many small engine manufacturers do not publish camshaft lobe height specifications.*

3 Measure the camshaft bearing journal diameters **(see illustration 11.3)**.

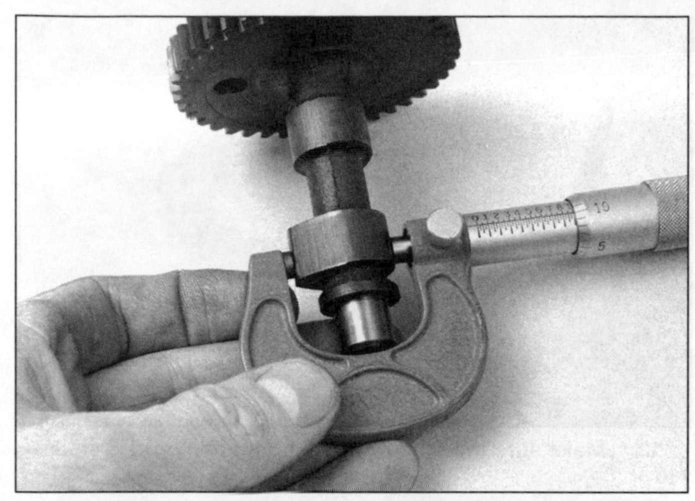

11.2 If the cam lobe height is less than specified, engine performance will suffer - install a new camshaft

11.3 Measure the camshaft bearing journal diameter with a micrometer to determine if excessive wear has occurred

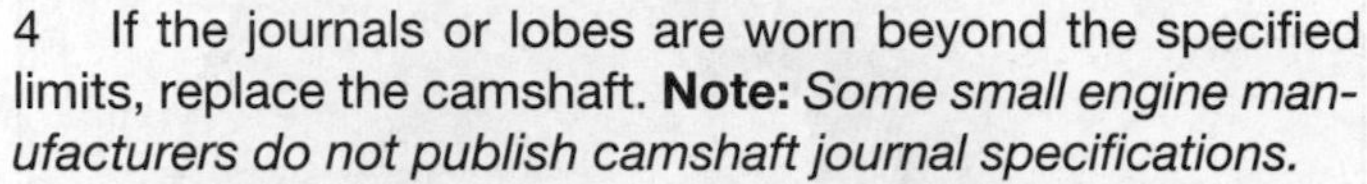

4 If the journals or lobes are worn beyond the specified limits, replace the camshaft. **Note:** *Some small engine manufacturers do not publish camshaft journal specifications.*

5 If an automatic spark advance mechanism is installed, check the weight for free movement and make sure the spring pulls it back. If it doesn't, and the weight isn't binding, install a new spring (if available separately).

6 If an automatic compression release mechanism is attached to the camshaft, check the components for binding and wear **(see illustration 11.6)**.

# 12 Piston/connecting rod inspection

1 If the cylinder must be rebored, there's no reason to check the piston, since a new (larger) one will have to be installed anyway. Before the inspection can be carried out, the piston/connecting rod assembly must be cleaned with solvent and the original piston rings removed from the piston. **Note:** *Always use new piston rings when the engine is reassembled - check with a dealer to ensure the correct ones are purchased and installed.* Using a piston ring installation tool, if available, or your fingers, remove the rings from the piston **(see illustration 12.1)**. Be careful not to nick or gouge the piston in the process.

2 Scrape all traces of carbon off the top of the piston **(see illustration 12.2)**. A hand-held brass wire brush or a piece of fine emery cloth can be used once the majority of deposits have been scraped away. Do not, under any circumstances, use a wire brush mounted in an electric drill to remove deposits from the piston - the piston material is soft and will be eroded by the wire brush.

3 Use a piece of broken piston ring to remove carbon

11.6 If equipped, inspect the compression-release mechanism on the camshaft; (A) flyweight, (B) springs - replace any worn components and install a new spring during any rebuild

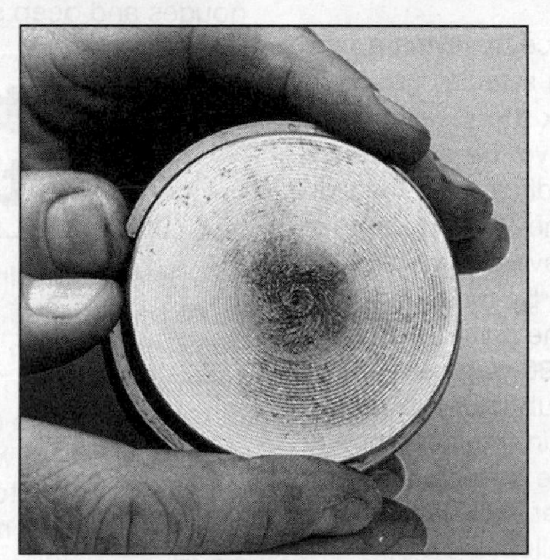

12.1 If you don't have the special tool, the rings can be removed from the piston by hand, but be careful not to break them (unless new ones are being installed)

12.2 Remove the carbon from the top of the piston with a scraper, then use fine emery cloth or steel wool and solvent to finish the job

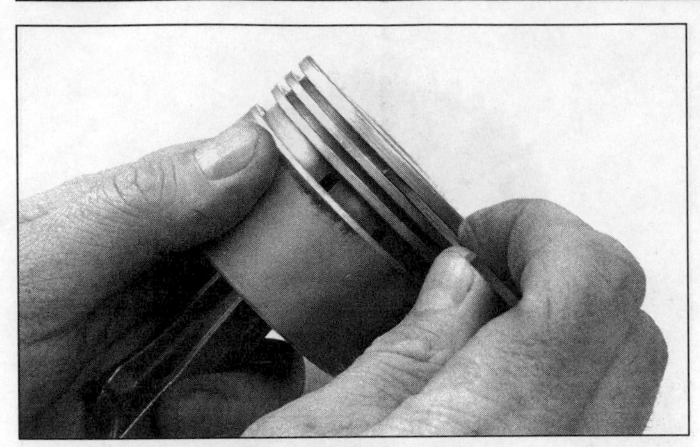

12.3 Clean the piston ring grooves with a ring groove cleaning tool or a section of broken piston ring

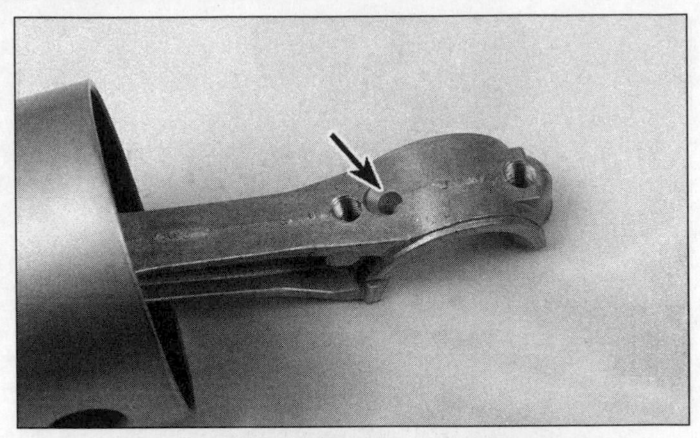

12.4 Make sure the oil hole in the connecting rod is clear

deposits from the ring grooves **(see illustration 12.3)**. Special tools are also available for this job. Be very careful to remove only the carbon deposits. Don't remove any metal and don't nick or scratch the sides of the ring grooves.

4 Once the deposits have been removed, clean the piston and connecting rod with solvent and dry them with compressed air (if available). **Warning:** *Always wear eye protection when using compressed air!* Make sure the oil return holes in the back side of the oil ring groove and the oil hole in the lower end of the rod are clear **(see illustration 12.4)**.

5 If the piston and cylinder aren't damaged or worn excessively, and if the engine block isn't rebored or replaced, a new piston won't be necessary. New piston rings, as mentioned above, should normally be installed when an engine is rebuilt.

6 Carefully inspect the piston for cracks around the skirt, at the pin bosses and at the ring lands.

7 Look for scoring and scuffing on the thrust faces of the skirt, holes in the piston crown and burned areas at the edge of the crown.

8 Measure the piston ring side clearance by laying a new piston ring in each groove and slipping a feeler gauge in beside it **(see illustration 12.8)**. Check the clearance at three or four locations around each groove. Be sure to use the correct ring for each groove; they are different. If the side clearance is greater than specified (see the Specifications in your engine Chapter), a new piston will have to be used.

9 Check the piston-to-bore clearance by measuring the bore (see Engine block inspection) and the piston diameter. Measure the piston across the skirt, at a 90-degree angle to the piston pin, near the lower edge. Subtract the piston diameter from the bore diameter to obtain the clearance (if applicable - not all manufacturers provide specifications). If it's greater than specified, the cylinder will have to be rebored and a new piston and rings installed.

10 Check the piston-to-rod clearance by twisting the piston and rod in opposite directions. Any noticeable play indicates excessive wear, which must be corrected by installing a new piston, connecting rod or piston pin (or all three.

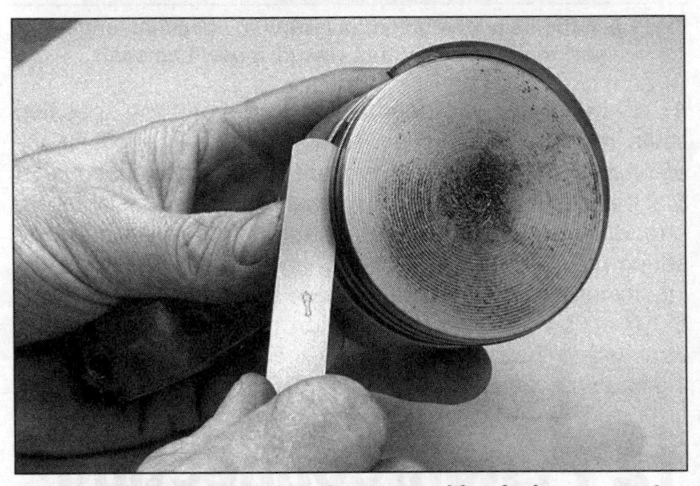

12.8 Check the ring side clearance with a feeler gauge at several points around the groove

11 Check the connecting rod for cracks and other damage. Clean and inspect the bearing surface for score marks, gouges and deep scratches.

# 13 Piston ring installation

1 Before installing the new piston rings, the ring end gaps must be checked. It's assumed the piston ring side clearance has been checked and verified correct (see Piston/connecting rod inspection above).

2 Insert the top (upper compression) ring into the cylinder and square it up with the cylinder wall by pushing it in with the top of the piston **(see illustration 13.2)**. The ring should be near the bottom of the cylinder, at the lower limit of ring travel.

3 Measure the end gap. To do this, slip feeler gauges between the ends of the ring until a gauge equal to the gap width is found **(see illustration 13.3)**. The feeler gauge should slide between the ring ends with a slight amount of

13.2 When checking piston ring and gap, the ring must be square in the cylinder bore - this is done by pushing it down with the top of a piston

13.3 Once the ring is at the lower limit of travel and square in the cylinder, measure the end gap with a feeler gauge

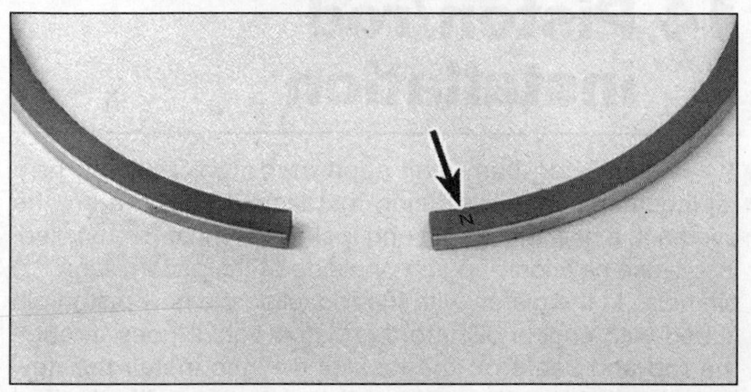

13.11 Piston rings are normally marked (arrow) to indicate the side that faces up, toward the top of the piston

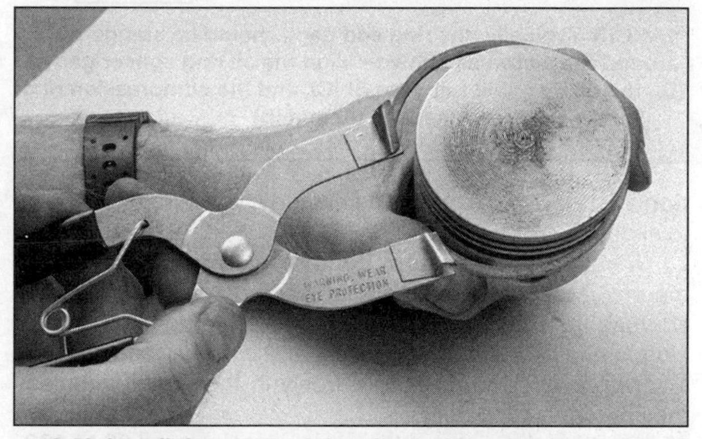


13.12 Install the compression rings with a ring expander - remember, the mark must face up!

drag. Compare the measurement to the Specifications in your engine Chapter. If the gap is larger or smaller than specified, double-check to make sure you have the correct rings before proceeding.

4 If the gap is too small, it must be enlarged or the ring ends may come in contact with each other during engine operation, which can cause serious damage. The gap can be increased by filing the ring ends very carefully with a fine file. Mount the file in a vise equipped with soft jaws, slip the ring over the file with the ends contacting the file face and slowly move the ring to remove material from the ends. When performing this operation, file only from the outside in. After filing any rings, use a fine stone or file to remove any burrs from the ends of the rings.

5 If the end gap is excessive double-check to make sure you have the correct rings for the engine.

6 Repeat the procedure for each ring.

7 Once the ring end gaps have been checked/corrected, the rings can be installed on the piston. **Note:** *Follow the instructions with the new piston rings if they differ from the information here.*

8 Install the piston rings. The oil control ring (lowest one on the piston) is installed first. On most engines it's composed of three separate components. Slip the spacer/expander into the groove first. If an anti-rotation tang is used, make sure it's inserted into the drilled hole in the ring groove.

9 Next, install the lower side rail. Don't use a piston-ring installation tool on the oil ring side rails - they may be damaged. Instead, place one end of the side rail into the groove between the spacer/expander and the ring land, hold it firmly in place and slide a finger around the piston while pushing the rail into the groove. Next, install the upper side rail in the same manner.

10 After the three oil ring components have been installed, check to make sure both the upper and lower side rails can be turned smoothly in the ring groove.

11 The lower compression ring is installed next. It usually will be stamped with a dot or mark which must face up, toward the top of the piston **(see illustration 13.11)**. **Note:** *Always follow the instructions printed on the ring package or box - different manufacturers may require different approaches. Don't mix up the upper and lower compression rings, as they have different cross sections.*

12 Use a piston ring installation tool and make sure the identification mark is facing the top of the piston, then slip the ring into the middle groove on the piston **(see illustra-**

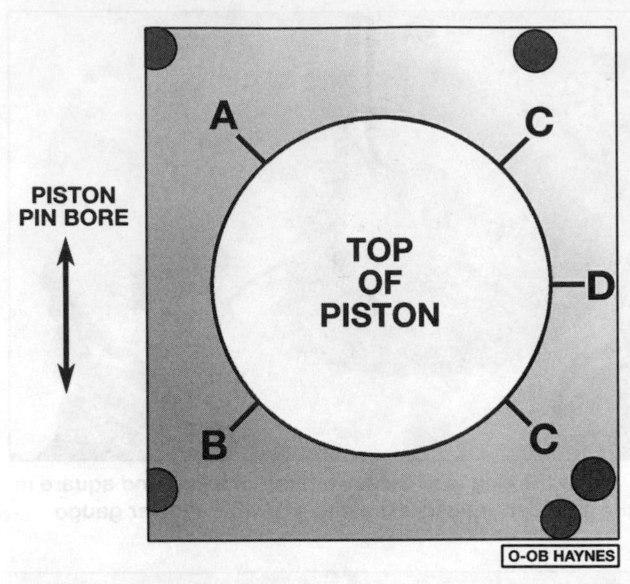


**13.15 Typically, the ring end gaps should be staggered around the piston as shown - align the oil ring spacer gap at (D), the oil ring side rail gaps at (C), and the compression ring gaps at (A) and (B)**

**tion 13.12)**. Don't expand the ring any more than necessary to slide it over the piston.

13 Install the upper (top) compression ring in the same manner. Make sure the mark is facing up. Be careful not to confuse the upper and lower compression rings. **Note:** *On Honda engines, the top ring is usually chrome-faced.*

14 Make sure the rings turn freely in the grooves (unless they're pinned in place).

15 Position the rings so the gaps are staggered 90 to 120-degrees apart **(see illustration 13.15)**. Do not align the end gaps over the piston pin bore. **Note:** *On some engines different ring gap spacing may be required. Refer to the appropriate engine Chapter for more information.*

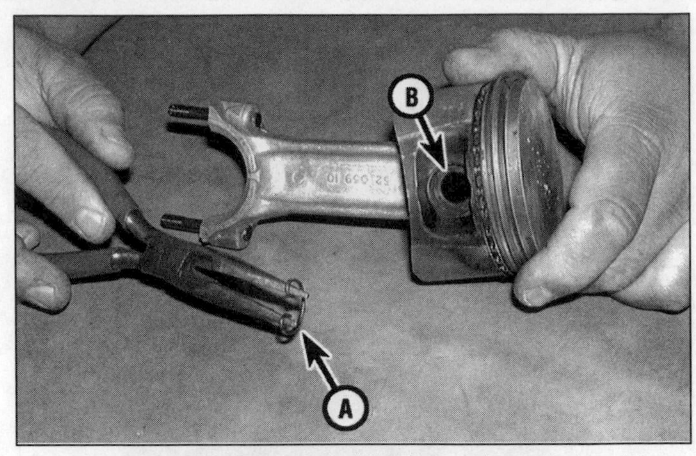


**14.1 With the new piston pin (B) pushed through the piston and rod, install the final lock-ring (A)**

# 14 Piston/rod installation

1 If the piston has been separated from the rod, or a replacement piston is being installed during the engine overhaul, a new piston pin and lock-rings must be installed. Install one new lock-ring on one side of the piston, align the pin holes in the piston with the rod, slide the new piston pin (lubed with engine oil) into the piston until it goes through the rod and seats on the lock-ring. Then install the new lock-ring on the other side of the piston **(see illustration 14.1)**. **Note:** *Always orient the piston to the connecting rod as originally installed.*

2 Lubricate the piston and rings with clean engine oil and attach a piston ring compressor to the piston **(see illustration 14.2)**. Leave the skirt protruding about 1/4-inch to guide the piston into the cylinder. The rings must be compressed until they're flush with the piston.

**14.2 Tighten the ring compressor tool around the piston/rings, leaving 1/4-inch of piston extending below the tool**

**14.4 Keep the ring compressor seated against the engine while tapping down the piston**

**15.2 Check the valve seats (arrow) in the engine block or head - look for pits, cracks and burned areas - L-head type engine shown here**

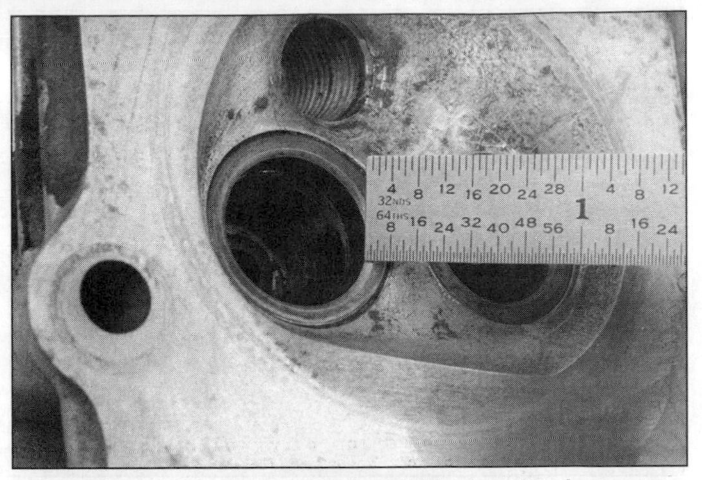


**15.3 Use a ruler to measure the width of each valve seat - OHV head shown**

3 Rotate the crankshaft until the connecting rod journal is at TDC (Top Dead Center) and apply a coat of engine oil to the cylinder walls. If the piston has an arrow in the top, it must face a certain way (see the engine Chapter for your model). Make sure the match marks on the rod and cap will be facing out when the rod/piston assembly is in place. Gently insert the piston/connecting rod assembly into the cylinder and rest the bottom edge of the ring compressor on the engine block. Tap the top edge of the ring compressor to make sure it's contacting the block around its entire circumference.

4 Carefully tap on the top of the piston with the end of a wooden or plastic hammer handle while guiding the end of the connecting rod into place on the crankshaft journal **(see illustration 14.4)**. The piston rings may try to pop out just before entering the bore, so keep some pressure on the ring compressor. Work slowly - if any resistance is felt as the piston enters the cylinder, stop immediately. Find the problem up and fix it before proceeding. Do not, for any reason, force the piston into the cylinder - you'll break a ring and/or the piston.

**15.4a A small hole gauge can be used to determine the inside diameter of the valve guide**

5 Install the connecting rod cap, a NEW lock plate (if used), the oil dipper (if used) and the bolts or washers and nuts. Make sure the marks you made on the rod and cap (or the manufacturer's marks) are aligned and facing out and the oil dipper is oriented correctly.

# 15 Valve/tappet inspection and servicing

## Inspection

1 If you're working on an OHV engine, inspect the head very carefully for cracks and other damage. If cracks are found, a new head is needed. Use a precision straightedge and feeler gauge(s) to check the head gasket surface for warpage. Lay the straightedge diagonally (corner-to-corner), intersecting the head bolt holes, and try to slip a 0.004-inch thick feeler gauge under it near each hole. Repeat the check with the straightedge positioned between each pair of holes along the sides of the head. If the feeler gauge will slip between the head surface and the straightedge, the head is warped. Check with a small engine repair shop or automotive machine shop about the possibility of resurfacing it.

2 Examine the valve seats **(see illustration 15.2)**. If they're pitted, cracked or burned, valve service that's beyond the scope of the home mechanic is required - take the engine or head to a small engine repair shop or automotive machine shop and have new valves and seats installed.

3 Measure each valve seat width and compare it to the Specifications in your engine Chapter **(see illustration 15.3)**. If it's not within the specified range, or if it varies around its circumference, valve seat service is required.

4 Clean the valve guides to remove any carbon buildup, then measure the inside diameters of the guides (at both

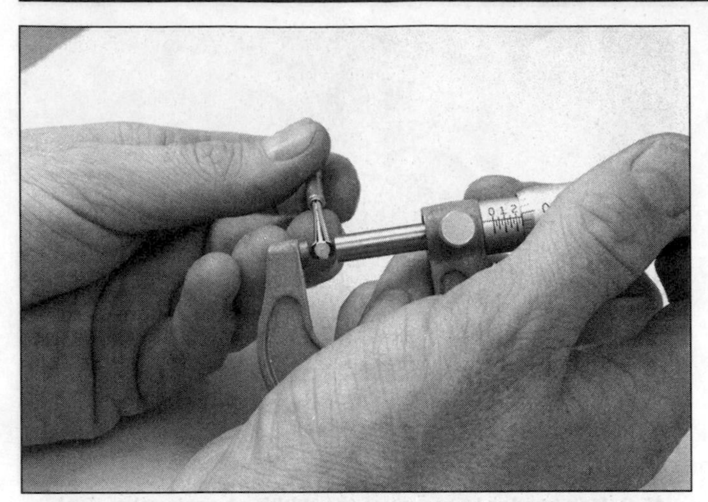

15.4b Measure the small hole gauge with a micrometer to obtain the actual size of the guide

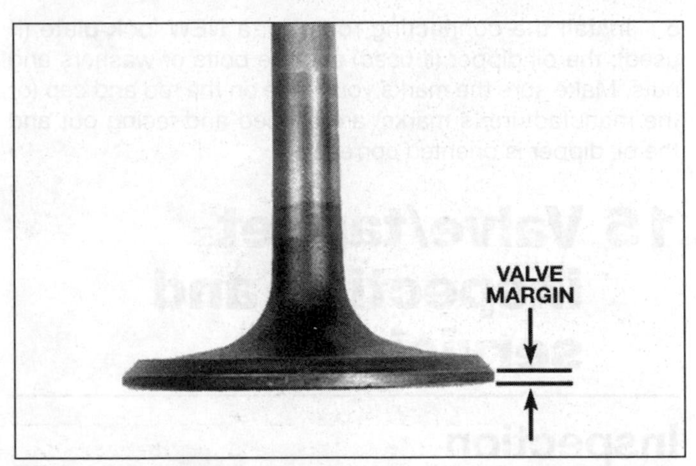


15.5b The margin width on each valve must be as specified - if no margin exists, the valve must be replaced

ends and the center of the guide). This is done with a small hole gauge and a 0-to-1 inch micrometer **(see illustrations 15.4a and 15.4b)**. Record the measurements for future reference. These measurements, along with the valve stem diameter measurements, will enable you to compute the valve-to-guide clearance. This clearance, when compared to the Specifications, will be one factor that will determine the extent of valve service work required. The guides are measured at the ends and at the center to determine if they're worn in a bell-mouth pattern (more wear at the ends). If they are, guide replacement or reconditioning is an absolute must. Some manufacturers don't publish valve-to-guide clearance specifications. Instead, they distribute special plug gauges that are inserted into the guides to determine how much wear has occurred in the guide. If no specifications are listed for your particular engine, have the guides checked and serviced by a small engine repair shop or automotive repair shop.

5 Carefully inspect each valve. Check the face (the area that mates with the seat) for cracks, pits and burned spots. Check the valve stem and the keeper groove or hole for

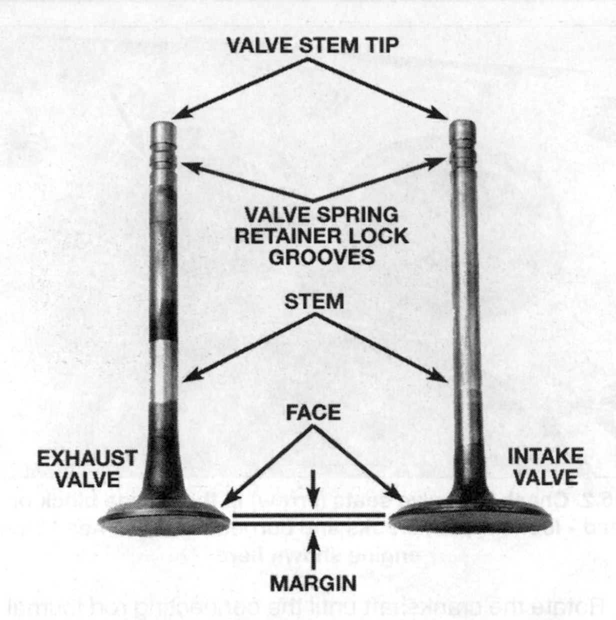


15.5a Check for valve wear the points shown here

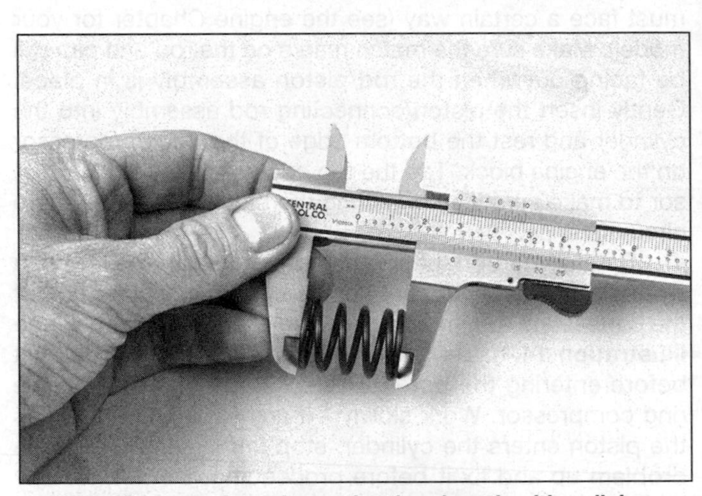

15.7 Measure the valve spring free length with a dial or vernier caliper

cracks. Rotate the valve and check for any obvious indication that it's bent. Check the end of the stem for pitting and excessive wear **(see illustrations 15.5a and 15.5b)**. The presence of any of the above conditions indicates the need for valve replacement.

6 Measure the valve stem diameter. By subtracting the stem diameter from the valve guide diameter, the valve-to-guide clearance is obtained. If the valve-to-guide clearance is greater than specified, the guides will have to be replaced and new valves may have to be installed.

7 Check the end of each valve spring for wear and pitting. Measure the free length and compare it to the Specifications, if applicable **(see illustration 15.7)**. Any springs that are shorter than specified have sagged and shouldn't be reused.

8 Stand the spring on a flat surface and check it for squareness **(see illustration 15.8)**.

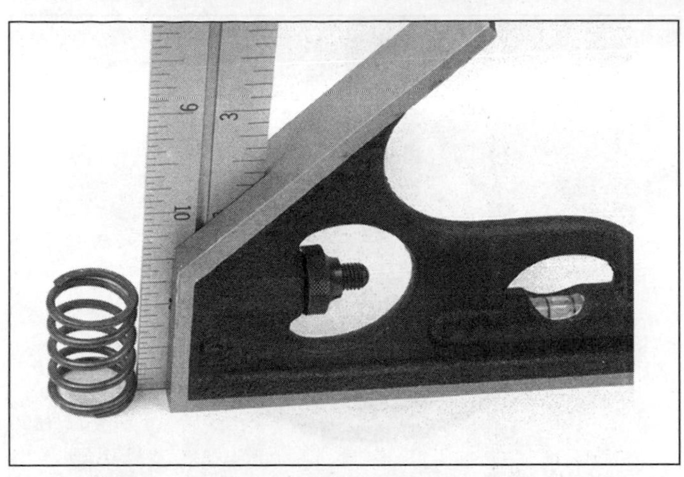

15.8 Check each valve spring for distortion with a square

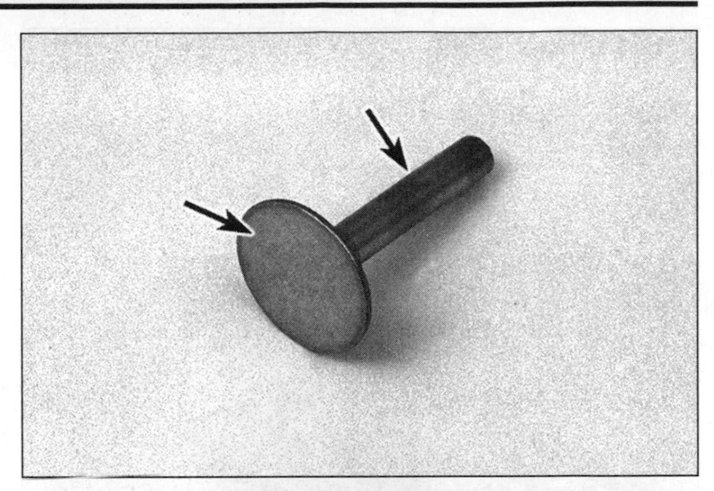

15.10 Check the tappet stems and ends (arrows) for wear and damage

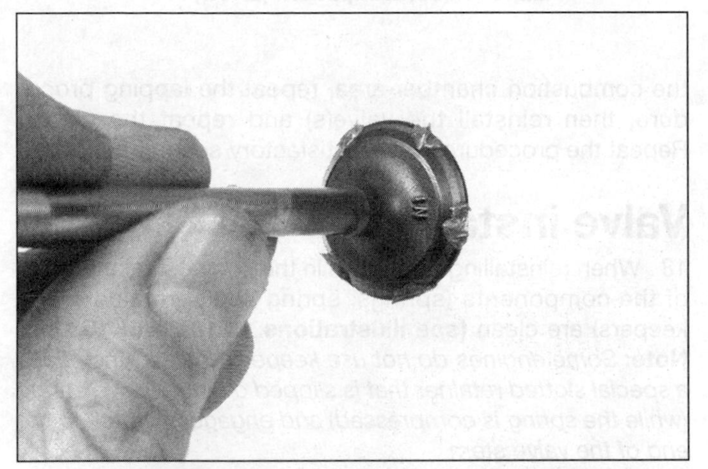

15.13 Apply the lapping compound very sparingly, in small dabs, to the valve face only

15.14a Rotate the lapping tool back-and-forth between the palms of your hands

9 Check the spring retainers and/or keepers or pins for obvious wear and cracks. Questionable parts should not be reused - extensive damage will occur in the event of failure during engine operation.

10 Check the tappets for wear, score marks and galling **(see illustration 15.10)**. Make sure they fit snugly in the holes and move freely without binding or catching. Some manufacturers publish specifications for tappet-to-tappet-bore clearances. Check the engine Chapter for your specific model.

## Valve lapping

11 If the inspection indicates that no service work is required, the valve components can be reinstalled in the engine block or head (see the appropriate engine chapter). **Note:** *If you have your engine torn down this far, it's wise to have the valves and seats ground at an automotive machine shop, to ensure long-lasting performance. A number of modern engines require a "three-angle" valve job to ensure a concentric and proper-width seat contact. This work is precise, and can't be duplicated by the home mechanic without experience and expensive equipment.*

12 Before reinstalling the valves, they should be lapped to ensure a positive seal between the faces and seats. If the valves and seats were not ground at a machine shop, lapping will provide a better seal, though not as long-lasting a seal as provided by a valve job. If the valves and seats were ground, the lapping procedure can be used simply as a final check of the concentricity and width of the seats. **Note:** *This procedure requires fine valve lapping compound (available at auto parts stores) and a valve lapping tool (see Chapter 1).*

13 Clean the valve and seat with lacquer thinner and apply a thin wipe of machinist's blue dye to the seat and allow it to dry. Apply a small amount of fine lapping compound to the valve face **(see illustration 15.13)**, then slip the valve into the guide. **Note:** *Make sure the valve is installed in the correct guide and be careful not to get any lapping compound on the valve stem.*

14 Attach the lapping tool to the valve and rotate the tool between the palms of your hands. Use a back-and-forth motion rather than a circular motion. Lift the valve off the seat at regular intervals to distribute the lapping compound evenly **(see illustrations 15.14a and 15.14b)**.

15.14b Lift the tool and valve periodically to redistribute the lapping compound on the valve face and seat

15.15a After lapping, the valve face should exhibit a uniform, unbroken contact pattern (arrow) . . .

15 Continue the lapping procedure until the valve face and seat contact area is uniform in width and unbroken around the entire circumference of the valve face and seat **(see illustrations 15.15a and 15.15b)**. When just checking the concentricity of a shop-performed valve job, only a little lapping is necessary to make the contact area show up. Where the machinist's blue dye is removed is an exact indication of the seat contact.

16 Carefully remove the valve from the guide and wipe off all traces of lapping compound. Use solvent to clean the valve and wipe the seat area thoroughly with a solvent-soaked cloth. Repeat the procedure for the remaining valve.

17 Once both valves have been lapped, check for proper valve sealing by pouring a small amount of solvent into each of the ports with the valves in place and held tightly against the seats. If the solvent leaks past the valve(s) into the combustion chamber area, repeat the lapping procedure, then reinstall the valve(s) and repeat the check. Repeat the procedure until a satisfactory seal is obtained.

## Valve installation

18 When reinstalling the valves in the engine, make sure all of the components (springs, spring seats, retainers and keepers) are clean **(see illustrations 15.18a and 15.18b)**. **Note:** *Some engines do not use keepers. Instead they have a special slotted retainer that is slipped over the valve spring (while the spring is compressed) and engages a notch in the end of the valve stem.*

19 Lubricate the components with engine oil and install the valves in their guides. Place the spring seat in place (if equipped) and the spring and retainer. Compress the spring

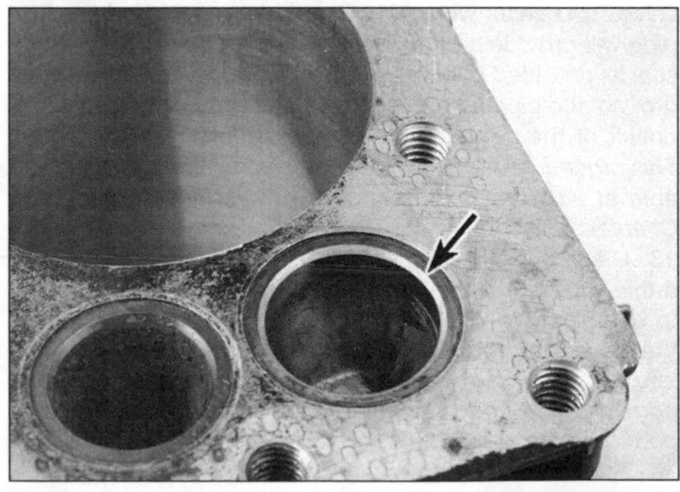

15.15b . . . and the seat should be the specified width (arrow), with a smooth, unbroken appearance

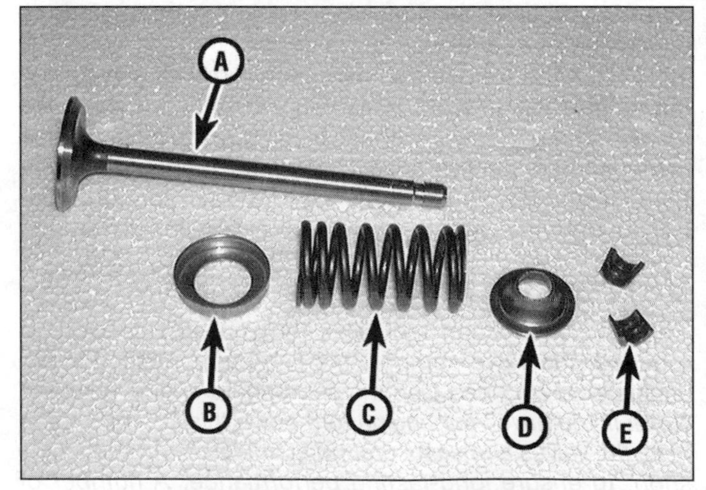


15.18a Valve components

*A* Valve
*B* Spring seat (not on all models)
*C* Spring
*D* Retainer
*E* Keepers

15.18b On some engines, the retainer is slotted (arrow) and it is fitted over the valve stem at the large end of the slot, then pushed in until the smaller end of the slot fits into a groove in the valve stem

15.19 With the spring compressor in place, install the keepers (arrow), making sure they are seated in the retainer before releasing the compressor tool

with a valve-spring compressor until the keepers can be installed with a small magnet or needlenose pliers **(see illustration 15.19)**. The keepers can be made to stay in place if they are coated with multi-purpose grease first. Slowly release the spring compressor, making sure the keepers stay seated in the retainer.

## Notes

# 6 Briggs & Stratton engines

## Contents

1.1 Typical Briggs & Stratton model number

## 1 Engine identification numbers/models covered

The engine model designation system used by Briggs & Stratton consists of a five or six digit number, normally found on the shroud, with the term "model" immediately under it **(see illustration 1.1)**. The model number can be used to determine the major features of the engine by comparing each digit to the accompanying chart **(see illustration 1.2)**. The digits in the model number can be explained generally as follows:

**BRIGGS & STRATTON MODEL NUMBER SYSTEM**

| A | B | C | D | E |
|---|---|---|---|---|
| | FIRST DIGIT AFTER DISPLACEMENT | SECOND DIGIT AFTER DISPLACEMENT | THIRD DIGIT AFTER DISPLACEMENT | FOURTH DIGIT AFTER DISPLACEMENT |
| CUBIC INCH DISPLACEMENT | BASIC DESIGN SERIES | CRANKSHAFT, CARBURETOR, GOVERNOR | PTO BEARING, REDUCTION GEAR, AUXILIARY DRIVE, LUBRICATION | TYPE OF STARTER |
| 6<br>8<br>9<br>10<br>11<br>12<br>13<br>16<br>17<br>18<br>19<br>22<br>23<br>24<br>25<br>26<br>28<br>29<br>30<br>32<br>35<br>40<br>42 | 0<br>1<br>2<br>3<br>4<br>5<br>6<br>7<br>8<br>9<br>A to Z | 0 - Horizontal Shaft Diaphragm Carburetor Pneumatic Governor<br>1 - Horizontal Shaft Vacu-Jet Carburetor Pneumatic Governor<br>2 - Horizontal Shaft Pulsa-Jet Carburetor Pneumatic or Mechanical Governor<br>3 - Horizontal Shaft Flo-Jet Carburetor Pneumatic Governor<br>4 - Horizontal Shaft Flo-Jet Carburetor Mechanical Governor<br>5 - Vertical Shaft Vacu-Jet Carburetor Pneumatic or Mechanical Governor<br>6 - Vertical Shaft<br>7 - Vertical Shaft Flo-Jet Carburetor Pneumatic or Mechanical Governor<br>8 - Vertical Shaft Flo-Jet Carburetor Mechanical Governor<br>9 - Vertical Shaft Pulsa-Jet Carburetor Pneumatic or Mechanical Governor | 0 - Plain Bearing/DU Non-Flange Mount<br>1 - Plain Bearing Flange Mounting<br>2 - Sleeve Bearing Flange Mounting Splash Lube<br>3 - Ball Bearing Flange Mounting Splash Lube<br>4 - Ball Bearing Flange Mounting Pressure Lubrication on Horizontal Shaft<br>5 - Plain Bearing Gear Reduction (6 to 1) CW Rotation Flange Mounting<br>6 - Plain Bearing Gear Reduction (6 to 1) CCW Rotation<br>7 - Plain Bearing Pressure Lubrication on Vertical Shaft<br>8 - Plain Bearing Auxiliary Drive (PTO) Perpendicular to Crankshaft<br>9 - Plain Bearing Auxiliary Drive Parallel to Crankshaft | 0 - Without Starter<br>1 - Rope Starter<br>2 - Rewind Starter<br>3 - Electric Starter Only 120 Volt Gear Drive<br>4 - Electric Starter/Generator 12 Volt Belt Drive<br>5 - Electric Starter Only 12 Volt Gear Drive<br>6 - Alternator Only<br>7 - Electric Starter 12 Volt Gear Drive With Alternator<br>8 - Vertical Pull Starter or Side Pull Starter |

Example: **MODEL** 92902

| 9 | 2 | 9 | 0 | 2 |
|---|---|---|---|---|
| 9 Cubic Inch | Design Series 2 | Vertical Crankshaft Pulsa-Jet Carburetor Pneumatic Governor | Plain Bearing/DU Non-Flange Mount | Rewind Starter |

**1.2 Use this chart to decipher the model number on Briggs & Stratton engines**
*Illustration courtesy of and with permission of Briggs & Stratton Corp.*

The first digit of a five-digit number, or the first two digits of a six-digit number, indicate the engine displacement in cubic inches. For example, a 17 would indicate a seventeen cubic inch engine and a 28 would mean a 28 cubic inch engine.

**Note:** *The information in this repair manual applies only to engines with a displacement of 16 cubic inches or more. If the model number has five digits, or if it has six digits and the first two digits are 13 or less, refer to Haynes Manual number 10340.*

The digit immediately after the displacement (the third digit from the left, if it's a six-digit model number) indicates the basic design series. It has to do with the cylinder type, ignition system and general engine configuration.

The second digit after the displacement (fourth digit from the left) indicates the crankshaft orientation (4 for horizontal and 7 for vertical) and the type of carburetor and governor installed on the engine.

The third digit after the displacement indicates the type of bearings used in the engine. It also will tell you if the engine is equipped with a reduction gear or auxiliary drive.

The last digit in the model number indicates the type of starter used on the engine.

The models covered by this manual are as follows:

**L-head engines**

| | |
|---|---|
| Single-cylinder | |
| 17000 series (horizontal and vertical) | 7 HP |
| 19000 series (horizontal and vertical) | 8 HP |
| 22000, 23000 and 24000 (horizontal cast iron) | 9, 10 HP |
| 22000 series (horizontal and vertical) | 10 HP |
| 25000 series (horizontal and vertical) | 11 HP |
| 28000 series (vertical) | 12, 12.5 HP |
| 30000 series (horizontal cast iron) | 12, 13 HP |
| 32000 series (horizontal cast iron) | 14, 16 HP |
| Twin-cylinder | |
| 40000 series (horizontal and vertical) | 14, 16 HP |
| 42000 series (horizontal and vertical) | 18 HP |

**Overhead Valve (OHV) engines**

| | |
|---|---|
| Single-cylinder | |
| 115400 and 117400 series (horizontal) | 5.5 HP |
| 161400 series (horizontal) | 9 HP |
| 185400 (horizontal) | 9 HP |
| 235400 series (horizontal) | 10 HP |
| 245400 series (horizontal) | 12 HP |
| 260700 and 261700 series (vertical) | 14 HP |
| 28E700, 28N700, 28P700, 28Q700 and 287700 series (vertical) | 16 HP |
| V-twin cylinder | |
| 290000 series (horizontal and vertical) | 12.5 HP |
| 303000 series (horizontal and vertical) | 16 HP |
| 350000 series (horizontal and vertical) | 18 HP |
| 351000 series (horizontal and vertical) | 20 HP |

# 2 Recoil starter service

## Rope replacement - all models

1 If the rope breaks, the starter doesn't have to be disassembled to replace it, but it may be a good idea to take the opportunity to do a thorough cleaning job and check the spring and drive mechanism.

2 There are two approaches you may be faced with when replacing the rope on a recoil starter. The method you use will depend on the starter type.

3 Remove the starter from the engine **(see illustration 2.3)**. Hold the shroud or recoil starter housing in a vise or clamp it to the workbench so it doesn't move around as you're working on the rope. Use soft jaws in the vise to prevent damage to the shroud or housing.

4 If you can't see the knot in the pulley end of the rope, the starter will have to be disassembled to install the new rope - the procedure is included later in this section. If the knot is visible **(see illustration 2.4)**, you should be able to replace the rope without disassembling the starter.

5 If the rope isn't broken, pull it all the way out.

**2.3 Remove the screws or bolts (arrows) and lift off the cover**

**2.4 If the knot is visible, the rope can be replaced easily without disassembling the recoil starter**

**2.6 Use a locking pliers or a C-clamp to restrain the pulley so it doesn't rewind**

6 Hold the pulley with locking pliers or a C-clamp so the spring won't rewind and the pulley is held in position for installing the rope **(see illustration 2.6)**.

2.7 Cut off the knot and pull the rope out

2.9 Turn the pulley to apply tension to the spring, then hold it in place while installing the rope

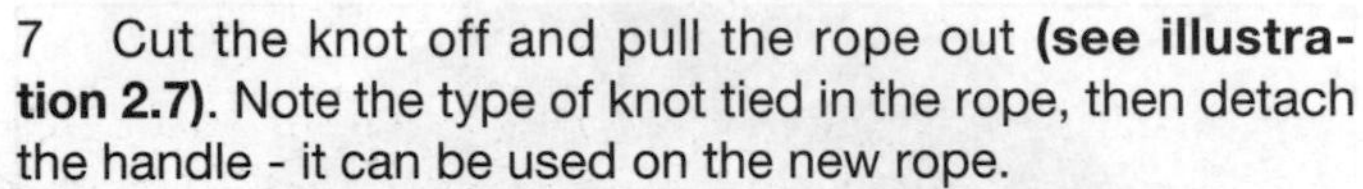

7 Cut the knot off and pull the rope out **(see illustration 2.7)**. Note the type of knot tied in the rope, then detach the handle - it can be used on the new rope.

8 If the old rope was broken, you'll have to wind up the recoil spring before installing the new rope.

9 Turn the pulley against spring tension until it stops completely **(see illustration 2.9)**, then back it off one full turn. This will prevent the spring from being wound too far when the rope is pulled out (which can break it off). **Note:** *A tool made from 3/4-inch square metal or hardwood stock will fit into the drive hole in the pulley and can be turned with a wrench.*

10 Cut a piece of new rope the same length and diameter as the original. Rope lengths may vary from three to four feet - if you don't have the exact replacement part, start with five feet of rope and cut it off if necessary when you see how it fills the pulley. The most common rope diameter is 5/32-inch. **Note:** *The rope should fill the pulley groove without binding.*

11 If the rope is made of nylon, melt the ends with a flame to prevent fraying.

2.12 The rope is difficult to thread into the pulley, so attach it to a piece of wire and use the wire to pull it into place

12 Turn the pulley so the opening for the rope is positioned as close to the opening in the housing as possible, then insert the rope into the housing opening and out through the pulley opening. This can be tricky - if the rope won't cooperate, hook a piece of wire through the end of the rope and bend it over with pliers, then thread the wire through the holes and use it to pull the rope into place **(see illustration 2.12)**. **Note:** *The rope must pass inside a guide lug on the old style metal pulley.*

13 Tie a knot in the rope and pull the knot tight against the hole in the pulley. On some models, the knot can be manipulated/pulled down into a cavity in the pulley. Make sure the knot doesn't contact the pulley retaining tangs.

14 Release the locking pliers or C-clamp while holding the rope, then allow the rope to rewind onto the pulley until the groove is full.

15 Pull the rope out slightly and attach the handle (make sure it's secure or the rope will disappear into the starter and you'll have to start over).

16 Check the starter for proper operation.

## Spring replacement - L-head engines

17 If the rope won't rewind and it isn't due to binding in the recoil starter, the spring may be broken.

18 Cut the knot at the pulley and remove the rope.

19 With the rope removed, grasp the outer end of the rewind spring with pliers **(see illustration 2.19)** and pull it out of the housing (if possible).

20 Bend one of the pulley retaining tangs up and lift out the pulley to disconnect the inner end of the spring **(see illustration 2.20)**. The housing has two spare tangs in case they break off.

21 Clean the rewind housing, pulley and spring with solvent and dry them with compressed air (if available) or a cloth. **Warning:** *Always wear eye protection when using compressed air!* Straighten the new spring so it doesn't tangle during installation.

2.19 Use pliers to grasp the outer end of the recoil starter spring and pull it out

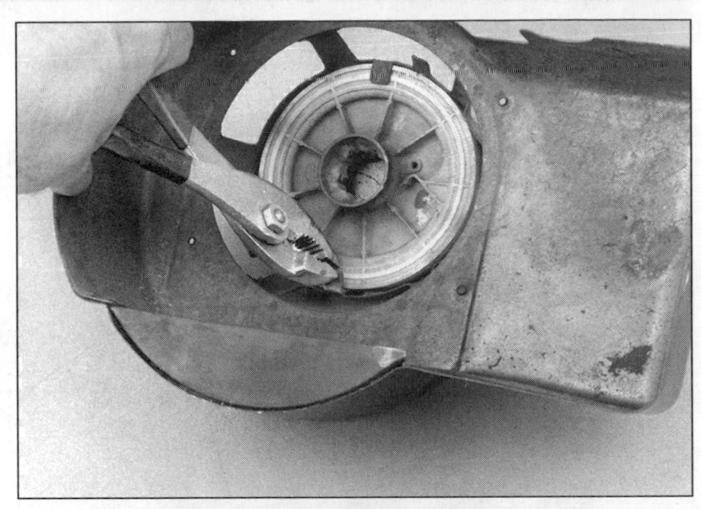

2.20 Bend up one of the tangs to remove the pulley - if it breaks off, use one of the spare tangs to hold the pulley in when it's reinstalled

22 Apply a light coat of oil to both sides of the spring.
23 If the pulley is made of steel, lubricate the end that contacts the housing and the spring face with grease.
24 Make sure the pulley, spring and housing are oriented correctly, then insert either end of the new spring through the housing opening and attach it to the pulley **(see illustration 2.24).**
25 Position the pulley in the housing and bend the tang down; if the tang breaks off, use the new ones to secure the pulley. The gap between the tang and pulley should be 1/16-inch (the pulley must be completely seated in the rewind housing when measuring the tang gap) **(see illustration 2.25). Note:** *Do not remove the nylon bumpers from the old style tangs when replacing a metal pulley with a nylon pulley. Replace the nylon bumpers if they're worn.*
26 Carefully wind the spring up by turning the pulley counterclockwise until the outer end of the spring can be locked in the housing slot **(see illustration 2.9 and the accompanying illustration 2.26).**

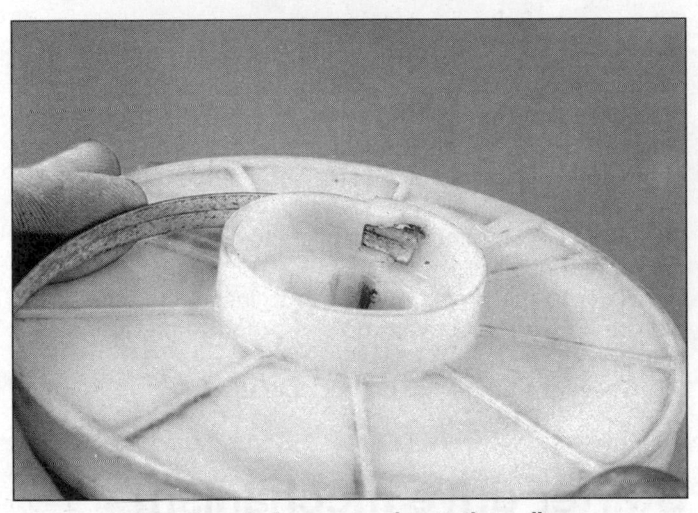

2.24 Attach the new spring to the pulley

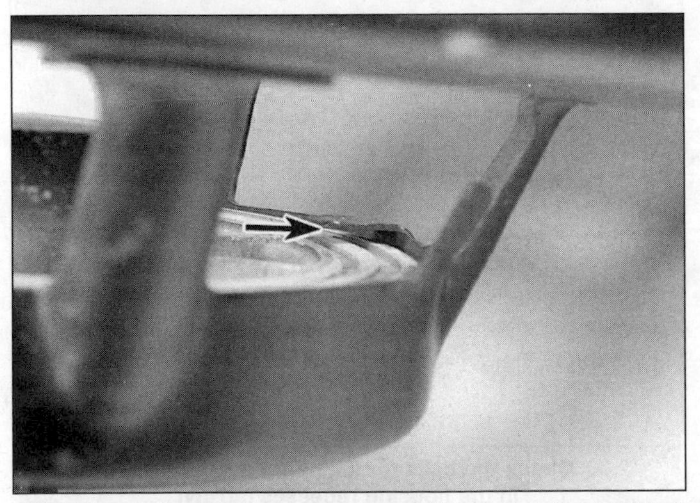

2.25 Check the pulley-to-tang gap so the pulley doesn't bind

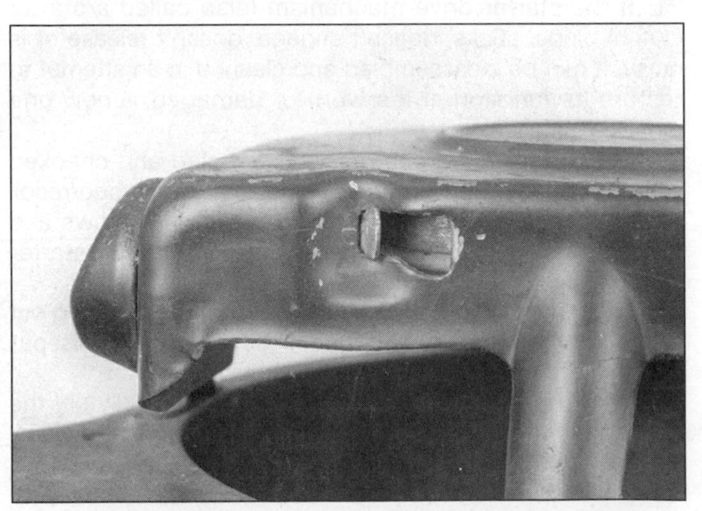

2.26 Make sure the outer end of the spring is securely locked in the narrow end of the housing slot

2.27 Wind up the spring until it's tight, then back it off one full-turn or until the rope holes line up

2.34 Use a small screwdriver or cold chisel and a hammer to pry the cover off the starter clutch

27 Keep turning the pulley until the spring is tight, then back it off one full turn or until the opening for the rope in the pulley and the opening in the housing are aligned **(see illustration 2.27)**. Check again to make sure the outer end of the spring is locked in the small end of the tapered slot in the housing.

28 Use a C-clamp or locking pliers to hold the pulley while the rope is installed.

29 Refer to the procedure above to install the rope. If the old one is worn or frayed, now is a good time to install a new one.

30 Check the starter for proper operation.

## Starter drive mechanism - L-head engines

31 If the starter drive mechanism (also called a starter clutch) binds, sticks, doesn't engage, doesn't release or is noisy, it can be disassembled and cleaned in an attempt to restore its function. If it's worn or damaged, a new one must be installed.

32 The starter drive can be disassembled and checked without removing it from the engine, but the shroud/recoil starter must be removed first. The two small screws and screen must also be detached. Disassemble the starter clutch as follows:

33 If you have an old style clutch, pry the retaining ring out of the housing groove with a small screwdriver and separate the ratchet and cover from the housing.

34 If you have a new style sealed clutch, carefully pry the cover off the housing with a small screwdriver or cold chisel and hammer **(see illustration 2.34)**.

35 Clean the components with solvent and dry them with compressed air (if available) or a clean cloth. **Warning:** *Always wear eye protection when using compressed air!* **Note:** *If you're working on a new style sealed clutch, clean the ratchet (the part that fits over the crankshaft) with a cloth only - don't submerge it in solvent.*

36 Check the balls for flat spots and the ratchet and housing for wear patterns caused by the balls **(see illustration 2.36)**.

37 Check inside the ratchet bore for rust and damage that could cause it to bind on the crankshaft **(see illustration 2.37)**. Check the crankshaft for nicks, burrs and a "mushroomed" end that could cause the drive mechanism to catch or bind.

38 If the clutch is a sealed type, make sure the seals are in place and in good condition **(see illustration 2.38)**.

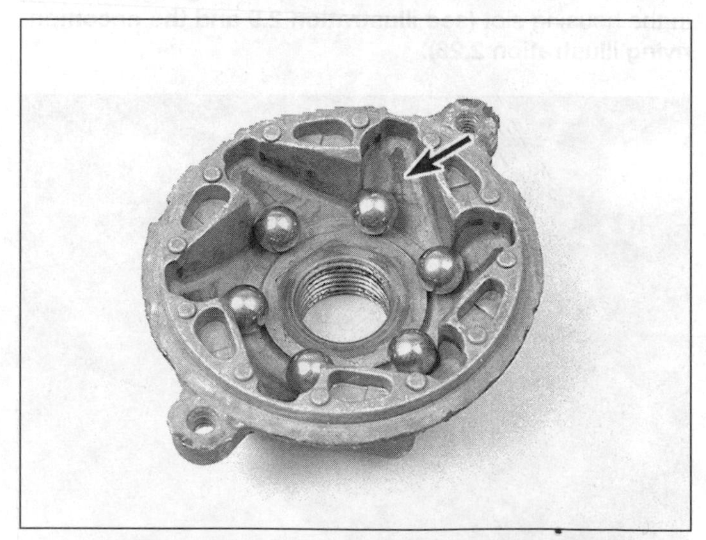

2.36 Check the balls for flat spots and look for wear in the housing recesses (arrow)

2.37 Check the ratchet bore (arrow) for rust and the end of the crankshaft for burrs and distortion that could cause the ratchet to bind on the shaft

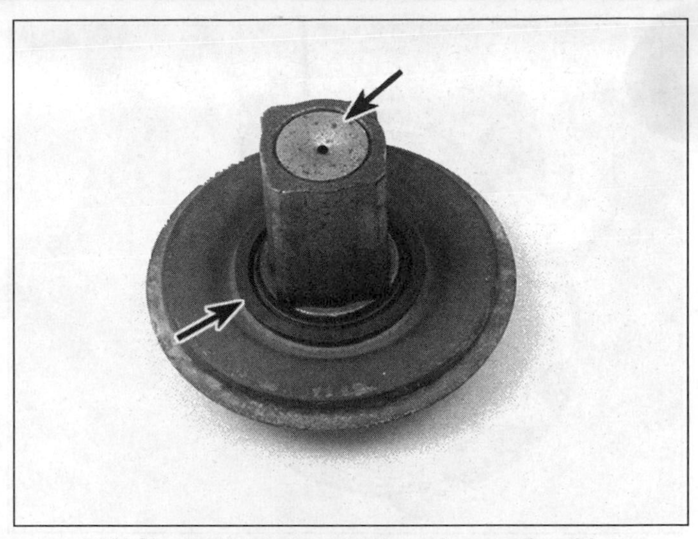

2.38 Check the starter clutch seals - one is rubber and one is felt (arrows)

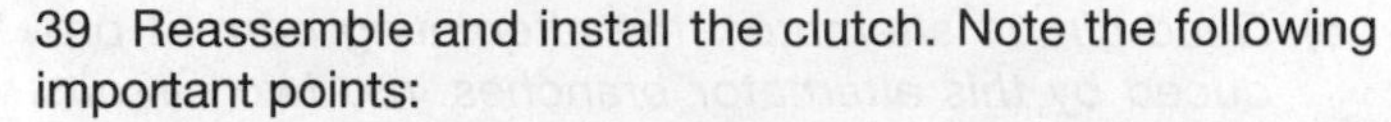

39 Reassemble and install the clutch. Note the following important points:

a) ***DO NOT*** *lubricate the ball cavity area! Starter clutch installation is part of Flywheel installation.*

b) *When installing a sealed starter clutch, apply one drop of engine oil only to the end of the crankshaft.*

c) *Tighten the starter clutch securely, but don't over-tighten and strip the threads.*

d) ***DO NOT*** *run the engine without the screen screws installed!*

**Note:** *A sealed clutch can be installed on older engines by modifying the recoil starter pulley and crankshaft. The old pulley can be made to fit the new clutch by cutting off the pulley hub until it's 1/2-inch tall. The crankshaft must be shortened 3/8-inch and the end chamfered with a file. A different screen (part no. 221661) is required with the new style starter clutch.*

## Spring and starter drive mechanism - OHV engines

40 Remove the shoulder screw and lift off the retainer **(see illustration 2.40)**. Remove the pawls and related parts.

41 If you're working on single-cylinder engines model 161400, 185400, 260700, or 261700, or an early-model twin-cylinder engine, remove the spring retainer, spring and cup from the starter housing.

42 If you're working on a single-cylinder engine other than the ones listed in Step 41, or a later model twin-cylinder engine, lift the pulley out of the starter housing **(see illustration 2.42a)**. Turn the pulley over and lift out the spring with needle-nose pliers **(see illustration 2.42b)**. **Warning:** *Release the spring slowly so it doesn't fly out and cause injury.*

43 Installation is the reverse of removal.

2.40 Remove the retainer, then lift off the pawls and springs

2.42a Lift the pulley out of the housing

2.42b Note the positions of the spring hooks, then secure the spring with locking pliers before disengaging it

3.3a The charge coil is secured by screws or bolts (arrows)

# 3 Charging and electric starting systems

## Charging system

1 The charging system on all models uses one or more charging coils (stators) positioned behind the flywheel. Magnets mounted on the flywheel rotate past the coil(s), producing alternating current in the stator windings. Different designs treat the current differently:

a) *Unregulated DC alternator: This charging system produces unregulated direct current which is used to charge a 12-volt battery. The alternating current produced by the alternator is rectified to direct current by a single diode in the wiring harness.*

b) *Unregulated AC alternator: This alternator produces unregulated alternating current which is used to operate lights. With this system, it's normal for the brightness of the lights to change as engine speed changes.*

c) *Dual-circuit alternator: This system is a combination of the unregulated DC and unregulated AC alternators, with a DC circuit for charging a 12-volt battery and an AC circuit for operating lights.*

d) *Tri-circuit alternator: The alternating current produced by this alternator branches into two circuits, each containing a diode which rectifies the current to direct current. One circuit provides 5 amps negative direct current to power lights; the other provides 5 amps positive direct current to charge the battery and operate the electric clutch (if equipped).*

e) *Quad-circuit alternator: the alternating current produced by this alternator branches into two circuits, each passing through the regulator-rectifier, which rectifies the alternating current to direct current. One circuit provides 5-amp unregulated current to power lights; the other provides 5-amp regulated current to charge the battery and operate the electric clutch (if equipped).*

f) *5 and 9-amp alternators: The alternating current produced by these alternators passes through a single wire to a regulator-rectifier, which regulates the current and rectifies it to direct current. Both designs use the same stator; the difference in output is due to more powerful flywheel magnets in the 9-amp system.*

g) *4, 10, 13- and 16-amp alternators: These are very similar to 5 and 9-amp alternators, but the alternating current passes through two wires on its way to the regulator-rectifier. Some 4 and 10-amp models use a pair of stators, rather than a single circular stator.*

## Component removal and installation

2 Refer to Section 8 or Section 9 and remove the flywheel.

3 Follow the wiring harness from the charge coil(s) to the electrical connector and disconnect it. Remove the coil mounting screws and take the coil(s) off **(see illustrations 3.3a and 3.3b)**.

4 If the charging system is equipped with a diode, unplug it.

5 If the charging system is equipped with a regulator/rectifier, disconnect its electrical connector **(see illustra tion 3.3b)**. Note how the regulator/rectifier is mounted, then remove the mounting bolt or screw and lift it off the engine.

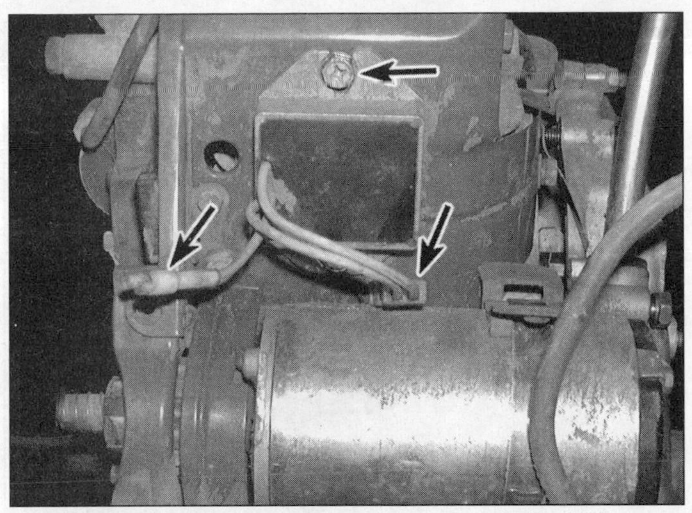

3.3b Disconnect the electrical connector(s) and undo the mounting screw to remove the regulator-rectifier

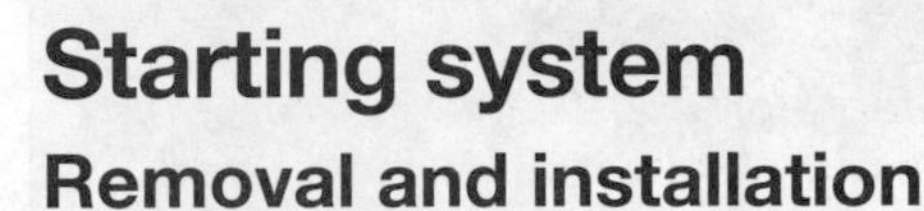

# Starting system

## Removal and installation

### Starter motor

6 Disconnect the cable from the negative terminal of the battery.

7 Pull back the rubber cover, remove the nut retaining the starter cable to the starter and disconnect the cable **(see illustration 3.7)**.

8 Remove the starter mounting bolts. On some engines, a stay bracket is secured by these bolts; be sure to note the location of the brackets so they can be reinstalled correctly.

9 Pull the starter away from the engine.

10 Install the starter by reversing the removal procedure.

### Starter solenoid

11 Disconnect the cable from the negative terminal of the battery.

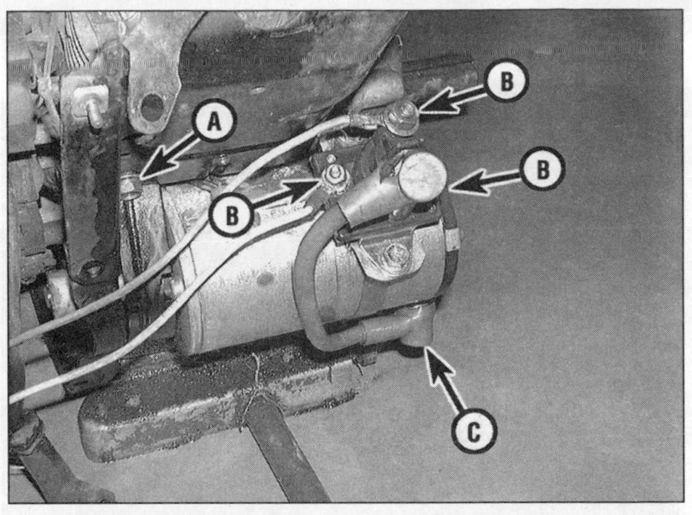


3.7 Typical starter installation

*A Upper mounting bolt (lower bolt hidden)*
*B Solenoid wire terminals*
*C Cable-to-starter terminal*

12 Pull back the rubber covers, remove the nuts retaining the starter and battery cables to the solenoid and disconnect the cables **(see illustration 3.7)**.

13 Disconnect the switch (thin) wire from the solenoid. Remove the solenoid mounting bolts or nuts and detach it from the starter.

14 Refer to Chapter 3 to check the solenoid.

15 Installation is the reverse of removal. Make sure the cables and wire are in good condition. Also make sure the solenoid mounting points are clean and free of corrosion, and tighten the bolts or nuts securely.

## Component replacement

16 Mark the position of the housing to each end cover. Remove the through-bolts and detach both end covers **(see illustration 3.16a)**. Drive out the roll pin and remove the drive clutch and pinion gear **(see illustration 3.16b)**.

3.16a Remove the through-bolts and end covers for access to the internal parts

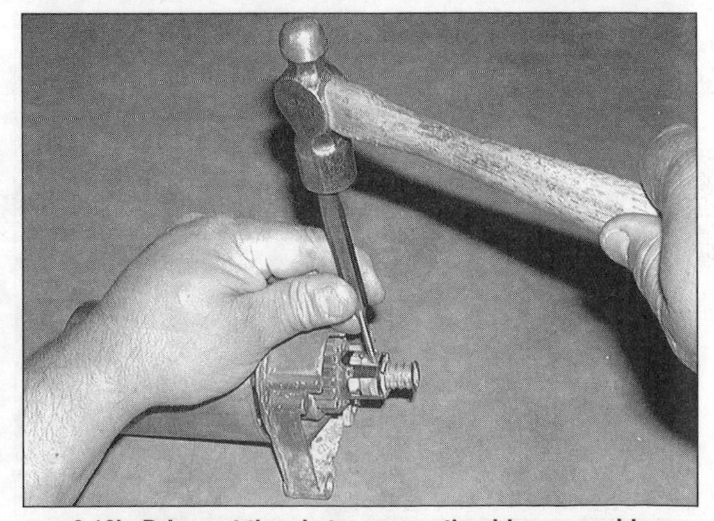

3.16b Drive out the pin to remove the drive assembly

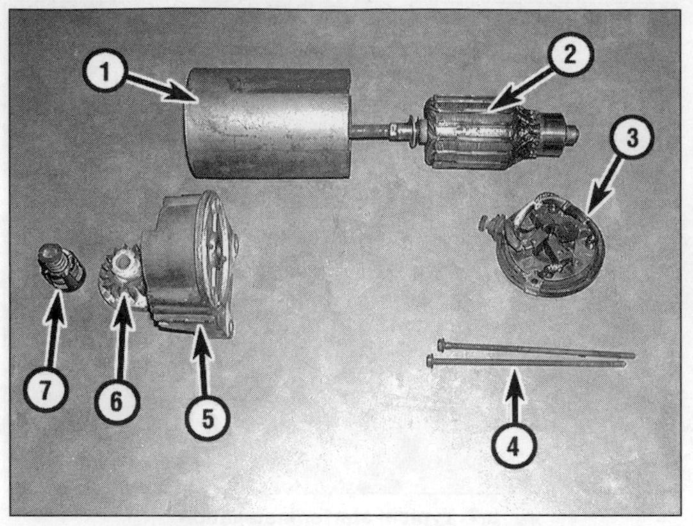


3.16c Typical starter components

1 Starter housing
2 Armature
3 End cover (with brushes and springs)
4 Through bolts
5 Drive housing
6 Pinion gear
7 Drive assembly

3.17a This type of brush plate uses flat coil springs

Pull the armature out of the housing and remove the brush holder **(see illustration 3.16c)**. Separate the brushes from the holder.

17 The parts of the starter motor that will most likely require regular attention are the brushes **(see illustrations 3.17a and 3.17b)**. Measure the length of the brushes (flat coil spring or compression spring type) or the width of the brushes (end-bearing type). Compare the results to the values listed in this Chapter's Specifications. If any of the brushes are worn beyond the specified limit, replace the brushes as a complete set. If the brushes are not worn excessively, cracked, chipped or otherwise damaged, they may be re-used.

3.17b This type of brush end plate uses compression-type coil springs

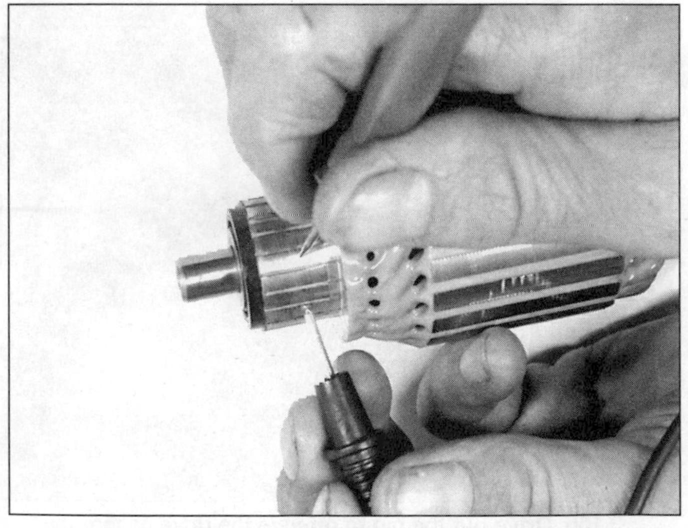

3.19 Continuity should exist between pairs of commutator bars

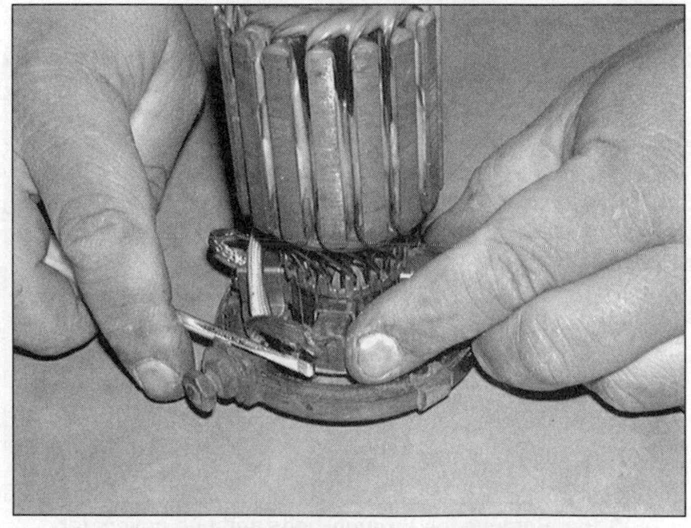

3.21 Pry back the brush springs so the commutator can be installed between the brushes

18 Inspect the commutator for scoring, scratches and discoloration. The commutator can be cleaned and polished with crocus cloth, but do not use sandpaper or emery paper. After cleaning, wipe away any residue with a cloth soaked in electrical contact cleaner or denatured alcohol. The commutator on starters with flat coil-type or compression-type brush springs can be turned on a lathe to the minimum diameter listed in this Chapter's Specifications. If this is done, clean the grooves between the commutator segments with a piece of hacksaw blade.
19 Using an ohmmeter or a self-powered test light, check for continuity between the commutator bars **(see illustration 3.19)**. There should be no continuity between the commutator and the shaft. If the checks indicate otherwise, the armature is defective.
20 Check for continuity between the pair of negative brushes and the pair of positive brushes. There should be continuity in both cases.
21 Assembly is the reverse of the disassembly procedure, with the following addition: If the starter has flat coil-type brush springs, push the brushes back so the commutator can be passed between them **(see illustration 3.21)**. Once this is done, position the springs so they push the brushes against the commutator.

# 4 Ignition system

## Points ignition systems

1 The point ignition system, installed on early models and smaller horsepower engines, is also called the "flywheel magneto" ignition system. This mechanical ignition system relies on the opening and the closing of the points to trigger the secondary voltage to the spark plug. The breaker points and the condenser are located under the flywheel. As the flywheel rotates, the magnet mounted on the flywheel passes the coil mounted on the engine case. The magnetic field from the flywheel magnet (magneto) causes a current to generate in the ignition coil primary circuit (winding). As the flywheel continues to rotate, it passes the last pole in the lamination stack on the coil. This produces a large change in the magnetic field ultimately producing high current in the primary circuit. This current change in the primary circuit builds until the points open and the secondary circuit is induced to flow to the spark plug and fire, releasing a high voltage spark. Points ignition systems include the ignition points, the coil, the condenser, the flywheel magneto, the ignition cam or plunger, the stator plate, and dust cover.

## Electronic ignition systems

2 The bulk of modern Briggs & Stratton engines use a "electronic magneto" ignition system. The electronic ignition system, also called "breakerless" ignition, incorporates solid state electronic components. There are two different types of breakerless systems on Briggs & Stratton engines; the first generation system is not actually a true "breakerless" system, it uses a set of magnetically actuated sealed points to trigger the secondary voltage similar to the points systems. The latter type incorporates an electronic module/trigger unit and is known as the "Magnatron" system.
3 This electronic ignition system is more durable than the old points system that commonly fail due to burning, oxidation and mechanical wear, and there is no timing to set. The electronic-magneto system operates similar to a points system but the method used to "break" the electronic signal differs. Flywheel rotation generates electricity in the primary circuit of the ignition coil assembly when the magnets rotate past the charge coil. As the flywheel rotates, it passes a trigger coil where a low voltage signal is produced. The low voltage signal allows the transistor to close allowing the energy stored in the capacitor to fire the secondary circuit or spark plug. The transistor acts as a primary circuit switching device in the same manner as the points in a mechanical (points) ignition system. Spark is produced whenever the engine is rotated, and a tab on the module is connected to either a kill switch or the OFF side of an ignition switch, to ground the module so the engine will stop.

## Ignition coil

### Removal

4 Remove the blower housing.
5 Disconnect the coil primary wire, stop switch wire and ground wire as necessary. On some models, it may be necessary to unsolder the stop switch wire from the terminal on the coil.
6 Unscrew the spark plug cap from the plug wire. Remove the coil mounting screws and remove the coil from the engine.

### Installation and adjustment

7 Position the flywheel so the magnet is not under the coil. Install the coil and tighten the mounting bolts hand-tight.
8 Rotate the flywheel until the magnet is under the coil. Refer to this Chapter's Specifications and place the appropriate size feeler gauge between the coil armature segments and the flywheel **(see illustration 4.8)**.
9 Loosen the coil mounting bolts, allow the magnet to pull the armature segment against the feeler gauge and tighten the mounting bolts securely. Rotate the flywheel to remove the feeler gauge. Be sure to check/adjust the clearance between each armature segment and the flywheel.
10 Connect the wiring and spark plug wire as necessary.

4.8 With the flywheel magnet positioned under the ignition coil armature, place the feeler gauge between the armature segment and the flywheel. Allow the magnet to pull the armature against the flywheel and tighten mounting bolts

4.12a Check the flywheel key (arrow) – if it's sheared off install a new one

## Breaker points

**Note:** *Models equipped with breaker points may have either internally or externally mounted breaker points. Internally mounted points are located under a cover behind the flywheel. Externally mounted points are located under a cover behind the blower housing. The flywheel need not be removed to replace the points on external models. This procedure describes the internal Type I breaker points. Internal Type II and Type III points and external points differ somewhat in appearance, but the procedure is basically the same.*

## Removal

11 On models equipped with internal breaker points, refer to Section 8 and remove the flywheel. On models equipped with external breaker points, remove the blower housing and the breaker point cover.

12 On models equipped with internal breaker points, check the flywheel key for damage, then remove the screws and lift off the breaker point cover **(see illustrations 4.12a and 4.12b)**.

13 Check the breaker points to see if they're burned, pitted, worn or covered with oil. If they're in good condition, they can be dressed with a point file, cleaned and readjusted (see Chapter 4). If their condition is at all doubtful, replace the points.

14 Check the crankshaft seal where the crankshaft protrudes from the engine. If oil has been leaking past the seal, it should be replaced (see Section 8).

15 On models equipped with internal breaker points, remove the screw and detach the condenser, then depress

4.12b Remove the screws (arrows) and lift off the breaker points cover

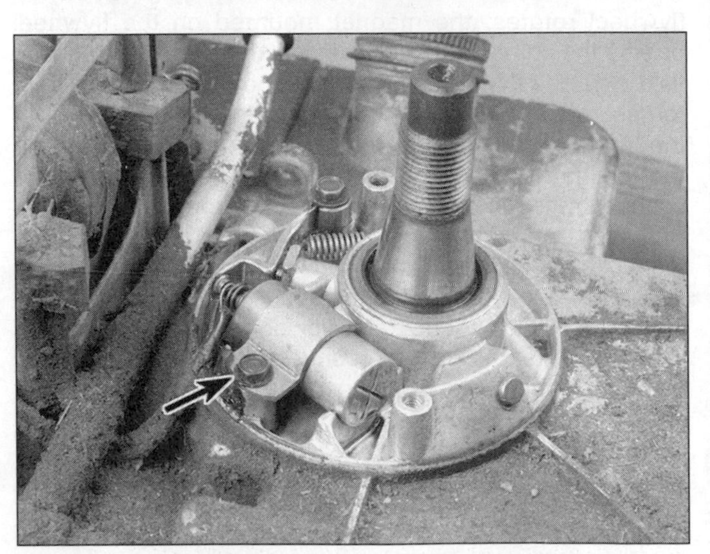

4.15a Remove the condenser screw . . .

4.15b . . . and the moveable point screw

4.17 Pull out the plunger and check it for wear

the small spring and release the primary wire from the terminal on the end of the condenser. Remove the screw and lift out the moveable point, return spring and post **(see illustrations 4.15a and 4.15b)**.

16 On models equipped with external breaker points, remove the condenser and mounting bracket screw, loosen the adjusting screw locknut and loosen the adjusting screw until the breaker points and condenser can be removed as an assembly.

17 Pull out the plunger and check it for wear **(see illustration 4.17)**. If it's worn to less than 0.870-inch in length (internal points) or 1.115-inch (twin-cylinder external points) or damaged in any way, install a new one. Take the old one with you to a small engine repair shop to make sure you get the correct replacement. **Note:** *If oil has been leaking past the plunger bore and fouling the points, the plunger bore is probably worn. Take the engine to a small engine repair shop and have the bore checked (a special gauge is available for this purpose). If it's worn, a small engine repair shop will ream it out and install a bushing to repair the bore.*

18 Clean the ignition point cavity with contact cleaner and wipe it out with a rag **(see illustration 4.18)**. Use the contact cleaner to remove oil and dirt from the faces of the new breaker points as well.

## Installation

19 Reinstall the plunger (with the grooved end out, toward the moveable point), the moveable point, the return spring and the post **(see illustration 4.19)**. Make sure the slot in the post engages the nub in the recess, that the moveable point arm is seated in the slot in the post and that the ground wire is under the screw.

20 On Type I models, attach the wire to the new condenser (the new points should have a little plastic tool designed to compress the spring that holds the wire on the condenser) and carefully clamp the condenser to the engine **(see illustration 4.20)**. Leave the mounting/adjusting screws just loose enough to adjust the point gap.

4.18 Clean the points cavity with contact cleaner

4.19 Install the plunger and moveable point

4.20 Install the condenser

4.24 Make sure the points open and close

## Adjustment

21 Rotate the crankshaft until the point gap is at it's widest opening.

22 Insert a 0.020-inch feeler gauge between the breaker point contacts (see Chapter 4). On single-cylinder models with internal Type I breaker points, pry the condenser forward or backward to adjust the gap. On single-cylinder models with internal Type II or III breaker points, pry the mounting bracket to adjust the point gap. On models equipped with external breaker points, back-off the locknut and turn the adjusting screw to adjust the point gap.

23 After the gap is properly set, tighten the mounting screws and locknut as applicable.

24 Turn the crankshaft and make sure the point contacts open and close **(see illustration 4.24)**.

25 Reinstall the cover and tighten the screws. If the cover is distorted, replace it with a new one or oil and moisture will foul the points.

26 Use RTV sealant to seal off the wires to prevent oil and moisture from getting to the points **(see illustration 4.26)**.

4.26 Seal the cover with RTV sealant where the wires enter

# Ignition timing

## L-head single-cylinder models 230000, 240000 and 320000 with electronic ignition

27 The ignition coil mounting bracket is slotted allowing for two timing settings depending on fuel type. When installing the coil place it in one of the two positions described below.

28 On gasoline fueled models, center the mounting bolts in the bracket slots and tighten the mounting bolts.

29 On kerosene fueled models, place the mounting bolts in the far left side of the slots and tighten the mounting bolts.

## L-head single-cylinder models 230000, 243400, 300000 and 320000 with external breaker points ignition

30 Check the breaker point gap and adjust if necessary.

31 Connect the positive lead of a self-powered continuity tester or digital multimeter to the breaker points primary lead, connect the negative lead to the breaker points mounting bracket.

32 Disconnect the ignition coil ground wire and isolate it from the coil.

33 Rotate the flywheel clockwise until continuity is indicated (points closed). Continue to rotate the flywheel clockwise until no continuity is indicated (points open).

34 Check the arrow on the flywheel, it should be aligned with the arrow on the ignition coil bracket. If the arrows are not aligned, remove the flywheel (without moving the crankshaft). Loosen the coil bracket bolts, temporarily install the flywheel and flywheel key and align the arrows. Remove the flywheel and tighten the coil bracket bolts. Install the flywheel and tighten the nut to the torque listed in this Chapter's Specifications.

5.1 Check the throttle shaft for wear

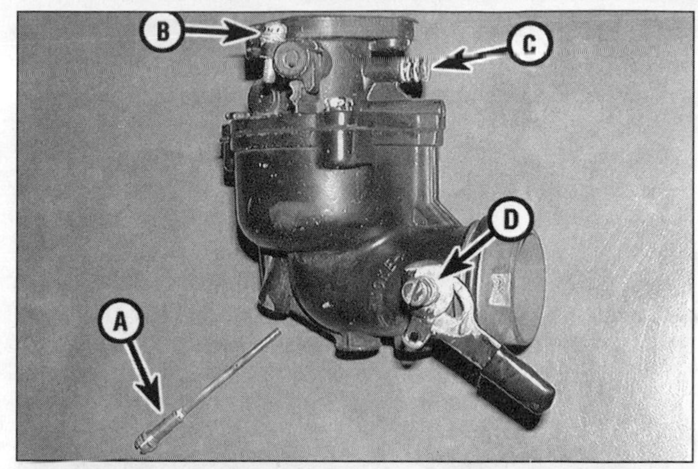


**5.2 Two-piece Flo-jet carburetor**

*A Main jet (main jet needle valve not shown)*
*B Idle speed screw*
*C Idle mixture screw*
*D Choke*

# 5 Carburetor disassembly and reassembly

Several carburetor types have been used on Briggs & Stratton engines. These include the two-piece Flo-jet and Briggs & Stratton/Walbro LMT. **Note:** *It's not always easy to determine which type of carburetor you have, since they're not marked, so be sure to take it with you when purchasing parts or a replacement carburetor.*

The following procedures describe how to disassemble and reassemble typical Briggs & Stratton carburetors so new parts can be installed. Read the sections in Chapter 5 on carburetor removal and overhaul before doing anything else.

In some cases it may be more economical (and much easier) to install a new carburetor rather than attempt to repair the original. Check with a small engine repair shop to see if parts are readily available and compare the cost of new parts to the price of a complete ready-to-install carburetor before deciding how to proceed.

## Two-piece Flo-jet carburetor (L-head single-cylinder engines)

### Disassembly

1 Move the throttle shaft back-and-forth **(see illustration 5.1)**. If you can feel side-to-side play, the shaft bushing is worn excessively. Replace it as described below.

2 Remove the main jet needle valve, then unscrew the main jet from the bore **(see illustration 5.2)**. On two-piece Flo-jet carburetors, the float bowl - which is actually the lower part of the carburetor housing - is attached to the upper carburetor housing with several screws. On these carburetors the main jet projects diagonally into a recess in the upper housing, so the main jet needle valve, packing nut and main jet must be taken out before removing the screws and separating the two parts of the carburetor.

**5.3a Push out the throttle lever pin . . .**

3 If you need to replace the throttle plate or throttle shaft bushings, push out the pin that retains the throttle stop **(see illustration 5.3a)**. Remove the throttle plate screws, then push out the shaft and remove the throttle plate **(see illustration 5.3b)**.

**5.3b . . . remove the throttle plate screw and pull out the shaft**

5.4 To remove shaft bushings, thread the bushing onto a tap and pull it out . . .

5.5 . . . press the new bushings in with a vise

4 Secure a 1/4-inch tap in a vise and thread it into the throttle shaft bushing **(see illustration 5.4)**. Pull the carburetor away from the vise to remove the bushing, then repeat the procedure to remove the other bushing.

5 Use the vise to press the new bushings into place **(see illustration 5.5)**. Temporarily slide the throttle shaft in and rotate it back-and-forth. If the throttle shaft doesn't move freely in the new bushings, ream them by hand with a 7/32-inch drill bit.

6 Check the choke shaft movement in the same way as the throttle shaft (Step 1). If it's loose, remove the choke plate screws **(see illustration 5.6a)**. Remove the choke screw, spring and washer, then pull out the choke shaft and remove the choke plate **(see illustrations 5.6b and 5.6c)**.

7 Remove the screws and separate the upper body and float from the lower body. Measure the float height on both sides **(see illustration 5.7a)**. It should be even (the float should be parallel with the surface of the carburetor).

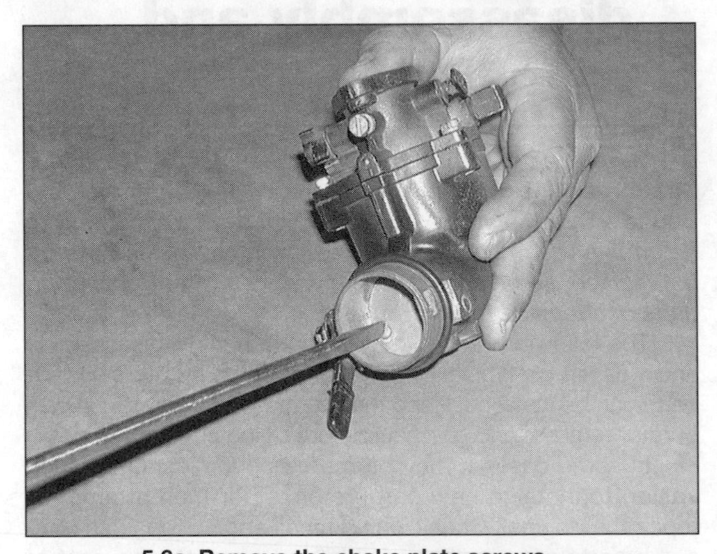

5.6a Remove the choke plate screws . . .

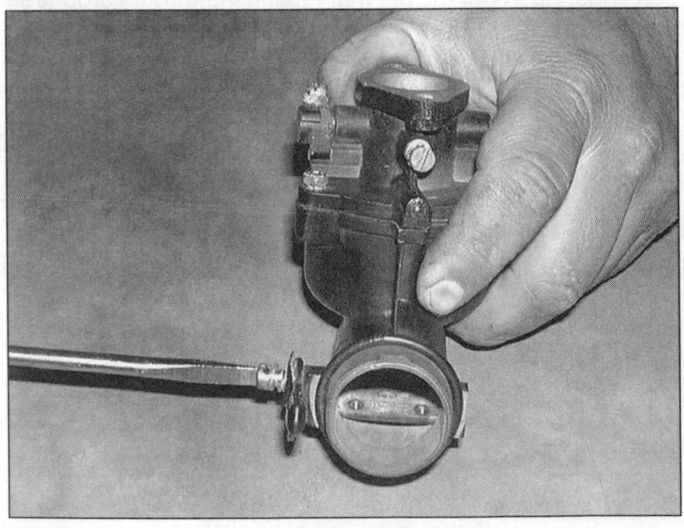

5.6b . . . and the shaft screw . . .

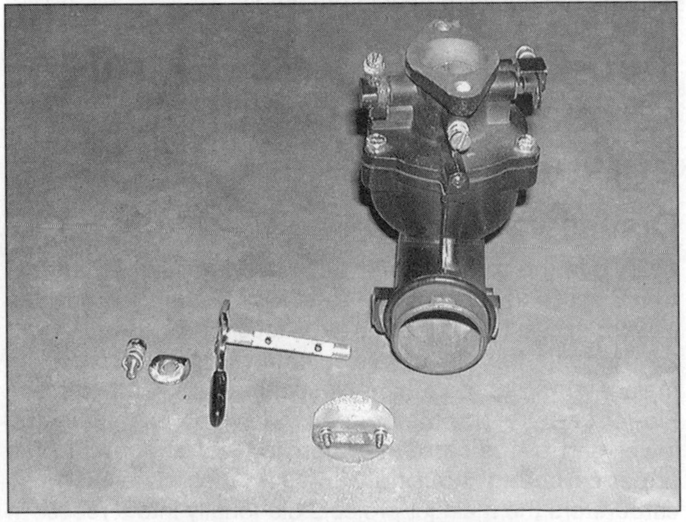

5.6c . . . and pull out the shaft

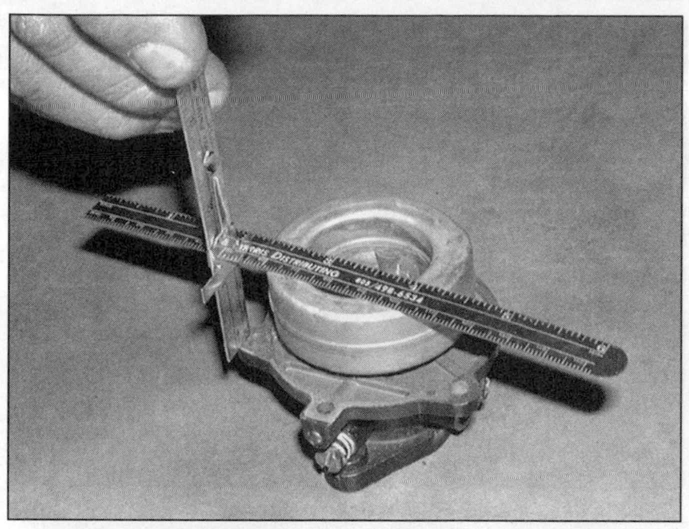
5.7a Measure float height with a steel rule or float gauge

5.7b If there's a tang on the float, bend it to change float level - some floats aren't adjustable

**Note:** *On some floats, the surface nearest the carburetor is flat and the surface farthest from the carburetor is not flat. Measure from the flat surface.* If there's no bendable tang, the float will have to be replaced if the height is incorrect. If there is a bendable tang, bend it to change the float height **(see illustration 5.7b)**.

8 To remove the float, pull out its pivot pin **(see illustration 5.8)**.

## Inspection

9 To do a thorough cleaning job, remove any Welsh plugs from the carburetor body.

Drill a small hole in the center of the plug and pry it out with a punch or thread a sheet metal screw into the hole, grasp the screw head with pliers and pull the plug out. **Note:** *Some two-piece Flo-jet carburetors have a large Welsh plug in the end of the throttle bore that should be removed only if the choke valve/shaft must be replaced with new parts.*

10 Refer to Chapter 5 and follow the cleaning/inspection procedures outlined under Carburetor overhaul. It's very important to get all sludge, varnish and other residue out of the carburetor passages.

11 Check the mixture adjusting screw tip for damage and distortion. The small taper should be smooth and straight.

12 Look for nicks and a pronounced groove or ridge on the tapered end of the needle valve **(see illustration 5.12)**. If there is one, a new needle and seat should be used when the carburetor is reassembled. **Note:** *Some inlet needle seats can be unscrewed, while others are a press fit.*

13 To remove a pressed-in seat, grip the hex-head of a self-tapping bolt in a vise and position the inlet seat bore over the threaded end of the bolt. Turn the carburetor body until the bolt threads in the seat bore and draw the seat out. A tap can be used in place of the bolt.

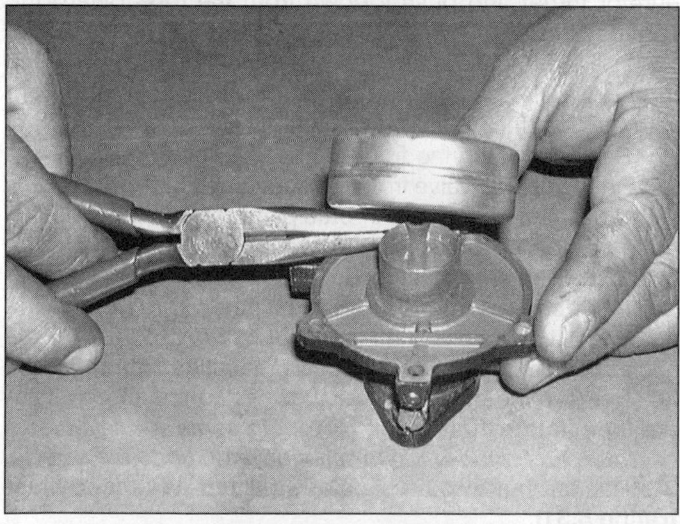
5.8 Pull out the float pin

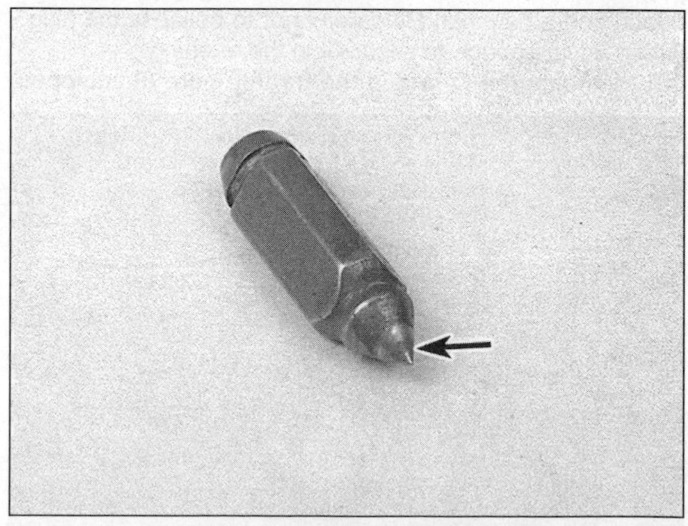
5.12 Check the needle valve for a groove or ridge in the tapered area (arrow)

5.24 Remove the bowl screw, washer, float bowl and O-ring

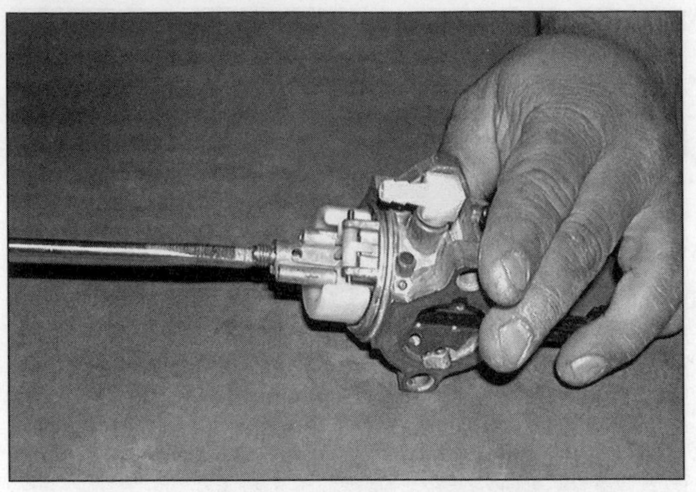

5.25 Remove the main nozzle

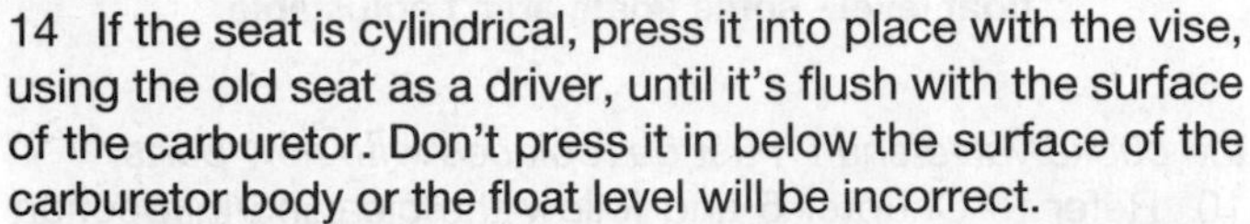

14 If the seat is cylindrical, press it into place with the vise, using the old seat as a driver, until it's flush with the surface of the carburetor. Don't press it in below the surface of the carburetor body or the float level will be incorrect.

15 Check the float pivot pin and bore for wear - if the pin fits loosely in the bores excessive amounts of fuel will enter the float bowl and flooding will occur.

16 Shake the float to see if there's gasoline in it. If there is, install a new one. Cork floats will absorb gasoline, but it's hard to tell if this has occurred.

## Assembly

17 Once the carburetor parts have been cleaned thoroughly and inspected, reassemble it by reversing the above procedure. Note the following important points:

18 Be sure to use new O-rings, gaskets and rubber diaphragms.

19 When installing new Welsh plugs, apply a small amount of non-hardening sealant to the outside edge and seat the plug in the bore with a 1/4-inch or larger diameter pin punch and a hammer. Be careful not to collapse the plug - flatten it just enough to secure it in the opening.

20 Lubricate the O-ring in the throttle bore (if equipped) with a small amount of engine oil so it won't be damaged when the carburetor is installed.

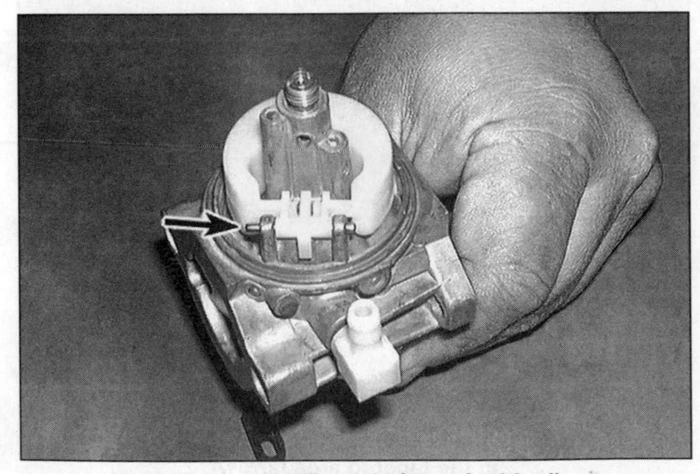

5.26 Pull out the float pin (arrow) with pliers

21 When the inlet needle valve assembly is installed, be sure the clip (if used) is attached to the float tab.

22 When fastening the housings together on a two-piece Flo-jet carburetor, tighten the screws in small increments to avoid distorting anything. Also, make sure the upper end of the fuel nozzle enters the recess in the upper body.

23 Turn the mixture adjusting screw(s) in until they bottom and back each one out the number of turns listed in this Chapter's Specifications.

# LMT carburetor (L-head and OHV single-cylinder engines)

24 Remove the float bowl screw and washer and detach the bowl and its gasket **(see illustration 5.24)**.

25 Unscrew the main nozzle **(see illustration 5.25)**. If the main nozzle won't come out of the bore, reach into the carburetor throat and push it down from the top, toward the float bowl.

26 Pull out the hinge pin with pliers **(see illustration 5.26)**. If there's a flat on one end of the pin, it goes on the right with the float facing away from you.

27 Remove the spring from the needle valve (if equipped). Detach the needle valve from the float **(see illustration 5.27)**.

28 Check the choke and throttle plates for wear by moving them back-and-forth in their bores. The bushings can be replaced if the shafts are loose.

29 If the choke plate is secured by screws, remove them. If not, pull it out of the shaft **(see illustration 5.29a)**. Pull the shaft and seal out of the carburetor **(see illustration 5.29b)**.

30 Remove the throttle plate screws (if equipped) or retainers **(see illustration 5.30)**. **Note:** *On some models, you'll need a Reed Prince 1/4-inch screwdriver to undo the screws.*

31 Pull out the throttle plate and related parts **(see illustration 5.31)**.

32 Unscrew the main air jet and idle mixture screw **(see illustrations 5.32a and 5.32b)**.

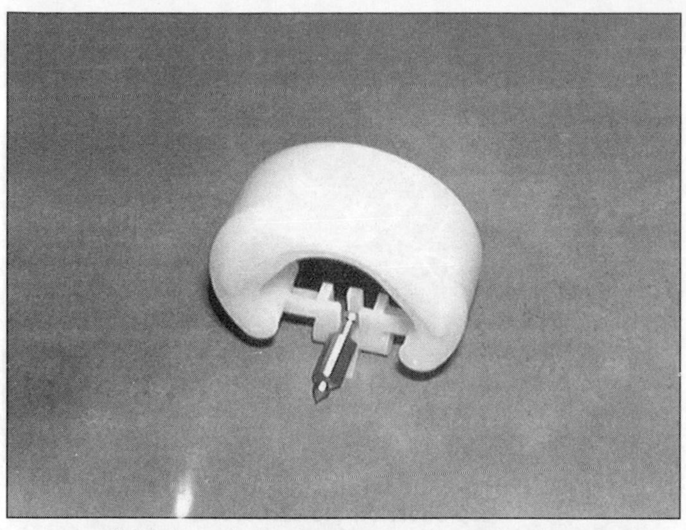

5.27 Detach the needle valve from the float

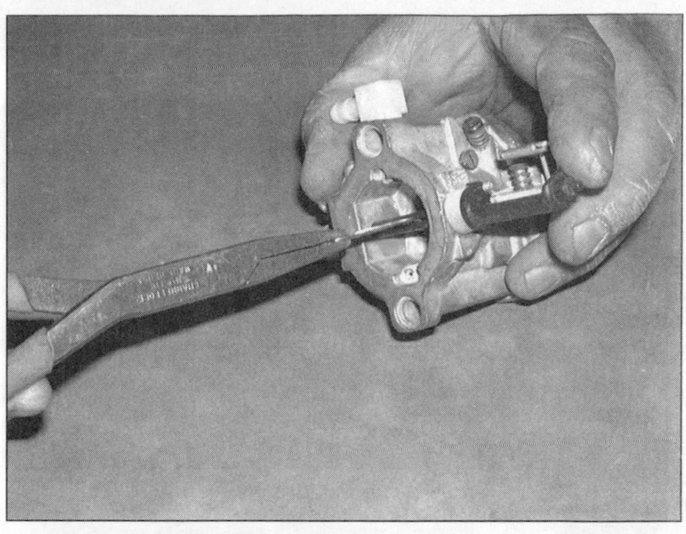

5.29a Pull out the choke plate . . .

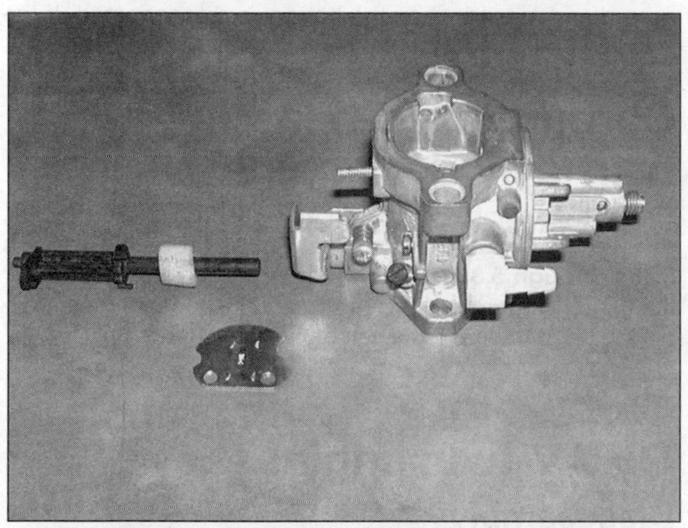

5.29b . . . then withdraw the shaft and seal

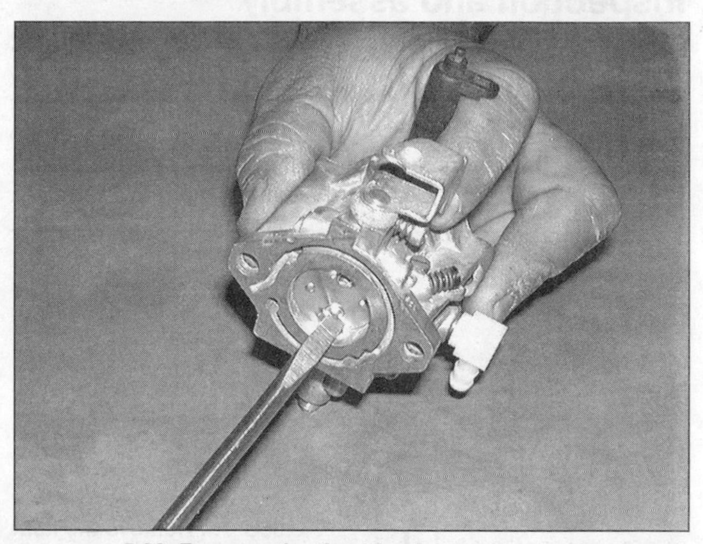

5.30 Remove the throttle plate screws . . .

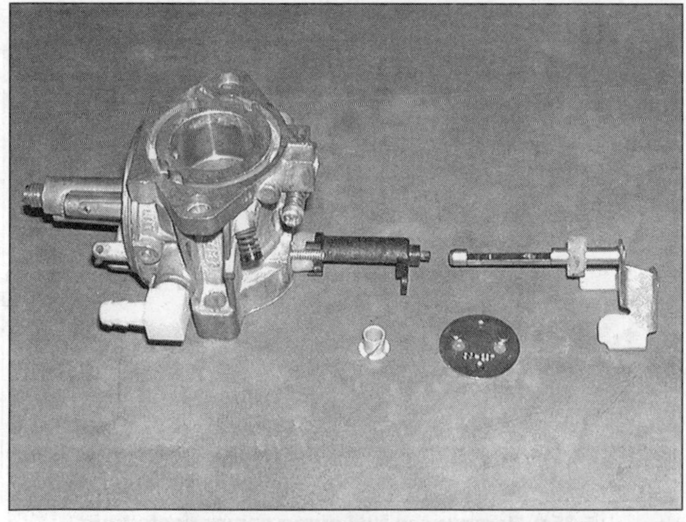

5.31 . . . pull out the shaft and remove the collar

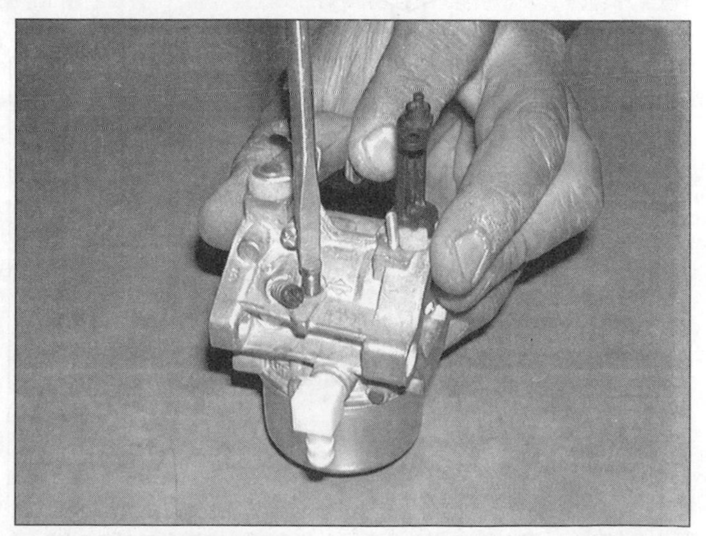

5.32a Unscrew the main air jet . . .

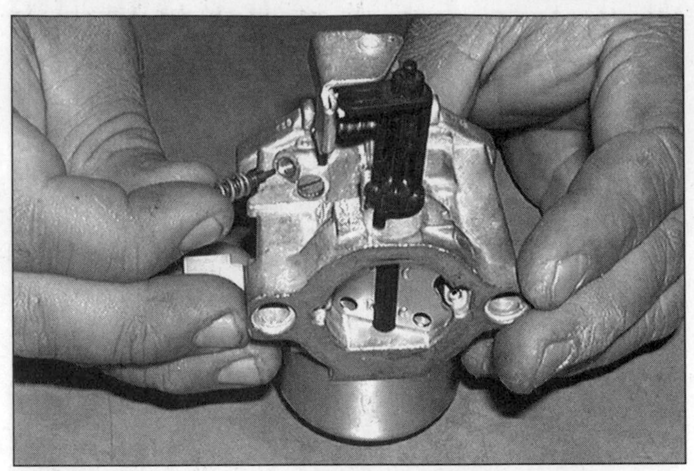

5.32b ... and the idle mixture screw and spring

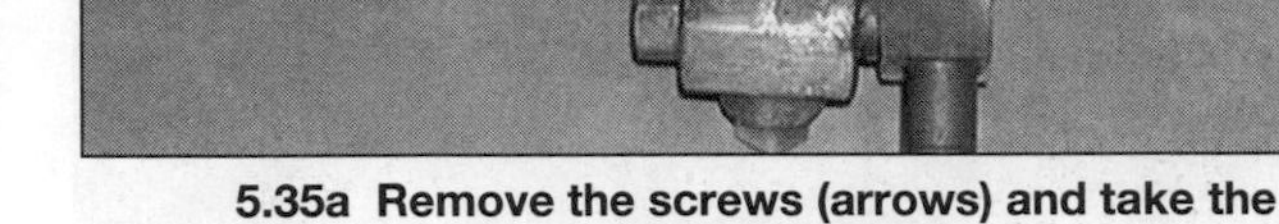

5.35a Remove the screws (arrows) and take the top cover off the carburetor

## Inspection and assembly

33 Procedures are the same as for the two-piece Flo-jet, described earlier.

34 If the carburetor has a solenoid, test it by connecting the solenoid terminal to a 9-volt battery with a short length of wire (the battery must be in new or nearly new condition). Connect the solenoid body to the battery negative terminal. The solenoid plunger should pull in when the battery is connected and extend when the battery is disconnected. If the solenoid doesn't retract or extend when it should, replace it.

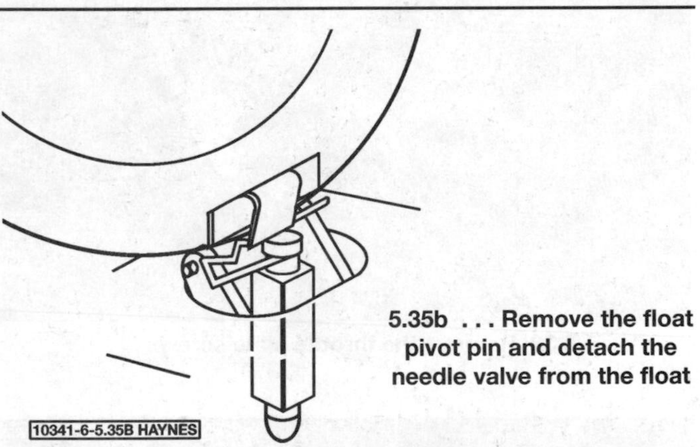


5.35b ... Remove the float pivot pin and detach the needle valve from the float

# Carburetor (L-head twin-cylinder engines)

35 To disassemble the carburetor used on these models, refer to the accompanying illustrations **(see illustrations 5.35a through 5.35i)**. Refer to the procedures described for the Flo-jet carburetor for cleaning and inspection. Reassemble the carburetor in the reverse order of disassembly.

# Carburetor - OHV V-twin engines

## Disassembly

38 Remove the upper body screws, lift off the upper body and remove the gasket **(see illustration 5.38)**. The emulsion tube will come away with the upper body.

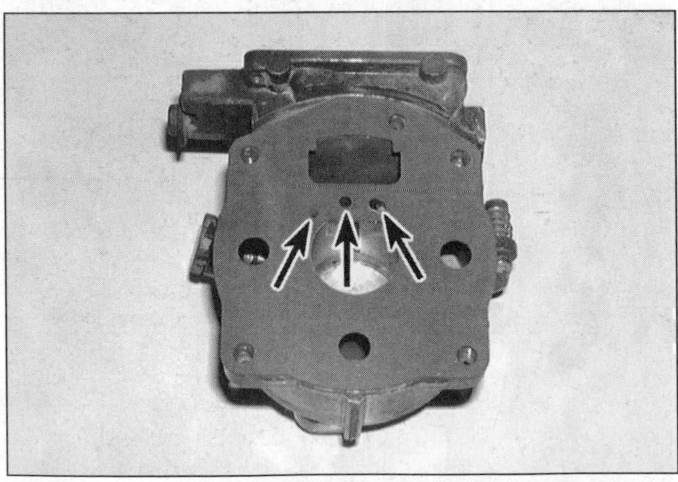

5.35c ... Remove the float chamber gasket; make sure the holes (arrows) line up on installation

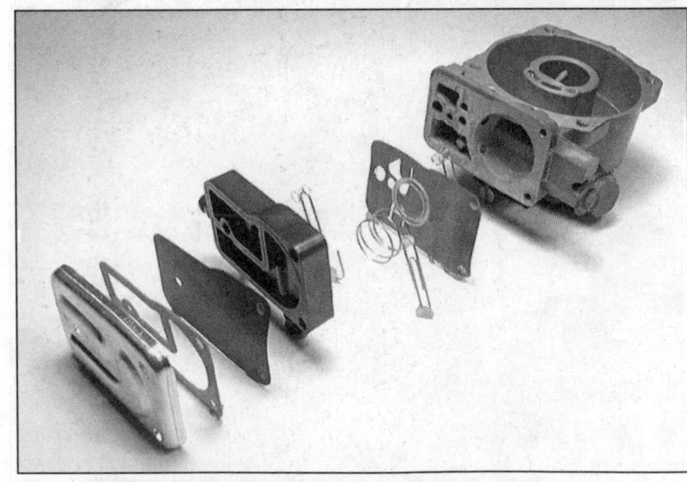

5.35d Remove the fuel pump components from the side of the carburetor

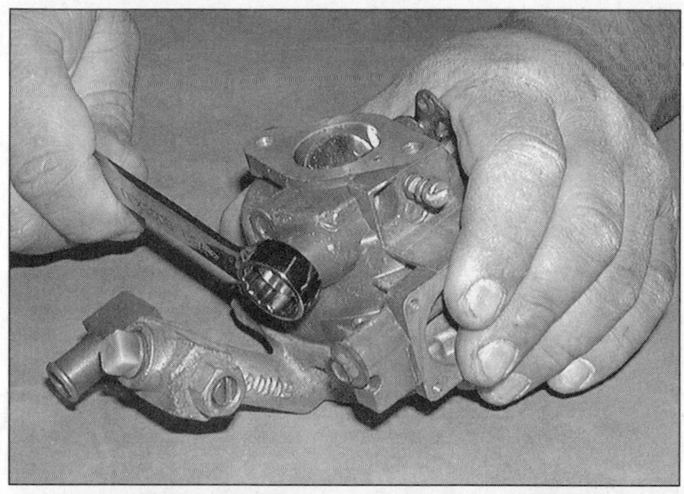

5.35e Remove the plug or solenoid from the carburetor body

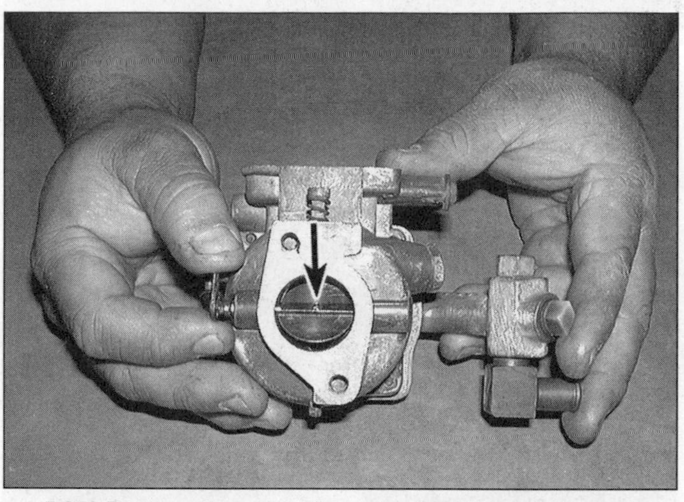

5.35f Remove the screw from the throttle plate (arrow)

5.35g Remove the throttle plate from the shaft

5.35h Pull the shaft out of the body and inspect the seal (remove the choke shaft in the same way if necessary)

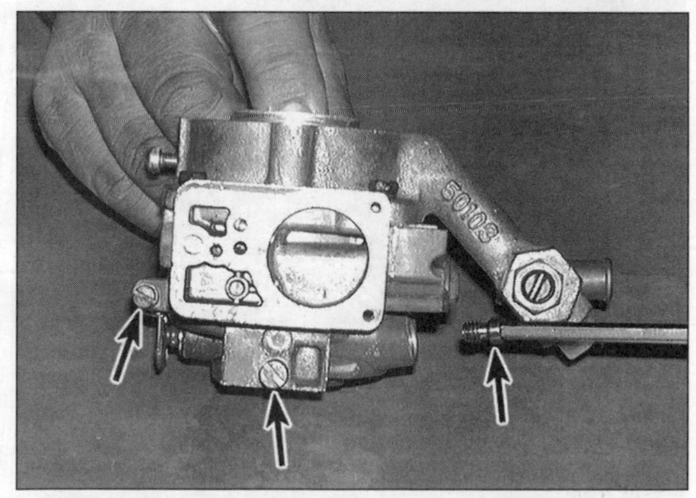


5.35i Remove the idle speed screw (left) and idle mixture screw (center); unscrew the fixed jet (right) with a 3/16-inch hex wrench

5.38 Loosen the cover screws evenly, then lift off the cover and emulsion tube

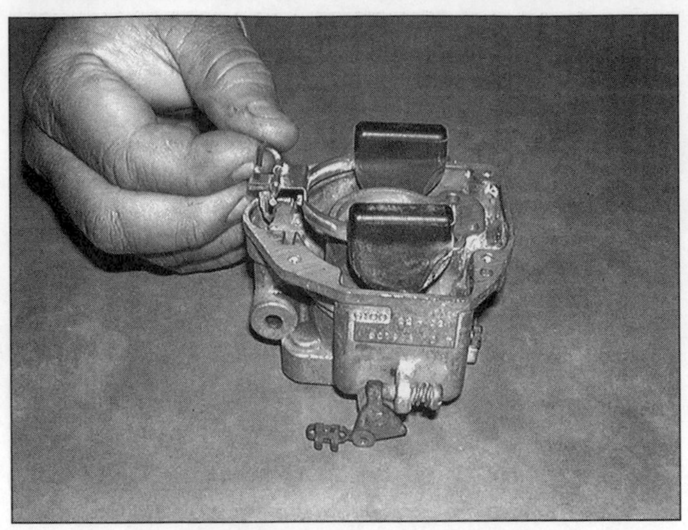
5.39 Lift off the float hinge and float

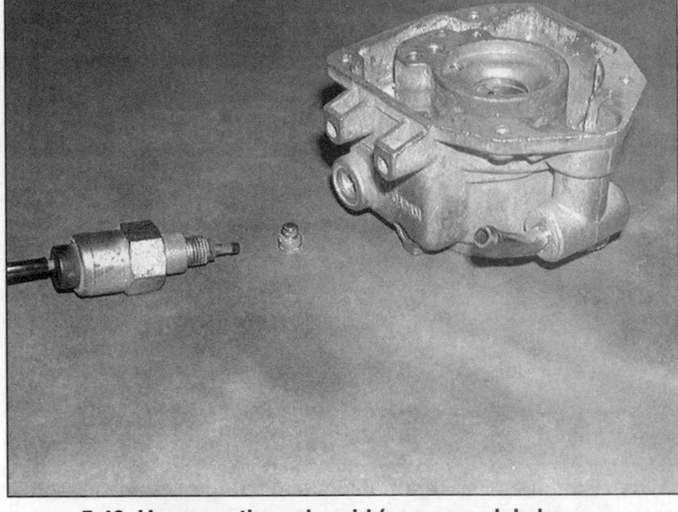
5.40 Unscrew the solenoid (some models have a plug instead), then remove the main jet

39 Lift the float hinge away from the lower body, together with the float and needle valve **(see illustration 5.39)**.

40 Unscrew the solenoid (if equipped) from the carburetor body, then unscrew the main jet **(see illustration 5.40)**. If the carburetor doesn't have a solenoid, there will be a threaded plug and washer in the hole.

41 Remove the idle mixture screw and spring **(see illustration 5.41)**.

42 Unscrew the idle jet (if equipped) **(see illustration 5.42)**. Some models have a nozzle in this hole.

43 Remove the throttle valve screw and take off the throttle valve **(see illustration 5.43)**.

44 Remove the choke plate screws and the choke plate **(see illustration 5.44)**. Pull the choke shaft out of the bore.

45 Drill-out the Welsh plug and pry it out of its bore with a small punch **(see illustration 5.45)**.

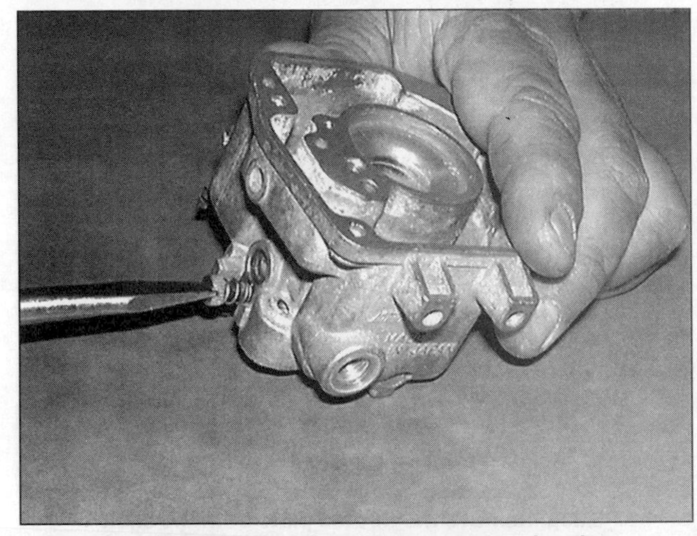
5.41 Remove the idle mixture screw and spring

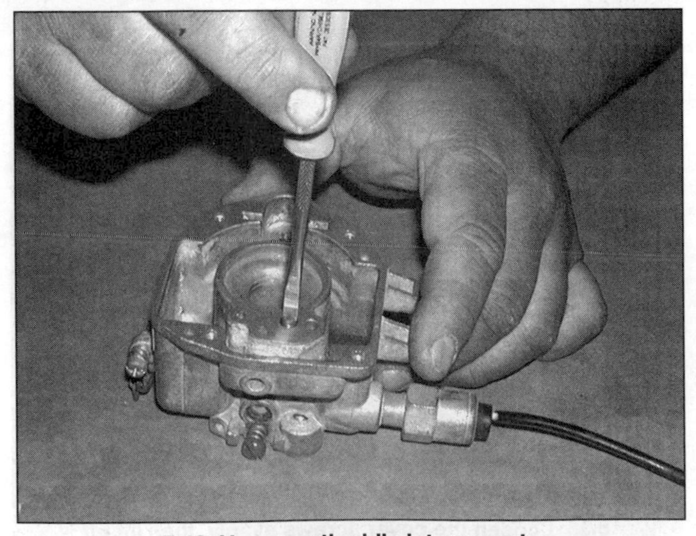
5.42 Unscrew the idle jet or nozzle

5.43 Remove the throttle plate screw, take off the plate and pull out the shaft

## Inspection

46 Slip the throttle shaft back into its bore. Place a 1/4-inch-thick flat surface, such as a piece of metal, on a flat bench top. Place the carburetor on the 1/4-inch-thick surface with the throttle lever above the bench top **(see illustration 5.46)**. Try to slip a 0.010-inch feeler gauge between the throttle lever and the bench top. If it will fit, there's too much play in the throttle shaft. Remove the throttle shaft and check it for worn spots where it rises in the bushings. If the throttle shaft is worn, it can be replaced. If the bushings are worn, the carburetor will have to be replaced.

47 The remainder of inspection is basically the same as for two-piece Flo-jet carburetors, described earlier.

48 Assembly is the reverse of removal and is generally the same as for two-piece Flo-jet carburetors.

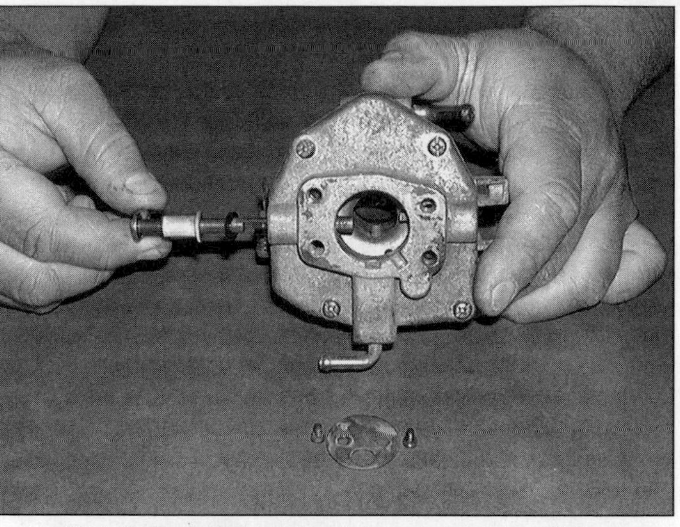

**5.44 Remove the choke plate screws, take off the plate and pull out the shaft**

# 6 Carburetor and governor adjustment

1 When making carburetor adjustments, the air cleaner must be in place and the fuel tank must be at least half full. For idle speed and mixture specifications, refer to Section 10.

## Carburetor

### Two-piece Flo-jet carburetors

2 Run the engine for five minutes to warm it up.

3 Place the speed control lever (if equipped) in the Fast position.

4 Turn the main jet needle valve (located on the underside of the float bowl) slowly inward until engine speed drops, then back it out slowly until the engine runs roughly.

5 Turn the screw back in to a point halfway between the points in Step 4.

6 Turn the throttle fully counterclockwise by hand. While holding it against the stop, set idle speed to the value listed in this Chapter's Specifications.

7 While still holding the throttle counterclockwise against the stop, adjust the idle mixture screw the same way you adjusted the main jet needle valve in Step 4. Recheck the idle speed and adjust it if necessary.

8 Release the throttle and check engine acceleration. If it won't accelerate smoothly, repeat the adjustment, trying a slightly richer mixture.

### LMT carburetors

9 Turn the idle mixture adjusting screw clockwise until it seats lightly, then back it out 1-1/2 turns. Run the engine for five minutes to warm it up.

10 Place the throttle control lever in the Slow position, then turn the idle speed adjusting screw until 1750 rpm is obtained.

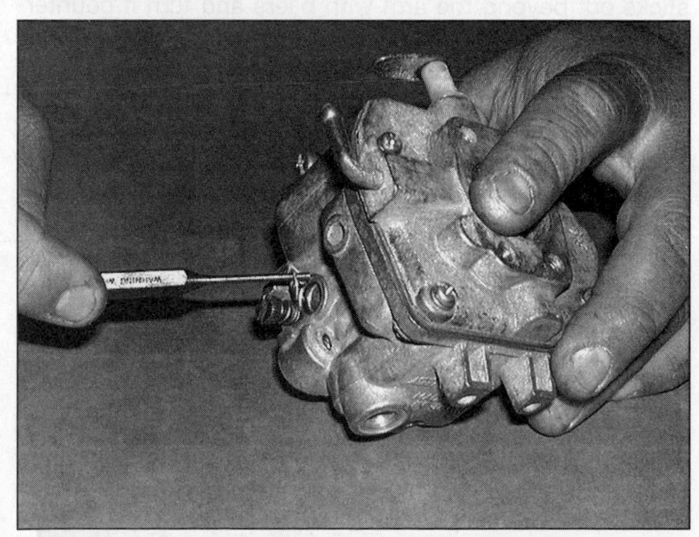

**5.45 Remove the Welsh plug with a small punch**

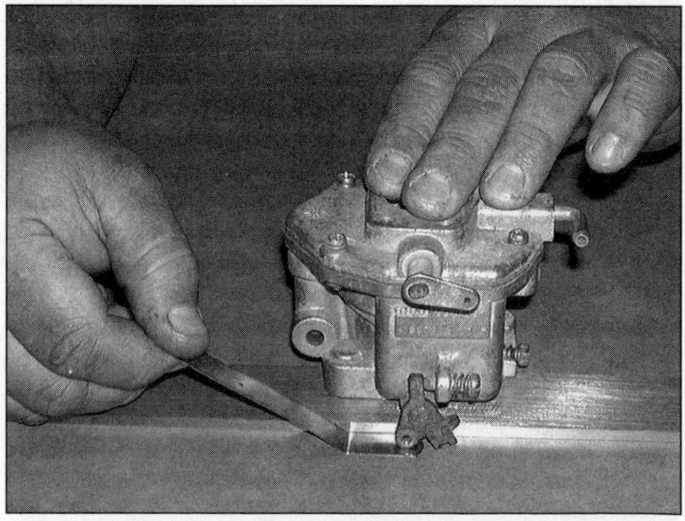

**5.46 Measure throttle shaft play with a feeler gauge**

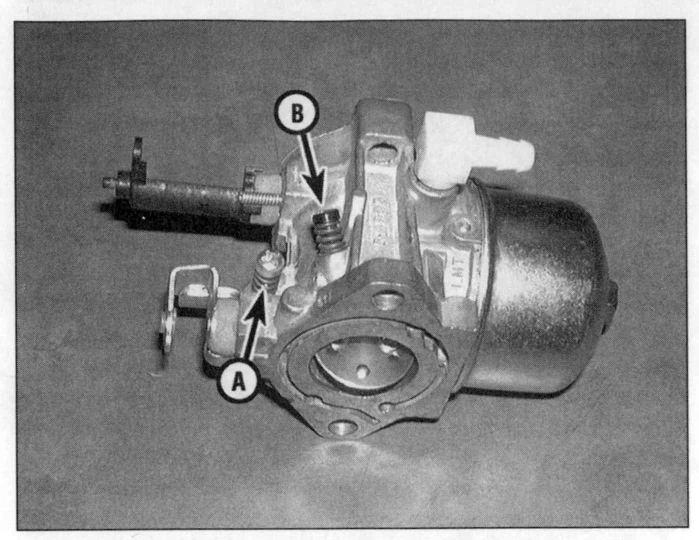


**6.11 Carburetor adjustment screws (LMT shown)**

*A Idle speed screw*
*B Idle mixture screw*

11 Slowly turn the idle mixture adjusting screw in until the engine begins to slow down **(see illustration 6.11)**.
12 Slowly turn the screw out until the engine begins to slow down.
13 Adjust the screw until it's halfway between the two points and the engine runs smoothly.

## Twin II carburetor

14 Turn the idle mixture adjusting screw clockwise until it seats lightly, then back it out 1-1/2 turns.
15 If the carburetor has a high speed mixture adjusting screw, back it out 1-1/2 turns from the seated position.
16 Run the engine for five minutes to warm it up.
17 Place the speed control lever in the idle position. Hold the throttle lever against the stop and set idle speed to 1300 rpm.
18 Turn the idle mixture adjusting screw in until the engine begins to slow down.
19 Slowly turn the screw out until the engine begins to run unevenly.
20 Adjust the screw until it's halfway between the two points and the engine runs smoothly.
21 Hold the throttle lever against the stop and set idle speed to 1000 rpm.
22 Release the throttle. Governed idle speed should be 1300 rpm.
23 If the carburetor has a high speed mixture screw, adjust it as described in Steps 18 through 20 with the throttle in the Fast position.
24 Move the speed control lever from SLOW to FAST - the engine should accelerate without hesitating or sputtering. If it doesn't, turn the high speed screw out in small increments.

## Carburetor (OHV V-twin engines)

25 Run the engine for five minutes to warm it up.
26 Place the speed control lever in the idle position. Hold the throttle lever against the stop and set idle speed to 1750 rpm.
27 Turn the idle mixture adjusting screw in until the engine begins to slow down.
28 Slowly turn the screw out until the engine begins to run unevenly.
29 Adjust the screw until it's halfway between the two points and the engine runs smoothly.
30 If the carburetor is equipped with a limiter cap, install it with its tang halfway between the stops.
31 If the engine has a governed idle, reset idle speed to 1200 rpm.

# Governor

32 It is best not to change the relationship of any linkages or other components in the governor system, unless a worn or broken component is to be replaced. If the engine is being overhauled, do not change any adjustments. If the governor was working suitably before the overhaul, and none of the connections are altered, it will continue to work properly. While governor adjustment should be made with a small-engine tachometer to avoid setting the engine to a higher rpm than it is designed for, some basic adjustments can be made by the home mechanic.
33 On most engines, a governor shaft sticks through the side of the crankcase, while the governor mechanism is protected inside the crankcase. An arm attaches to this shaft with a clamp-bolt/nut. The basic initial adjustment on begins with loosening the nut/bolt where the arm attaches to the shaft. Pull the arm back away from the carburetor to its limit and hold it there. Grip the end of the shaft that sticks out beyond the arm with pliers and turn it counter-clockwise to its limit. Tighten the bolt/nut on the arm while the arm and shaft are in this relationship.
34 Sensitivity is the other common adjustment for governors. If the engine seems like it is hunting for consistent speed as the load changes, the sensitivity is too sensitive and must be lowered. If the engine speed drops too much when the engine encounters a load, the governor adjustment isn't sensitive enough.
35 The governor sensitivity is adjusted by moving the position of the governor spring, where it hooks into the lower portion of the governor arm. A series of holes are provided. To increase the sensitivity, raise the spring into the next higher hole in the arm. To decrease sensitivity, hook the spring into a hole lower than its original position. The center of this arm is the factory position, but varying equipment and loads are accommodated by the extra adjustment holes.

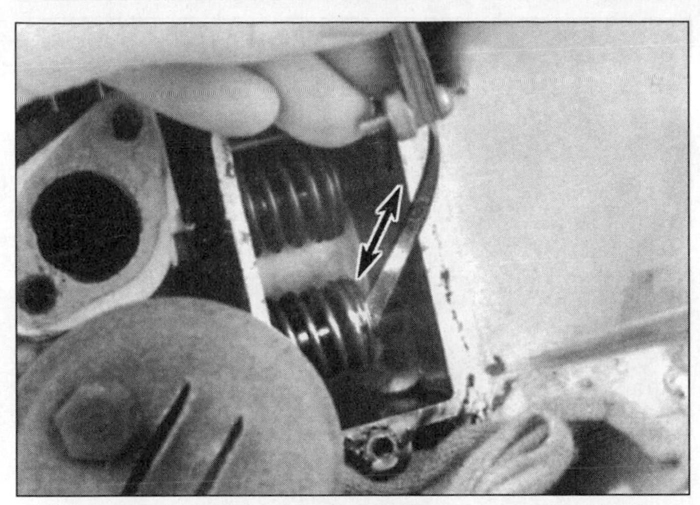

7.8 If the clearance is correct, the feeler gauge will fit between the valve stem and tappet with a slight drag

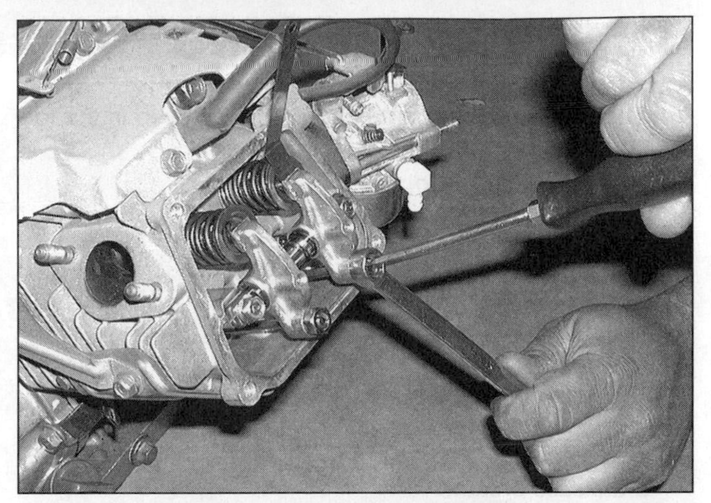

7.14 Loosen the locknut and turn the adjusting screw to adjust the valve clearance

# 7 Valve clearance check and adjustment

1 Correct valve tappet clearance is essential for efficient fuel use, easy starting, maximum power output, prevention of overheating and smooth engine operation. It also ensures the valves will last as long as possible.

## L-head engines

2 When the tappet is closed, clearance should exist between the end of the valve stem and the tappet. The clearance is not adjustable; if it's incorrect, the valves will have to be removed and the stem ends ground down carefully and lapped to provide more clearance. This is a major job, covered in Section 8 and Chapter 5. If the clearances are too great, new valves will have to be installed (again, a major repair procedure).
3 A feeler gauge with a thickness equal to the valve clearance will be needed for the measurement. Valve clearances are listed in this Chapter's Specifications.
4 Disconnect the wire from the spark plug and ground it to the engine.
5 Remove the bolts and detach the tappet cover plate or the crankcase breather assembly to provide access to the tappets.
6 Turn the crankshaft by hand and watch the valves to see if either one sticks in its guide.
7 Turn the crankshaft until the intake valve is wide open, then turn it another 360-degrees (one complete turn). This will ensure that the valves are completely closed for the clearance check.
8 Select a feeler gauge thickness equal to the specified valve clearance and slip it between the valve stem end and the tappet **(see illustration 7.8)**.
9 If the feeler gauge can be moved back-and-forth with a slight drag, the clearance is correct. If it's loose, the clearance is excessive; if it's tight (watch the valve to see if it's forced open slightly), the clearance is too small.
10 If the clearance is incorrect, refer to Section 8 and Chapter 5 for valve service procedures.
11 Reinstall all removed components.

## Overhead Valve (OHV) engines

12 Detach the cylinder head cover from the engine (see Chapter 9).
13 Remove the spark plug and place your thumb over the plug hole, then slowly turn the crankshaft (using the recoil starter, if equipped) until you feel pressure building up in the cylinder. Use a flashlight to look into the spark plug hole and see if the piston is at the top of its stroke. Continue to turn the crankshaft until it is. If it's too hard to see into the spark plug hole, insert a piece of stiff wire or a long screwdriver blade against the piston and use it to gauge piston movement. **Note 1:** *Be sure to check for compression buildup as you turn the engine. The piston also comes to top dead center on its exhaust stroke.*
**Note 2:** *The piston must be exactly at top dead center or the compression release may open the exhaust valve part way and give an inaccurate reading.*
14 Select a feeler gauge thickness equal to the specified valve clearance and slip it between the valve stem end and the rocker arm **(see illustration 7.14)**.
15 If the feeler gauge can be moved back-and-forth with a slight drag, the clearance is correct. If it's loose, the clearance is excessive; if it's tight (watch the valve to see if it's forced open slightly), the clearance is too small.
16 To adjust the clearance, loosen the rocker arm locknut and turn the pivot in or out as required.
17 Hold the pivot with a wrench and tighten the locknut securely, then recheck the clearance.
18 Reinstall all removed components.

# 8 L-head engines

The engine components should be removed in the following general order:

*Engine cover (if used)*
*Cooling shroud/recoil starter*
*Carburetor/fuel tank*
*Muffler*
*Cylinder head*
*Small shroud around cylinder fins (if used)*
*Flywheel*
*Flywheel brake components (if equipped)*
*Ignition components*
*Intake manifold*
*Crankcase breather*
*Oil sump (with balancer gears and counterweights if equipped)*
*Oil slinger/governor*
*Crankshaft*
*Camshaft*
*Tappets*
*Piston/connecting rod assembly*
*Valves*

**Note:** *These procedures refer to the "magneto side" and "drive side" of the engine. The magneto side of the engine is also called the flywheel side. This is the side with the flywheel, charging coil and ignition coil. The drive side of the engine is also called the power takeoff (PTO) side.*

# Single-cylinder engines

## Disassembly

1 For shroud/recoil starter, carburetor and muffler removal, refer to Chapters 4 and 5 as necessary. To remove the cylinder head, loosen the bolts in the opposite order of the tightening sequence, then carefully pry it off without damaging the gasket surfaces **(see illustrations 8.57a and 8.57b)**. The remaining components can be removed to

**8.2 Remove the screws (arrows) and the debris screen (if equipped)**

complete engine disassembly as described below. **Note:** *If the engine is cast-iron or has a cast-iron sleeve, use a ridge reamer to remove the carbon/wear ridge from the top of the cylinder bore after the cylinder head is off. Follow the manufacturer's instructions included with the tool. If the engine is made of aluminum, the ridge does not have to be removed.*

2 Remove the small bolts (arrows) and lift off the debris screen (if equipped) **(see illustration 8.2)**.

3 Hold the flywheel with the special tool and remove the starter clutch **(see illustration 8.3)** or the large nut and washer. If you don't have the special tool, be very careful not to damage the flywheel.

4 In most cases, a special puller (available from the engine manufacturer) will be needed for removing the flywheel **(see illustration 8.4)**. In this example, the puller body is slipped over the end of the crankshaft, the bolts are threaded into the flywheel holes (they may have to cut their own threads if the flywheel has never been removed before), the lower nuts are tightened against the flywheel and the upper nuts are tightened in 1/4-turn increments until the flywheel pops off the shaft taper.

5 Remove the flywheel key and place in a safe location (so it won't be lost).

**8.3 Hold the flywheel with a special tool or a strap wrench while loosening the nut**

**8.4 Use a puller to remove the flywheel**

**8.8a Remove the intake manifold and crankcase breather**

**8.8b Check the breather with a spark plug gauge**

6 Refer to Chapter 4 and remove the ignition points and plunger (if equipped), then note how the wires to the coil are routed (it's a very good idea to draw a simple sketch). Refer to Section 4 and detach the coil/spark plug wire assembly.

7 If it's still in place, remove the bolts and separate the intake manifold/tube from the engine. Remove the gasket and discard it.

8 Remove the mounting bolts and detach the crankcase breather assembly and gasket from the engine **(see illustrations 8.8a and 8.8b)**. Clean it with solvent, then try to slip a 0.045-inch wire-type spark plug gauge into the space between the fiber disc valve and the breather body (at several points) as shown here. DO NOT apply any force to the valve as this check is done. If the gauge fits into the space, install a new breather assembly when the engine is reassembled.

9 Use emery cloth to remove any rust and burrs from the drive end of the crankshaft so the bearing in the oil sump, crankcase cover or bearing support can slide over it **(see illustration 8.9)**.

10 The crankcase opening is covered by an oil sump, crankcase cover or bearing support, depending on model. Model 230000, 240000, 300000 and 320000 engines have a bearing support on the drive side as well as on the magneto side.

11 On aluminum model 250000 horizontal crankshaft engines, the crankcase cover contains two counterbalance weights. These are gear-driven by the crankshaft.

12 On cast iron engines with a horizontal crankshaft, there's a counterbalance weight mounted in the end cover on the magneto side of the engine. The weight rotates around the crankshaft and is driven by gears, which cause it to rotate in the opposite direction from the crankshaft. Model 300400 and 320400 cast iron engines have a second counterweight, mounted in a cover on the drive side of the engine.

13 On cast iron engines with external balancer gears, remove them (some models have the gears only on the magneto side of the engine; others have them on both sides of the engine). The small gear is secured by a snap-ring; the large gear is secured by a bolt.

14 Loosen the bolts that secure the oil sump or crankcase cover to the engine block in 1/4-turn increments to avoid warping the sump or cover, then remove them **(see illustration 8.14)**. If you're working on an engine with a cover on the drive side, loosen its bolts as well.

**8.9 Clean the end of the crankshaft before you remove the cover**

**8.14 Loosen the cover bolts evenly . . .**

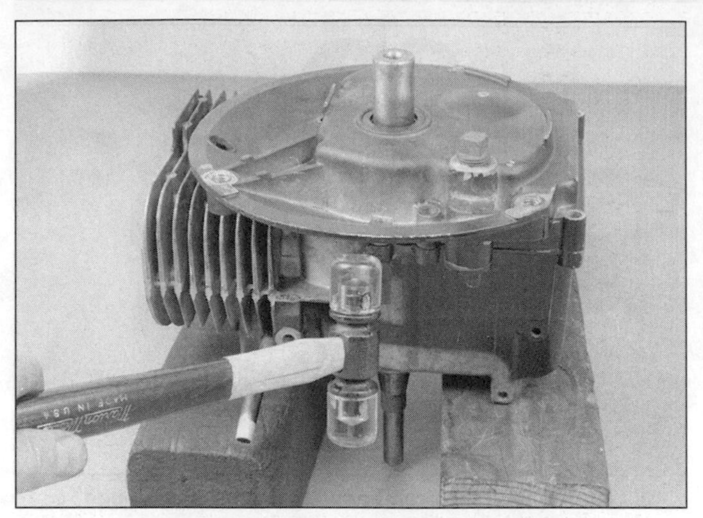

**8.15 . . . then carefully tap the cover off**

**8.16 Lift the oil slinger/governor off the camshaft (aluminum engines)**

15 Tap the sump or cover with a soft-face hammer to break the gasket seal, then separate it from the engine block and crankshaft **(see illustration 8.15)**. If it hangs up on the crankshaft, continue to tap on it with the hammer, but be very careful not to crack or distort it if it's made of aluminum. If thrust washers are installed on the crankshaft or camshaft, slide them off and set them aside.

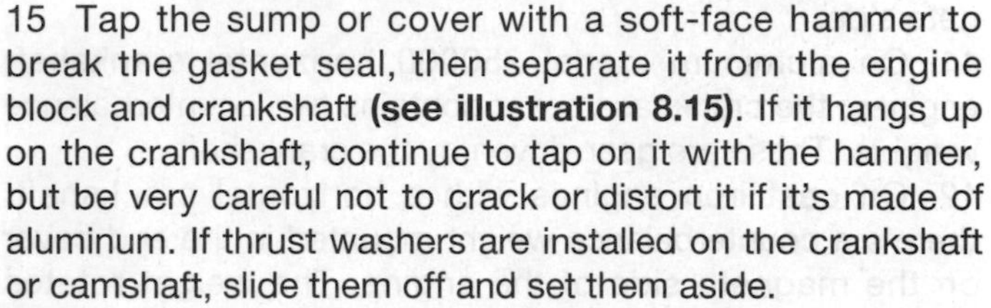

16 On aluminum block engines, lift the oil slinger/governor assembly off the end of the camshaft **(see illustration 8.16)**.

17 Mark the side of the connecting rod and cap that faces out and note how the oil dipper (if used) is installed **(see illustration 8.17)**. The parts must be reassembled in the exact same relationship to the crankshaft.

18 Flatten the locking tabs on the connecting rod bolts with a punch and hammer, then loosen the bolts in 1/4-turn increments until they can be removed by hand **(see illustrations 8.18a and 8.18b)**. Separate the cap from the connecting rod, move the end of the rod away from the crankshaft journal and push the piston/rod assembly out through the top of the bore.

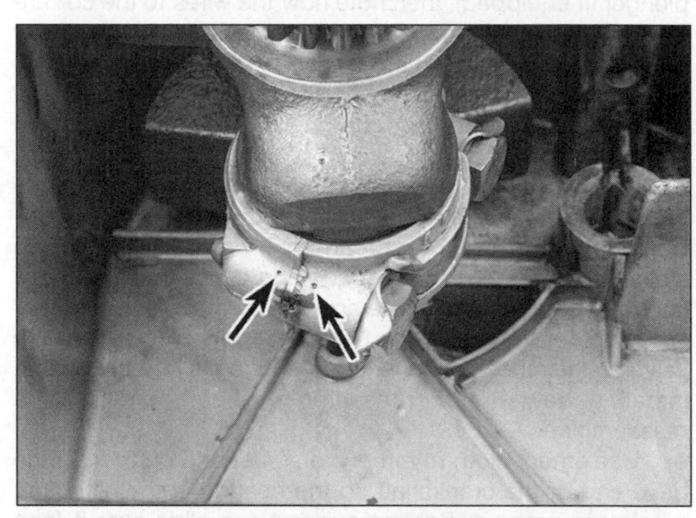

**8.17 Mark the outward-facing sides of the connecting rod and cap (arrows); if there's an oil dipper (not shown), mark it too**

**8.18a Bend back the lockwasher tabs . . .**

**8.18b . . . and remove the connecting rod bolts**

**8.19 Look for the camshaft and crankshaft alignment marks (arrows) - these are the marks used on plain bearing aluminum engines**

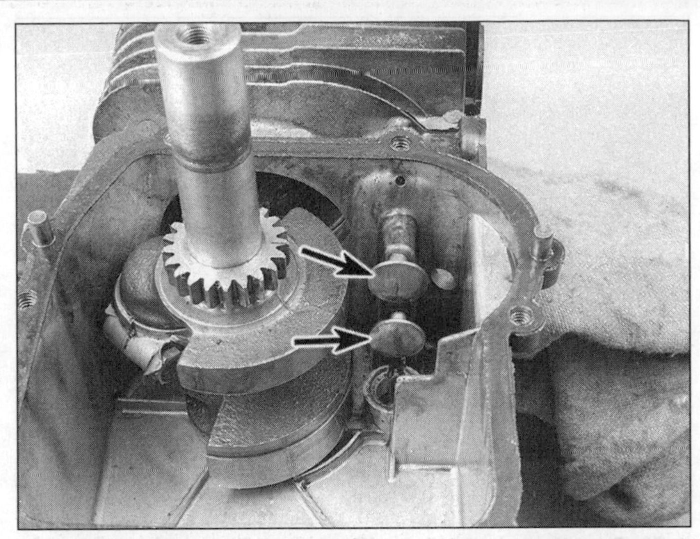

**8.24 With the camshaft removed, pull out the tappets (arrows)**

## Aluminum engines with plain bearings

19 On aluminum block engines, turn the crankshaft until the marks on the timing gears are aligned, then lift out the camshaft **(see illustration 8.19)**.

## Aluminum engines with ball bearings

20 The ball bearing on these models blocks the view of the crankshaft gear, so the crankshaft timing mark is on the counterweight. Align the punch mark on the camshaft gear with the punch mark on the crankshaft counterweight (the crankshaft mark will be visible between two of the teeth on the camshaft gear).

21 With the timing marks aligned, lift out the camshaft and crankshaft together.

## Cast iron engines

22 Align the timing marks on the camshaft and crankshaft **(see illustration 8.19)**. If there's a bearing support for the camshaft support shaft bolted to the magneto side of the engine, unbolt it.

23 Tap the camshaft support shaft out of the cylinder block from the drive side with a hammer and punch. On models with a support shaft plug in the magneto side of the engine, this will drive the plug out. Lift the crankshaft out, then the camshaft. If you're working on a vertical crankshaft engine, the oscillating counterbalance mechanism is removed together with the crankshaft.

24 Once the camshaft is removed, the tappets can be pulled from their bores **(see illustration 8.24)**. Store them in marked containers so they can be returned to their original locations.

25 Three methods have been used to hold valve spring retainers in place: Pins, slotted retainers (one per valve) and split-type keepers (two per valve). If the engine you're working on has pins or keepers, insert the valve spring compressor jaw between the retainer and the valve chamber wall. If it has slotted retainers, insert the compressor jaw between the spring and retainer and position the remaining jaw on the outside of the valve chamber **(see illustration 8.25)**.

26 Compress the intake valve spring with the special tool and remove the pin, keepers or retainer, then withdraw the valve through the top of the engine. Pull out the spring (and retainer if necessary), then repeat the procedure for the exhaust valve.

27 On some engines, the mechanical governor housing is bolted to the engine; on others, the mechanical governor is integrated into the oil slinger assembly that slips over the end of the camshaft. The lever is attached to the governor shaft with a roll pin or clamped to it with a bolt or bolt/nut. If the governor components are worn or damaged, install a new assembly and have the shaft bushing replaced by a small engine repair shop.

**8.25 Remove the valve keepers**

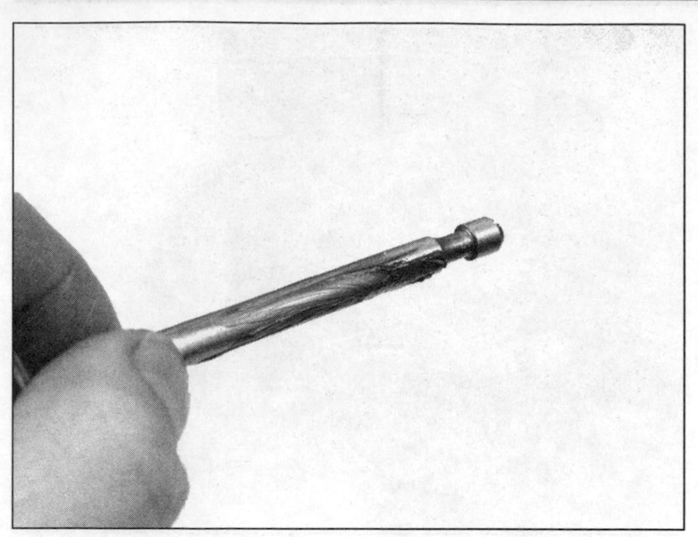

8.30 Coat the valve stems with engine oil or engine assembly lube before installing them

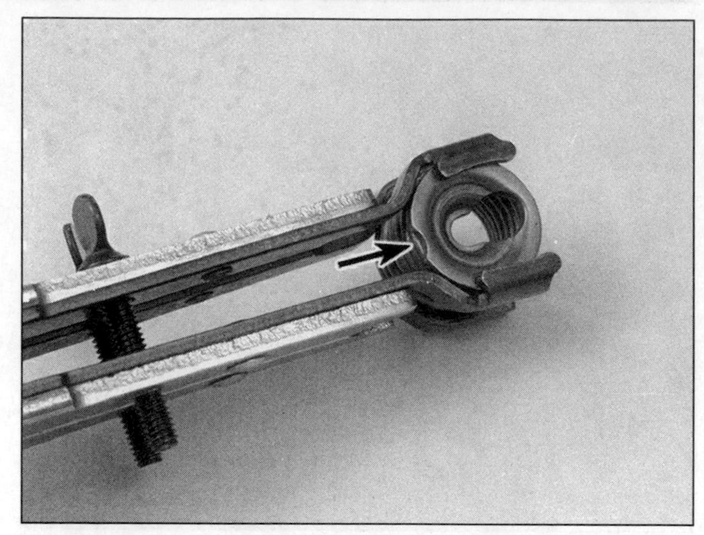

8.31 The retainer cutout (arrow) faces outward on installation

## Inspection of components

28 If you're working on a vertical crankshaft cast iron engine, disassemble the oscillating counterbalancer from the crankshaft. Two designs are used:

a) *If you're working on an early model, bend back the lockwashers tabs from the bolt heads and remove the bolts and spacers. Separate the weights (drive side and magneto side) and dowel pins and detach the link. Lay the parts in order so they can be reassembled correctly. Label the drive side and magneto side counterweights so they don't get mixed up.*

b) *Later models are basically the same as early ones, but have only one bolt (with no lockwasher or spacer) and one dowel pin.*

29 After the engine has been completely disassembled, refer to Chapter 5 for the cleaning, component inspection and valve lapping procedures. **Note:** *Special test equipment is needed to check the ignition coil/electronic ignition module. If you suspect the coil/module is causing ignition problems and the troubleshooting procedures in Chapter 3 and Section 4 don't solve the problem, have it checked by a small engine repair shop* Once you've inspected and serviced everything and purchased any necessary new parts, which should always include new gaskets and seals, reassembly can begin. Begin by reinstalling the PTO and mechanical governor components (if used) in the crankcase, then proceed as follows:

8.34 Lubricate the crankshaft seal before installing the cover

## Reassembly

30 Coat the intake valve stem with clean engine oil or engine assembly lube, then reinstall it in the block **(see illustration 8.30)**. Make sure it's returned to its original location.

31 Compress the spring with the retainer in place **(see illustration 8.31)**. The small cutout in the edge of the slotted retainer should face out to facilitate retainer installation. Pull the valve out enough to position the spring. Push the valve back in and install the pin/keepers or slotted retainer. Release the compressor and make sure the retainer is securely locked on the end of the valve. Repeat the procedure for the exhaust valve.

32 Lubricate the tappets and install them in their bores **(see illustration 8.24)**.

33 If you're working on a cast iron engine with a vertical crankshaft, reassemble the oscillating counterbalancer onto the crankshaft (see Step 28 above). The rounded side of the free end of the link (the end that fits over the pivot post in the crankcase) faces the drive side of the engine. Tighten the bolt(s) to the torque listed in this Chapter's Specifications. If the counterbalancer uses a lockplate, bend it against the bolts after tightening them.

34 Lubricate the crankshaft magneto side oil seal lip **(see illustration 8.34)**, the plain bearing (if applicable) and the

8.39 Compress the rings with a ring compressor

8.40 Coat the cylinder wall with oil

connecting rod journal with clean engine oil, then reposition the crankshaft in the crankcase. If a ball-bearing is used on the magneto side, lubricate it with clean engine oil.

35 Lubricate the camshaft lobes, journals and shaft bore with engine assembly lube.

**Note:** *On ball-bearing equipped aluminum block engines, the crankshaft and camshaft must be installed as an assembly. Align the timing marks on the camshaft gear and crankshaft - this is very important! The crankshaft timing mark is on the counterweight.*

36 On cast-iron engines with plain bearings, install the camshaft and support shaft before the crankshaft.

37 On cast-iron engines with ball bearings, position the camshaft gear in the crankcase recess, install the crankshaft, then install the camshaft/support shaft.

38 Check the timing marks on the camshaft and crankshaft to make sure they're aligned properly. Apply sealant to the cam support shaft hole plug and press or drive it into the opening in the flywheel side of the crankcase. Camshaft end play must be as specified (it should be okay unless the flywheel side cam bearing was replaced or a new camshaft was installed.)

39 Before installing the piston/connecting rod assembly, the cylinder bore must be perfectly clean and the top edge of the bore must be chamfered slightly so the rings don't catch on it. Position the piston ring end gaps 120-degrees apart (see Chapter 5). Lubricate the piston and rings with clean engine oil, then attach a ring compressor to the piston **(see illustration 8.39)**. Leave the skirt protruding about 1/4-inch. Tighten the compressor until the piston cannot be turned, then loosen it until the piston turns in the compressor with resistance.

40 Rotate the crankshaft until the connecting rod journal is at TDC (Top Dead Center - top of the stroke) and apply engine oil to the cylinder walls **(see illustration 8.40)**. If the piston has a notch in the top, it must face the magneto side of the engine. Make sure the mark you made on the rod will be facing out when the rod/piston assembly is in place. Gently insert the piston/connecting rod assembly into the cylinder and rest the bottom edge of the ring compressor on the engine block. Tap the top edge of the ring compressor to make sure it's contacting the block around its entire circumference.

41 Carefully tap on the top of the piston with the end of a wooden or plastic hammer handle while guiding the end of the connecting rod into place on the crankshaft journal **(see illustration 8.41)**. The piston rings may try to pop out just before entering the bore, so keep some pressure on the ring compressor. Work slowly - if any resistance is felt as the piston enters the cylinder, stop immediately. Find out what's hanging up and fix it before proceeding. Do not, for any reason, force the piston into the cylinder - you'll break a ring and/or the piston.

8.41 Tap the piston into the bore with a hammer handle

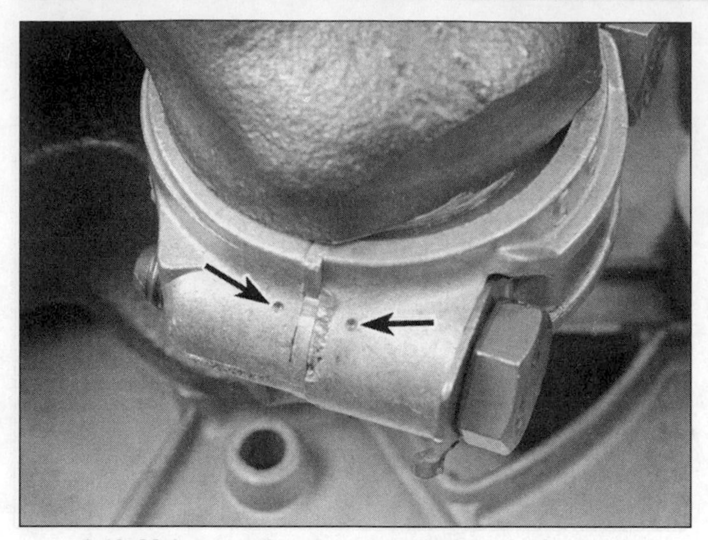
8.42 Make sure the connecting rod marks (arrows) both face outward

8.43 Tighten the bolts evenly to the specified torque

42 Install the connecting rod cap, a NEW lock plate, the oil dipper (if used) and the bolts **(see illustration 8.42)**. Make sure the marks you made on the rod and cap are aligned and facing the direction they were originally and the oil dipper is oriented correctly. **Note:** *Some replacement connecting rods are packaged with a thick washer under each bolt head - remove and discard them. If a lock plate is installed, one, two or no thin washers may be used. If no oil dipper is used, use two thin washers. If the oil dipper is held by one bolt, use one thin washer under the bolt not holding the dipper. If the dipper is attached by two bolts, don't use any washers.*

43 Tighten the bolts in three steps to the torque listed in this Chapter's Specifications **(see illustration 8.43)**. Temporarily install the camshaft (if not already in place) and turn the crankshaft through two complete revolutions to make sure the rod doesn't hit the cylinder or camshaft. If it does, the piston/connecting rod or camshaft is installed incorrectly.

44 If nothing binds or contacts anything, bend up the locking tabs to keep the connecting rod bolts from loosening **(see illustration 8.44)**.

45 Install the oil slinger/governor assembly on the end of the camshaft (aluminum block engines). Some engines also have a spring washer that must be slipped over the end of the camshaft after the oil slinger is in place.

46 Lubricate the crankshaft main bearing journal and the lip on the oil seal in the sump (or crankcase cover/bearing support) with clean engine oil or engine assembly lube **(see illustration 8.46)**.

47 Make sure the dowel pins are in place, then position a new gasket on the crankcase (the dowel pins will hold it in place) **(see illustration 8.47)**. A very thin coat of non-hardening sealant can be used if desired. Carefully lower the oil sump (or crankcase cover/bearing support) into place over the end of the crankshaft until it seats on the crankcase. If a

8.44 Bend up the locktabs to secure the bolts

8.46 Lubricate the cover seal

8.47 Install a new cover gasket over the dowels

8.48 Tighten the cover bolts evenly, in a criss-cross pattern

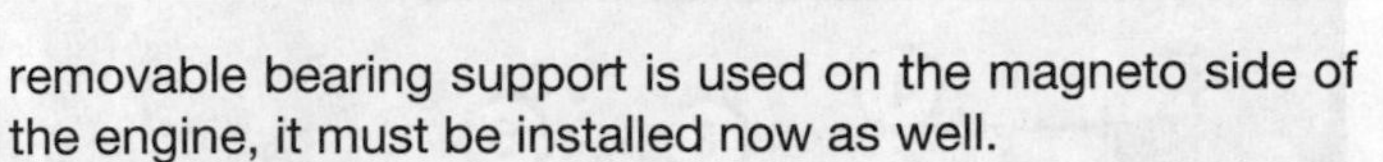

removable bearing support is used on the magneto side of the engine, it must be installed now as well.

48 Install the bolts and tighten them to the torque listed in this Chapter's Specifications. Follow a criss-cross pattern and work up to the final torque in three equal steps to avoid warping the oil sump or cover **(see illustration 8.48)**.

49 Measure crankshaft end play with a dial indicator **(see illustration 8.49)**. The crankshaft end play must be checked and adjusted as follows:

50 On aluminum engines, it must be 0.002 to 0.008-inch with a 0.015-inch thick gasket (standard). If the end play is less than specified, use additional gaskets in various combinations to correct it (they're available in 0.005, 0.009 and 0.015-inch thickness). If the end play is greater than specified, a thrust washer is available for installation over the drive end of the crankshaft to reduce play (additional or different thickness gaskets may be needed along with the thrust washer). **Note:** *The thrust washer cannot be used on engines with two ball-bearings - replace worn parts instead.*

8.49 Measure crankshaft end play with a dial indicator

51 On cast-iron engines (models 230000 and 240000), the crankshaft end play must be 0.002 to 0.008-inch with a 0.015-inch thick gasket (standard) under the bearing support plate. If the end play is less than specified, use additional gaskets in various combinations to obtain the correct end play (they are available in various thickness).
**Note:** *The thrust washer cannot be used on engines with two ball-bearings - replace worn parts instead.*

52 On cast iron engines with a bearing support at each end (models 300000 and 320000), end play doesn't need to be checked unless the crankshaft or one of the bearing supports has been replaced. Adjust it if necessary by adding or removing shims in various thickness.

53 Install the crankcase breather and intake manifold. Use new gaskets and tighten the bolts securely.

54 Install the ignition coil/spark plug wire assembly and pneumatic governor vane (if applicable), but don't tighten the bolts completely - just snug them up. The ignition coil bolt holes are slotted; refer to Section 4 and adjust the gap between the flywheel and coil before snugging up the bolts. Be sure to reroute the wires from the ignition coil properly.

55 Install the ignition points (if equipped), then make sure the tapered portion of the crankshaft and the inside of the flywheel hub are clean and free of burrs. Position the Woodruff key in the crankshaft keyway and install the flywheel. Install the washer and tighten the large nut or the starter clutch to the torque listed in this Chapter's Specifications. Install the debris screen (if equipped) and tighten the screws.

56 If a flywheel brake is used, reinstall it now.

57 Install the cylinder head, using a new gasket. Following the recommended tightening sequence, tighten the head bolts evenly, in several stages, to the torque listed in this Chapter's Specifications **(see illustrations 8.57a and 8.57b)**.

58 To install the remaining components, refer to Chapters 4 and 5 as necessary. **Caution:** *Be sure to fill the crankcase to the correct level with the specified oil before attempting to start the engine.*

**8.57a Cylinder head TIGHTENING sequence - aluminum engines**

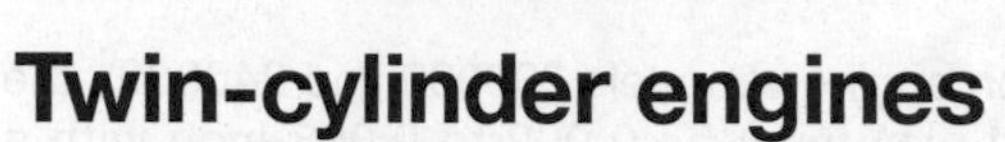

# Twin-cylinder engines

59 Disassembly and assembly procedures for twin-cylinder L-head models are basically the same as for singles, with the following additions:

a) *There's a separate torque sequence for each cylinder head **(see illustrations 8.59a and 8.59b).***

b) *The governor gear and oil slinger on vertical crankshaft models are mounted inside the crankcase cover **(see illustrations 8.59c and 8.59d).***

c) *The crankshaft and camshaft timing marks on plain bearing models are visible when the crankcase cover is removed **(see illustration 8.59e).** On ball bearing models, the ball bearing blocks the view of the crankshaft gear, so the crankshaft mark is on the crankshaft counterweight instead of the gear.*

d) *When assembling the connecting rods, position the oil spray hole for no. 1 cylinder away from the camshaft and the oil spray hole for no. 2 cylinder toward the camshaft.*

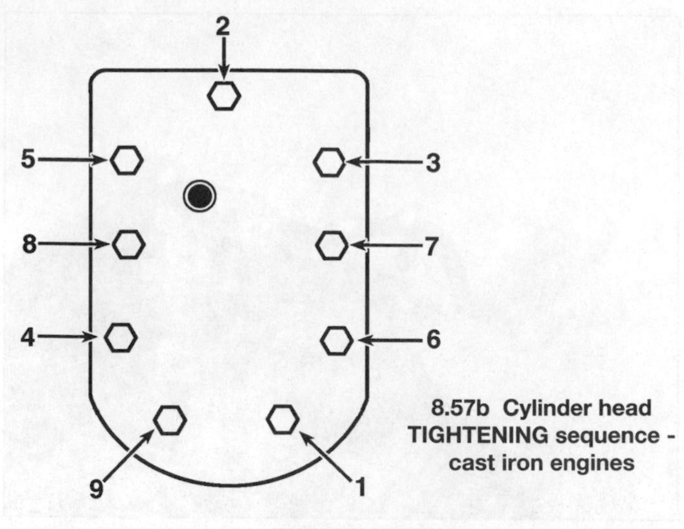


**8.57b Cylinder head TIGHTENING sequence - cast iron engines**

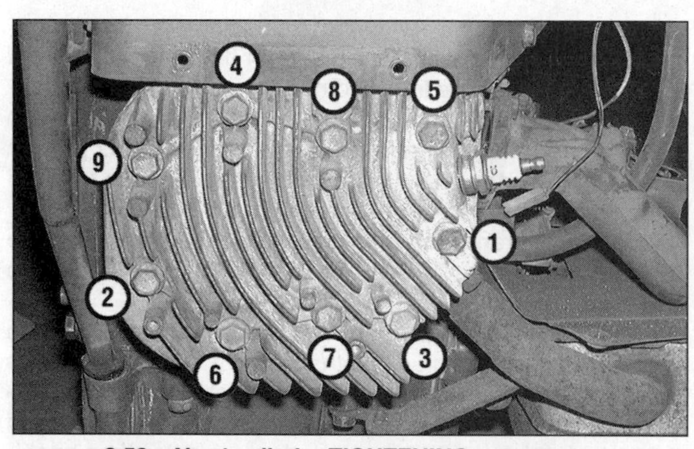


**8.59a No. 1 cylinder TIGHTENING sequence - L-head twin engines**

e) *To disassemble and inspect the oil pump (if equipped), remove the cover screws and lift off the cover **(see illustrations 8.59f and 8.59g).** The rotors can then be lifted out.*

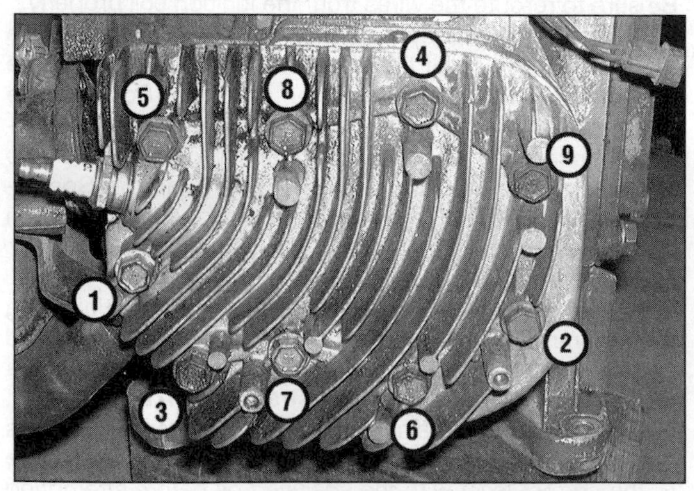


**8.59b No. 2 cylinder TIGHTENING sequence - L-head twin engines**

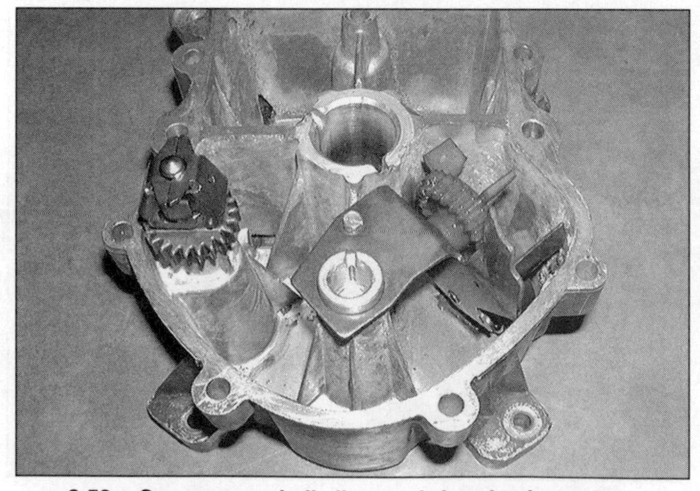

**8.59c Governor and oil slinger - L-head twin engines**

8.59d Governor gears - L-head twin engines

8.59e Camshaft and crankshaft timing marks (arrows) - L-head twin engines (plain bearings)

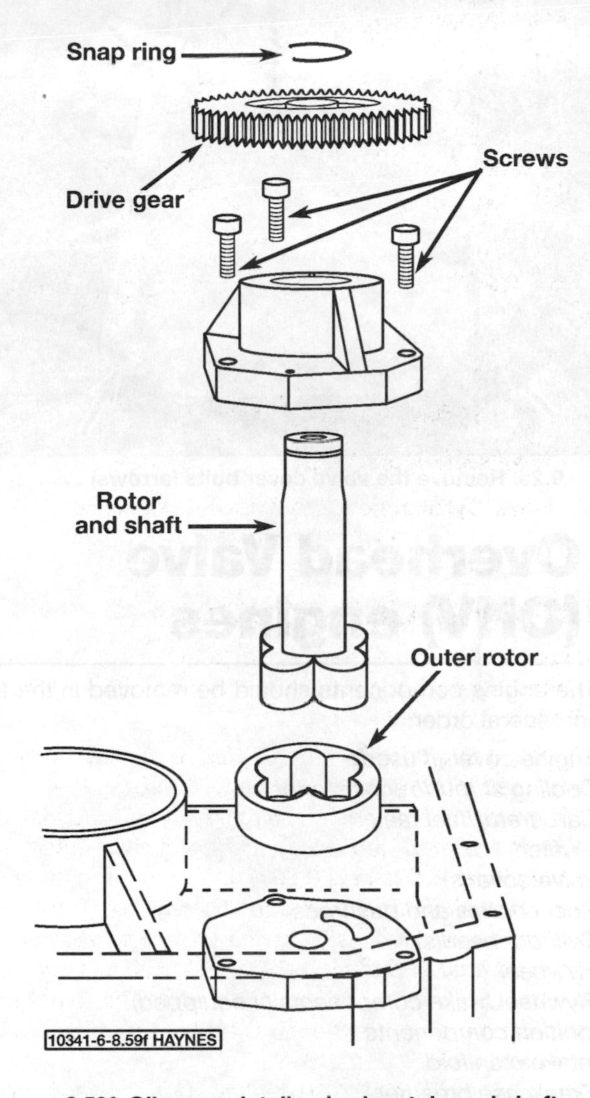


8.59f Oil pump details - horizontal crankshaft L-head twin engines

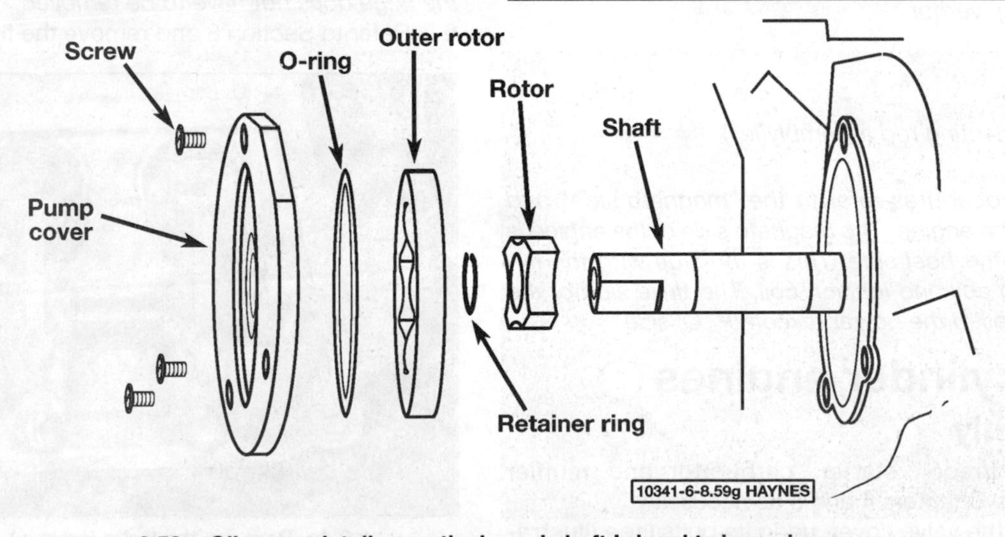


8.59g Oil pump details - vertical crankshaft L-head twin engines

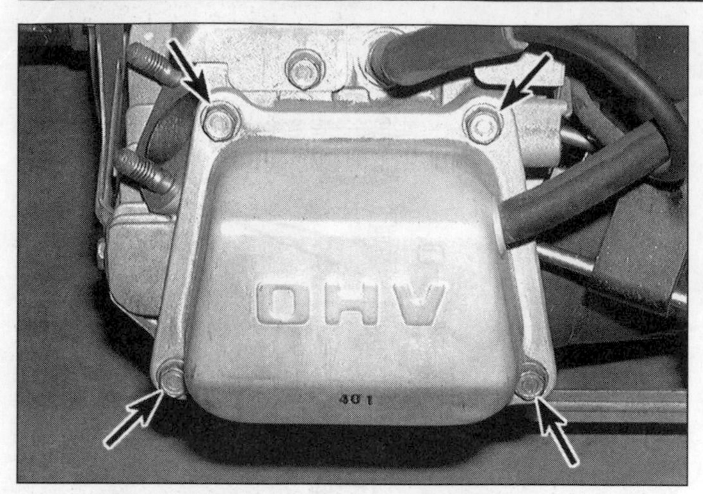

9.2a Remove the valve cover bolts (arrows) . . .

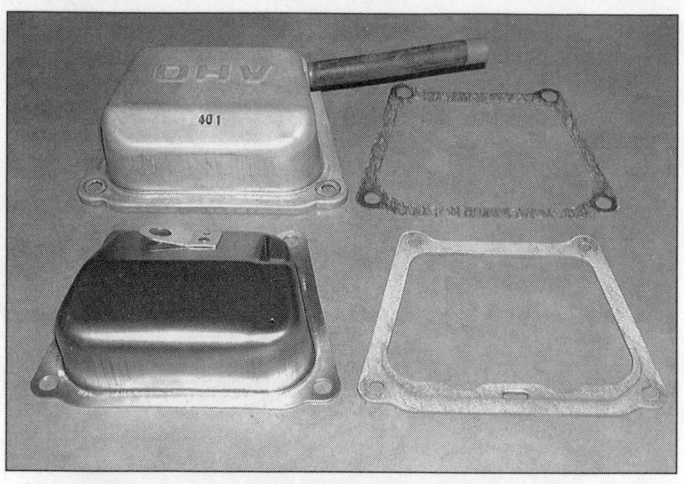

9.2b . . . and lift off the cover, gasket, breather and second gasket

# 9 Overhead Valve (OHV) engines

The engine components should be removed in the following general order:

*Engine cover (if used)*
*Cooling shroud/recoil starter*
*Carburetor/fuel tank*
*Muffler*
*Valve cover(s)*
*Rocker arms and pushrods*
*Cylinder head(s)*
*Flywheel*
*Flywheel brake components (if equipped)*
*Ignition components*
*Intake manifold*
*Crankcase breather*
*Oil sump (with balancer gears and counterweights if equipped)*
*Oil slinger/governor*
*Crankshaft*
*Camshaft*
*Tappets*
*Piston/connecting rod assembly(ies)*
*Valves*

**Note:** *These procedures refer to the "magneto side" and "drive side" of the engine. The magneto side of the engine is also called the flywheel side. This is the side with the flywheel, charging coil and ignition coil. The drive side of the engine is also called the power takeoff (PTO) side.*

## Single-cylinder engines

### Disassembly

1 For shroud/recoil starter, carburetor and muffler removal, refer to Chapters 4 and 5 as necessary.

2 To remove the valve cover, undo its bolts **(see illustration 9.2a)**. Take off the cover, gasket, breather and second gasket **(see illustration 9.2b)**. **Note:** *If the cover is stuck, tap it from the side with a rubber mallet or dead blow hammer. Don't pry it or the gasket surfaces may be damaged.*

3 On all models except 235400 and 245400, remove the rocker nut from above each rocker arm, then lift the rocker arms off the pivots.

4 On 235400 and 245400 engines, remove the rocker shaft bolts **(see illustration 9.4a)**. Lift off the rocker assembly, remove the snap-ring from each end and take the rocker arms off the shaft **(see illustration 9.4b)**.

5 To remove the cylinder head, loosen the bolts in the opposite order of the tightening sequence **(see illustrations 9.53a, 9.53b, and 9.53c)**. Tap the head with a rubber mallet or dead blow hammer to free it, then lift it off and remove the gasket. Don't pry the head off or the gasket surfaces may be damaged. **Note:** *If the engine is cast-iron or has a cast-iron sleeve, use a ridge reamer to remove the carbon/wear ridge from the top of the cylinder bore after the cylinder head is off. Follow the manufacturer's instructions included with the tool. If the engine is made of aluminum, the ridge does not have to be removed.*

6 Refer to Section 8 and remove the flywheel and its key.

9.4a Remove the bolts (arrows) and lift off the rocker assembly

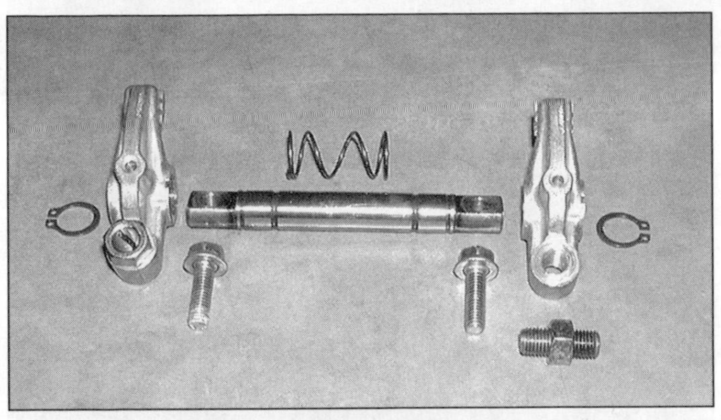

9.4b Rocker assembly details

9.11 Typical crankcase cover bolts (A) and governor crank (B)

7 Refer to Section 4 and detach the coil/spark plug wire assembly.

8 If it's still in place, remove the bolts and separate the intake manifold from the engine. Remove the gasket and discard it.

9 Remove the mounting bolts and detach the crankcase breather assembly and gasket from the engine. Clean it with solvent, then try to slip a 0.045-inch wire-type spark plug gauge into the space between the fiber disc valve and the breather body (at several points) **(see illustration 8.8b)**. DO NOT apply any force to the valve as this check is done. If the gauge fits into the space, install a new breather assembly when the engine is reassembled.

10 Use emery cloth to remove any rust and burrs from the drive end of the crankshaft so the bearing in the oil sump, crankcase cover or bearing support can slide over it **(see illustration 8.9)**.

11 The crankcase opening is covered by a crankcase cover. The shape and specific cover bolt locations differ from model to model, but in all cases, the bolts should be loosened in stages in a criss-cross pattern to prevent warping the cover **(see illustration 9.11)**. **Note:** *On some models, one of the bolts has a coat of sealant on the threads. Label this bolt, clean the threads and apply a new coat of non-hardening sealant on assembly.*

12 Tap the sump or cover with a soft-face hammer to break the gasket seal, then separate it from the engine block and crankshaft. If it hangs up on the crankshaft, continue to tap on it with the hammer, but be very careful not to crack or distort it if it's made of aluminum. If thrust washers are installed on the crankshaft or camshaft, slide them off and set them aside.

13 On 28E700, 28N700, 28P700, 28Q700 and 287700 engines, lift the oil slinger/governor assembly off the end of the camshaft **(see illustration 8.16)**.

14 Some horizontal crankshaft models use a counterbalance shaft, which resembles a camshaft, mounted in the engine block on the opposite side of the crankshaft from the camshaft. The counterbalance shaft is gear-driven in the opposite direction from the crankshaft to cancel out vibrations. Before removing the shaft, locate the alignment marks on camshaft and counterbalance shaft **(see illustration 9.14)**. Pull out the counterbalance shaft, then pull out the camshaft.

15 Once the camshaft is removed, the tappets can be pulled from their bores. Store them in marked containers so they can be returned to their original locations.

16 Mark the sides of the connecting rod and cap that face out **(see illustration 9.16)**. The parts must be reassembled in the exact same relationship to each other and to the crankshaft.

9.14 Align the timing marks on the crankshaft with those on the camshaft and the counterbalancer (if equipped)

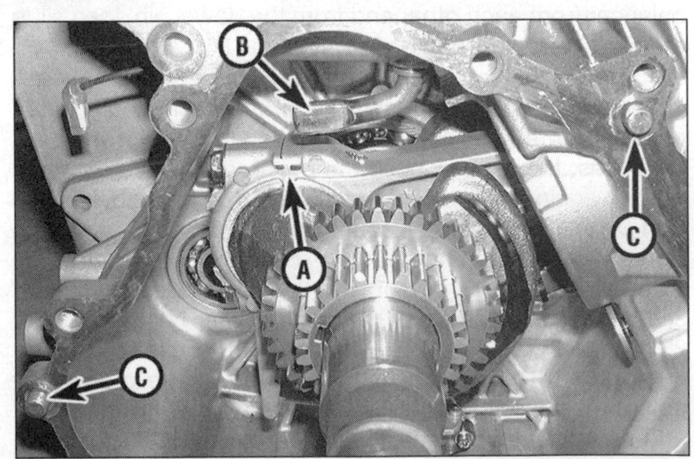


9.16 Make alignment marks on the rod and cap (A); check the governor crank arm for wear (B); locate the cover dowels (C)

**9.17 The governor gear (A) and crankshaft timing gear (B) can be pried off if necessary; loosen the rod cap nuts with a ratchet and socket**

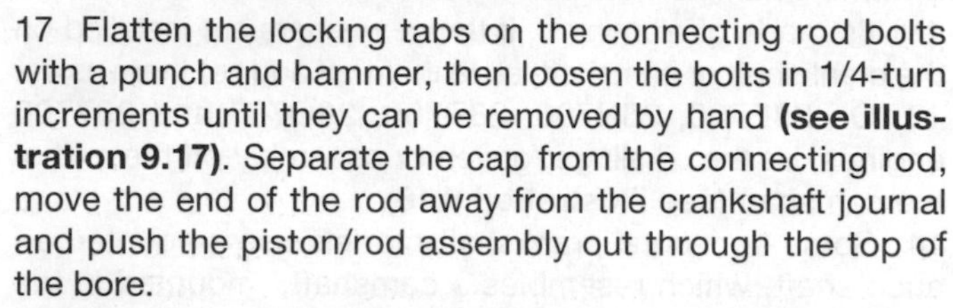

17 Flatten the locking tabs on the connecting rod bolts with a punch and hammer, then loosen the bolts in 1/4-turn increments until they can be removed by hand **(see illustration 9.17)**. Separate the cap from the connecting rod, move the end of the rod away from the crankshaft journal and push the piston/rod assembly out through the top of the bore.

18 If you're working on a vertical crankshaft engine with an oscillating counterbalancer, pry off the crankshaft gear with two screwdrivers.

19 Lift the crankshaft out of the crankcase.

# Inspection of components

20 If you're working on a model 260700 or 261700 vertical-crankshaft engine, remove the snap-ring and link pin and detach the counterweight from the connector link.

21 If you're working on a model 28E700, 28N700, 28P700, 28Q700 or 287700 vertical-crankshaft engine, remove the counterbalancer bolt. Separate the halves of the counterbalancer from each other and from the dowel pin, then slide them off the crankshaft and remove the link.

22 . The mechanical governor is mounted in the crankcase cover. It consists of a pushrod and three centrifugal levers mounted on a gear. When the rotational speed of the gear increases, the levers cause the pushrod to extend. It then pushes against the arm of the governor crank **(see illustration 9.16)**. The governor crank controls the throttle setting to regulate engine speed. If the governor components are worn or damaged, pry out the governor gear with two screwdrivers and remove the thrust washer beneath it.

23 Models 260700, 261700, 28P700 and 28Q700 are equipped with an oil pump, mounted in the crankcase cover. Remove the three cover bolts and lift off the cover and O-ring. Lift out the rotors (on models 260700 and 261700 the inner rotor is secured by a snap-ring) and check them for scoring and wear. Replace the rotors as a set if

**9.24 Note the direction of the punch mark on the piston top (arrow)**

they're worn. Check with a small engine repair shop; you may have to buy a new cover as well. Use a new snap-ring on 260700 and 261700 models.

24 If a new piston will be installed, look for a mark on its crown that indicates the direction of installation **(see illustration 9.24)**. If you don't see a mark, make your own with a scribe or sharp punch.

25 Pry the circlip out of one side of the piston with needle-nosed pliers or a pointed tool **(see illustration 9.25)**. **Warning:** *Wear eye protection when removing the circlip!*

26 Push the piston pin out from the opposite side to free the piston from the rod **(see illustration 9.26a)**. You may have to deburr the area around the groove to enable the pin to slide out (use a triangular file for this procedure). If the pin won't come out, you can fabricate a piston pin removal tool from a long bolt, a nut, a piece of tubing and washers **(see illustration 9.26b)**.

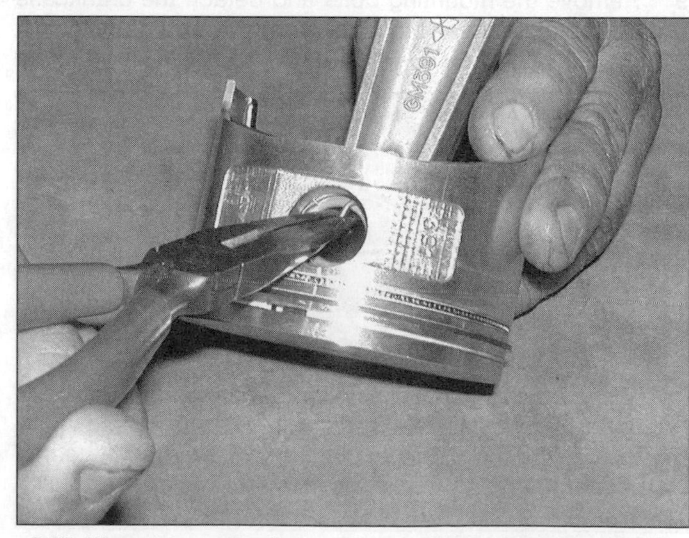

**9.25 Wear eye protection and remove the piston pin circlips**

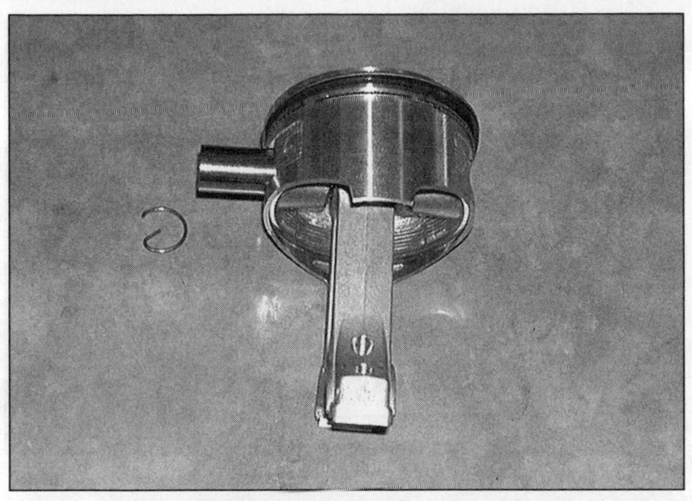

**9.26a Push the pin out part way and take the piston off the rod**

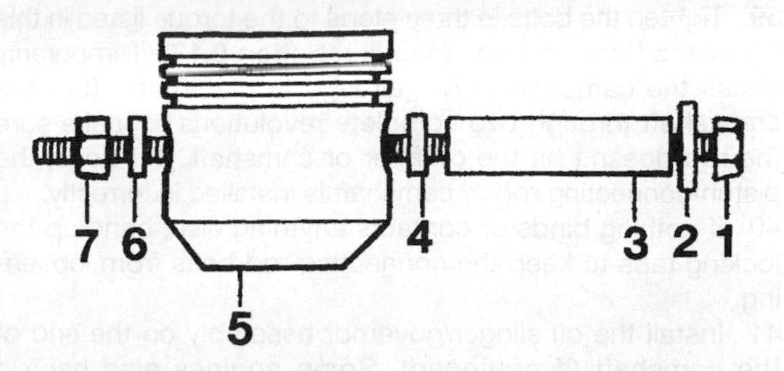


**9.26b If the pin is stuck, you can make this removal tool**

1 *Bolt*
2 *Washer*
3 *Pipe (A)*
4 *Padding (A)*
5 *Piston*
6 *Washer (B)*
7 *Nut (B)*
A *Large enough for piston pin to fit inside*
B *Small enough to fit through piston pin bore*

27 After the engine has been completely disassembled, refer to Chapter 5 for the cleaning, component inspection and valve lapping procedures. **Note:** *Special test equipment is needed to check the ignition coil/electronic ignition module. If you suspect the coil/module is causing ignition problems and the troubleshooting procedures in Chapter 3 don't solve the problem, have it checked by a small engine repair shop.* Once you've inspected and serviced everything and purchased any necessary new parts, which should always include new gaskets and seals, reassembly can begin. Begin by reinstalling the PTO and mechanical governor components (if used) in the crankcase and cover, then proceed as follows:

## Reassembly

28 If the valves were removed from the cylinder head, refer to Chapter 5 and install them.
29 Lubricate the tappets and install them in their bores.
30 If you're working on an engine with a two-piece oscillating counterbalancer (28E700, 28N700, 28P700, 28Q700, 287700), assemble it onto the crankshaft. The rounded edge of the free end of the link faces the drive side of the crankshaft. Tighten the bolt to the torque listed in this Chapter's Specifications.
31 Lubricate the crankshaft magneto side oil seal lip, the plain bearing (if applicable) and the connecting rod journal with clean engine oil or engine assembly lube, then reposition the crankshaft in the crankcase. If a ball-bearing is used on the magneto side, lubricate it with clean engine oil. **Note:** *If the crankshaft timing gear was removed, install it with its timing mark facing outward so you can see it.*
32 If the engine has a one-piece oscillating counterbalancer (260700, 1261700), install the link on the crankshaft, then install the dowel pin, counterweight and pivot shaft.
33 If the engine has a shaft-type counterbalancer, install it. Make sure the counterbalancer timing mark aligns with the mark on the crankshaft.
34 Lubricate the camshaft lobes, journals and shaft bore with engine assembly lube. Install the camshaft, aligning its timing mark with the crankshaft mark.
35 Before installing the piston/connecting rod assembly, the cylinder bore must be perfectly clean and the top edge of the bore must be chamfered slightly so the rings don't catch on it. Position the piston ring end gaps 120-degrees apart (see Chapter 5). Lubricate the piston and rings with clean engine oil, then attach a ring compressor to the piston **(see illustration 8.39)**. Leave the skirt protruding about 1/4-inch. Tighten the compressor until the piston cannot be turned, then loosen it until the piston turns in the compressor with resistance.
36 Rotate the crankshaft until the connecting rod journal is at BDC (Bottom Dead Center - bottom of the stroke) and apply engine oil to the cylinder walls **(see illustration 8.40)**. Be sure the notch in the top of the piston faces in the proper direction **(see illustration 9.24)**. If the word MAG is cast in the rod, it faces the magneto side of the engine. Make sure the mark you made on the rod will be facing out when the rod/piston assembly is in place. Gently insert the piston/connecting rod assembly into the cylinder and rest the bottom edge of the ring compressor on the engine block. Tap the top edge of the ring compressor to make sure it's contacting the block around its entire circumference.
37 Carefully tap on the top of the piston with the end of a wooden or plastic hammer handle while guiding the end of the connecting rod into place on the crankshaft journal **(see illustration 8.41)**. The piston rings may try to pop out just before entering the bore, so keep some pressure on the ring compressor. Work slowly - if any resistance is felt as the piston enters the cylinder, stop immediately. Find out what's hanging up and fix it before proceeding. Do not, for any reason, force the piston into the cylinder - you'll break a ring and/or the piston.
38 Install the connecting rod cap, a NEW lock plate, the oil dipper (if used) and the bolts. Make sure the marks you made on the rod and cap are aligned and facing the direction they were originally and the oil dipper is oriented correctly.

39 Tighten the bolts in three steps to the torque listed in this Chapter's Specifications **(see illustration 9.17)**. Temporarily install the camshaft (if not already in place) and turn the crankshaft through two complete revolutions to make sure the rod doesn't hit the cylinder or camshaft. If it does, the piston/connecting rod or camshaft is installed incorrectly.

40 If nothing binds or contacts anything else, bend up the locking tabs to keep the connecting rod bolts from loosening.

41 Install the oil slinger/governor assembly on the end of the camshaft (if equipped). Some engines also have a spring washer that must be slipped over the end of the camshaft after the oil slinger is in place.

42 Lubricate the crankshaft main bearing journal and the lip on the oil seal in the sump (or crankcase cover/bearing support) with clean engine oil or engine assembly lube.

43 Make sure the dowel pins are in place, then position a new gasket on the crankcase (the dowel pins will hold it in place) **(see illustration 9.16)**. A very thin coat of non-hardening sealant can be used if desired. Carefully lower the oil sump (or crankcase cover/bearing support) into place over the end of the crankshaft until it seats on the crankcase.

44 Install the bolts and tighten them to the torque listed in this Chapter's Specifications. Follow a criss-cross pattern and work up to the final torque in three equal steps to avoid warping the oil sump or cover.

46 Measure crankshaft end play with a dial indicator. The crankshaft end play must be checked and adjusted as follows:

47 Models 161400, 260700, 261700, 28E700, 28N700, 28P700, 28Q700 and 287700 - If the end play is less than specified, use additional gaskets in various combinations to correct it (they're available in 0.005, 0.009 and 0.015-inch thickness). If the end play is greater than specified when using only the standard gasket, replace the end cover with a new one.

48 Models 185400, 235400 and 245400 - If the end play is greater than specified, install one or more thrust washers on between the drive end bearing and the crankcase cover to reduce play.

49 Install the crankcase breather and intake manifold. Use new gaskets and tighten the bolts securely.

9.53a Cylinder head TIGHTENING sequence - four-bolt head

50 Install the ignition coil/spark plug wire assembly but don't tighten the bolts completely - just snug them up. The ignition coil bolt holes are slotted; refer to Section 4 and adjust the gap between the flywheel and coil before snugging up the bolts. Be sure to reroute the wires from the ignition coil properly.

51 Make sure the tapered portion of the crankshaft and the inside of the flywheel hub are clean and free of burrs. Position the Woodruff key in the crankshaft keyway and install the flywheel. Install the washer and tighten the large nut or the starter clutch to the torque listed in this Chapter's Specifications. Install the debris screen (if equipped) and tighten the screws.

52 If a flywheel brake is used, reinstall it now.

53 Install the cylinder head, using a new gasket. Following the recommended tightening sequence, tighten the cylinder head bolts evenly, in several stages, to the torque listed in this Chapter's Specifications **(see illustrations 9.53a, 9.53b and 9.53c)**.

54 To install the remaining components, refer to Chapters 4 and 5 as necessary. **Caution:** *Be sure to fill the crankcase to the correct level with the specified oil before attempting to start the engine.*

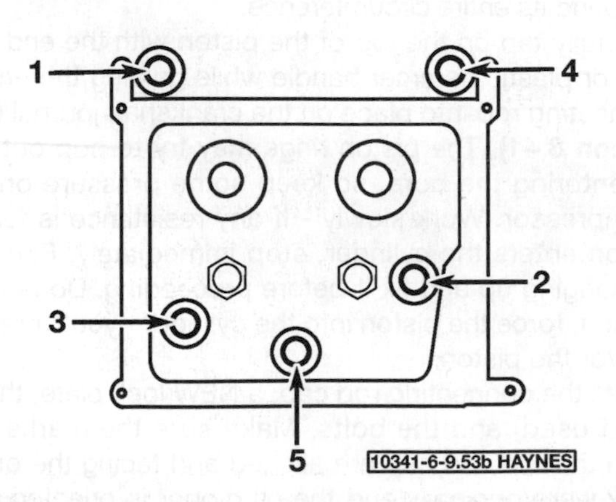

9.53b Cylinder head TIGHTENING sequence - five-bolt head

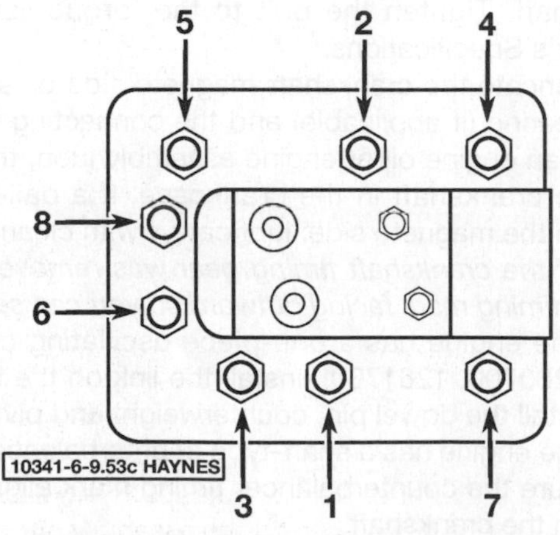

9.53c Cylinder head TIGHTENING sequence - eight-bolt head

**9.56 Remove the valve cover nuts (arrows) and take the cover off**

**9.57a Loosen the rocker arm studs evenly to relieve valve spring pressure . . .**

# V-twin engines

## Disassembly

55 For shroud/recoil starter, carburetor and muffler removal, refer to Chapters 4 and 5 as necessary.

56 To remove a valve cover, undo its nuts **(see illustration 9.56)**. Take off the cover and gasket. **Note:** *If the cover is stuck, tap it from the side with a rubber mallet or dead blow hammer. Don't pry it or the gasket surfaces may be damaged.*

57 Loosen the rocker arm nuts evenly to relieve valve spring pressure, then undo the nuts and take the rocker assembly off **(see illustrations 9.57a, 9.57b and 9.57c)**. Pull out the pushrods. **Note:** *The exhaust pushrods are aluminum (gray color). Don't mix them up with the intake pushrods.*

58 To remove the cylinder head, loosen the bolts in the opposite order of the tightening sequence **(see illustration 9.53a)**. Tap the head with a rubber mallet or dead blow hammer to free it, then lift it off and remove the gasket. Don't pry the head off or the gasket surfaces may be damaged. **Note:** *If the engine is cast-iron or has a cast-iron sleeve, use a ridge reamer to remove the carbon/wear ridge from the top of the cylinder bore after the cylinder head is off. Follow the manufacturer's instructions included with the tool. If the engine is made of aluminum, the ridge does not have to be removed.*

59 Refer to Section 8 and remove the flywheel and its key.

60 Refer to Section 4 and detach the coil/spark plug wire assembly.

61 Remove the mounting bolts and detach the crankcase breather assembly and gasket from the engine **(see illustration 9.61)**. Inspect it as described in Step 9.

**9.57b . . . then lift the assembly off**

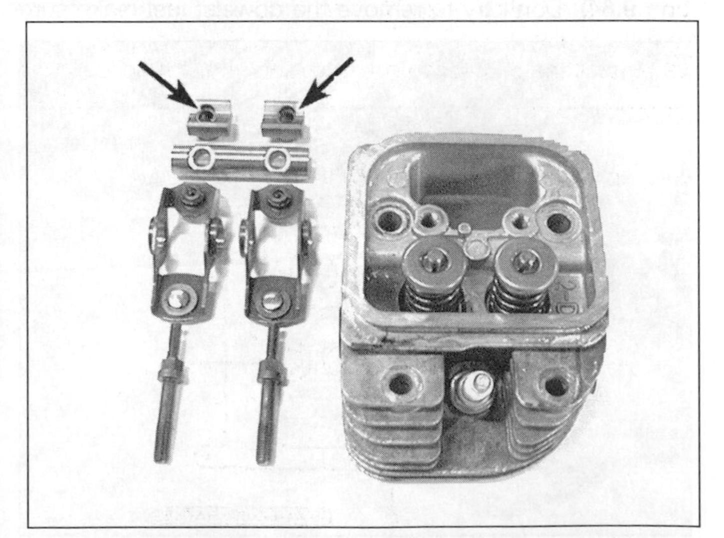

**9.57c Rocker assembly details; the support bolt holes (arrows) face away from each other**

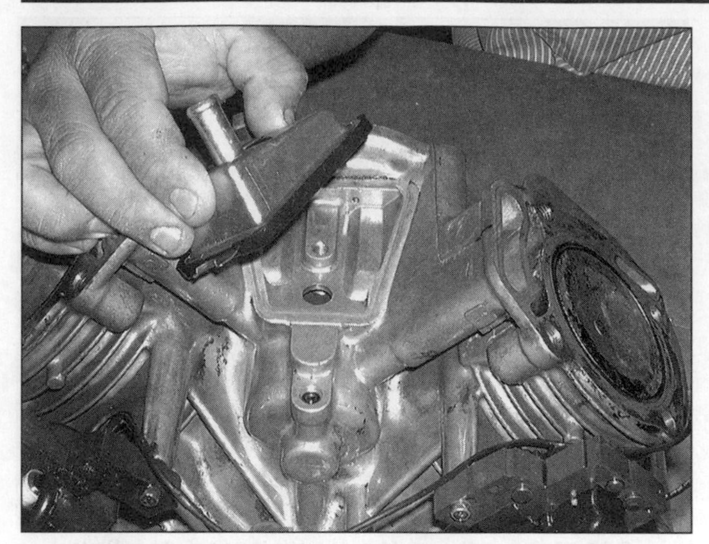

9.61 Unbolt the breather and remove its gasket

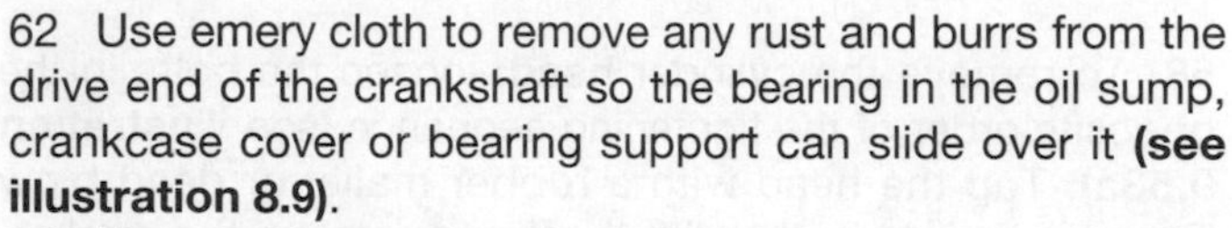

9.63 Crankcase cover bolts (A) and oil pump bolts (B) (horizontal crankshaft OHV twin)

62 Use emery cloth to remove any rust and burrs from the drive end of the crankshaft so the bearing in the oil sump, crankcase cover or bearing support can slide over it **(see illustration 8.9)**.

63 The crankcase opening is covered by a crankcase cover. The shape and specific cover bolt locations differ from model to model, but in all cases, the bolts should be loosened in stages in a criss-cross pattern to prevent warping the cover **(see illustration 9.63)**. If you're planning to remove the oil pump from the cover, it's easier to loosen its bolts now, while the cover is still bolted to the engine.

64 Tap the sump or cover with a soft-face hammer to break the gasket seal, then separate it from the engine block and crankshaft. If it hangs up on the crankshaft, continue to tap on it with the hammer, but be very careful not to crack or distort it if it's made of aluminum. If thrust washers are installed on the crankshaft or camshaft, slide them off and set them aside. Locate the dowels and the small O-ring **(see illustration 9.64)**. Don't try to remove the dowels; just make sure they're there. Set the O-ring aside where it won't be lost.

65 Lift out the camshaft, together with the governor **(see illustration 9.65)**.

66 Once the camshaft is removed, the tappets can be pulled from their bores. Store them in marked containers so they can be returned to their original locations.

67 Look for the mark on the side of each connecting rod that faces out (toward the drive side of the engine). The connecting rod for number one cylinder (the cylinder that's closest to the magneto) should be labeled OUT 1. The connecting rod for number two cylinder should be labeled OUT 2. There should also be an alignment mark on each rod and cap, so the rod and cap will be reassembled in the same relationship to each other. Finally, there is a notch on the top of each piston that faces the magneto side of the engine. The parts must be reassembled in the exact same relationship to the crankshaft.

68 The remainder of connecting rod and crankshaft removal is the same as for single-cylinder engines.

9.64 Don't try to remove the dowels (arrow); locate the O-ring and be sure to reinstall it

9.65 Remove the governor (arrow) and camshaft

**9.69 Rotate the governor crank (arrow) clockwise as you install the crankcase cover**

**9.72a Unbolt the windage tray . . .**

# Inspection of components

69 The mechanical governor is mounted on the end of the camshaft. It consists of a pushrod and three centrifugal levers mounted on a gear. When the rotational speed of the gear increases, the levers cause the pushrod to extend. It then pushes against the arm of the governor crank **(see illustration 9.69)**. The governor crank controls the throttle setting to regulate engine speed. If the governor components are worn or damaged, replace them.

70 All models are equipped with an oil pump, mounted in the crankcase cover.

71 If you're working on a vertical crankshaft model, remove the outer rotor and oil screen.

72 If you're working on a horizontal crankshaft model, remove the windage plate and oil pickup screen **(see illustrations 9.72a, 9.72b and 9.72c)**.

73 Check the rotors and gear for scoring and wear **(see illustration 9.73)**. Replace the pump if they're worn.

**9.72b . . . and the oil pickup . . .**

**9.72c . . . remove the pickup and O-ring from the pump**

**9.73 Replace the pump if the rotors or gear are damaged**

74 If a new piston will be installed, look for a mark on its crown that indicates the direction of installation **(see illustration 9.24)**. If you don't see a mark, make your own with a scribe or sharp punch.

75 Refer to Steps 25 and 26 in this Section to remove the pistons from the connecting rods.

76 After the engine has been completely disassembled, refer to Chapter 5 for the cleaning, component inspection and valve lapping procedures. **Note:** *Special test equipment is needed to check the ignition coil/electronic ignition module. If you suspect the coil/module is causing ignition problems, have it checked by a small engine repair shop.*

## Reassembly

77 Once you've inspected and serviced everything and purchased any necessary new parts, which should always include new gaskets and seals, reassembly can begin. Begin by reinstalling the PTO and mechanical governor components (if used) in the crankcase and cover.

78 The remainder of assembly is basically the same as for single-cylinder OHV engines, but there's no counterbalancer. Tighten all fasteners to the torque's listed in this Chapter's Specifications.

# 10 Specifications

### General

| | |
|---|---|
| Engine oil | |
| Type | API grade SH or better high quality detergent oil |
| Viscosity | |
| Above 40-degrees F | SAE 30 |
| Zero to 40-degrees F | SAE 5W-30 or 10W-30 |
| Below zero | SAE 5W-20 or 5W-30 synthetic oil |
| Capacity | |
| L-head single-cylinder engines | |
| 170000 | |
| Vertical crankshaft | 36 fluid ounces |
| Horizontal crankshaft | 44 fluid ounces |
| 190000 through 286700 aluminum | 48 fluid ounces |
| 230000 through 320000 cast iron | 64 fluid ounces |
| L-head twin-cylinder engines | |
| Without oil filter | 3 pints* |
| With oil filter | 3-1/2 pints |
| OHV single-cylinder engines | |
| 161400 | 40 fluid ounces |
| 185400, 2235400, 245400 | 41 fluid ounces |
| 260700, 261700 | |
| Without oil filter | 64 fluid ounces |
| With oil filter | 72 fluid ounces |
| 28P700, 28Q700 | 56 fluid ounces |
| 28E700, 28N700, 287700 | 48 fluid ounces |
| OHV twin-cylinder engines | |
| Without oil filter | 3 pints |
| With oil filter | 3-1/2 pints |

**Some early production models have a capacity of 3-1/2 pints. Check the dipstick.*

### Fuel system

| | |
|---|---|
| Float level | |
| All except 290400, 294400, 303400 | Parallel to mounting surface |
| 290400, 294400, 303400 | |
| With fuel pump | 1/16-inch below bowl surface* |
| Without fuel pump | 3/32 inch above bowl surface* |

| | |
|---|---|
| Initial mixture setting (from lightly seated position) | |
| L-head single-cylinder engines | |
| Two-piece Flo-jet | 1-1/4 turns |
| Crossover Flo-jet | 1-1/2 turns |
| L-head twin-cylinder engines | 1-1/2 turns |
| OHV single-cylinder engines | 1-1/4 turns |
| OHV twin-cylinder engines | |
| 290500 through 350000 | 1-1/4 turns |
| 351400 and 351700 | 3/4 turn |
| Idle speed | |
| L-head single-cylinder engines | |
| LMT Flo-jet carburetor | 1750 rpm ungoverned |
| Two-piece Flo-jet | |
| Aluminum engines | 1750 rpm ungoverned |
| Cast iron engines | 1200 rpm ungoverned |
| Crossover Flo-jet | 1750 rpm (governor in idle position) |
| L-head twin-cylinder engines | 1000 rpm ungoverned |
| OHV single-cylinder engines | |
| Model 161400, 260700, 261700 | 1200 rpm ungoverned |
| Model 235400, 245400, 28E700, 28N700, 28Q700, 287700 | 1750 rpm** |
| OHV twin-cylinder engines | |
| Red governed idle spring | 1200 rpm |
| White governed idle spring | 900 rpm |

*Carburetor held just past vertical position so needle valve is closed.*
***Reset to 1200 after mixture adjustment if engine has a governed idle.*

**Ignition system**

| | |
|---|---|
| Spark plug type (Champion) | |
| L-head single and twin-cylinder engines | |
| Standard | J19LM |
| Resistor plug | RJ19LM |
| Optional resistor plug | RJ12 |
| OHV single-cylinder engines | RC12YC |
| OHV V-twin engines | |
| Standard | RC12YC |
| Optional | RC14YC |
| Spark plug gap | 0.030-inch |
| Ignition point gap | 0.020-inch |
| Ignition coil-to-flywheel gap | |
| L-head single-cylinder engines | |
| Three-leg coil | |
| Aluminum engines | 0.012 to 0.016 inch |
| Cast iron engines | 0.022 to 0.026 inch |
| Two-leg coil | 0.010 to 0.014 inch |
| L-head twin-cylinder engines (electronic ignition) | 0.008 to 0.012 inch |
| OHV single-cylinder engines | |
| 161400, 235400, 245400, 260700, 261700 | 0.008 to 0.012 inch |
| 185400 | 0.012 to 0.020 inch |
| 280000 | 0.010 to 0.014 inch |
| OHV twin-cylinder engines | 0.008 to 0.012 inch |

**Starting system (electric)**

| | |
|---|---|
| Brush wear limits | |
| End-bearing type | 1/4 inch beyond brush lead |
| Flat coil spring and compression coil spring type | 1/8 inch beyond brush lead |
| Commutator diameter limit | 1.230 inch |

**Engine**

Cylinder bore diameter

| | |
|---|---|
| L-head single-cylinder engines | |
| 170000, 190000, 230000 | 2.999 to 3.000 inches |
| 220000, 250000, 280000 | 3.4365 to 3.4375 inches |
| 240000 | 3.0615 to 3.0625 inches |
| 300000 | 3.4365 to 3.4375 inches |
| 320000 | 3.5615 to 3.5625 inches |
| L-head twin-cylinder engines | 3.4365 to 3.4375 inches |
| OHV single-cylinder engines | |
| 185400 | 3.1496 to 3.1504 inches |
| 235400, 245400 | 3.5047 to 3.5039 inches |
| 260700, 261700, 28E700, 27N700, 28P700, 28Q700, 287700 | 3.4365 to 3.4375 inches |
| OHV twin-cylinder engines | |
| 290000 through 303000 | 2.677 to 2.678 inches |
| 350000 | 2.835 to 2.836 inches |

Piston pin diameter limit

| | |
|---|---|
| L-head single-cylinder engines | |
| 170000, 190000, 240000 | 0.671 inch |
| 220000, 250000, 280000, 300000, 320000 | 0.799 inch |
| 230000 | 0.734 inch |
| L-head twin-cylinder engines | 0.799 inch |
| OHV single-cylinder engines | |
| 161400, 260700, 261700 | 0.800 inch |
| 185400 | 0.7072 inch |
| 235400, 245400 | 0.7996 inch |
| 28E700, 28N700, 28P700, 28Q700, 287700 | 0.799 inch |
| OHV twin-cylinder engines | 0.6718 inch |

Piston pin bore limit

| | |
|---|---|
| L-head single-cylinder engines | |
| 170000, 190000, 240000 | 0.673 inch |
| 220000, 250000, 280000, 300000, 320000 | 0.801 inch |
| 230000 | 0.736 inch |
| L-head twin-cylinder engines | 0.802 inch |
| OHV single-cylinder engines | |
| 161400, 260700, 261700 | 0.801 inch |
| 185400 | 0.7102 inch |
| 235400, 245400 | 0.8028 inch |
| 28E700, 28N700, 28P700, 28Q700, 287700 | 0.801 inch |
| OHV twin-cylinder engines | 0.6735 inch |

Piston ring gap limits

| | |
|---|---|
| L-head engines | |
| Aluminum cylinder bores | |
| Compression rings | 0.035 inch |
| Oil ring | 0.045 inch |
| Cast iron cylinder bores | |
| Compression rings | 0.030 inch |
| Oil ring | 0.035 inch |
| OHV single-cylinder engines | |
| 161400 | |
| Compression rings | 0.030 inch |
| Oil ring rails | 0.065 inch |
| 185400, 235400, 245400, 260700, 261700 (all rings) | 0.030 inch |
| 28E70, 28N700, 28P700, 28Q700 | |
| Top compression ring | 0.025 inch |
| Second compression ring, oil ring | 0.030 inch |

| | |
|---|---|
| 287700 | |
| Compression rings | 0.030 inch |
| Oil ring | 0.035 inch |
| OHV V-twin engines | 0.030 inch |
| Piston ring side clearance limits (with new rings) | |
| L-head engines | 0.009 inch |
| Single-cylinder OHV engines | |
| 161600, 260700, 261700 | |
| Compression rings | 0.004 inch |
| Oil ring | 0.008 inch |
| 185400, 235400, 245400 (all rings) | 0.007 inch |
| 28E700, 28N700, 28P700, 28Q700, 287700 (all rings) | 0.006 inch |
| V-twin OHV engines | 0.008 inch |
| Crankpin journal diameter limit | |
| L-head single-cylinder engines | |
| 170000 | 1.090 inch |
| 190000 | 1.122 inch |
| 220000, 250000, 280000 | 1.247 inch |
| 230000 | 1.1844 inch |
| 240000, 300000, 320000 | 1.3094 inch |
| L-head twin-cylinder engines | 1.622 inch |
| OHV single-cylinder engines | |
| 161400 | 1.427 inch |
| 185400 | 1.3368 inch |
| 235400, 245400 | 1.4953 inch |
| 260700, 261700 | 1.622 inch |
| 28E700, 28N700, 28P700, 28Q700, 287700 | 1.247 inch |
| OHV twin-cylinder engines | 1.455 inch |
| Main bearing journal diameter limit | |
| L-head single-cylinder engines | |
| 170000, 190000 | |
| Magneto side | |
| With counterbalancer | 1.179 inch |
| Without counterbalancer | 0.997 inch |
| Drive side | 1.179 inch |
| 220000, 250000, 280000 | 1.367 inch |
| 230000 | 1.3679 inch |
| 240000, 300000, 320000 | Ball bearings |
| L-head twin-cylinder engines | 1.367 inch |
| OHV single-cylinder engines | |
| 161400 | |
| Magneto side | 1.497 inch |
| Drive side | Ball bearing |
| 185400, 235400, 245400 | Ball bearings |
| 260700, 261700 | 1.622 inch |
| 28E700, 28N700, 28P700, 28Q700, 287700 | 1.376 inch |
| OHV twin-cylinder engines | |
| Magneto side | 1.179 inch |
| Drive side | 1.375 inch |
| Crankshaft end play | |
| L-head single-cylinder engines | 0.002 inch |
| L-head twin-cylinder engines | |
| Horizontal crankshaft | 0.004 inch |
| Vertical crankshaft | 0.002 inch |

**Engine (continued)**

| | |
|---|---|
| Crankshaft end play | |
| OHV single-cylinder engines | |
| 161400, 260700, 261700, 28E700, 28N700, 28P700, 28Q700, 287700 | 0.002 inch |
| 185400, 235400, 245400 | 0.001 inch |
| OHV twin-cylinder engines | 0.003 to 0.015 inch |
| Camshaft end play (L-head cast iron single-cylinder engines) | 0.002 inch |
| Valve clearance (cold) | |
| L-head aluminum single-cylinder engines (except 286700)* | |
| Intake | 0.005 to 0.007 inch |
| Exhaust | 0.009 to 0.011 inch |
| L-head aluminum single-cylinder engine model 286700* | |
| Intake | 0.004 to 0.006 inch |
| Exhaust | 0.009 to 0.011 inch |
| L-head cast iron single-cylinder engines | |
| Intake | 0.007 to 0.009 inch |
| Exhaust | 0.017 to 0.019 inch |
| L-head twin-cylinder engines | |
| Intake | 0.004 to 0.006 inch |
| Exhaust | 0.007 to 0.009 inch |
| OHV single-cylinder engines | |
| 161400 | 0.003 to 0.005 inch |
| 185400, 235400, 245400 | 0.002 to 0.004 inch |
| 260700, 261700 | 0.003 to 0.005 inch |
| 28E700, 28N700, 28P700, 28Q700, 287700 | |
| Intake | 0.003 to 0.005 inch |
| Exhaust | 0.005 to 0.007 inch |
| OHV twin-cylinder engines | 0.004 to 0.006 inch |

**Including aluminum engines with cast iron cylinder liners.*

**Torque specifications**

| | |
|---|---|
| Oscillating counterbalancer bolt(s) (single-cylinder engines only) | |
| L-head engines | |
| Two bolts | 80 in-lbs |
| One bolt | 115 in-lbs |
| OHV engines | 115 in-lbs |
| Connecting rod cap nuts/bolts | |
| L-head single-cylinder engines | |
| 170000 | 165 in-lbs |
| 190000, 220000, 250000 | 185 in-lbs |
| 230000, 240000, 300000, 320000 | 190 in-lbs |
| 280000 | |
| Both fasteners the same size | 185 in-lbs |
| One small, one large fastener | |
| Small fastener (tighten first) | 160 in-lbs |
| Large fastener | 260 in-lbs |
| L-head twin-cylinder engines | 190 in-lbs |
| OHV single-cylinder engines | |
| 161400, 260700, 261700 | 200 in-lbs |
| 185400, 235400, 245400 | 175 in-lbs |
| 28E700, 28N700, 28P700, 28Q700, 287700 | |
| One small, one large fastener | |
| Small fastener (tighten first) | 130 in-lbs |
| Large fastener | 260 in-lbs |
| Both fasteners the same size | 185 in-lbs |
| OHV V-twin engines | 115 in-lbs |

| | |
|---|---|
| Crankcase cover/oil pan bolts | |
| L-head single-cylinder engines | |
| 170000, 190000, 220000, 250000 | 140 in-lbs |
| 280000 | |
| With separate lockwasher, no thread sealant | 140 in-lbs |
| With separate lockwasher and thread sealant | 200 in-lbs |
| With integral washer and machined tip | 200 in-lbs |
| Crankcase cover/oil pan bolts | |
| L-head single-cylinder engines (continued) | |
| 230000, 240000, 300000, 320000 | |
| Magneto side | 90 in-lbs |
| Drive side | 190 in-lbs |
| L-head twin-cylinder engines | |
| Horizontal crankshaft | |
| Crankcase cover | 225 in-lbs |
| Base | 27 ft-lbs |
| Vertical crankshaft | |
| Steel crankcase cover | 250 in-lbs |
| Aluminum crankcase cover | 27 ft-lbs |
| Oil pan | 225 in-lbs |
| OHV single-cylinder engines | |
| 161400, 260700, 261700 | 250 in-lbs |
| 185400, 235400, 245400 | 175 in-lbs |
| 28E700, 28N700, 28P700, 28Q700, 287700 | 140 in-lbs |
| OHV twin-cylinder engines | 170 in-lbs |
| Cylinder head bolts | |
| L-head single-cylinder engines | |
| Aluminum | 165 in-lbs |
| Cast iron | 190 in-lbs |
| L-head twin-cylinder engines | 160 in-lbs |
| OHV single-cylinder engines | |
| 115400, 117400 | 220 in-lbs |
| 161400 | 75 in-lbs |
| 185400 | 300 in-lbs |
| 235400, 245400 | 35 ft-lbs |
| 260700, 261700 | 225 in-lbs |
| 28E700, 28N700, 28P700, 28Q700, 287700 | 220 in-lbs |
| OHV twin-cylinder engines | 165 in-lbs |
| Flywheel nut | |
| L-head single-cylinder engines | |
| Aluminum engines | 65 ft-lbs |
| Cast iron engines | 145 ft-lbs |
| L-head V-twin engines | 150 ft-lbs |
| OHV single-cylinder engines | |
| 161400, 28E700, 28N700, 28P700, 28Q700, 287700 | 65 ft-lbs |
| 185400, 235400, 245400 | 60 ft-lbs |
| 260700, 261700 | 125 ft-lbs |
| 28E700, 28N700, 28P700, 28Q700, 287700 | 65 ft-lbs |
| Rocker assembly (OHV engines) | |
| Rocker arm studs | |
| 161400, 260700, 261700 | 140 in-lbs |
| 185400 | 175 in-lbs |
| 28E700, 28N700, 28P700, 28Q700, 287700 | 85 in-lbs |
| V-twin engines | 140 in-lbs |

**Torque specifications (continued)**

Rocker shaft bolts (235400, 245400) ............................ 85 in-lbs
Valve cover fasteners
    161400, 260700, 261700.......................................... 55 in-lbs
    260700, 261700...................................................... 55 ft-lbs
    28E700, 28N700, 28P700, 28Q700, 287700............. 60 ft-lbs
    V-twin engines........................................................ 25 in-lbs

# 7 Tecumseh/Craftsman engines

## Contents



## 1 Engine identification numbers/models covered

**Note:** *For Craftsman engine identification, refer to the cross reference charts at the end of this Chapter and convert the number to a Tecumseh model number.*

1 The engine designation system used by Tecumseh consists of a model and serial number, normally found on the shroud. It may also be located on a tag attached to the crankcase **(see illustrations 1.1a and 1.1b)**. The number can be used to determine the major features of the engine by comparing each digit to the key. The letters/digits in the model number can be explained generally as follows:

2 The first letter or group of letters in a model number indicates the basic engine type:

*ECH = Exclusive Craftsman Horizontal*
*ECV = Exclusive Craftsman Vertical*
*H = Horizontal Shaft*
*HH = Horizontal Heavy Duty (cast-iron)*
*HHM = Horizontal Heavy Duty (cast-iron) Medium Frame*
*HM = Horizontal Medium Frame*

**1.1a Identification number location on a L-head engine with a vertical recoil starter (arrow). This Craftsman number must be converted to a Tecumseh number using the conversion tables at the end of this Chapter**

1.1b Tecumseh engine decal information

*HMSK = Horizontal Medium Frame Snow King*
*HS = Horizontal Small Frame*
*HSK = Horizontal Snow King*
*HSSK = Horizontal Small Frame Snow King*
*LAV = Lightweight Aluminum Vertical*
*OH = Overhead Valve Heavy Duty (cast iron)*
*OHH = Overhead Valve Horizontal*
*OHM = Overhead Valve Horizontal Medium Frame*
*OHSK = Overhead Valve Horizontal Snow King*
*OHV = Overhead Valve Vertical*
*OVM = Overhead Valve Vertical Medium Frame*
*OVRM = Overhead Valve Vertical Rotary Mower*
*OVXL = Overhead Valve Vertical Medium Frame Extra Life*
*TNT = Toro N' Tecumseh*
*TVM = Tecumseh Vertical (medium frame)*
*TVS = Tecumseh Vertical Styled*
*TVXL = Tecumseh Vertical Extra Life*
*V = Vertical shaft*
*VH = Vertical Heavy Duty (cast-iron)*
*VM = Vertical Medium Frame*

**Note:** *New short blocks are identified by a tag that is wired to the case. Letters SBH indicate Short Block Horizontal or SBV indicate Short Block Vertical. The short block serial number is also listed on this tag.*

**L-head light duty engines** include:
H60, HH60, HSK60, V60, VH60, TVM140 .............. 6 HP
H70, HH70, HM70, HMSK70, HSK70, V70, VH70, VM70, TVM170 .............. 7 HP
H80, HM80, HHM80, HMSK80, V80, VM80, TVM195, TVXL195 .............. 8 HP
HM100, HMSK100, VM100, TVM220, TVXL220 .............. 10 HP

**L-head heavy duty engines** include:
HH80, VH80 .............. 8 HP
HH100, VH100 .............. 10 HP
HH120 .............. 12 HP

**OHV light duty engines** include:
OHH55, OHSK55 .............. 5.5 HP
OVRM60 .............. 6 HP
OHM, OHSK, OVM, OVXL, OHV .............. 11 to 13 HP
OHV .............. 13.5 to 17 HP

**OHV heavy duty engines** include:
OH120, OH140, OH150, OH160, OH180 .............. 12 to 18 HP

3 The numbers following the letter prefixes (model designations) indicate a cubic inch displacement or horsepower rating. First, determine the correct engine model using the letter key. To verify the correct horsepower rating of the engine, refer to the above chart and cross reference the model number. A typical model designation plate on a "V" type or vertical engine will indicate a letter "V" within the engine code (example: **TVM195-57010B serial 3105C**) and more specifically a Tecumseh Vertical Style medium frame engine. Then 195 for a 19.5 cubic inch displacement, followed by 57010B for the parts identification number, followed by the serial number which includes the numeral 3 (1993) for the year of manufacture, then 106 (April 16) for the calendar day of manufacture, and last, the C for the line and shift location at the factory. Follow the chart to determine that this Tecumseh engine is rated at 8 hp.

4 These Tecumseh/Craftsman engines are distinctly separated by the types of engine design and the metallurgy of the engine components. There are basically two types of engines, either the horizontal shaft design or the vertical shaft design. The horizontal engine positions the crankshaft horizontally (vertical-stroke piston) while the vertical crankshaft engine design positions the crankshaft vertically (horizontal-stroke piston). Early models use the L-head design which places the valves in the block, requiring the air/fuel mixture to enter the combustion chamber in an L direction. Later models are equipped with overhead valves (OHV) which places the valves in the cylinder head directly over the combustion chamber. Vertical-crankshaft engines are commonly used in lawnmowers while horizontal crankshaft designs are used in pumps, snowblowers, etc. Metallurgy changes with engine designs also. Small frame engines are equipped with aluminum blocks and cylinder bores. Medium frame engines are equipped with aluminum blocks with cast iron sleeves. Heavy frame engines are equipped with cast iron blocks and cylinder heads. There are heavy duty versions of the L-head engines as well as the Overhead Valve (OHV) engines.

5 There are several different types of oiling systems used on these engines. Smaller horsepower horizontal engines are equipped with the "splash oiling system." A dipper is installed onto the end of the connecting rod to agitate the oil bath in the sump causing oil to splash over the cylinder walls and the bearing assembly. Small horsepower vertical engines are equipped with a plunger style oil pump located on the end of the crankshaft. The camshaft eccentric moves the barrel back and forth on the plunger forcing oil through the hole in the camshaft and out of an oiling port

located between the crankshaft and camshaft endcaps in the case. The oil is then sprayed out under pressure over the camshaft and crankshaft bearing assemblies and the cylinder wall. Larger horsepower engines are equipped with a rotary oil pump (lobed gear type) that forces the pressurized oil onto the bearing surfaces, the piston, cylinder and connecting rod. The plunger style and the rotary style oil pumps are both referred to as "a pressurized oiling system."

6 Some of these models are equipped with an automatic decompression mechanism to allow for a much easier "pull" on the recoil starter The automatic decompression assembly is mounted on the camshaft. The automatic decompression system releases the compression of the engine by lifting the exhaust valve while the engine is cranking. There are basically two different type systems; a flyweight system and a camshaft lobe system. The flyweight system incorporates a release lever that activates by centrifugal motion, lifting the exhaust valve slightly. After the engine starts, the rpm levels overcome the weight of the flyweight (release lever) to restore full compression. The camshaft lobe system incorporates a slightly raised profile on the intake lobe of the camshaft that allows compression to leak out during slow cranking. These systems are mostly incorporated on larger horsepower L-head engines with vertical crankshafts.

7 These engines are also equipped with governor systems to ensure constant operation at the selected speed against load variations. The governor assembly consists of a centrifugal flyweight design. The governor assembly is located in the crankcase cover or the oil pan depending upon the type of crankshaft arrangement; vertical or horizontal. As the engine speed increases, the weights on the governor assembly move outward. The shape of the weights force the governor spool to lift, thereby forcing the governor linkage (using lever action) to close the throttle. As the rpm decreases, the governor spool will fall, allowing the governor shaft to retract and increase throttle demand. The governor acts as a throttle and rpm limiter.

# 2 Recoil starter service

## Rope replacement

1 Some larger heavy duty engines with vertical pull (vertical engagement) starters with V notches in the bracket and heavy duty (cast iron blocks) with horizontal pull recoil starters are equipped with recoil starters that DO NOT require complete disassembly to install a new rope. In the event the starter rope has broken off inside the recoil starter, it will be necessary, in most cases, to disassemble the recoil starter for replacement. **Note:** *Vertical pull recoil starters with vertical engagement systems are equipped with gears to transfer torque to a vertical crankshaft; this system is not common. Most vertical pull starters engage a horizontal crankshaft, and most horizontal recoil starters engage a vertical crankshaft.*

2 Starter related problems will require the starter to be removed from the engine and visually inspected. Refer to Section 3 for information on the starting systems. Rope replacement should be done with the correct size and specified length rope to ensure easy starting and pulling.

**Rope diameter selections**
#4-1/2 rope = 9/64-inch diameter
#5 rope = 5/32-inch diameter
#6 rope = 3/16-inch diameter

**Rope length standards**
54-inch (standard stamped steel starter)
61-inch (vertical pull horizontal engagement type)
65-inch (vertical pull vertical engagement type)
85-inch (extended handlebar rope start)

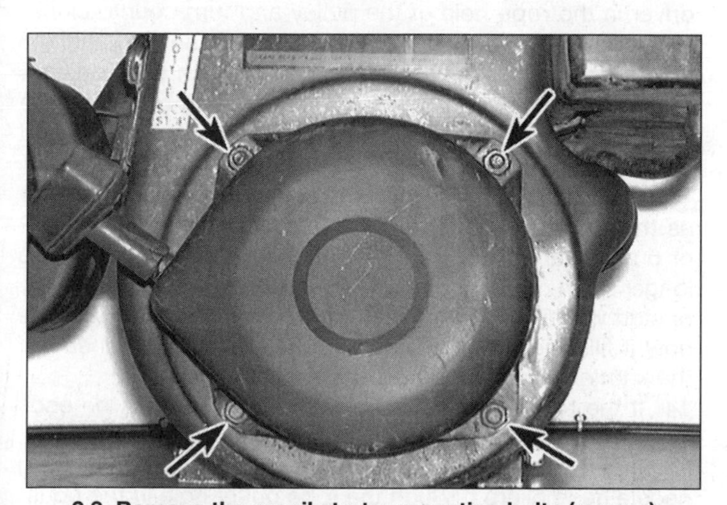

2.3 Remove the recoil starter mounting bolts (arrows)

3 Remove the starter assembly mounting bolts and separate the starter assembly from the engine **(see illustration)**.

4 If the rope breaks and if the starter doesn't have to be disassembled to replace it, it may be a good idea to take the opportunity to do a thorough cleaning job and check the spring and dog mechanism. There are two approaches you may be faced with when replacing the rope on a recoil starter. The method you use will depend on the starter type. The following procedure covers only those recoil starters that DO NOT require complete disassembly for rope replacement. Refer to the recoil starter service procedures for all others.

2.8 Hold the pulley in position using a clamp while inserting the rope end into the hole

2.9 Use locking pliers or a clamp to restrain the pulley so it doesn't rewind

## Vertical pull (vertical engagement) with V notch (L-head engines)

5 Rotate the pulley until the staple in the pulley lines up with the "V" notch. Pry the staple out using a small screwdriver.

6 Remove the rope from the recoil assembly.

7 Rotate the pulley counterclockwise to fully rewind the starter return spring until it is tight. Carefully "unwind" or "back-off" the starter until the "V" notch lines up with the hole in the pulley.

8 Hold the pulley in position and insert the new rope into the hole **(see illustration 2.8)**. Tie a left hand knot in the rope end to secure it in position. Make sure the rope end knot does not protrude from the cavity and inhibit the rotation of the pulley.

## Horizontal pull (vertical engagement) OHV heavy duty engines

9 If the old rope was broken, you'll have to wind up (preload) the recoil spring before installing the new rope. If the old rope is not broken, you'll have to wind up the recoil spring, align the knot with the housing and cut the knotted end of the rope. With the recoil assembly mounted in a vise, turn the pulley against spring tension until it stops completely (approximately 7 turns). Back it off one full turn. This will prevent the spring from being wound too far when the rope is pulled out (which can break it off). Restrain the pulley with the clamp **(see illustration 2.9)**. **Note:** *The hole in the pulley should line up with the hole in the housing.*

10 If it isn't broken, pull the rope all the way out of the assembly leaving the knotted end in the recoil assembly.

11 Pull the knot out of the cavity with a pair of needle-nose pliers, then cut the knot off and pull the rope out **(see illustration 2.11)**. Note the type of knot used in the rope, then detach the handle - it can be used on the new rope.

12 If the recoil assembly has not been preloaded (broken rope), hold the shroud or recoil starter housing in a vise or clamp it to the workbench so it doesn't move around as you're working on the rope. Use soft jaws in the vise to prevent damage to the shroud or housing. Position a screwdriver in the rope hole of the pulley and turn counterclockwise until tight, then allow the coil to slowly unwind until the hole in the pulley for the rope lines up with the housing. Clamp the pulley into position using locking pliers and be sure to not damage the painted surfaces of the pulley and housing.

2.11 Cut off the knot and pull the rope out (OHV heavy duty engine)

13 Cut a piece of new rope the same length and diameter as the original. Standard rope size (diameter) is no. 4-1/2 or 5 and standard length is 65-inches, although some are longer - if in doubt, make it the same length as the old rope or start with 54-inches and cut off any excess when you see how it fills the pulley. Be sure to install a washer ahead of the knotted end to secure it against the pulley.

14 If the rope is made of nylon, cauterize (melt) the ends with a match to prevent fraying.

15 Insert the rope into the pulley hole and guide it using needle nose pliers through the rope outlet hole in the housing until it bottoms at the end of the rope (knot end). Secure

2.22a Drive the roll pin retainer with a punch down through the socket

2.22c Remove the screw (arrow) and detach the retainer cup

the handle and double knot the handle end of the rope. Hold the rope with tension by the handle and release the pliers or clamp from the pulley assembly. Allow the rope to slowly wind onto the pulley as the spring tension activates.

16 Before mounting the starter assembly, place the hub into the cup and pull the rope slightly until the starter dogs engage equally in the teeth. Tighten the starter assembly to 40 to 80 inch lbs. Check to make sure the mounting holes are aligned properly and the blower housing is not bent.

17 Check the starter for proper operation.

# Recoil assembly service

18 **Note:** *The following procedures are broken down into categories that describe the type of recoil starter cover (stamped steel, stylized, aluminum, etc.) and from there, each type is categorized by the type of retaining mechanism. In order to completely distinguish the type of recoil starter on your system, it will be necessary to compare the photos with the entire working components of the recoil starter.*

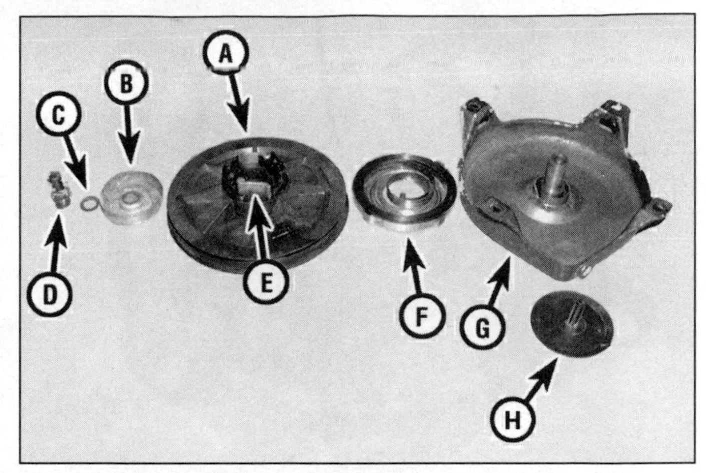


2.22b Exploded view of a stylized recoil starter with a roll pin retainer

A *Reel*
B *Retainer*
C *Washer*
D *Brake spring*
E *Starter dog(s)*
F *Recoil spring*
G *Recoil housing*
H *Cap*

## Horizontal type recoil assemblies

### Stylized recoil starter and stamped steel starter

19 The stylized and stamped steel starter assembly has a distinct teardrop or oval shaped cover and is easily distinguished by the elongated section near the handle. All of these types are equipped with the dog-type recoil starter system.

20 Release the tension of the recoil spring by removing the knot from the rope handle and allowing the rope to wind itself around the starter assembly.

22 Separate the pulley from the ratchet mechanism. There are several variations of the ratcheting mechanism. Be careful not to lose or damage the starter dogs, springs and pins when disassembling.

a) *On* ***stylized recoil starters with roll pin retainer****, install a 1-inch diameter deep socket under the retainer. Set the recoil assembly on the bench supported by the socket. Use a 5/16 inch or 1/4 inch roll pin punch to drive out the center pin.* **Note:** *There are some stamped steel housing recoil assemblies that have the roll pin retainer. These special stamped steel recoil starters use the smaller (1/4 inch) roll pin. Drive the pin from the inside, down through the socket and out the bottom. Remove the brake spring, the retainer, washers and pulley assembly. Note the exact position of each of these components for ease of reassembly* **(see illustrations 2.22a and 2.22b). Warning:** *The recoil spring is under extreme tension while it is installed in the housing. Handle the recoil spring with caution. Be sure to wear eye protection and gloves when handling this assembly.*

b) *On* ***stamped steel recoil starters with screw retainer****, remove the center screw and the retainer cup. Then remove the starter dogs, the dog springs and the brake springs* **(see illustrations 2.22c and 2.22d).**

2.22d Lift out the starter dog and brake spring (arrows)

2.30 Remove the screws and detach the cover (vertical pull starter)

**Note:** *There are some stylized housing recoil starters that are equipped with the retainer screw. Follow this disassembly procedure. Note the exact position of these components for ease of reassembly.*

c) *On **stylized recoil starters with plastic retainers**, use a small screwdriver to pry the retainer legs apart and lift out the retaining wedge. Pinch the legs of the retainer together and pull on the head of the retainer to remove it from the housing. Remove the pulley assembly from the recoil housing. Inspect the components and replace the damaged components with new ones. Note the exact position of each of these components for ease of reassembly.*

23 Installation is the reverse of removal. Be sure to note that the starter dogs face OUT on the stamped steel starter or the dogs face IN on the stylized recoil starter.

24 Install the recoil spring and keeper assembly into pulley by rotating the spring until it locks into position. The spring should have a light coating of oil on it. Place the pulley assembly into the recoil starter housing. **Note:** *Some models are not equipped with a keeper assembly therefore, install the recoil spring directly onto the pulley.*

25 Install the brake spring, the starter dogs and the dog return spring.

26 Install the remaining recoil assembly components:

a) *On **stylized recoil starters with the roll pin retainer**, install a new center pin into the assembly. Be sure to place two new washers between the center leg of the starter and the retainer. Discard the old plastic washers and the center spring pin. Place the recoil assembly onto a flat surface and drive the center pin in until it is approximately 1/8-inch from the starter assembly. Be careful to NOT drive the center pin too far. The retainer will bend and the starter dogs will not engage the starter cup.* **Note:** *On the stamped steel starter assemblies with the roll pin, drive the center pin in until it contacts the shoulder in the starter housing.*

b) *On **stamped steel recoil starters with the screw retainer**, install a new retainer screw into the assembly. Replace the retainer cup . Tighten the retainer screw to 65 to 75 in-lbs. (stamped steel covers) or 115 to 135 in-lbs. (cast aluminum covers).* **Note:** *On Snow King engines, the starter dog posts should be lubricated with SAE 30 engine oil.* **Note:** *In the event the retainer screw on older models has been stripped or enlarged, consult a parts retailer or small engine repair shop for a specialized replacement screw that will fit into the pulley after it has been drilled and tapped.*

c) *On **stylized recoil starters with plastic retainers**, pinch the two legs of the plastic retainer together and install it into the center shaft hole of the pulley. Rotate the retainer so that the two tabs on the bottom part of the retainer lock between the dog and pulley hub. Push in until the legs of the plastic retainer pop out of the center shaft. Turn the starter over until the locking tabs lock between the retainer legs.*

27 To replace the rope, wind the starter pulley counterclockwise four or five turns to preload the recoil spring with the new rope and then insert the rope into the starter housing and tie a knot in the rope end after the handle is installed. Release the pulley and allow the rope to wind over the pulley.

28 Installation is the reverse of removal.

## Vertical-pull starters

### Vertical pull/ horizontal engagement

29 Remove the handle and relieve the spring tension (if necessary) by allowing the rope to slip past the rope clip.

30 Remove the two small screws and detach the spring cover **(see illustration 2.30)**.

31 Carefully remove the spring **(see illustration 2.31)**.

32 Remove the screw and detach the center hub **(see illustration 2.32)**.

33 Detach the gear and pulley assembly. Disassemble the components by removing the snap-ring and washer.

2.31 Grasp the spring securely when removing it and be prepared for sudden release of spring tension

2.32 Remove the screw and detach the center hub

34 The rope can now be removed from the pulley.

35 Attach the new rope. Use number 4-1/2 or 5 braided rope. Cauterize (melt) the ends by burning them with a match and wiping them with a cloth while hot. Standard rope length is 61-inches, although some applications require a longer rope. Check the old one if in doubt and make it the same length.

36 Assemble the gear and pulley and install the washer and snap-ring. **Caution:** *The brake spring must fit snugly in the gear groove. DO NOT lubricate the brake spring or the spiral on the pulley.*

37 Lubricate the center shaft with a small amount of grease.

38 Place the gear and pulley in position and make sure the brake spring loop is positioned over the metal tab on the bracket. The rope clip must fit tightly on the bracket. The raised spot on the clip fits into the hole in the bracket.

39 Install the center hub and screw. Tighten the screw to 44 to 55 in-lbs. If the screw is loose, it'll prevent the rope from retracting.

40 Install the spring (new springs are confined in a retainer). Lay the spring and retainer over the receptacle and push the spring out of the retainer into position - make sure the ends are positioned correctly.

41 Install the cover and screws.

42 Wind the rope onto the pulley by slipping it past the rope clip. When the rope is completely wound onto the pulley, turn the pulley two more turns to put tension on the spring. Tie a knot in the end of the rope so it doesn't rewind completely into the pulley.

43 When installing the starter on the engine, adjust it so the head of the tooth is no closer than 1/16-inch to the base of the flywheel gear tooth.

44 Thread the rope through the guide and install the handle, then check the starter for proper operation.

## Vertical-pull vertical engagement

45 Pull the rope out far enough to lock it in the V-shaped cutout in the bracket.

46 If you have to remove the handle, pry out the staple with a small screwdriver.

47 Place the starter bracket on top of a deep socket large enough to receive the head of the center pin, then drive out the pin.

48 Rotate the spring capsule strut until it's aligned with the legs of the brake spring. Insert a pin or nail no longer than 3/4-inch through the hole in the strut so it catches in the gear teeth. This will keep the capsule in a wound position.

49 Slip the sheave out of the bracket. **Caution:** *Do not attempt to remove the spring capsule from the sheave assembly unless it's completely unwound.*

50 Squeeze and hold the spring capsule firmly against the gear sheave with your thumb at the outer edge of the capsule.

51 Carefully remove the retainer pin from the strut and slowly relieve your grip on the assembly so the spring capsule rotates in a controlled manner to unwind completely.

52 Take the spring capsule off the gear sheave. If the rope is being removed, pry the staple up with a small screwdriver. **Note:** *Do not lubricate any of the starter parts. The starter uses number 4-1/2 braided rope. Standard rope length with the handle mounted on the shroud is 65-inches. If the handle is mounted in any other position, measure from the shroud to the handle and add the additional length. Cauterize (melt) the rope ends by burning them with a match and wiping them with a cloth while hot.*

53 Insert the rope end through the hole of the gear sheave opposite the staple platform and tie a left-hand knot. Pull the knot back into the cavity, making sure the rope end doesn't protrude from the cavity.

54 Wind the rope onto the sheave clockwise, as seen from the gear side of the gear sheave.

55 Reinstall the brake spring - be careful not to spread it more than necessary.

56 Install the spring capsule. Make sure the starter spring end hooks on the gear hub.

57 Wind the spring up four full turns and position the strut between the two brake spring legs. Insert the pin into

the strut.

58 If the starter is equipped with a locking pawl or delay pawl and spring, make sure they're in place, then grasp the gear and spring capsule assembly and slide it into the bracket, making sure the legs of the brake spring are positioned in the slots of the bracket.

59 Feed the rope end under the rope guide and hook it into the V-notch. Remove the pin, the strut will rotate clockwise against the bracket.

60 Insert the new center pin by carefully pressing or driving it firmly into place.

# 3 Charging and electric starting systems

## Charging system components

1 The electrical system consists of three main components - the battery, the starting circuit and charging circuit. The battery functions within both the starting circuit and the charging circuit. Be sure the battery is checked and replaced, if necessary, before attempting to diagnose any charging circuit or starting circuit malfunction. **Note:** *There are some applications where a 120 volt A/C current is plugged into the power equipment to start the engine but generally, a 12 volt battery and starter are used on electric start engines. The following diagnostic procedures do not cover the 120 volt A/C systems.*

2 The charging system consists of alternator stator (charge coils), diodes (A/C to D/C switch), a voltage regulator/rectifier assembly, an ignition switch, flywheel magnets and the battery. Some early models with battery and charging systems will only possess some of these particular components. The charging system works independently of the starting circuit and other control circuits that regulate lighting, accessories, etc. The engine must be rotating at speed to produce electrical current flow. Most small engine alternators produce A/C current when flywheel magnets rotate over the stator (conductor) and cut the magnetic field thereby generating a fluctuating A/C current. From there, it is necessary to convert the alternating waveform from A/C to D/C voltage to charge the battery. A rectifier or diode uses only one half of the A/C voltage signal thereby allowing the conversion process from alternating (A/C) to direct current (D/C).

3 Consult a parts retailer or small engine repair shop for the exact type of system installed on your equipment. **Note:** *There are many variations of the Tecumseh charging system. Some heavy duty applications produce 18 or 35 watt power for the lighting system on tractors, lawnmowers, generators, etc. Some light duty applications produce a 3 or 5 amp trickle charge for the battery through a diode inline with the harness. Other applications include a 7 amp charging system with a voltage regulator mounted under the blower housing. Consult a Tecumseh parts distributor for the exact type of charging system included on your application. The parts distributor will need the engine identification number to make the correct selection.*

## Removal and installation

### Regulator

4 Disconnect the negative battery cable.

5 Remove the blower housing. **Note:** *Most regulator/rectifier assemblies are mounted under the blower housing on these models. Refer to the appropriate engine section for blower housing removal. All others are mounted externally.*

6 To remove the regulator, mark and disconnect the harness connector from the regulator. **Note:** *On some engines, all three wires are in a single connector that snaps off the regulator/rectifier assembly.*

7 Remove the bolts holding the regulator to the engine shroud or equipment and remove the regulator

### Stator

8 To remove the stator, refer to the appropriate engine section and remove the flywheel. These models will require a special knock-off tool to separate the flywheel from the crankshaft.

9 Disconnect the leads from the stator.

10 Remove the screws and remove the stator from the engine's front cover.

11 Installation (regulator and stator) is the reverse of the removal procedure.

## Starting system

12 The starting system consists of a 12 volt battery, battery cables, a starter and ignition switch, safety switches and a starter solenoid. There are two different types of starters, each available in 12V or 120V models; the "exposed shaft starters" and the "cap assembly starters". Although similar in construction, the cap assembly starter is equipped with a cover over the drive gear and the drive nut while the exposed shaft starter is not covered. Starter solenoids are mounted externally. **Note:** *The starters labeled "CSA" cannot be repaired but instead, they must be replaced as a single unit. Starters labeled "UL Listed" can be serviced.*

## Removal and installation

### Solenoid

13 Disconnect the negative battery cable, then remove the cable from the battery to the starter solenoid.

14 Remove any other connectors from the solenoid, then remove the mounting screw(s).

15 Separate the starter from the equipment.

3.19 Make matchmarks on the starter case for alignment purposes before removing the end cap

3.24 Some brush assemblies attach with a cable end (A) while others attach with a solder joint (B). After the brushes have been disconnected, pull the spring (C) over the brush assembly and separate the brush from the holder

### Starter motor

16 There are several different makes of starter motors used on the various Tecumseh engine models depending upon engine size and application.
17 Disconnect the negative battery cable at the battery, and the positive cable from the starter motor (the cable from the solenoid).
18 Remove the mounting bolts.
19 Make matching marks on the case and both ends before disassembling the starter **(see illustration 3.19)**.
20 Installation is the reverse of the removal procedure.

### Brush replacement

21 There are two problem areas concerning starter motors. If the motor has been abused (cranked regularly for more than 15 seconds at a time without a cooling off period) or has been in service for many years, the armature or field coils may be damaged. Examination of the starter with the end cap off will indicate it the commutator is worn or damaged. These problems are generally not user-serviced, and a rebuilt or new replacement starter is in order.
22 More common is the general wear of the brushes that ride on the commutator, and this is a normal replacement procedure that can be handled by the home mechanic.
23 The commutator end is the opposite end from the starter drive. Remove the long through-bolts and remove the commutator end cap.
24 The brushes are contained within the end cap **(see illustration 3.24)**. A new brush kit will include the brushes and brush springs. **Note:** *All four brushes must be replaced at the same time*.
25 When all the brushes and springs are back in place in the commutator end cap, they must be kept in position while the end cap is slipped over the starter shaft. Insert the through-bolts, align the case markings and tighten the bolts with nuts on the drive end.
26 Starter installation is the reverse of the removal procedure.

# 4 Ignition systems

## Breaker points ignition systems

1 The point ignition system, installed on early models and smaller horsepower engines, is also called the "flywheel magneto" ignition system. This mechanical ignition system relies on the opening and the closing of the points to trigger the secondary voltage to the spark plug. The breaker points and the condenser are located under the flywheel. As the flywheel rotates, the magnet mounted on the flywheel passes the coil mounted on the engine case. The magnetic field created from the flywheel magnet (magneto) passing near the coil mounted on crankcase, causes a current to generate in the ignition coil primary circuit (winding). As the flywheel continues to rotate, it passes the last pole in the lamination stack on the coil. This produces a large change in the magnetic field ultimately producing high current in the primary circuit. This current change in the primary circuit builds until the points open and the secondary circuit is induced to flow to the spark plug and fire, releasing a high voltage spark.
2 Most points ignition systems include the ignition points, the coil, the condenser, the flywheel magneto, the ignition cam, the stator plate (dust cover, the cam wiper). The stator plate consists of the points, the cam wiper, the condenser and laminations. The laminations are strips of iron riveted together to form an iron core. If corrosion is severe, the laminations will be hampered by the excess deposits requiring the stator plate to be replaced.
3 There are two different type points ignition systems on these models; the standard point system (internal coil) and the fixed timed system (external coil). Both systems require a dial indicator to set the ignition timing.

4.4 Remove the flywheel key (arrow)

4.5 Release the retainer clip and lift off the point cover and gasket

4.6 Remove the nut (arrow) and detach the primary wires from the point terminal

4.7 Slide the movable point up and off the post, then remove the spring and terminal insulator

4.8 Remove the screw (arrow) and lift out the fixed point

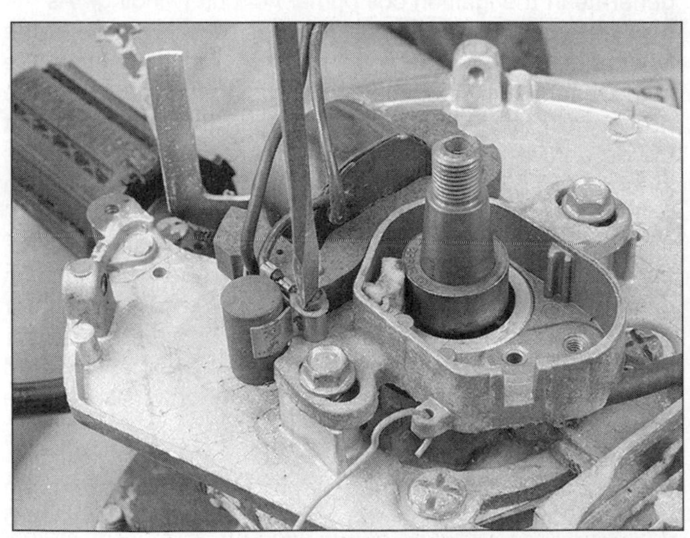

4.9 Remove the mounting screw and detach the condenser

## Breaker points

### Removal

4 Remove the flywheel. Refer to the appropriate section in this Chapter. Check the flywheel key. If it's sheared off, install a new one **(see illustration 4.4)**. Check the contact points to see if they're burned, pitted, worn down or covered with oil. Once you've gone to the trouble of removing the flywheel, new points should be installed.

5 Release the retainer clip and lift off the point cover and gasket; look for oil leaking past the crankshaft seal **(see illustration 4.5)**.

6 Remove the nut and detach the primary wires from the point terminal **(see illustration 4.6)**. When installing a new condenser, you'll have to cut the original wire at the terminal (the new one will have a terminal that fits over the post).

7 Slide the movable point up, off the post, and remove the spring and the terminal and insulator **(see illustration 4.7)**.

4.10 Clean the ignition point cavity with contact cleaner, then wipe it with a rag

4.11 Install the new fixed point, leaving the screw loose enough to allow movement of the plate

4.13 Slip the new movable point over the post and position the insulator in the cutout

8 Remove the screw and lift out the fixed point **(see illustration 4.8)**.

9 Remove the mounting screw and detach the condenser, then install the new one in its place and route the wire over to the point terminal **(see illustration 4.9)**.

10 Clean the ignition point cavity with contact cleaner and wipe it out with a rag **(see illustration 4.10)**.

## Installation

11 Install the new fixed point - leave the screw loose enough to allow movement of the plate **(see illustration 4.11)**.

12 Slip the new movable point over the post and position the insulator in the cutout - slip the primary and condenser wires onto the terminal and install the nut.

13 Turn the crankshaft very slowly until the cam opens the movable point as far as possible - if the cam was removed to replace the oil seal, make sure it's installed with the correct side out **(see illustration 4.13)**.

14 Insert a clean feeler gauge - 0.020-inch thick - between the contact points and move the fixed point very carefully with a screwdriver until the gap between the points is the same thickness as the feeler gauge (be careful not to change the position of the movable point as this is done).

15 Turn the crankshaft and make sure the movable arm opens and closes.

16 Install the gasket and point cover and snap the retainer clip into place.

## Ignition timing

17 All points type systems require that the engine be timed. Most systems will not require a precise setting using a dial indicator unless the engine was disassembled for rebuilding purposes or the stator plate mounting bolts were loosened by mistake. In most cases, a simple points replacement with the proper feeler gauge setting will time the engine.

18 Remove the cylinder head (see Section 8) and rotate the crankshaft to bring the piston to the top of its travel, then turn the crankshaft slightly counterclockwise (viewed from the magneto side). Attach a dial indicator to the engine, with the plunger of the dial indicator contacting the top of the piston, perpendicular to it.

19 Find TDC by rotating the crankshaft clockwise when looking at the magneto end of the crankshaft. Turn the crankshaft until the indicator needle stops moving to indicate that the piston has reached the top of its stroke. Zero the dial indicator. This setting is TDC.

20 Rotate the crankshaft counterclockwise (magneto side) until the dial indicator is 0.010-inch past the specified timing setting (refer to the Specifications listed at the end of this Chapter). For example, if the setting is 0.070-inch, rotate the crankshaft counterclockwise until the dial indicator reads 0.080-inch, then back again to 0.070 inch. This will take out any extra slack between the piston, connecting rod and crankshaft. **Note:** *It's a good idea to temporarily reinstall the flywheel (without moving the crankshaft) and making matchmarks on the flywheel and the engine. That way, the next time the ignition timing must be set, the cylinder head will not have to be removed (the marks can simply be lined up).*

21 Next, set the ignition points at this crankshaft setting. Connect an ohmmeter to the ignition point lead and ground, then loosen the two bolts holding the stator assembly and rotate the assembly until the ohmmeter indicates that there is a break (open) in the circuit. A static timing light can be used for this check also. Tighten the stator assembly bolts in this position.

22 Reconnect the leads to the points terminal and tighten the securing nut making sure the leads do not touch the flywheel. Be sure to adjust the air gap between the ignition coil and the flywheel on externally mounted ignition coil system (see Steps 32 and 33). The specifications for air gap on externally mounted coils is 0.0125 inch.

23 Clean the cylinder head and engine block mating surfaces, then install the cylinder head using a new gasket. **Note:** *If the cylinder head and piston crown are covered with carbon deposits, decarbonize the head and piston crown before installing the head.*

# Electronic ignition systems

24 The electronic ignition system also called the "solid state ignition" system (SSI) incorporates solid state electronic components with no mechanical features. These systems are equipped with flywheel magnets (magneto), charge coil (external or internal), capacitor, silicon controlled rectifier, pulse transformer, trigger coil and spark plug. The charge coil, trigger coil, pulse transformer, capacitor and rectifier assembly are all contained within the ignition coil/module assembly.

25 This electronic ignition system is more durable than the conventional points system that commonly fail due to burning, oxidation and mechanical wear. The SSI system operates similar to a points system but the method used to "break" the electronic signal differs. Flywheel rotation generates electricity in the primary circuit of the ignition coil assembly when the magnets rotate past the charge coil. As the flywheel rotates, it passes a trigger coil where a low voltage signal is produced. The low voltage signal allows the transistor to close allowing the energy stored in the capacitor to fire the secondary circuit or spark plug. The transistor acts as a primary circuit switching device in the same manner as the points in a mechanical (points) ignition system.

26 These systems do not require an ignition timing procedure.

## Coil/module assembly

### Removal

27 Disconnect the spark plug lead from the plug. On models with battery systems, disconnect the negative cable from the battery. **Caution:** *Never connect battery voltage to any wire or component of the electronic-magneto ignition system, or the module will be damaged.*

28 On electronic ignition systems that mount the ignition coil inside the flywheel, remove the flywheel. Refer to the appropriate engine section.

**4.29 Ignition coil mounting bolts (arrows)**

29 At the module, pull the connector from the kill lead tab on the module. Remove the two module mounting screws and remove the coil/module **(see illustration 4.29)**.

### Installation

30 Installation is the reverse of the removal procedure.

31 When installing the module (external), the gap between the flywheel magnet and the module must be set.

32 Turn the flywheel so that the magnet is directly under the module. Loosen the mounting screws slightly and insert a clean non-magnetic feeler gauge of the correct thickness (0.0125 inch). The magnet will pull the module down to squeeze the feeler gauge between module and magnet. Tighten the module mounting screws.

33 Rotate the flywheel a few revolutions to make sure that the magnet never touches the module's legs. Use the feeler gauge to check that both legs of the module are the same distance from the magnet. Tighten the mounting bolts.

# 5 Carburetor disassembly and reassembly

## General information

1 There are two basic types of carburetors installed on these engines. One type of carburetor uses a hollow float to maintain the operating level of fuel in the carburetor. This is called the FLOAT type carburetor. Another type uses a rubber diaphragm and it is called the DIAPHRAGM type carburetor. This type exposes one side to intake manifold pressure and the other side to atmospheric pressure. The diaphragm acts as a float, maintaining the proper fuel level in the carburetor. The diaphragm type carburetor allows the engine to operate while it is tilted at various angles; this type is installed on portable equipment.

**5.2 Location of the manufacturing number on the carburetor (arrow)**

2 There are two different manufacturers for these carburetors, Tecumseh and Walbro. Tecumseh carburetors are identified by the manufacturing numbers and date code stamped onto the flange on the side of the carburetor **(see illustration 5.2)**. Walbro carburetors can also be distinguished using the manufacturing numbers stamped onto the side of the carburetor body. Some carburetors are equipped with an adjustable main jet (main mixture screw) easily identified by the screw on the bottom of the float chamber. This adjustment screw is termed the "high-speed mixture adjustment screw." The "low speed mixture adjustment screw" is located near the throttle plate. Others are equipped with fixed main jets where NO adjustment is required. All carburetors are equipped with an idle mixture screw and an idle speed screw. **Note:** *Tecumseh carburetors use series and letter designations to indicate the specific type carburetor; Series I, III, IV (L-head engines), Series III, IV, VI (dual system), VII (OHV light-duty) and HH80-100-120 and VH100 (OHV heavy duty engines). Walbro carburetors use letter and series to designate carburetor types; Model LMH, LME, Series VI, VIII, IX (L-head and OHV) and HH80-100 and VH80-100 (OHV heavy duty).*

# Disassembly

3 The following procedures describe how to disassemble and reassemble the carburetor so new parts can be installed. Read the sections in Chapter 5 on carburetor removal and overhaul before doing anything else. In some cases it may be more economical (and much easier) to install a new carburetor rather than attempt to repair the original. Check with a parts retailer or small engine repair shop to see if parts are readily available and compare the cost of new parts to the price of a complete ready-to-install (standard service) carburetor before deciding how to proceed. The carburetor model number and date code are stamped on the edge of the mounting flange.

## Float type carburetors

4 While counting the number of turns, carefully turn the idle mixture adjusting screw in until it bottoms, then remove it along with the spring. Repeat this procedure if the carburetor is equipped with a high-speed mixture adjusting screw (located at the bottom of the fuel bowl).

5 Remove the idle speed adjustment screw and spring from the carburetor body. Counting and recording the number of turns required to bottom the screws will enable you to return them to their original positions and minimize the amount of adjustment required after reassembly.

6 Detach the float bowl from the carburetor body. On some carburetors the float bowl is held in place with a bolt **(see illustration 5.6a)**, while on others, it's held in place with the high speed mixture adjusting screw fitting **(see illustration 5.6b)**. Be sure to note the locations of any gaskets/washers used.

7 If you're working on a Walbro carburetor, note how the float spring is positioned before removing the float - it may be a good idea to draw a simple sketch to refer to during reassembly. Push the float pivot pin out of the carburetor body **(see illustration 5.7)**.

8 Remove the float assembly, the inlet needle valve and inlet needle seat **(see illustration 5.8)**. Note how the inlet needle valve is attached to the float. The retainer clip, if

**5.6a The float bowl may be held in place with a bolt (arrow) . . .**

**5.6b . . . or a high speed mixture adjusting screw fitting**

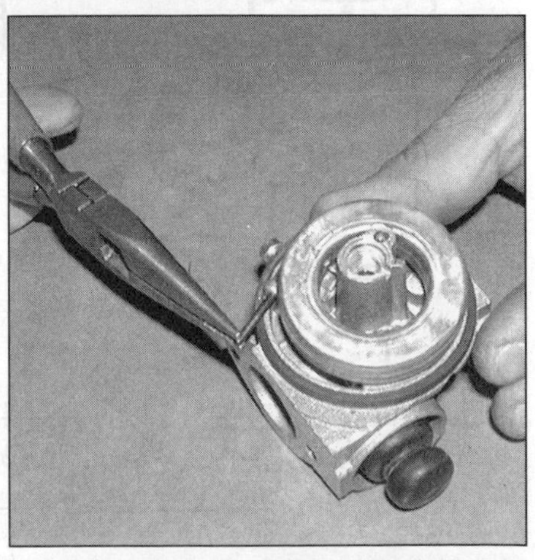

**5.7 Remove the float pin from the hinge**

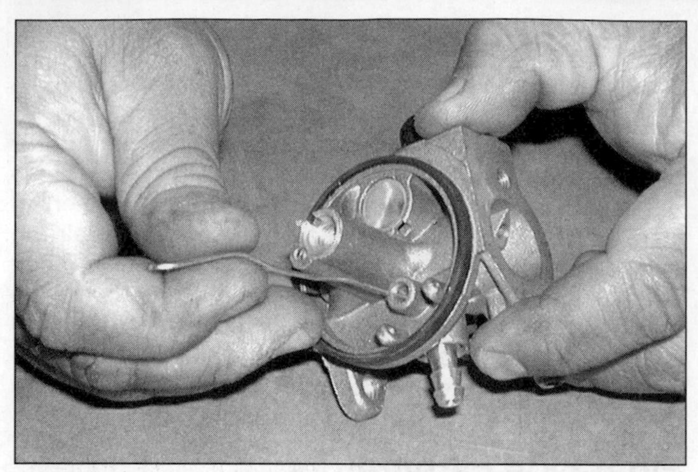

5.8 Remove the inlet needle seat using a wire or paper clip with a slight hook

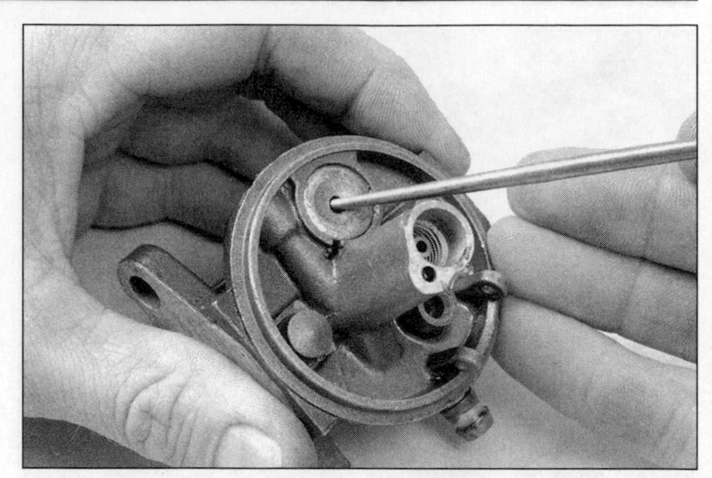

5.10a To remove a Welch plug, drill a 1/8 inch hole in the plug and use an awl or other tool to pry it out - don't damage the bore in the process

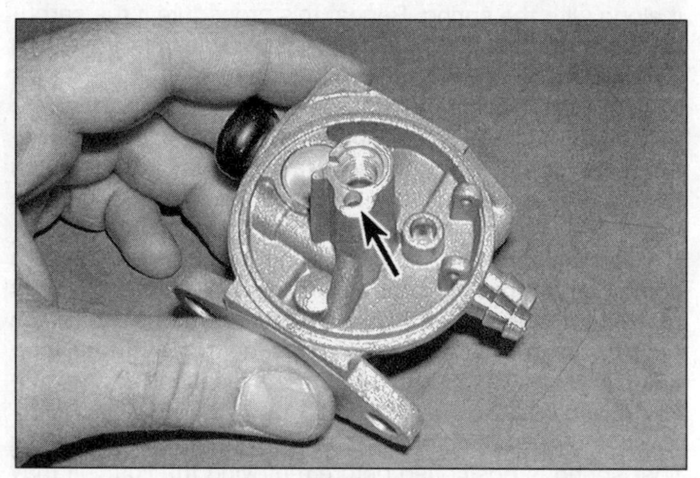

5.10b DO NOT remove the ball plugs (arrows) from the carburetor body!

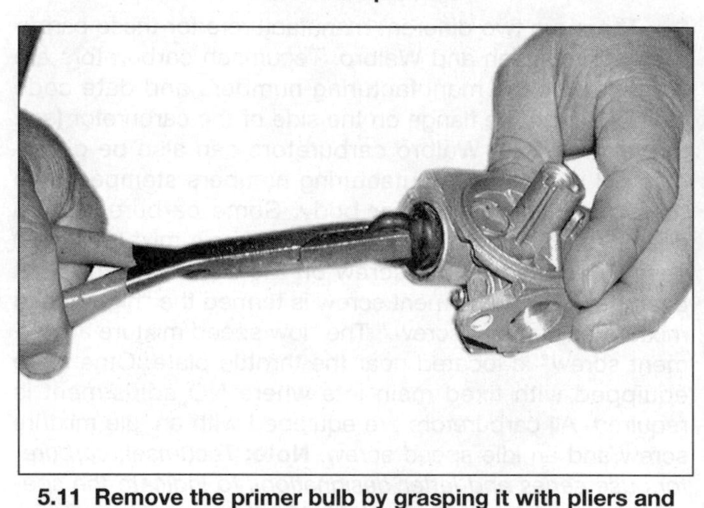

5.11 Remove the primer bulb by grasping it with pliers and twisting it out of the body

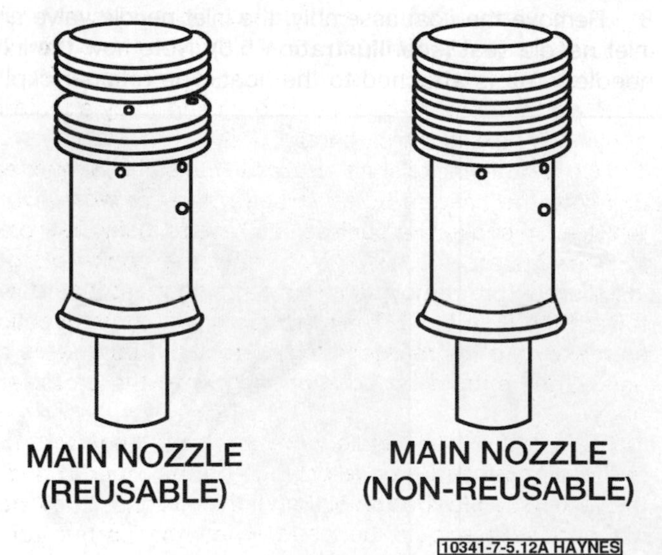


5.12 Details of the Walbro carburetor main nozzle

used, must be positioned the same way during reassembly.

9 Remove the float bowl gasket.

10 To do a thorough cleaning job, remove any Welch plugs from the carburetor body **(see illustration 5.10a)**. **Note:** *Do not remove any brass cup or ball plugs (if used)* **(see illustration 5.10b)**. *One may be located near the inlet needle seat cavity to seal off the idle air bleed. Another one may be located in the base, where the float bowl mounting bolt or high speed adjustment screw fitting seals the idle fuel passage. A third plug may be located on the side of the main nozzle casting, sealing the idle fuel passage.* Drill a small hole in the center of the Welch plug and pry it out with a punch or thread a sheet metal screw into the hole, grasp the screw head with a pliers and pull the plug out.

11 If the carburetor has a primer bulb, it can be removed with pliers **(see illustration 5.11)**. First, remove the inner circlip with a small screwdriver tip and then grasp it securely and twist and pull to detach it from the carburetor body. The retainer can be pried out with a screwdriver. If the original primer is removed, discard it and install a new one when reassembling the carburetor. A 3/4-inch deep socket can be

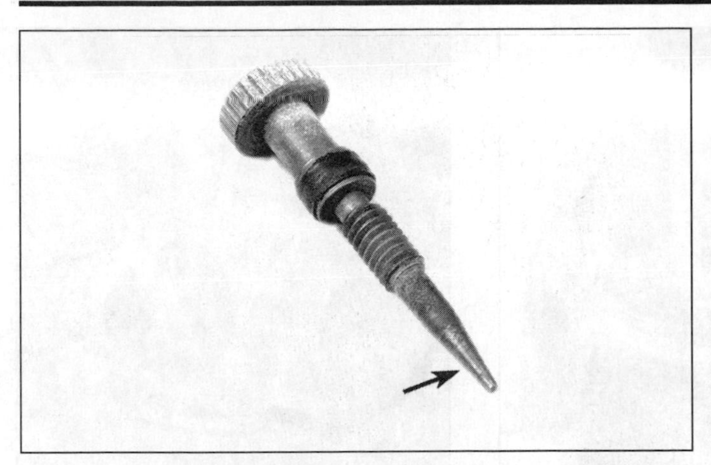

5.14 Check the main mixture adjusting screw tip for damage and distortion - if it is blunted, bent or has a groove worn in it, install a new one

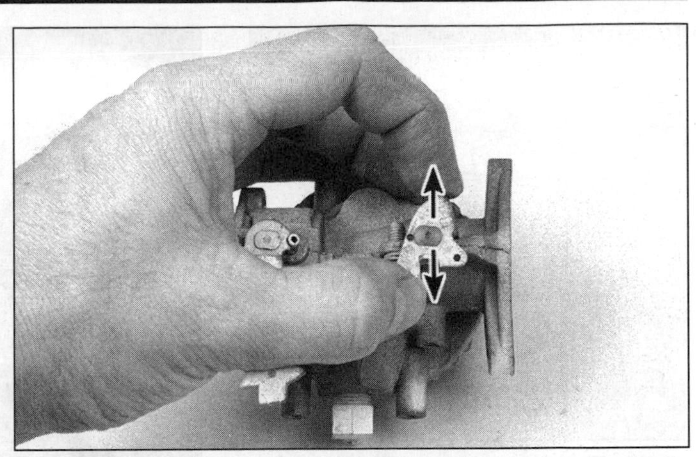

5.15 Check for wear in the throttle shaft or bores by moving the shaft back-and-forth

5.16 The line on the throttle valve plate (arrow) must be in the 12 o'clock position when it is reinstalled

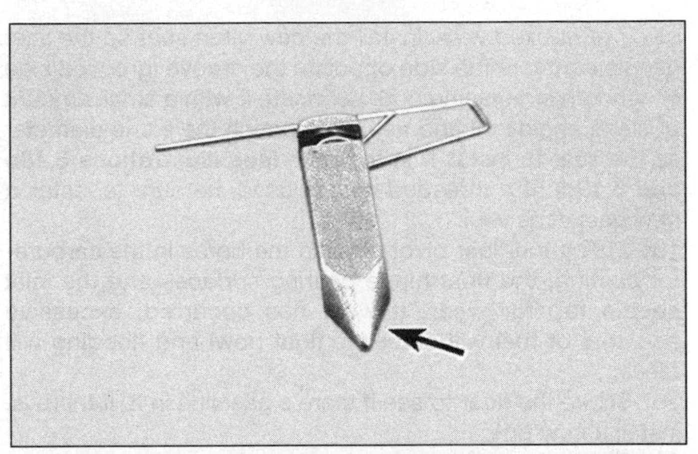

5.18a Check the inlet needle valve for a groove or ridge in the tapered area (arrow)

used to seat the new bulb/retainer in the cavity.

12 Do not attempt to remove the main nozzle on a Tecumseh model carburetor. On Walbro carburetors, the main nozzle may be removed for cleaning if necessary. Be sure to use the correct size screwdriver. If the main nozzle is the original nozzle with full threads it must be replaced with a service replacement nozzle **(see illustration 5.12)**.

13 Refer to Chapter 5 and follow the cleaning/inspection procedures outlined under Carburetor Overhaul. **Note:** *Don't soak plastic or rubber parts in carburetor cleaner. If the float is made of cork, don't puncture it or soak it in carburetor cleaner.*

14 Check each mixture adjusting screw tip for damage and distortion **(see illustration 5.14)**. The small taper should be smooth and straight.

15 Check the throttle plate shaft for wear by moving it back-and-forth **(see illustration 5.15)**.

16 Check the throttle plate fit in the carburetor bore. If there's play in the shaft, the bore is worn excessively, which may mean a new carburetor is required. On some carburetors, replacing the throttle plate shaft or plate, or both, may cure the problem. Check with a parts retailer or small engine repair shop. The throttle plate and shaft don't have to be removed unless new parts are required. If they are removed, make sure the line or number on the plate is facing out and the line is in the 12 o'clock position when reinstalled **(see illustration 5.16)**. If the throttle binds after the parts are reinstalled, loosen the screw and reposition the plate on the shaft. If dust seals are used, they should be positioned next to the carburetor body.

17 Check the choke shaft for play in the same manner and examine the linkage holes to see if they're worn. Don't remove the choke shaft unless you have to install new parts to compensate for wear. Note how the choke plate is installed before removing it; the flat side must face down, toward the float bowl. They will operate in either direction, so make sure it's reassembled correctly. If dust seals are used, they should be positioned next to the carburetor body.

18 Check the inlet needle valve and seat. Look for nicks and a pronounced groove or ridge on the tapered end of the valve **(see illustration 5.18a)**. If there is one, a new needle and seat should be used when the carburetor is reassembled. They are normally installed as a matched set. **Note:** *Some inlet needle seats can be unscrewed, while others are a viton plastic insert that can be pulled out with a*

5.18b The viton inlet needle valve seat must be installed with the grooved side in

5.18c Use a pin punch to push the viton seat in until it is bottomed in the bore

5.22 The float spring (arrow) must face the choke end of the carburetor

*piece of hooked wire.* Install the new viton seat so the inlet needle contacts the side opposite the groove (grooved side in, smooth side facing out). Lubricate it with a small amount of clean engine oil and use a pin punch the same diameter as the seat to install it in the bore **(see illustrations 5.18b and 5.18c)**. If a threaded seat is used, be sure to install a new gasket as well.

19 Check the float pivot pin and the bores in the carburetor casting, the float hinge bearing surfaces and the inlet needle tab for wear. If wear has occurred, excessive amounts of fuel will enter the float bowl and flooding will result.

20 Shake the float to see if there's gasoline in it. If there is, install a new one.

21 Check the fuel inlet fitting to see if it's clean and unobstructed. If it's damaged or plugged, a new one can be installed. Twist and pull on the old one to remove it. When installing the new one, insert it into the opening in the carburetor body, then apply a non-hardening thread locking compound to the exposed part of the shank. Push it in until the shoulder on the fitting contacts the carburetor. Make sure the fitting points in the same direction as the original.

22 Once the carburetor parts have been cleaned thoroughly and inspected, reassemble it by reversing the above procedure. **Note:** *Make sure all fuel and air passages in the carburetor body, main nozzle, inlet needle seat and float bowl mounting bolt are clean and clear. Be sure to use new gaskets, seals and O-rings. Whenever an O-ring or seal is installed, lubricate it with a small amount of grease or oil. Don't overtighten small fasteners or they may break off.* When installing new Welch plugs, apply a small amount of non-hardening sealant (such as Permatex no. 2) to the outside edge and seat the plug in the bore with a 1/4-inch or larger diameter pin punch and a hammer. Be careful not to collapse the plug - flatten it just enough to secure it in the opening. When the inlet needle valve assembly is installed, be sure the retaining clip is attached to the float tab. **Note:** *On Walbro carburetors only, when attaching the float to the carburetor, position the spring between the float hinges with the long spring end pointing toward the choke end of the carburetor* **(see illustration 5.22)**. *Wind the spring back to put tension on it, then set the float in place, release the tension and install the pin.*

5.23 A drill bit can be used to gauge the specified float level on Walbro carburetors

23 Invert the carburetor and check the float level: On Walbro carburetors and OHV light-duty engines equipped with a Tecumseh carburetor, the gap between the float and the carburetor body on the side directly opposite the pivot pin is the float measurement. A drill bit of the correct size can be used as a gauge **(see illustration 5.23)**. The float should touch the drill bit when the bit is flush with the edge of the float. If the float is too LOW or HIGH, bend the tab near the float hinge until the correct height is attained. Refer to the Specifications listed at the end of this Chapter for the correct setting.

24 On L-head engines equipped with a Tecumseh carburetor, special gauges are required to check the float level. Obtain the gauge from a Tecumseh parts distributor and follow the instructions provided with the tool. **Note 1:** *All Tecumseh Series VII and Walbro LMK carburetors are equipped with a fixed float height that is non-adjustable.* **Note 2:** *If float sticking occurs due to deposits, or when the*

*fuel tank is filled for the first time, loosen the carburetor bowl nut one complete turn then spin the bowl 1/4 turn in either direction to free-up the float and finally, return the bowl to the original position and tighten the nut.*

25 Install the idle speed adjustment screw with the carburetor in an upright position (not upside-down or sideways). This will prevent damage to carburetors with a metering rod in the idle circuit.

26 Turn the mixture adjusting screw(s) in until they bottom and back each one out the number of turns required to restore them to their original positions. Refer to the chart in Section 6.

## Diaphragm type carburetors

27 While counting the number of turns, carefully screw the high-speed mixture adjusting screw in until it bottoms, then remove it along with the spring. **Note:** *The high-speed mixture adjusting screw is the screw furthest from the engine (the low-speed mixture adjusting screw is the screw closer to the engine).*

28 Remove the low speed mixture adjustment screw and spring from the fitting in the same manner. Counting and recording the number of turns required to bottom the screws will enable you to return them to their original positions and minimize the amount of adjustment required after reassembly.

29 Detach the diaphragm cover from the carburetor body. Be sure to note the locations of any gaskets/washers used. Mark the diaphragm to insure correct installation.

30 Use a 9/32-inch thin wall socket to unscrew and remove the inlet needle and seat assembly from the carburetor body.

31 Mark the direction of the inlet fitting on the carburetor body. Some carburetors are equipped with a fuel strainer as an integral part of the fuel fitting.

32 To do a thorough cleaning job, remove any Welch plugs from the carburetor body. **Note:** *Do not remove any brass cup or ball plugs (if used). One may be located near the inlet needle seat cavity to seal off the idle air bleed.* Drill a small hole in the center of the Welch plug and pry it out with a punch or thread a sheet metal screw into the hole, grasp the screw head with a pliers and pull the plug out.

33 Refer to Chapter 5 and follow the cleaning/inspection procedures outlined under Carburetor overhaul. **Caution:** *Don't soak plastic or rubber parts in carburetor cleaner.*

34 Check each mixture adjusting screw tip for damage and distortion **(see illustration 5.14)**. The small taper should be smooth and straight.

35 Check the throttle plate shaft for wear by moving it back-and-forth **(see illustration 5.15)**.

36 Check the throttle plate fit in the carburetor bore. If there's play in the shaft, the bore is worn excessively, which may mean a new carburetor is required. On some carburetors, replacing the throttle plate shaft or plate, or both, may cure the problem; check with a parts retailer or small engine repair shop. The throttle plate and shaft don't have to be removed unless new parts are required. If they are removed, make sure the line or number on the plate is facing out and the line is in the 12 o'clock position when reinstalled **(see illustration 5.16)**. If the throttle binds after the parts are reinstalled, loosen the screw and reposition the plate on the shaft. If dust seals are used, they should be positioned next to the carburetor body.

37 Check the choke shaft for play in the same manner and examine the linkage holes to see if they're worn. Don't remove the choke shaft unless you have to install new parts to compensate for wear. Note how the choke plate is installed before removing it.

38 Check the inlet needle valve and seat. Look for nicks and a pronounced groove or ridge on the tapered end of the valve. If there is one, a new needle and seat should be used when the carburetor is reassembled. They are normally installed as a matched set. Install the new seat. Be sure to install a new gasket as well.

39 Install the diaphragm assembly. The rivet head must always face toward the inlet needle valve. On carburetors with an "F" cast into the carburetor flange, the diaphragm goes next to the carburetor body. Most other diaphragm carburetors position the gasket between the diaphragm and the carburetor body. Install the cover mounting screws and the cover.

40 Once the carburetor parts have been cleaned thoroughly and inspected, reassemble it by reversing the above procedure. Note the following important points:

a) *Make sure all fuel and air passages in the carburetor body, main nozzle, inlet needle seat and float bowl mounting bolt are clean and clear.*
b) *Be sure to use new gaskets, seals and O-rings.*
c) *Whenever an O-ring or seal is installed, lubricate it with a small amount of grease or oil.*
d) *Don't overtighten small fasteners or they may break off.*
e) *When installing new Welch plugs, apply a small amount of non-hardening sealant (such as Permatex no. 2) to the outside edge and seat the plug in the bore with a 1/4-inch or larger diameter pin punch and a hammer. Be careful not to collapse the plug. Flatten it just enough to secure it in the opening.*

41 Installation is the reverse of removal.

# 6 Carburetor and governor adjustment

## Carburetor

1 When making carburetor adjustments, the air cleaner must be in place and the fuel tank should be at least half full.

2 If not already done, turn the low and high speed FUEL/AIR MIXTURE adjusting screws clockwise until they seat lightly, then back them out the same number of turns as you recorded during disassembly. If you didn't record the settings, the accompanying chart of suggested initial settings should be a good starting point before making running adjustments:

| L-head and OHV light-duty engines | Main mixture screw | Idle mixture screw |
|---|---|---|
| Tecumseh carburetors | | |
| Float type carburetors | 1-1/2 turns | 1 turn |
| Diaphragm type carburetors | 1 turn | 1 turn |
| Walbro carburetors | | |
| LMH | 1 1/2 turns | 1-1/2 turns |
| WHG and LME | 1-1/4 turns | 1 1/4 turns |
| LMK | Non-adjustable | 1 turn |
| **L-head and OHV heavy duty engines** | | |
| OHV engines | 1 turn | 1 turn |
| **L-head engines** | | |
| Tecumseh carburetors | 1-3/4 turn | 1-1/4 turn |
| Walbro carburetors | 1-1/2 turn | 1-1/4 turn |

3 Start the engine and allow it to reach operating temperature before making adjustments.
4 With all carburetor mixture adjustments, the ideal setting point is halfway between the rich and lean positions.
5 Start with the main or high-speed fuel mixture (on models with adjustable main jets) **(see illustration)**. The engine should be run at full throttle, and preferably under some load, such as mowing, pumping, snow-blowing, etc. In small increments, turn the main fuel screw out until the engine runs slower - this is the rich position. Adjust the screw inward until the engine speed increases again (to where it started) and then begins to decrease. This is the lean position. Set the screw to halfway between the rich and lean positions.
6 Place the throttle control lever in the SLOW or IDLE position, then adjust the low speed idle mixture screw in the same manner as in Step 5, except not under load **(see illustration)**. Leave the low speed idle mixture screw halfway between the rich and lean positions.
9 If the engine smokes under load, this indicates the high-speed mixture is too rich, and if it backfires under load, this indicates the high-speed mixture is too lean. If the engine was adjusted in warm weather, but stops frequently in cold weather, try adjusting the main fuel mixture slightly richer.
10 Move the speed control lever from SLOW to FAST - the engine should accelerate without hesitating or sputtering.
11 If the engine dies, it's too lean - turn the adjusting screws out in small increments. If the engine sputters and runs rough before picking up the load, it's too rich - turn the adjusting screws in slightly.
12 If the adjustments are "touchy", check the float level and make sure it isn't sticking.
13 Turn the idle adjusting screw to provide a satisfactory idle **(see illustration)**.

# Governor

14 It is best not to change the relationship of any linkages or other components in the governor system, unless a worn or broken component is to be replaced. If the engine is being overhauled, do not change any adjustments. If the governor was working suitably before the overhaul, and none of the connections are altered, it will continue to work properly. While governor adjustment should be made with a small-engine tachometer to avoid setting the engine to a higher rpm than it is designed for, some basic adjustments can be made by the home mechanic.
15 On most engines, a governor shaft sticks through the

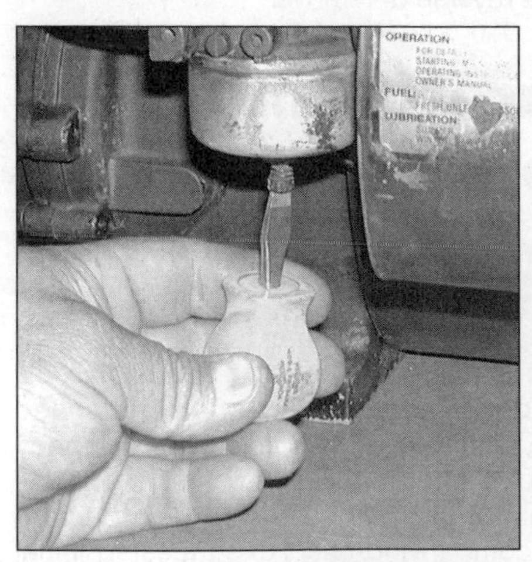

6.5 Adjusting the high speed mixture screw

6.6 Adjusting the low speed mixture screw

6.13 Turn the idle speed screw to set the idle to the lowest, smoothest idle rpm possible

7.9a Checking the valve clearance adjustment on an L-head engine

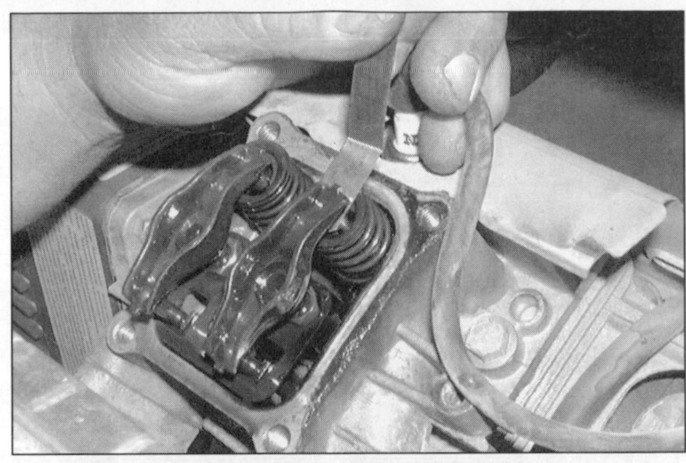

7.9b Checking the valve clearance adjustment on an OHV engine

side of the crankcase, while the governor mechanism is protected inside the crankcase. An arm attaches to this shaft with a clamp-bolt/nut. The basic initial adjustment on Tecumseh governors begins with loosening the nut/bolt where the arm attaches to the shaft **(see illustration 8.3)**. Pull the arm back away from the carburetor to its limit and hold it there (counterclockwise on vertical and clockwise on horizontal crankshaft engines). Grip the governor lever which is connected to the throttle and force it to wide open throttle (WOT). Tighten the bolt/nut on the clamp in this position.

16 Sensitivity is the other common adjustment for the governors. If the engine seems like it is hunting for consistent speed as the load changes, the sensitivity is too sensitive and must be lowered. If the engine speed drops too much when the engine encounters a load, the governor adjustment isn't sensitive enough.

17 On some governor assemblies, the governor sensitivity is adjusted by moving the position of the governor spring, where it hooks into the lower portion of the governor arm. A series of holes are provided. To increase the sensitivity, raise the spring into the next higher hole in the arm. To decrease sensitivity, hook the spring into a hole lower than its original position. The center of this arm is the factory position, but varying equipment and loads are accommodated by the extra adjustment holes.

# 7 Valve clearance - check and adjustment

1 Correct valve tappet clearance is essential for efficient fuel use, easy starting, maximum power output, prevention of overheating and smooth engine operation. It also ensures the valves will last as long as possible.

2 When the valve is closed, clearance should exist between the end of the stem and the tappet. The clearance is very small - measured in thousandths of an inch - but it's very important. The recommended clearance values are listed in the Specifications at the end of this Chapter. Note that intake and exhaust valves often require different clearance. **Note:** *The engine must be cold when the clearance is checked.* A feeler gauge with a blade thickness equal to the valve clearance will be needed for this procedure.

3 On L-head engines, if the clearance is too small, the valves will have to be removed and the stem ends ground down carefully and lapped to provide more clearance. This is a major job, covered in the overhaul and repair procedures. If the clearance is too great, new valves will have to be installed (again, a major repair procedure). On OHV engines, the tappets are adjustable, and with two wrenches, the clearance can be set to Specifications.

4 Disconnect the wire from the spark plug and ground it on the engine.

5 On L-head engines, remove the bolts and detach the tappet cover plate or the crankcase breather assembly **(see illustration 8.11)**. On OHV engines, remove the bolts and remove the valve cover **(see illustration 9.9a)**.

6 Turn the crankshaft by hand and watch the valves to see if they stick in the guide(s).

7 Turn the crankshaft until the intake valve is wide open, then turn it an additional 360-degrees (one complete turn). This will ensure the valves are completely closed for the clearance check. L-head engines are designed so that the valves are easily viewed with the tappet cover off. On OHV engines, make sure the engine is positioned on TDC. Refer to the timing procedure in Section 4.

8 Select a feeler gauge thickness equal to the specified valve clearance and slip it between the valve stem end and the tappet. Refer to the Specifications listed at the end of this Chapter.

9 If the feeler gauge can be moved back-and-forth with a slight drag, the clearance is correct **(see illustration 7.9a)**. If it's loose, the clearance is excessive; if it's tight (watch the valve to see if it's forced open slightly), the clearance is inadequate. On OHV engines with adjustable tappets, use two wrenches on the tappet to either extend or retract it to

8.2 Remove the dipstick tube mounting screw from the blower housing

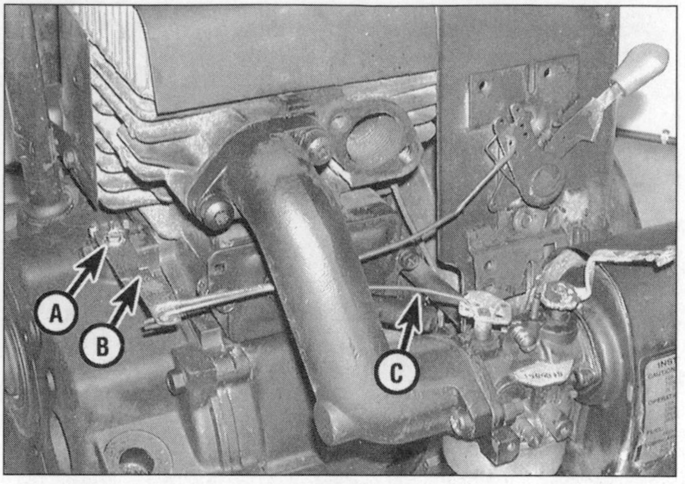

8.3 Governor linkage details

*A Governor lever clamp* *B Lever* *C Governor linkage*

achieve the proper valve clearance **(see illustration 7.9b)**. **Note:** *OHV engines are equipped with a variety of adjusting mechanisms on the rocker arms. Some require an Allen wrench and crowfoot wrench to lock the adjustment. Others use off-size adjusting and locknuts. Some use an adjustable screw and locknut arrangement.*

10 If the clearance is incorrect on L-head engines, refer to Chapter 5 for valve service procedures.

11 Reinstall the crankcase breather or tappet cover plate, using a new gasket if necessary.

# 8 L-head light duty engines

## Disassembly

The engine components should be removed in the following general order:

*Engine cover (if used)*
*Fuel tank*
*Cooling shroud/recoil starter*
*Carburetor/intake manifold*
*Muffler*
*Cylinder head*
*Flywheel*
*Flywheel brake components (if equipped)*
*Ignition components*
*Crankcase breather assembly*
*Oil sump/crankcase cover*
*Crankshaft*
*Camshaft*
*Tappets*
*Piston/connecting rod assembly*
*Valves*
*Governor components*

1 For fuel tank, carburetor/intake manifold, muffler and oil draining procedures, refer to Chapters 4 and 5 as necessary. For the shroud and recoil assembly removal, refer to the beginning of this chapter.

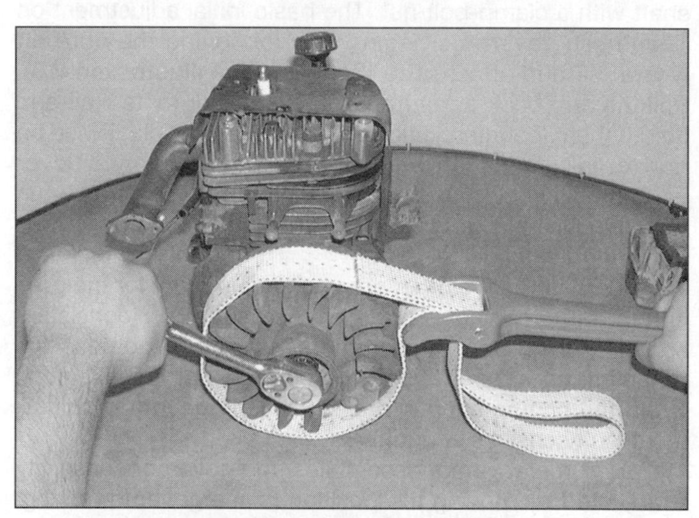
8.4 Remove the flywheel nut while locking the flywheel in position with a strap wrench

2 Remove the blower housing by removing the dipstick tube mounting screw **(see illustration 8.2)** and unscrewing the dipstick tube from the engine.

3 Disconnect the governor linkage from the carburetor. Be sure to make notes or a drawing of the linkage on your model to insure correct reassembly **(see illustration 8.3)**.

4 Prevent the flywheel from turning by using a strap wrench **(see illustration 8.4)** and remove the large nut, washer and starter cup. After the nut is removed, the starter cup and debris screen can be detached.

5 Thread a flywheel knock-off tool onto the flywheel shaft (check with your local power equipment dealer or small engine repair shop for tool availability). Thread the tool in until it bottoms out, then unthread it one complete turn **(see illustration 8.5)**.

8.5 Tap the knock-off tool while prying up on the flywheel with a flat bladed screwdriver or prybar

8.8 Note which side of the point cam faces UP (TOP)

8.10 Remove the coil/points assembly mounting bolts (arrows)

8.11 Remove the crankcase breather

6 Using a hammer, tap the knock-off tool while lifting the flywheel using a flat-bladed screwdriver or prybar to break the flywheel loose from the engine. Rotate the flywheel 180-degrees and pry at a different area if the flywheel is difficult to remove. DO NOT hammer on the end of the crankshaft and DO NOT use a jaw-type puller that applies force to the outer edge of the flywheel! As the tool is hit with the hammer, the flywheel should pop off the shaft taper. **Note:** *If a flywheel brake is used, remove the brake and related components.*

7 Place the flywheel upside-down on a wooden surface and check the magnets by holding a screwdriver at the extreme end of the handle while moving the tip toward one of the magnets - when the screwdriver tip is about 3/4-inch from the magnet, it should be attracted to it. If it doesn't, the magnets may have lost their strength and ignition system performance may not be up to par. Remove the flywheel key - if it's sheared off, install a new one. See Chapter 5 for checking procedures on the ignition and charging systems.

8 First remove the flywheel key, stator and baffle plate. Refer to Chapter 4 and remove the ignition points and related parts (if equipped), then note which side of the point cam is facing out and slide it up, off the end of the crankshaft (it should be marked TOP) **(see illustration 8.8)**.

9 Remove the cylinder head bolts, loosening them a little at a time ,working in an order opposite that of the tightening sequence **(see illustrations 8.46a and 8.46b)**. Remove the cylinder head.

10 Note how the wires to the coil are routed. If more than one wire is present, mark them with pieces of tape. Mark the coil bracket and engine bosses with a scribe or center punch, then remove the bolts and detach the coil/spark plug wire assembly **(see illustration 8.10)**.

11 If it's still in place, remove the bolts and separate the intake manifold from the engine - remove the gasket and discard it. Remove the mounting bolts and detach the crankcase breather assembly **(see illustration 8.11)** and gasket from the engine.

8.12 Remove the crankcase cover (vertical crankshaft engine shown)

8.16 Remove the eccentric for the plunger style oil pump

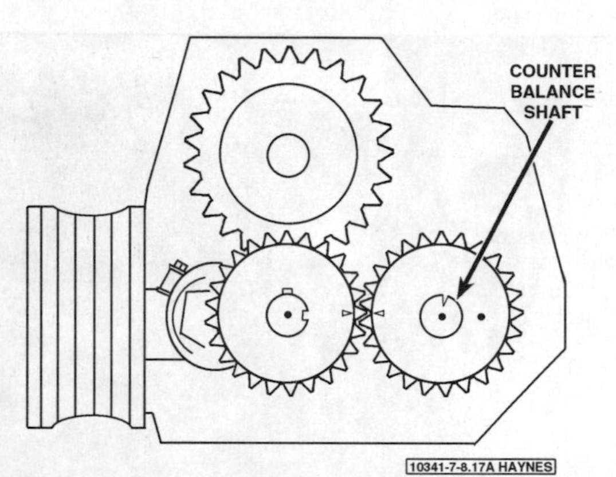


8.17a Be sure to align the balance shaft gears before disassembly

8.17b Align the camshaft and crankshaft timing gears (arrows) and paint the marks

12 Remove the crankcase cover **(see illustration 8.12)**. On engines equipped with a gear reduction assembly, turn the crankshaft to roll the reduction shaft gear off the crankshaft worm gear when removing the crankcase cover.

13 Loosen the oil sump-to-engine block bolts in 1/4-turn increments to avoid warping the sump, then remove them. Some engines have a crankcase cover (horizontal engine) instead of an oil sump (vertical crankshaft engines). They're attached to the engine block with several bolts and removal is similar.

14 Tap the sump/crankcase cover with a soft-face hammer to break the gasket seal, then separate it from the engine block and crankshaft. If it hangs up on the crankshaft, continue to tap on it with the hammer, but be very careful not to crack or distort it (especially if it's made of aluminum). If thrust washers are installed on the crankshaft or camshaft, slide them off and set them aside.

15 Use a ridge reamer to remove the carbon/wear ridge from the top of the cylinder bore. Follow the manufacturer's instructions included with the tool.

16 Vertical crankshaft engines are equipped with a plunger-type oil pump that's driven by an eccentric on the camshaft - it can be lifted out before the camshaft is removed **(see illustration 8.16)**. **Note:** *There are several different types of oiling systems installed on these engines. Smaller horsepower horizontal engines are equipped with the "splash oiling system." A dipper is installed on the end of the connecting rod to splash the oil in the sump onto the cylinder walls and the connecting rod bearing. Small horsepower vertical engines are equipped with a plunger style oil pump located on the end of the camshaft. The camshaft eccentric moves the barrel back and forth on the plunger, forcing oil through the hole in the camshaft and out of an oiling port located between the crankshaft and camshaft end caps in the case. The oil is then sprayed out under pressure over the camshaft and crankshaft bearing assemblies and the cylinder wall. Larger horsepower engines are equipped with a rotary oil pump (lobed gear type) that forces the pressurized oil onto the bearing surfaces, the piston, cylinder and connecting rod. The plunger style and the rotary style oil pumps are both referred to as "a pressurized oiling system."*

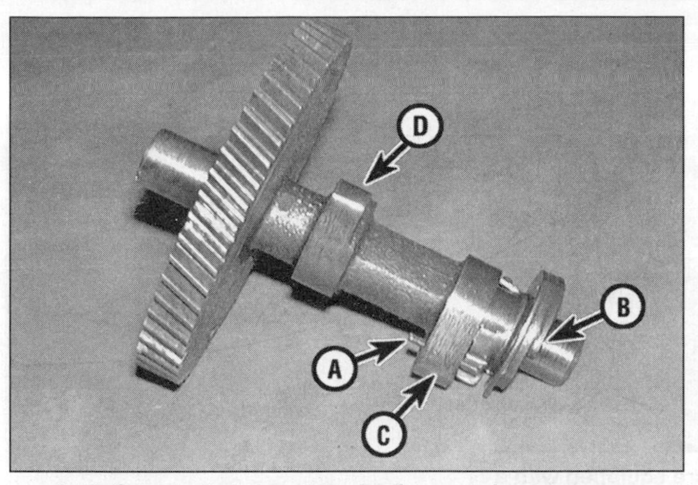


**8.18 Camshaft decompression lever and components**

*A Roll pin*
*B Spring*
*C Exhaust cam lobe*
*D Intake cam lobe*

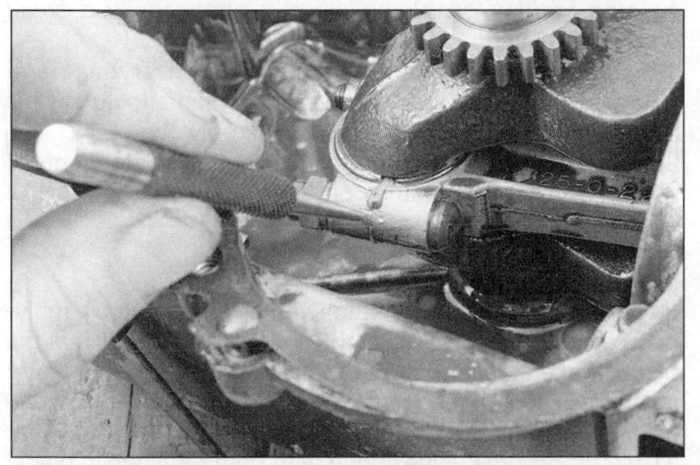

**8.20 Look for matchmarks on the connecting rod and cap - if you can't see any, use a punch to make your own marks**

17 Align the timing marks **(see illustrations 8.17a and 8.17b)**. On engines equipped with a balance shaft system, it will be necessary to make two sets of timing marks; one for the counterbalance shaft gears and the other set for the camshaft and crankshaft gears. Remove the counterbalance shaft gears to access the camshaft and crankshaft gears. **Note:** *Some engines may be equipped with an Ultra-Balance Counterbalance system. This system uses a single weighted shaft that drives off the crankshaft. This shaft functions as a counterweight to the weights on the crankshaft that cause excessive engine vibration. Be sure to paint the timing marks before disassembly and make notes or pictures to help the reassembly procedure.* **Note:** *On VM70, HHM80, HM80, HM100, TVM195 and TVM220 engines, rotate the camshaft clockwise three teeth past the aligned position to allow the compression release mechanism to clear the exhaust valve tappet allowing the camshaft to be removed.*

18 Remove the camshaft and balance shaft gears (if equipped). **Note:** *Remove the compression relief mechanism and check to make sure the roll pin is locked into the plunger* **(see illustration 8.18)**.

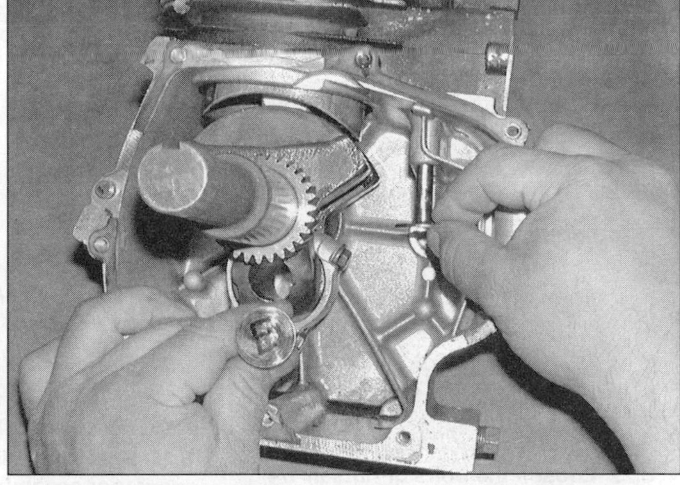

**8.19 After the camshaft is removed, pull out the tappets and store them in marked containers to aid in reassembly**

**8.21 Turn the crankshaft so the rod journal is at the bottom of the stroke (Bottom Dead Center). Flatten the locking tabs (if used) on the connecting rod bolts with a punch and hammer**

19 After the camshaft is removed, pull out the tappets and store them in marked containers so they can be returned to their original locations **(see illustration 8.19)**.

20 Look for match marks on the connecting rod and cap. If you can't see any, mark the side of the connecting rod and cap that faces OUT and note how the oil dipper (if used) is installed (the parts must be reassembled in the exact same relationship to the crankshaft) **(see illustration 8.20)**. **Note:** *Some models use offset piston and connecting rods. Be sure the match marks on the connecting rods are facing OUT when installing the connecting rods on all types of engines.*

21 Turn the crankshaft so the rod journal is at the bottom of its stroke (Bottom Dead Center). Flatten the locking tabs, if equipped, on the connecting rod bolts with a punch and hammer **(see illustration 8.21)**.

22 Loosen the bolts or nuts in 1/4-turn increments until they can be removed by hand **(see illustration 8.22)**. Separate the cap (and washers, if used) from the connecting rod,

8.22 Loosen the bolts or nuts in 1/4 turn increments until they can be removed by hand

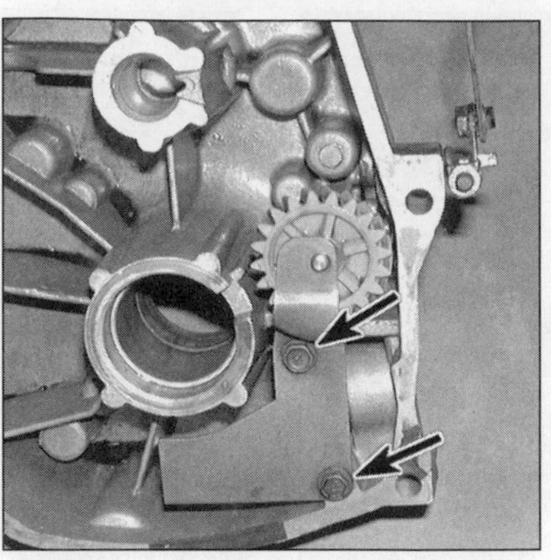

8.25 Some models are equipped with a governor assembly that is attached to the cover with a bracket and two bolts (arrows)

8.32a Install the washer and gear assembly and make sure the retaining ring is secured in the shaft groove

move the end of the rod away from the crankshaft journal and push the piston/rod assembly out through the top of the bore. The crankshaft can now be lifted out.

23 Compress the intake valve spring and remove the retainer, then withdraw the valve through the top of the engine. Pull out the spring, then repeat the procedure for the exhaust valve. Some engines have a retainer at the base of the spring as well.

24 Do not separate the lever from the governor shaft unless new parts are needed - the lever mount will be damaged during removal.

25 The governor assembly can be withdrawn from the gear shaft after the retaining rings are removed (one on each side of the spool), or on other models, after the bolts and cover are removed **(see illustration 8.25)**. Note how the parts fit together to simplify reassembly (a simple sketch would be helpful). Check the governor parts for wear and damage. If the gear shaft must be replaced, measure how far it protrudes before removing it and don't damage the crankcase boss. Clamp the shaft in a vise and tap the crankcase boss with a soft-face hammer to extract the shaft from the hole. **Caution:** *DO NOT twist the shaft with pliers or the mounting hole will be enlarged and the new shaft won't fit into it securely.*

26 When installing the new shaft, coat the serrated end with stud and bearing mount liquid after the shaft has been started in the hole with a soft-face hammer. Use a vise or press to finish installing the shaft and make sure it protrudes the same amount as the original (or the distance specified on the instruction sheet included with the new part). Wipe any excess stud and bearing mount liquid off the shaft and mounting boss flange.

## Inspection of components

27 After the engine has been completely disassembled, refer to Chapter 5 for the cleaning, component inspection, cylinder honing and valve lapping procedures. **Note:** *Special test equipment is needed to check the ignition coil/electronic ignition module. If you suspect the coil/module is causing ignition problems, have it checked by a small engine repair shop.*

28 Once you've inspected and serviced everything and purchased any necessary new parts, which should always include new gaskets and seals, reassembly can begin.

29 Begin by reinstalling the gear reduction components (if used) in the crankcase, then proceed as follows:

# Reassembly

30 Coat the intake valve stem with clean engine oil, moly-base grease or engine assembly lube, then reinstall it in the block. Make sure it's returned to its original location. **Note:** *If a seal is used on the intake valve, always install a new one when the engine is reassembled.*

31 Compress the valve spring with both retainers in place, then pull the valve out enough to position the spring and install the slotted retainer. Release the compressor and make sure the retainer is securely locked on the end of the valve. Repeat the procedure for the exhaust valve. **Note:** *Some models are equipped with dampening coils (tight wound coils) on the end of the valve spring. Position the dampening coils near the engine block away from the valve spring retainers.*

32 Install the washer and governor gear assembly **(see illustration 8.32a)**, make sure the retaining ring is secured in the shaft groove, then install the spool and the outer retaining ring **(see illustration 8.32b)**.

33 Lubricate the crankshaft magneto side oil seal lip **(see illustration 8.33)**, the plain bearing (if applicable) and the connecting rod journal with clean engine oil, moly-base grease or engine assembly lube, then reposition the crankshaft in the crankcase. If a ball-bearing is used on the magneto side, lubricate it with clean engine oil.

34 Before installing the piston/connecting rod assembly, the cylinder must be perfectly clean and the top edge of the bore must be chamfered slightly so the rings don't catch on it. Stagger the piston ring end gaps and make sure they're positioned opposite the valve seats when the piston is

8.32b Install the spool and the outer retaining ring

8.33 Lubricate the crankshaft (magneto side) oil seal lip

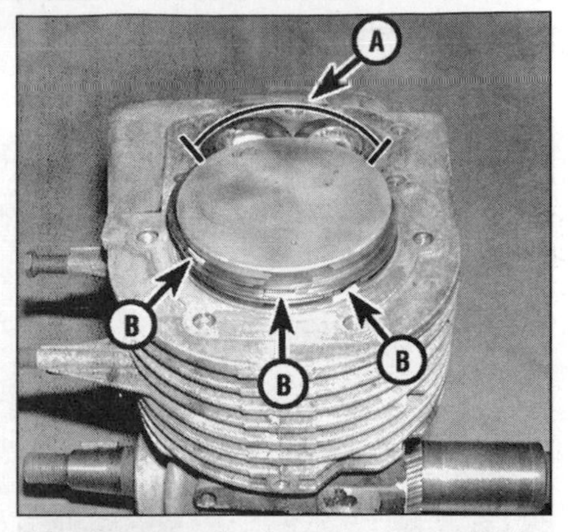

8.34 Stagger the ring gaps to avoid the "trenched" area

*A Trenched area*
*B Ring gap*

8.37 Make sure the marks on the rod and cap align properly (arrows)

8.38 Tighten the bolts or nuts to the specified torque

installed. Lubricate the piston and rings with clean engine oil, then attach a ring compressor to the piston (see Chapter 5). Leave the skirt protruding about 1/4-inch. Tighten the compressor until the piston cannot be turned, then loosen it until the piston turns in the compressor with resistance. **Note:** *The cylinder block on H60, HHM80, HM100, TVM125, TVM140 and TVM220 models are equipped with "trenching". This design incorporates a slightly recessed edge (or "trench") in the combustion chamber around the valves and near the cylinder walls. Take caution when gapping the rings to avoid this area as the rings may get caught on this recessed lip when installing the piston into the cylinder* **(see illustration 8.34).**

35 Apply a coat of engine oil to the cylinder walls. If the piston has an arrow in the top, it must face the valve seat side of the engine (if possible) or to the right when looking at the engine with the connecting rod pointing down. Make sure the match marks on the rod and cap will be facing out when the rod/piston assembly is in place. Gently insert the piston/connecting rod assembly into the cylinder and rest the bottom edge of the ring compressor on the engine block. Tap the top edge of the ring compressor to make sure it's contacting the block around its entire circumference. **Note:** *The piston(s) used on these models will have either an arrow stamped above the piston pin hole, a cast number inside the piston or an arrow stamped on the top of the piston. The piston must be installed with these marks toward the valves (front). If the piston is not marked with one of the above designations, the piston can be installed in either direction.*

36 Carefully tap on the top of the piston with the end of a wooden or plastic hammer handle (see Chapter 5) while guiding the end of the connecting rod into place on the crankshaft journal. The piston rings may try to pop out just before entering the bore, so keep some pressure on the ring compressor. Work slowly - if any resistance is felt as the piston enters the cylinder, stop immediately. Find out what's hanging up and fix it before proceeding. Do not, for any reason, force the piston into the cylinder - you'll break a ring and/or the piston.

37 Install the connecting rod cap, a NEW lock plate, the oil dipper (if used) and the bolts or washers and nuts. Make sure the marks you made on the rod and cap **(see illustration 8.37)** (or the manufacturer's marks) are aligned and facing out and the oil dipper is oriented correctly.

38 Tighten the bolts or nuts to the torque listed in the

8.39 Apply clean engine oil, moly-base grease or engine assembly lube to the tappets, then reinstall them

8.41 Install the plunger type oil pump onto the camshaft. Make sure the chamfered side of the pump body faces the camshaft and the plunger ball is seated in the recess in the oil sump after the sump is in place

8.42 Lubricate the crankshaft main bearing journal

8.43a Make sure the dowel pins are in place, then position a new gasket on the crankcase

8.43b Coat the lip of the seal in the sump with clean engine oil or assembly lube

Specifications at the end of this Chapter **(see illustration 8.38)**. Note that Durlock bolts (used without locking tabs) and regular bolts have different torque values. Work up to the final torque in three steps. Temporarily install the camshaft and turn the crankshaft through two complete revolutions to make sure the rod doesn't hit the cylinder wall or camshaft. If it does, the piston/connecting rod is installed incorrectly.

39 Apply clean engine oil, moly-base grease or engine assembly lube to the tappets, then reinstall them - make sure they're returned to their original locations **(see illustration 8.39)**.

40 If it's not already in place, install the camshaft. Apply clean engine oil, moly-base grease or engine assembly lube to the lobes and bearing journals and align the timing marks on the gears. This is very important **(see illustration 8.17b)**! The marks are usually dimples/lines or beveled teeth (or a combination of them) near the outer edge of the gears. On many engines, the camshaft gear mark must be aligned with the keyway for the crankshaft gear - no mark is included on the crankshaft gear itself. Install the balance shaft gears if equipped. Make sure the alignment marks are correctly aligned **(see illustration 8.17a)**. **Note:** *Some models are equipped with the Mechanical Compression Release (MCR) camshafts that are equipped with a locking pin in the camshaft that extends over the exhaust cam lobe to lift the valve to relieve engine compression during cranking. After the engine has started, centrifugal force moves the weight outward dropping the pin down and out of the way of the exhaust valve, thereby bringing the engine compression back to normal. Be sure this pin is locked into the correct position after installation.* **Note:** *Some models are equipped with a Bump Compression Release (BCR) camshaft. This type of camshaft is equipped with a small bump on the exhaust lobe to allow slight compression release when the engine is cranking slowly during start-up. Make sure the correct type of camshaft is installed in the particular engine by verifying the part numbers and camshaft types with a small engine parts distributor.*

41 Install the plunger-type oil pump (if used) after lubricat-

8.44 Apply thread cement to the bolt threads and tighten them to the specified torque

8.45 Checking the crankshaft endplay with a dial indicator

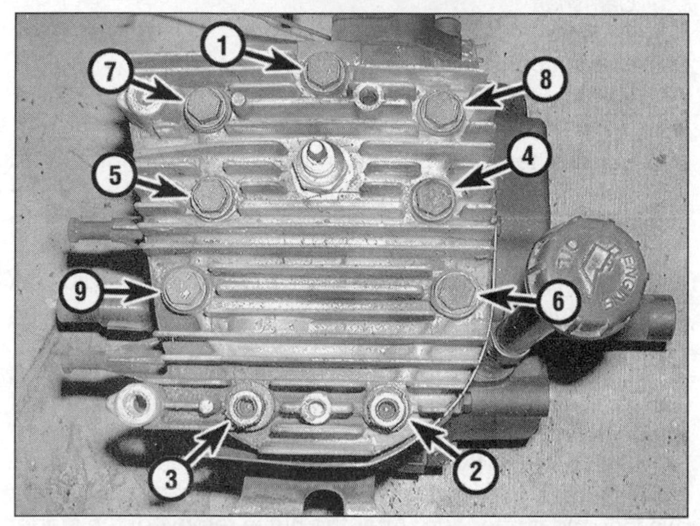

8.46a Bolt tightening sequence for TVM195, TVM220, VM and HM models (also applies to L-head heavy duty engines)

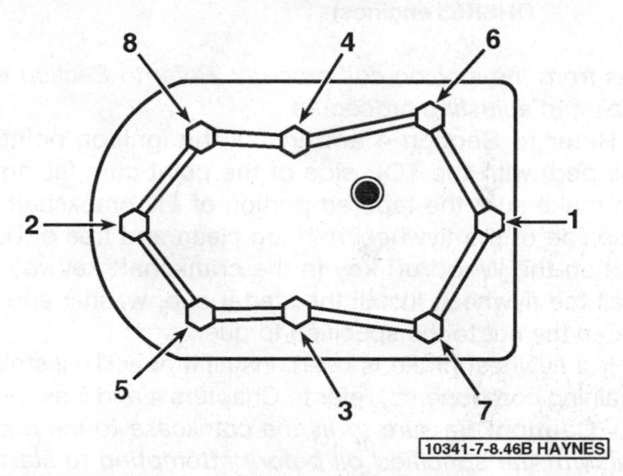

8.46b Bolt tightening sequence for L-head light duty engines except TVM195, TVM220, VM and HM models

ing it with clean engine oil **(see illustration 8.41)**. Be absolutely certain the chamfered side of the pump body faces the camshaft and the plunger ball is seated in the recess in the oil sump after the sump is in place.

42 Lubricate the crankshaft main bearing journal and the lip on the oil seal in the sump (or crankcase cover) with clean engine oil, moly-base grease or engine assembly lube **(see illustration 8.42)**.

43 Make sure the dowel pins are in place, then position a new gasket on the crankcase (the dowel pins will hold it in place) **(see illustrations 8.43a and 8.43b)**. Carefully lower the oil sump (or crankcase cover) into place over the end of the crankshaft until it seats on the crankcase. DO NOT damage the oil seal lip or leaks will result! The governor shaft must match up with the spool end and the oil pump shaft ball end must be engaged in the recess.

44 Apply a non-hardening thread locking compound to the bolt threads, then install and tighten them to the specified torque. Follow a criss-cross pattern and work up to the final torque in three equal steps to avoid warping the oil sump/cover **(see illustration 8.44)**.

45 The crankshaft endplay must be checked with a dial indicator and compared with the Specifications at the end of this Chapter. If it's excessive, check with a small engine repair shop regarding the availability of shims to correct it **(see illustration 8.45)**.

46 Install the cylinder head, using a new gasket. Tighten the bolts a little at a time, in the correct sequence **(see illustrations)**, to the torque listed in the Specifications at the end of this Chapter.

47 Install the crankcase breather (with the small hole down) and intake manifold (unless the carburetor and manifold were removed as an assembly). Use new gaskets and tighten the bolts securely.

48 Install the ignition coil/spark plug wire assembly and align the marks on the coil bracket and crankcase bosses, then tighten the coil mounting bolts securely. **Note:** *If the coil is mounted on the outside of the flywheel, don't tighten the bolts completely - just snug them up. The bolt holes are slotted; move the coil as far away from the flywheel as possible before snugging up the bolts. Be sure to reroute the*

**9.9a Remove the valve cover bolts (arrows) (OHH, OVRM and OHSK55 engines)**

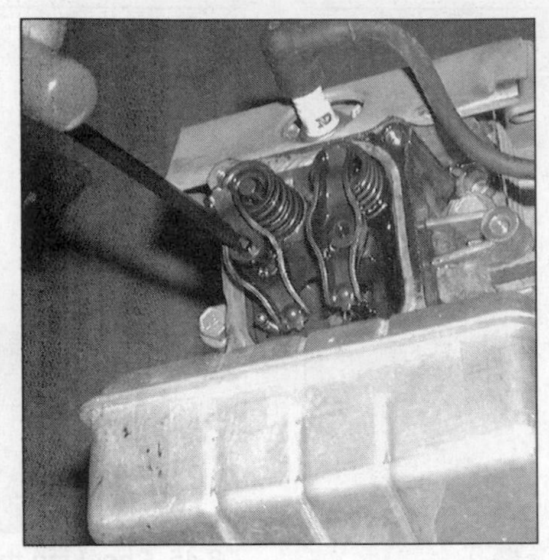
**9.9b Remove the rocker arms from the cylinder head**

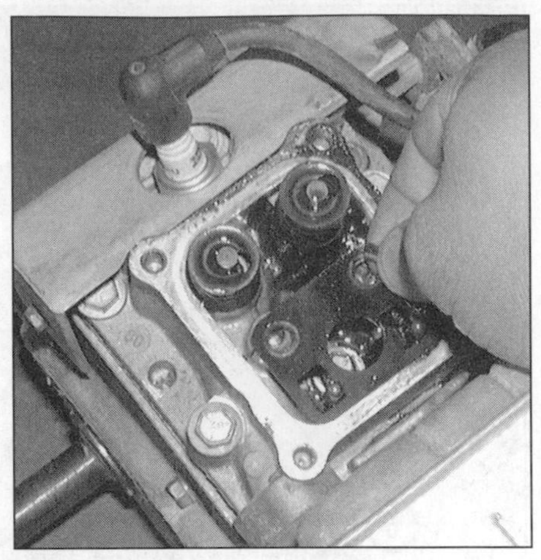
**9.9c Remove the pushrod guide plate from the cylinder head**

*wires from the ignition coil properly. Refer to Section 4 for the air gap adjusting procedure.*

49 Refer to Section 4 and install the ignition points (if equipped) with the TOP side of the point cam facing up, then make sure the tapered portion of the crankshaft and the inside of the flywheel hub are clean and free of burrs. Position the Woodruff key in the crankshaft keyway and install the flywheel. Install the starter cup, washer and nut. Tighten the nut to the specified torque.

50 If a flywheel brake is used, install it now. To install the remaining components, refer to Chapters 4 and 5 as necessary. **Caution:** *Be sure to fill the crankcase to the correct level with the specified oil before attempting to start the engine.*

# 9 Overhead Valve (OHV) light duty engines

## Disassembly

The engine components should be removed in the following general order:

*Engine cover (if used)*
*Fuel tank*
*Cooling shroud/recoil starter*
*Carburetor/intake manifold*
*Muffler*
*Flywheel*
*Flywheel brake components (if equipped)*
*Ignition components*
*Cylinder head*
*Crankcase breather assembly*
*Oil sump/crankcase cover*
*Balance shaft(s) (if equipped)*
*Crankshaft*
*Camshaft*
*Tappets*
*Piston/connecting rod assembly*
*Valves*
*Governor components*

1 For fuel tank, carburetor/intake manifold, muffler and oil draining procedures, refer to Chapters 4 and 5 as necessary. For the shroud and recoil assembly removal, refer to the beginning of this chapter.

2 Remove the blower housing by removing the dipstick tube mounting screw **(see illustration 8.2)** and unscrewing the dipstick tube from the engine.

3 Disconnect the governor linkage from the carburetor. Be sure to make notes or a drawing of the linkage on your model to insure correct reassembly **(see illustration 8.3)**.

4 Prevent the flywheel from turning by using a strap wrench **(see illustration 8.4)** and remove the large nut, washer and starter cup. After the nut is removed, the starter cup and debris screen can be detached.

5 Thread a flywheel knock-off tool onto the flywheel shaft (check with your local power equipment dealer or small engine repair shop for tool availability). Thread the tool until it bottoms out and unthread it one complete turn **(see illustration 8.5)**.

6 Using a hammer, tap the knock-off tool while lifting the flywheel using a flat- bladed screwdriver to break the flywheel loose from the engine. Rotate the flywheel 180-degrees and pry at a different area if the flywheel is difficult to remove. DO NOT hammer on the end of the crankshaft and DO NOT use a jaw-type puller that applies force to the outer edge of the flywheel! As the tool is hit with the hammer, the flywheel should pop off the shaft taper. **Note:** *If a flywheel brake is used, remove the brake and related components. Make detailed notes of the system to insure correct reassembly.*

7 Place the flywheel upside-down on a wooden surface and check the magnets by holding a screwdriver at the end of the handle while moving the tip toward one of the magnets - when the screwdriver tip is about 3/4-inch from the

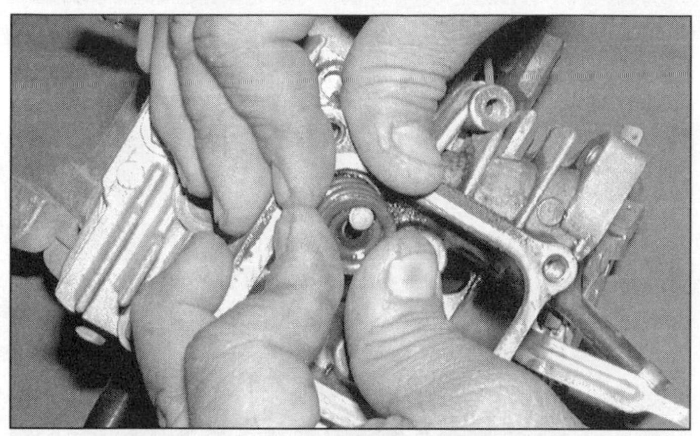

9.10 Using the tips of your thumbs and index fingers, depress the valve spring and remove the retainers (OHM, OHSK110, OHSK120, OVXL and OHV engines)

magnet, it should be attracted to it. If it doesn't, the magnets may have lost their strength and ignition system performance may not be up to par. Remove the flywheel key. If it's sheared off, install a new one. See Chapter 5 for checking procedures on the ignition and charging systems.

8 Remove the stator and baffle plate if equipped with a charging system.

9 There are two types of OHV engines that are classified according to valve train disassembly. On OHH, OVRM and OHSK55 engines, remove the rocker arm assembly and cylinder head with the valve assemblies intact. On all other overhead valve engines, refer to Step 10. Remove the valve cover bolts **(see illustration 9.9a)**. Remove the rocker arms from the cylinder head **(see illustration 9.9b)**. Remove the rocker arms, bearings (if equipped), nuts, rocker arm studs, retainer screw and guide plate **(see illustration 9.9c)** and push rods. **Note:** *On most engines, the exhaust push rod must be removed after the cylinder head is removed. Keep this in mind for reassembly purposes.*

10 On OHM, OHSK110, OHSK120, OVXL and OHV engines, the valve springs must be removed from the cylinder head to access the head bolts. Turn the crankshaft to position the piston down in the cylinder bore (to prevent damage to the valve and piston crown). Knock the valve cap loose using a deep socket and depress the valve springs with your thumbs **(see illustration 9.10)**. Remove the retainer, valve cap, retainers and springs from both the intake and exhaust valves. **Note:** *On most engines, the exhaust push rod must be removed after the cylinder head is removed. Keep this in mind for reassembly purposes.*

11 Remove the cylinder head bolts, loosening them a little at a time and working in an order opposite the tightening sequence **(see illustration 9.46 and 9.47)**. **Note:** *Some smaller horsepower engines are equipped with regular hex-head bolts, while larger horsepower engines will require an Allen wrench to remove the head bolts.*

12 If it's still in place, remove the bolts and separate the intake manifold from the engine. Remove the gasket and discard it. Remove the mounting bolts and detach the crankcase breather assembly and gasket from the engine.

9.13 Remove the crankcase cover

13 On horizontal shaft engines, remove the crankcase cover **(see illustration 9.13)**. Use an oil seal protector tool to avoid damaging the crankshaft seal. On some larger engines, remove the snap-ring and the oil seal before removing the crankcase cover. On engines equipped with a gear reduction assembly, turn the crankshaft to roll the reduction shaft gear off the crankshaft worm gear when removing the crankcase cover.

14 On vertical shaft engines, loosen the oil sump-to-engine block bolts in 1/4-turn increments to avoid warping the sump, then remove them. Tap the sump/cover with a soft-face hammer to break the gasket seal, then separate it from the engine block and crankshaft. If it hangs up on the crankshaft, continue to tap on it with the hammer, but be very careful not to crack or distort it (especially if it's made of aluminum). If thrust washers are installed on the crankshaft or camshaft, record their positions, slide them off and set them aside.

15 Turn the crankshaft to align the timing marks **(see illustrations 9.15a and 9.15b)**. On engines equipped with a bal-

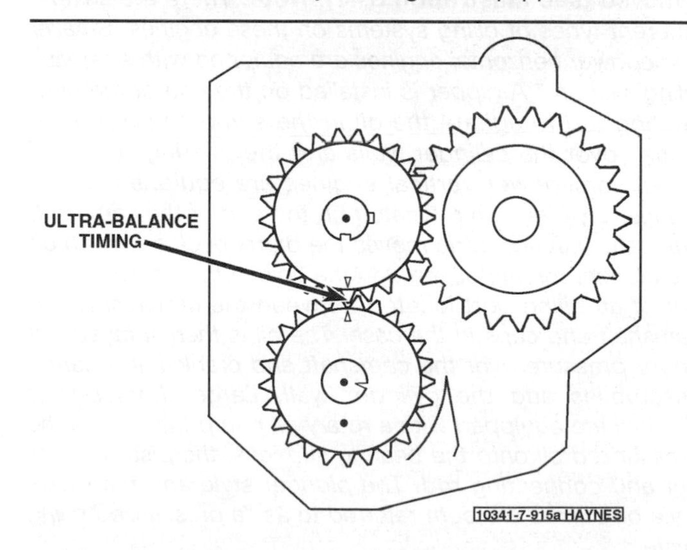


9.15a Align the timing marks on the balance shaft gears (arrows)

9.15b **Align the timing marks on the crankshaft and camshaft gears (arrows)**

9.16 **Remove the camshaft and tappets**

*A Camshaft B Tappets*

ance shaft system, it will be necessary to align the marks on the counterbalance shaft gears as well as the camshaft and crankshaft gears. Remove the counterbalance shaft gears to access the camshaft and crankshaft gears. **Note:** *It's a good idea to highlight the timing marks with paint before disassembly (to make them easier to see) and make notes or pictures to help during the reassembly procedure.*

16 Remove the camshaft **(see illustration 9.16)** and the balance shafts and gears, if the engine is equipped with the balance shaft system.

17 Use a ridge reamer to remove the carbon/wear ridge from the top of the cylinder bore after the cylinder head is off (see Chapter 5). Follow the manufacturer's instructions included with the tool.

18 Vertical crankshaft engines are equipped with a plunger-type oil pump that's driven by an eccentric on the camshaft; it can be lifted out before the camshaft is removed **(see illustration 8.41)**. **Note:** *There are several different types of oiling systems on these engines. Smaller horsepower horizontal engines are equipped with a "splash oiling system." A dipper is installed on the end of the connecting rod to agitate the oil in the sump causing oil to splash over the cylinder walls and the bearing surfaces. Small horsepower vertical engines are equipped with a plunger style oil pump located on the end of the camshaft. The camshaft eccentric moves the barrel back and forth on the plunger forcing oil through the hole in the camshaft and out of an oiling port located between the crankshaft and camshaft end caps in the case. The oil is then sprayed out under pressure over the camshaft and crankshaft bearing assemblies and the cylinder wall. Larger horsepower engines are equipped with a rotary oil pump that forces the pressurized oil onto the bearing surfaces, the piston, cylinder and connecting rod. The plunger style and the rotary style oil pumps are both referred to as "a pressurized oiling system."*

19 On horizontal crankshaft engines, note the location of the oil dipper attached to the end of the crankshaft and disconnect the oil dipper from the connecting rod. Some models are equipped with oil dippers attached to the rod cap by the connecting rod nuts, while others are cast into the connecting rod.

20 After the camshaft is removed, pull out the tappets and store them in marked containers so they can be returned to their original locations.

21 Look for match marks on the connecting rod and cap **(see illustration 9.21)**. If you can't see any, mark the side of the connecting rod and cap that faces OUT and note how the oil dipper (if used) is installed (the parts must be reassembled in the exact same relationship to the crankshaft). **Note:** *Some models use offset piston and connecting rods. Be sure the match marks on the connecting rods are facing OUT when installing the connecting rods on all types of engines.*

22 Turn the crankshaft so the rod journal is at the bottom of its stroke (Bottom Dead Center). Flatten the locking tabs (if used) on the connecting rod bolts with a chisel and hammer.

23 Loosen the bolts or nuts in 1/4-turn increments until they can be removed by hand. Separate the cap (and washers, if used) from the connecting rod, move the end of the rod away from the crankshaft journal and push the piston/rod assembly out through the top of the bore. The crankshaft can now be lifted out.

24 Do not separate the lever from the governor shaft unless new parts are needed. The lever mount will be damaged during removal.

25 Remove the governor assembly. Some are equipped with retaining clips, others are retained by a plate attached with bolts **(see illustrations 8.25, 8.32a and 8.32b)**. Note how the parts fit together to simplify reassembly (a simple sketch would be helpful). Check the governor parts for wear and damage. If the gear shaft must be replaced, measure how far it protrudes before removing it and don't damage the crankcase boss. Clamp the shaft in a vise and tap the crankcase boss with a soft-face hammer to extract the

9.21 Check for matchmarks on the connecting rods (arrows)

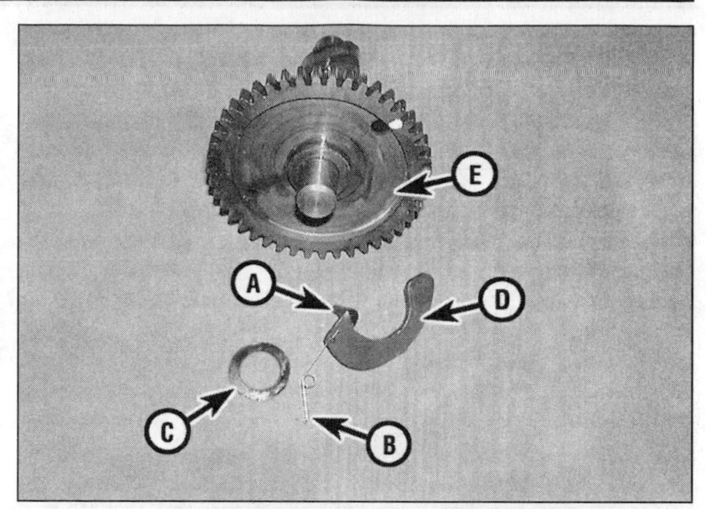


9.32 Compression release components

| | | | |
|---|---|---|---|
| *A* | *Spring end "A"* | *D* | *Compression release lever* |
| *B* | *Spring end "B"* | *E* | *Camshaft* |
| *C* | *Thrust washer* | | |

shaft from the hole. **Caution:** *DO NOT twist the shaft with pliers or the mounting hole will be enlarged and the new shaft won't fit into it securely.*

26 When installing the new shaft, coat the serrated end with stud and bearing mount liquid after the shaft has been started in the hole with a soft-face hammer. Use a vise or press to finish installing the shaft and make sure it protrudes the same amount as the original (or the distance specified on the instruction sheet included with the new part). Wipe any excess stud and bearing mount liquid off the shaft and mounting boss flange. At this point, the gear reduction components can be removed (if equipped). Be sure to note how the parts fit together. A simple sketch may prove helpful during reassembly. Check the gears and shaft for wear and damage. Install new parts if necessary. Replace the shaft oil seal even if the original shaft is reinstalled.

## Inspection of components

27 If you're working on a OHH, OVRM or OHSK55 model, compress the intake valve spring and remove the retainer (see Chapter 5).

28 After the engine has been completely disassembled, refer to Chapter 5 for the cleaning, component inspection, cylinder honing and valve lapping procedures. **Note:** *Special test equipment is needed to check the ignition coil/electronic ignition module. If you suspect the coil/module is causing ignition problems, have it checked by a small engine repair shop.*

29 Once you've inspected and serviced everything and purchased any necessary new parts, which should always include new gaskets and seals, reassembly can begin. Begin by reinstalling the gear reduction components, if equipped, in the crankcase, then proceed as follows:

# Reassembly

30 If you're working on a OHH, OVRM or OHSK55 model, coat the intake valve stem with clean engine oil, moly-base grease or engine assembly lube, then reinstall it in the block. Make sure it's returned to its original location. **Note:** *Remember, on OHM, OHSK110, OHSK120, OVXL and OHV engines, the valve springs and retainers must be installed after the cylinder head has been installed.*

31 Compress the valve spring with both retainers in place, then pull the valve out enough to position the spring and install the slotted retainer. Release the compressor and make sure the retainer is securely locked on the end of the valve. Repeat the procedure for the exhaust valve. **Note:** *Some models are equipped with dampening coils (tight wound coils) on the end of the valve spring. Position the dampening coils near the engine block away from the valve spring retainers.*

32 On OHSK55 models, assemble the compression release components onto the camshaft **(see illustration 9.32)**. Install spring end A through the release from the pin side, insert the end of spring B through the camshaft gear and slide the release pin into the small hole near the center of the cam gear. Install the thrust washer onto the camshaft next to the compression release mechanism. **Note:** *Some models are equipped with a Mechanical Compression Release (MCR) camshaft that is equipped with a locking pin in the camshaft that extends over the exhaust cam lobe to lift the valve to relieve engine compression during cranking. Be sure this pin is locked into the correct position after installation.*

33 Lubricate the crankshaft magneto side oil seal lip **(see illustration 8.33)**, the plain bearing (if applicable) and the connecting rod journal with clean engine oil, moly-base grease or engine assembly lube, then reposition the crankshaft in the crankcase. If a ball-bearing is used on the magneto side, lubricate it with clean engine oil.

34 Before installing the piston/connecting rod assembly, the cylinder must be perfectly clean and the top edge of the bore must be chamfered slightly so the rings don't catch on it. Install new piston rings on the piston and stagger the piston ring end gaps (see Chapter 5). Lubricate the piston and rings with clean engine oil, then attach a ring compressor to

9.35 Lubricate the cylinder with clean engine oil

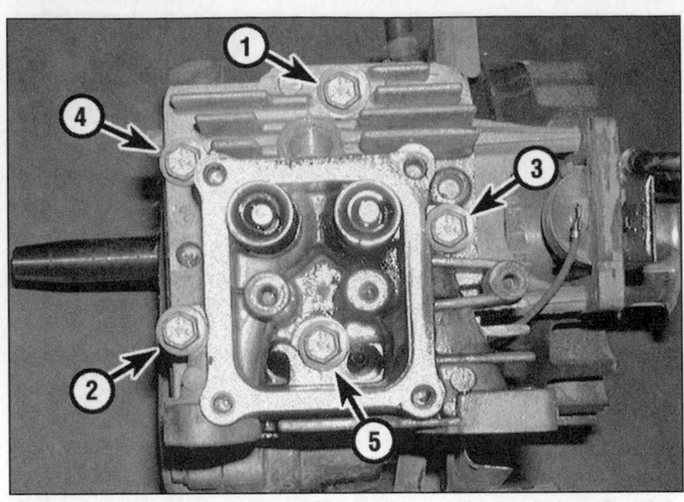


9.46 Cylinder head tightening sequence on OVRM, OHH and OHSK engines

the piston. Leave the skirt protruding about 1/4-inch. Tighten the compressor until the piston cannot be turned, then loosen it until the piston turns in the compressor with resistance.

35 Rotate the crankshaft until the connecting rod journal is at TDC (Top Dead Center - top of the stroke) and apply a coat of engine oil to the cylinder walls **(see illustration 9.35).** If the piston has an arrow in the top, it must face OUT of the engine (if possible) or toward the carburetor when looking at the engine with the connecting rod pointing down. Make sure the match marks on the rod and cap will be facing out when the rod/piston assembly is in place. Gently insert the piston/connecting rod assembly into the cylinder and rest the bottom edge of the ring compressor on the engine block. Tap the top edge of the ring compressor to make sure it's contacting the block around its entire circumference. **Note:** *The piston(s) used on these models will have either an arrow stamped above the piston pin hole, a cast number inside the piston or an arrow stamped on the top of the piston. The piston must be installed with these marks toward the carburetor. If the piston is not marked with one of the above designations, the piston can be installed in either direction.*

36 Carefully tap on the top of the piston with the end of a wooden or plastic hammer handle while guiding the end of the connecting rod into place on the crankshaft journal. The piston rings may try to pop out just before entering the bore, so keep some pressure on the ring compressor. Work slowly - if any resistance is felt as the piston enters the cylinder, stop immediately. Find out what's hanging up and fix it before proceeding. Do not, for any reason, force the piston into the cylinder - you'll break a ring and/or the piston.

37 Install the connecting rod cap, a NEW lock plate, the oil dipper (if used) and the bolts or washers and nuts. Make sure the marks you made on the rod and cap (or the manufacturer's marks) are aligned and facing out and the oil dipper is oriented correctly.

38 Tighten the bolts or nuts to the specified torque listed at the end of this Chapter. Note that Durlock bolts (used without locking tabs) and regular bolts have different torques. Work up to the final torque in three steps. Temporarily install the camshaft and turn the crankshaft through two complete revolutions to make sure the rod doesn't hit the cylinder or camshaft. If it does, the piston/connecting rod is installed incorrectly.

39 Apply clean engine oil, moly-base grease or engine assembly lube to the tappets, then reinstall them - make sure they're returned to their original locations.

40 Apply clean engine oil, moly-base grease or engine assembly lube to the lobes and bearing journals of the camshaft. Align the timing marks on the camshaft and crankshaft gears - this is very important! The marks are usually dimples/lines or beveled teeth (or a combination of them) near the outer edge of the gears **(see illustrations 9.15a and 9.15b)**. On many engines, the camshaft gear mark must be aligned with the keyway for the crankshaft gear - no mark is included on the crankshaft gear itself.

41 Install the plunger-type oil pump (if used) after lubricating it with clean engine oil. Be absolutely certain the chamfered side of the pump body faces the camshaft and the plunger ball is seated in the recess in the oil sump after the sump is in place **(see illustration 8.41)**. **Note:** *On rotary type oil pumps, install the oil pump drive shaft into the slot in the end of the camshaft. Install the oil pump rotor and ring into the sump, but the sump must be installed later.*

42 Lubricate the crankshaft main bearing journal and the lip on the oil seal in the sump (or crankcase cover) with clean engine oil, moly-base grease or engine assembly lube.

43 Install the counterbalance gears. Set the engine to TDC (see Section 4) and install the thinner counterbalance gear onto the far left boss and the thicker gear onto the far right boss. Make sure the alignment marks are correct **(see illustrations 9.15a and 9.15b)**.

44 Make sure the dowel pins are in place, then position a new gasket on the crankcase (the dowel pins will hold it in

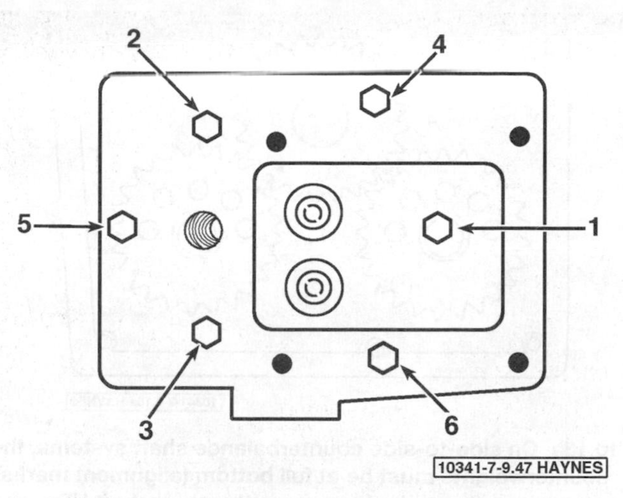


**9.47 Cylinder head tightening sequence on OHV11-17, OHM, OHSK110 and OHSK120**

place). Carefully lower the oil sump (or crankcase cover) into place over the end of the crankshaft until it seats on the crankcase. DO NOT damage the oil seal lip or leaks will result! The governor shaft must match up with the spool end and the oil pump shaft ball end must be engaged in the recess. Apply thread cement to the bolt threads, then install and tighten them to the specified torque. Follow a criss-cross pattern and work up to the final torque in three equal steps to avoid warping the oil sump/cover.

45 The crankshaft endplay must be checked with a dial indicator **(see illustration 8.45)** and compared to the value listed in this Chapter's Specifications. If it's excessive, check with a small engine repair shop regarding the availability of shims to correct it.

46 On conventional cylinder heads, install the cylinder head gasket and cylinder head, then tighten the bolts to the torque values listed in the Specifications at the end of this Chapter. Follow the correct tightening sequence **(see illustration 9.46)**.

47 Install the valve train components on non-conventional cylinder head systems. **Note:** *There are certain non-conventional engines that require valve spring assembly after the cylinder head is installed. On OHM, OHSK110, OHSK120, OVXL and OHV engines, the valve springs must be installed after the cylinder head bolts are tightened.*

a) First, install the pushrods with the cupped ends of the rods on top of the tappets. Be sure to install lubricated O-rings onto both sides of the pushrods.
b) Install the new valves into the head and place the cylinder head gasket and head onto the engine crankcase. Tighten the cylinder head bolts to the torque specifications listed at the end of this Chapter. Follow the correct tightening sequence **(see illustration 9.47)**.
c) Install the rocker arm housing, the push rod guide plate (with the legs facing up), new rocker arm studs and O-rings and rocker arm guide plate.
d) Lift each valve up until it contacts the seat and hold the valve in this position. Install a special air line adapter into the spark plug hole and apply compressed air into the cylinder chamber to force the valves against the valve seats under pressure. ***Note:*** If compressed air is not available, use a piece of nylon rope and feed it into the cylinder through the spark plug hole with the piston at Bottom Dead Center. Raise the piston until the rope touches the valves and forces the valves against the seat in the cylinder head. Perform the necessary engine operations and then remove the rope.
e) Install the new O-rings, valve spring caps, the valve springs and the valve spring retainers. The retainers should position the larger opening away from the engine. Use new O-rings and install the white retainers on the exhaust guide.
f) Use your thumbs to compress the valve springs and force the springs down to make clearance for the valve retainers. Compress the valve spring and insert the two keepers into the retainer.

48 Install the crankcase breather (with the small hole down) and intake manifold (unless the carburetor and manifold were removed as an assembly). Use new gaskets and tighten the bolts securely.

49 Adjust the valves with the engine cold (see Section 7).

50 Install the valve cover with a new gasket. Be sure to install new O-rings onto the cover bolts, if equipped.

51 Install the governor assembly. Refer to the carburetor and governor adjustment procedure in Section 6 and make sure the governor linkage and carburetor is adjusted properly.

52 Install the ignition coil/spark plug wire assembly and align the marks on the coil bracket and crankcase bosses, then tighten the coil mounting bolts securely. **Note:** *If the coil is mounted on the outside of the flywheel, don't tighten the bolts completely - just snug them up. The bolt holes are slotted; move the coil as far away from the flywheel as possible before snugging up the bolts. Check the coil/module adjusting procedure in Chapter 4. Be sure to reroute the wires from the ignition coil properly.*

53 Refer to Chapter 4 and install the ignition points (if equipped) with the TOP side of the point cam facing up, then make sure the tapered portion of the crankshaft and the inside of the flywheel hub are clean and free of burrs. Position the Woodruff key in the crankshaft keyway and install the flywheel. Install the starter cup, washer and nut. Tighten the nut to the specified torque. **Note:** *If the ignition coil is mounted outside the flywheel, turn the flywheel so the magnets are facing away from the coil assembly, then insert a feeler gauge and adjust the air gap. Follow the procedure in Section 4 for the ignition coil/module air gap adjustment.*

54 If a flywheel brake is used, install it now. To install the remaining components, refer to Chapters 4 and 5 as necessary. **Caution:** *Be sure to fill the crankcase to the correct level with the specified oil before attempting to start the engine.*

# 10 L-head and Overhead Valve (OHV) heavy duty engines

## Disassembly

The engine components should be removed in the following general order:

*Engine cover (if used)*
*Fuel tank*
*Blower shroud/recoil starter*
*Carburetor/intake manifold*
*Muffler*
*Flywheel*
*Flywheel brake components (if equipped)*
*Ignition components*
*Rocker arm housing*
*Cylinder head*
*Crankcase breather assembly*
*Oil sump/crankcase cover*
*Balance shaft(s) (if equipped)*
*Camshaft*
*Crankshaft*
*Tappets*
*Piston/connecting rod assembly*
*Valves*
*Governor components*

1 For fuel tank, carburetor/intake manifold, muffler and oil draining procedures, refer to Chapters 4 and 5 as necessary. For the shroud and recoil assembly removal, refer to the beginning of this chapter.

2 Remove the blower housing by removing the dipstick tube mounting screw and unscrewing the dipstick tube from the engine **(see illustration 8.5)**.

3 Disconnect the governor linkage from the carburetor. Be sure to make notes or a drawing of the linkage on your model to insure correct reassembly **(see illustration 8.3)**.

4 Hold the flywheel using a strap wrench **(see illustration 8.4)** and remove the large nut, washer and starter cup. After the nut is removed, the starter cup and debris screen can be detached.

5 Thread a flywheel knock-off tool onto the flywheel shaft (check with your local power equipment dealer or small engine repair shop for tool availability). Thread the tool until it bottoms out and unthread it one complete turn **(see illustration 8.5)**.

6 Using a hammer, tap the knock-off tool while lifting the flywheel using a flat- bladed screwdriver to break the flywheel loose from the engine. Rotate the flywheel 180-degrees and pry at a different area if the flywheel is difficult to remove. DO NOT hammer on the end of the crankshaft and DO NOT use a jaw-type puller that applies force to the outer edge of the flywheel! As the tool is hit with the hammer, the flywheel should pop off the shaft taper. **Note:** *If a flywheel brake is used, remove the brake and related components. Make detailed notes of the system to insure correct reassembly.*

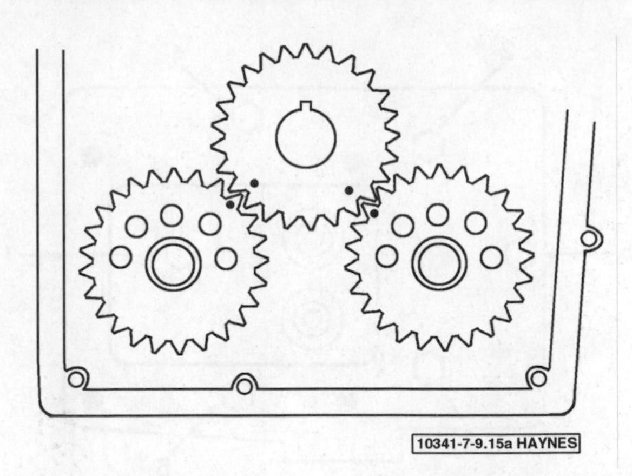


**10.13a On side-to-side counterbalance shaft systems, the counterweights must be at full bottom (alignment marks aligned) and the keyway on the crankshaft UP**

7 Place the flywheel upside-down on a wooden surface and check the magnets by holding a screwdriver at the end of the handle while moving the tip toward one of the magnets - when the screwdriver tip is about 3/4-inch from the magnet, it should be attracted to it. If it doesn't, the magnets may have lost their strength and ignition system performance may not be up to par. Remove the flywheel key - if it's sheared off, install a new one. See Chapter 5 for checking procedures on the ignition and charging systems.

8 First remove the flywheel key, stator and baffle plate and then refer to Section 4 and remove the coil/module assembly (electronic ignition assembly) and related parts (if equipped).

9 On OHV engines, remove the rocker arm assembly and cylinder head with the valve assemblies intact. First, remove the valve cover bolts **(see illustration 9.9a)**. Remove the rocker arms from the cylinder head **(see illustration 9.9b)**. Remove the rocker arm retaining clips, rocker arm and bearings (if equipped). **Note:** *Refer to Section 8 for L-head engine cylinder head removal procedures.*

10 On OHV engines, the valve springs must be removed from the cylinder head to access the head bolts. First, turn the piston down in the cylinder bore to prevent damage to the valve and piston crown. Next, knock the valve cap loose using a deep socket and depress the valve springs with the thumbs **(see illustration 9.10)**. Remove the keepers, valve cap, retainer and springs from both the intake and exhaust valves. If the valve springs cannot be compressed without the valves dropping down into the combustion chamber, install a special adapter into the spark plug hole and apply compressed air.

11 Remove the cylinder head bolts in an order opposite that of the tightening sequence **(see illustration 10.42** for OHV engines and **8.46a for L-head engines)**. **Note:** *Bolts numbered 1 and 5 are 1-3/8 inch long while bolts numbered 2, 3 and 4 are 1-3/4 inch long.*

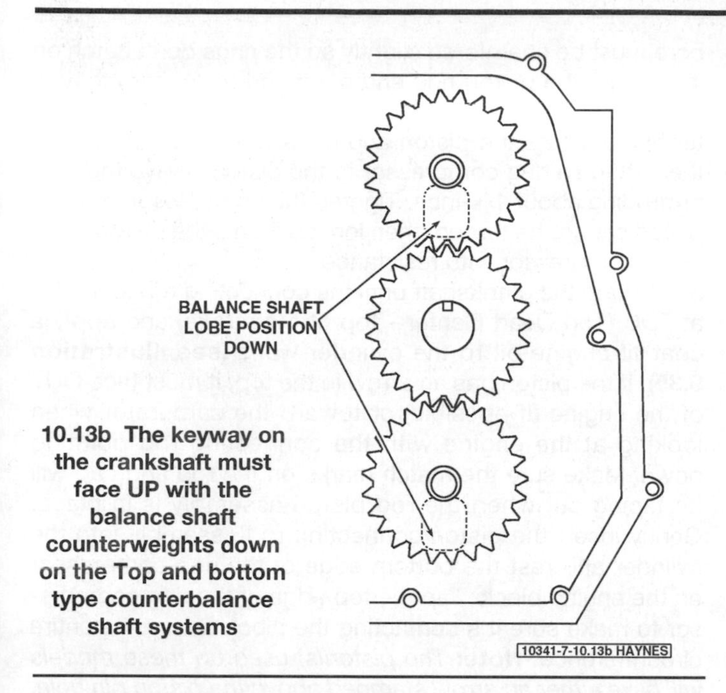


**10.13b The keyway on the crankshaft must face UP with the balance shaft counterweights down on the Top and bottom type counterbalance shaft systems**

12 Set the engine to TDC (refer to Section 4). Remove the pipe fittings and check to make sure the counterbalance shaft alignment slots are directly behind the pipe fitting openings **(see illustrations 10.40a and 10.40b)**. Remove the snap-ring and the oil seal, then remove the crankcase cover. On engines equipped with a gear reduction assembly, turn the crankshaft to roll the reduction shaft gear off the crankshaft worm gear when removing the crankcase cover. Return the piston to TDC.

13 Align the timing marks **(see illustrations 10.13a and 10.13b)**. If you can't see any marks, make your own corresponding marks on the camshaft gear and crankshaft gear. If equipped with counterbalance shafts, it will be necessary to mark the relationship of the balance shaft gears to the crankshaft gear. **Note:** *With the piston at TDC on the compression stroke, the counterbalance shaft weights should be at the bottom of their travel.* Remove the counterbalance shaft gears to access the camshaft and crankshaft gears. **Note:** *Some engines may be equipped with the counterbalance shafts mounted next to each other (side-to-side). On other models, the are arranged on opposite sides the crankshaft gear (180 degrees apart - top-and-bottom). These counterbalance systems require careful reassembly to insure correct timing. Be sure to make the timing marks before disassembly and make notes or drawings to help during the reassembly procedure.*

14 Vertical crankshaft engines are equipped with a plunger-type oil pump that's driven by an eccentric on the camshaft; it can be lifted out before the camshaft is removed. **Note:** *There are several different types of oiling systems on these engines. Smaller horsepower horizontal engines are equipped with a "splash oiling system." A dipper is installed onto the end of the connecting rod to agitate the oil in the sump, causing oil to splash over the cylinder walls and the bearing assembly. Vertical engines are equipped with a plunger style oil pump located on the end of the camshaft. The camshaft eccentric moves the barrel back and forth on the plunger forcing oil through the hole in the camshaft and out of an oiling port located between the crankshaft and camshaft end caps in the case. The oil is then sprayed out under pressure over the camshaft and crankshaft bearing assemblies and the cylinder wall. The plunger style oil pumps are referred to as "a pressurized oiling system".*

15 Remove the balance shafts and gears and cap and the camshaft **(see illustration 9.16)**.

16 After the camshaft is removed pull out the tappets and store them in marked containers so they can be returned to their original locations.

17 Use a ridge reamer to remove the carbon/wear ridge from the top of the cylinder bore after the cylinder head is off. Follow the manufacturer's instructions included with the tool (see Chapter 5).

18 On horizontal crankshaft engines, note the location of the oil dipper attached to the end of the crankshaft and disconnect the oil dipper from the connecting rod. Some models are equipped with oil dippers attached to the rod cap by the connecting rod nuts while others are cast into the connecting rod. **Note:** *Engines equipped with vertical connecting rods (horizontal crankshaft) cannot be replaced with connecting rods from horizontal systems (vertical crankshaft).*

19 Look for match marks on the connecting rod and cap - if you can't see any, mark the side of the connecting rod and cap **(see illustration 9.21)** that faces OUT and note how the oil dipper (if used) is installed (the parts must be reassembled in the exact same relationship to the crankshaft). **Note:** *Be sure the match marks on the connecting rods are facing OUT when installing the connecting rods on all types of engines.*

20 Turn the crankshaft so the rod journal is at the bottom of its stroke (Bottom Dead Center). Flatten the locking tabs (if used) on the connecting rod bolts with a chisel and hammer.

21 Loosen the bolts or nuts in 1/4-turn increments until they can be removed by hand. Separate the cap (and washers, if used) from the connecting rod, move the end of the rod away from the crankshaft journal and push the piston/rod assembly out through the top of the bore. The crankshaft can now be lifted out.

22 Do not separate the lever from the governor shaft unless new parts are needed - the lever mount will be damaged during removal.

23 The governor assembly can be withdrawn from the gear shaft after the spool and washer is removed from the governor gear shaft. Note how the parts fit together to simplify reassembly (a simple sketch would be helpful). Check the governor parts for wear and damage. If the gear shaft must be replaced, measure how far it protrudes before removing it and don't damage the crankcase boss. Clamp the shaft in a vise and tap the crankcase boss with a soft-face hammer to extract the shaft from the hole. **Caution:** *DO NOT twist the shaft with pliers or the mounting hole will be enlarged and the new shaft won't fit into it securely.*

24 When installing the new shaft, coat the serrated end with stud and bearing mount liquid after the shaft has been

started in the hole with a soft-face hammer. Use a vise or press to finish installing the shaft and make sure it protrudes the same amount as the original (or the distance specified on the instruction sheet included with the new part). Wipe any excess stud and bearing mount liquid off the shaft and mounting boss flange. **Note:** *At this point, the gear reduction components can be removed (if equipped). Be sure to note how the parts fit together - a simple sketch may prove helpful during reassembly. Check the gears and shaft for wear and damage. Install new parts if necessary - replace the shaft oil seal even if the original shaft is reinstalled.*

## Inspection of components

25 After the engine has been completely disassembled, refer to Chapter 5 for the cleaning, component inspection, cylinder honing and valve lapping procedures. **Note:** *Special test equipment is needed to check the ignition coil/electronic ignition module. If you suspect the coil/module is causing ignition problems, have it checked by a small engine repair shop.*

26 Once you've inspected and serviced everything and purchased any necessary new parts, which should always include new gaskets and seals, reassembly can begin. Begin by reinstalling the gear reduction components, if equipped, in the crankcase, then proceed as follows: **Note:** *These heavy duty engines are equipped with tapered roller bearings on the crankshaft. These bearings must be installed by a machine shop equipped with the necessary tools. Be sure to install new camshaft gear if the crankshaft gear is damaged. Do not replace one gear only.*

# Reassembly

27 Working on OHV cylinder heads, coat the intake valve stem with clean engine oil, moly-base grease or engine assembly lube, then reinstall it over the valve in the cylinder head. Remember, the cylinder head will be installed later with the valves inside the head. The valve springs and retainers will be compressed onto the valves after the cylinder head bolts and rocker arm housing are installed. **Note:** *Refer to Section 8 for L-head valve reassembly procedures.*

28 If equipped, assemble the compression release components onto the camshaft **(see illustration 9.32)**. **Note:** *Some models are equipped with a Mechanical Compression Release (MCR) camshaft that is equipped with a locking pin in the camshaft that extends over the exhaust cam lobe to lift the valve to relieve engine compression during cranking. Be sure this pin is locked into the correct position after installation.*

29 Lubricate the crankshaft magneto side oil seal lip **(see illustration 8.33)**, the plain bearing (if applicable) and the connecting rod journal with clean engine oil, moly-base grease or engine assembly lube, then reposition the crankshaft in the crankcase. If a ball-bearing is used on the magneto side, lubricate it with clean engine oil.

30 Before installing the piston/connecting rod assembly, the cylinder must be perfectly clean and the top edge of the bore must be chamfered slightly so the rings don't catch on it. Stagger the piston ring end gaps and make sure they're positioned correctly when the piston is installed (see Chapter 5). Lubricate the piston and rings with clean engine oil, then attach a ring compressor to the piston. Leave the skirt protruding about 1/4-inch. Tighten the compressor until the piston cannot be turned, then loosen it until the piston turns in the compressor with resistance.

31 Rotate the crankshaft until the connecting rod journal is at TDC (Top Dead Center - top of the stroke) and apply a coat of engine oil to the cylinder walls **(see illustration 9.35)**. If the piston has an arrow in the top, it must face OUT of the engine (if possible) or toward the carburetor when looking at the engine with the connecting rod pointing down. Make sure the match marks on the rod and cap will be facing out when the rod/piston assembly is in place. Gently insert the piston/connecting rod assembly into the cylinder and rest the bottom edge of the ring compressor on the engine block. Tap the top edge of the ring compressor to make sure it's contacting the block around its entire circumference. **Note:** *The piston(s) used on these models will have either an arrow stamped above the piston pin hole, a cast number inside the piston or an arrow stamped on the top of the piston. The piston must be installed with these marks toward the carburetor. If the piston is not marked with one of the above designations, the piston can be installed in either direction.*

32 Carefully tap on the top of the piston with the end of a wooden or plastic hammer handle while guiding the end of the connecting rod into place on the crankshaft journal. The piston rings may try to pop out just before entering the bore, so keep some pressure on the ring compressor. Work slowly - if any resistance is felt as the piston enters the cylinder, stop immediately. Find out what's hanging up and fix it before proceeding. Do not, for any reason, force the piston into the cylinder - you'll break a ring and/or the piston.

33 Install the connecting rod cap, a NEW lock plate, the oil dipper (if used) and the bolts or washers and nuts. Make sure the marks you made on the rod and cap (or the manufacturer's marks) are aligned and facing out and the oil dipper is oriented correctly.

34 Tighten the bolts or nuts to the specified torque listed at the end of this Chapter. Work up to the final torque in three steps. Temporarily install the camshaft and turn the crankshaft through two complete revolutions to make sure the rod doesn't hit the cylinder wall or camshaft. If it does, the piston/connecting rod is installed incorrectly.

35 Apply clean engine oil, moly-base grease or engine assembly lube to the tappets, then reinstall them. Make sure they're returned to their original locations.

36 Install the camshaft. Apply clean engine oil, moly-base grease or engine assembly lube to the lobes and bearing journals and align the timing marks on the gears - this is very important! The marks are usually dimples/lines or beveled teeth (or a combination of them) near the outer edge of the gears. On many engines, the camshaft gear mark must be aligned with the keyway for the crankshaft gear - no mark is included on the crankshaft gear itself.

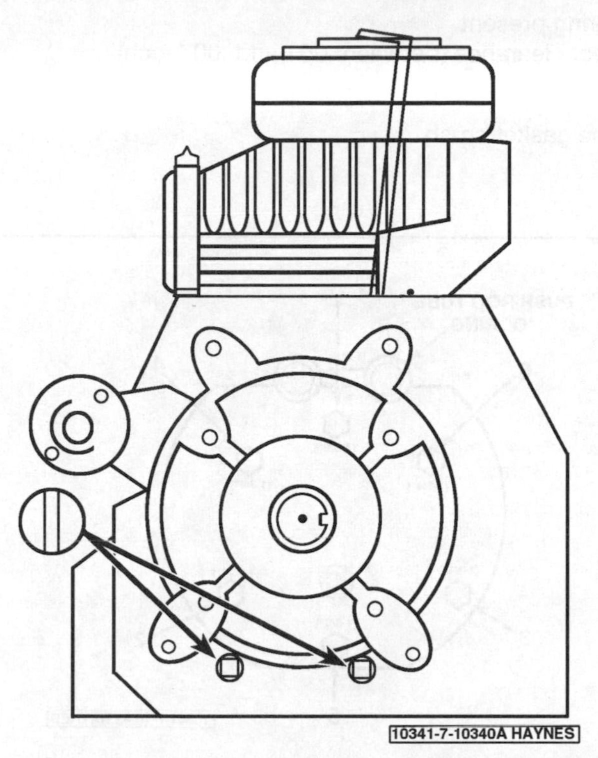


**10.40a Align the slots behind the fittings to align the counterbalance shafts (side-to-side counterbalance system)**

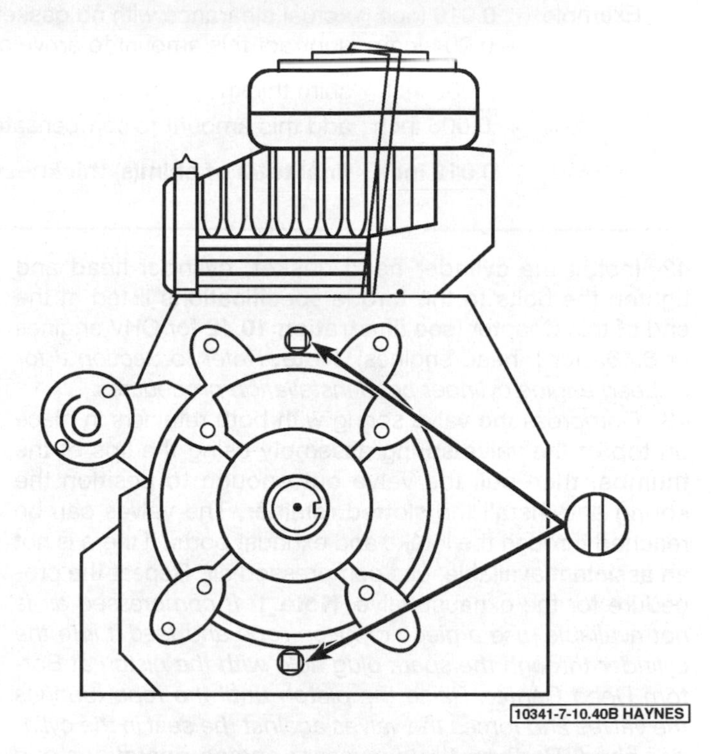


**10.40b Alignment slots for the top-to-bottom counterbalance system**

37 Install the plunger-type oil pump (if used) after lubricating it with clean engine oil. Be absolutely certain the chamfered side of the pump body faces the camshaft and the plunger ball is seated in the recess in the oil sump after the sump is in place **(see illustration 8.41)**.

38 Lubricate the crankshaft main bearing journal and the lip on the oil seal in the sump (or crankcase cover) with clean engine oil, moly-base grease or engine assembly lube.

39 Install the counterbalance gears. Set the engine to TDC (see Section 4). On systems equipped with DynaStatic counterbalance shafts, make sure the balance shafts are set properly. Make sure the alignment marks are correct **(see illustrations 10.13a or 10.13b)**.

40 Make sure the dowel pins are in place, then position a new gasket on the crankcase (the dowel pins will hold it in place). Carefully lower the oil sump (or crankcase cover) into place over the end of the crankshaft until it seats on the crankcase. DO NOT damage the oil seal lip or leaks will result! The governor shaft must match up with the spool end. Apply thread locking compound to the bolt threads, then install and tighten them to the specified torque. Follow a criss-cross pattern and work up to the final torque in three equal steps to avoid warping the oil sump/cover. Also, make sure the slotted alignment indicators for the counterbalance shafts are visible in the holes in the crankcase cover **(see illustrations 10.40a and 10.40b)** before tightening the bolts. This step is extremely important for counterbalance shaft alignment.

41 Once the crankcase cover has been installed over the end of the crankshaft and the roller bearing assembly is fitted to the case, the smaller crankcase cover must be installed and the endplay must be checked using a feeler gauge **(see illustration 10.41)**. Determine the gap between the cover and the machined surface using a feeler gauge without gaskets or O-ring installed between the assembly. If the measurement is less than 0.007, then the addition of the O-ring and gasket will reduce the actual measurement leaving the correct endplay setting. If the measurement is over 0.007 inch, then a special shim must be installed to compensate for the excess endplay. To find the correct size shim, take the actual measured endplay and add the variables (gasket and O-ring thickness). Follow this example:

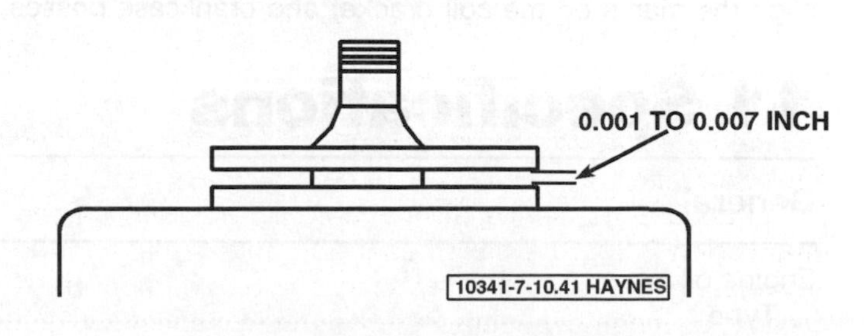


**10.41 Check the gap with the gasket removed from the end cover and install the correct size shim only if the clearance exceeds 0.007 inch**

| Example | 0.010 inch | actual clearance with no gasket or O-ring present |
|---|---|---|
| | – 0.004 inch | subtract this amount to arrive at correct clearance (between .001 and .007 inch) |
| | 0.006 inch | shim thickness required |
| | + 0.006 inch | add this amount to compensate for the gasket crush |
| | **0.012 inch** | **final total of shim(s) thickness** |

42 Install the cylinder head gasket, cylinder head and tighten the bolts to the torque specifications listed at the end of this Chapter (**see illustration 10.42** for OHV engines or **8.46a** for L-head engines). **Note:** *Refer to Section 8 for L- head engine cylinder head installation procedures.*

43 Compress the valve spring with both retainers in place on top of the valve spring assembly using the tips of the thumbs, then pull the valve out enough to position the spring and install the slotted retainer. The valves can be reached through the intake and exhaust ports. If there is not an assistant available, use compressed air. Repeat the procedure for the exhaust valve. **Note 1:** *If compressed air is not available, use a piece of nylon rope and feed it into the cylinder through the spark plug hole with the piston at Bottom Dead Center. Raise the piston until the rope touches the valves and forces the valves against the seat in the cylinder head. Perform the necessary engine operations and then remove the rope.* **Note 2:** *A white Teflon O-ring is used under the valve spring retainer on the exhaust valve and a black O-ring is used on the intake valve. Do not interchange the O-rings or damage may result.* **Note 3:** *Some models are equipped with dampening coils (tight wound coils) on the end of the valve spring. Position the dampening coils near the engine block away from the valve spring retainers.*

44 Install the crankcase breather with the small hole(s) down and intake manifold (unless the carburetor and manifold were removed as an assembly). Use new gaskets and tighten the bolts securely.

45 Adjust the valves with the engine cold (see Section 7).

46 Install the valve cover with a new gasket. Be sure to install new O-rings onto the cover bolts, if equipped.

47 Install the governor assembly **(see illustration 8.3)**. Refer to the carburetor and governor adjustment procedure in Section 6 and make sure the governor linkage and carburetor is adjusted properly.

48 Install the ignition coil/spark plug wire assembly and align the marks on the coil bracket and crankcase bosses, then tighten the coil mounting bolts securely. **Note:** *If the coil is mounted on the outside of the flywheel, don't tighten the bolts completely - just snug them up. The bolt holes are slotted; move the coil as far away from the flywheel as possible before snugging up the bolts. Check the coil/module air gap adjusting procedure in Section 4. Be sure to reroute the wires from the ignition coil properly.*

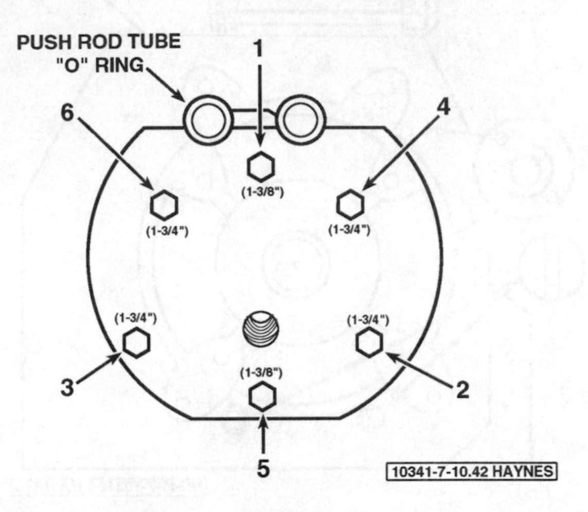


**10.42 OHV cylinder head tightening sequence - make sure the correct length bolt is installed into the correct position**

49 Position the Woodruff key in the crankshaft keyway and install the flywheel. Install the starter cup, washer and nut. Tighten the nut to the specified torque listed at the end of this Chapter.

50 If a flywheel brake is used, install it now. To install the remaining components, refer to Chapters 4 and 5 as necessary. **Caution:** *Be sure to fill the crankcase to the correct level with the specified oil before attempting to start the engine.*

# 11 Specifications

## General

Engine oil

| | |
|---|---|
| Type | API grade SH or better high quality detergent oil |
| Viscosity | |
| Above 32-degrees F | SAE 30 |
| Zero to 32-degrees F | SAE 5W-30 |
| Below zero-degrees F | SAE 0W-30 |

Engine oil capacity
- L-head light duty engines
  - H60, HH60, HSK60, H70, HH70, HSK70 .......... 19 ounces
  - V60, VH60, V70, VH70, TVM140 .......... 27 ounces
  - VM70, VM80, VM100, TVM170, TVM195, TVM220, TVXL195, TVXL220 .......... 32 ounces
  - HM70, HM80, HM100, HMSK70, HMSK80, HMSK100 .......... 26 ounces
- OHV light duty engines
  - OHH55, OHSK55, OVRM60 .......... 21 ounces
  - OHM, OHSK, OVM, OVXL, OVH11-13 .......... 32 ounces
  - OHV13.5-17
    - without oil filter .......... 55 ounces
    - with 2-1/4 inch filter .......... 62 ounces
    - with 2-5/8 inch filter .......... 64 ounces
- L-head and OHV heavy duty engines .......... Not available

## Ignition system

Spark plug type (Champion)
- TVS and TVXL engines .......... RJ19LM
- OHV and OVR, OH180, OHH55, OHSK55, OVXL120 (nos. 203000 on), OVXL125 (nos. 203000 on) .......... RN4C
- OVM120, OVXL120 (nos. 202700 to 202999), OVXL125 (nos. 202700 to 202999), OHM120, OHSK110, OHSK120 .......... RL86C
- OH120, OH140, OH160 .......... RL82C
- All others .......... RJ8C

Spark plug gap (all) .......... 0.030 inch
Ignition point gap .......... 0.020 inch
Ignition coil-to-flywheel gap .......... 0.0125 inch

## Fuel system

Float level
- L-head engines (Walbro carburetors only)
  - V80, VM80, H80, HM80, VH80, HH80, HHM80, TVM195 .......... 5/64 to 7/64 inch
  - All others .......... 7/64 to 1/8 inch
- OHV engines
  - Tecumseh carburetors .......... 11/64 inch
  - Walbro carburetors .......... 5/16 inch

## Torque specifications

Cylinder head bolts
- L-head light duty engines .......... 200 in-lbs
- OHV light duty engines .......... 19 ft-lbs
- L-head and OHV heavy duty engines .......... 20 ft-lbs

Connecting rod bolts
- L-head light duty engines
  - V60, VH60, VH70, HH60, HH70 .......... 105 in-lbs
  - TVM140, HSK60, H60 .......... 170 in-lbs
  - V70, HSK70, H70, VM70, VM80, VM100, TVM170, TVM195, TVM220, TVXL170, TVXL195, TVXL220, HMSK70, HMSK80, HMSK100, HM70, HM80, HM100, HHM80 .......... 210 in-lbs
- OHV light duty engines
  - OVRM60, OHH55, OHSK55 .......... 105 in-lbs
  - OVXL, OHM, OVM, OHSK, OHV11-13, OHV13.5-17 .......... 210 in-lbs
- L-head and OHV heavy duty engines .......... 110 in-lbs

**Torque specifications (continued)**

Crankcase cover
  L-head light duty engines
    TVM140 (except Power-Loc bolts), HSK60, H60, V70, HSK70, H70 .......... 115 in-lbs
    VM70, VM80, VM100, TVM140 (with Power-Loc bolts), TVM170, TVM195, V60, VH60, VH70, HH60, HH70, TVM220, TVXL170, TVXL195, TVXL220,HMSK70, HMSK80, HMSK100, HM70, HM80, HM100, HHM80 .......... 125 in-lbs
  OHV light duty engines
    OVRM60, OHH55, OHSK55 .......... 115 in-lbs
    OVXL, OHM, OVM, OHSK, OHV11-13, OHV13.5-17 .......... 125 in-lbs
  L-head and OHV heavy duty engines .......... 130 in-lbs
Flywheel nut
  L-head light duty engines with internal ignition .......... 39 ft-lbs
  L-head light duty engines with external ignition .......... 52 ft-lbs
  OHV light duty engines
    OVRM60, OHH55, OHSK55 .......... 35 ft-lbs
    OVXL, OHM, OVM, OHSK, OHV11-13, OHV13.5-17 .......... 58 ft-lbs
  L-head OHV heavy duty engines .......... 55 ft-lbs
Spark plug .......... 180 in-lbs
Ignition mounting bolts
  Direct to cylinder
    HH60, HH70, V70, VH60, VH70 .......... 90 in-lbs
    All others .......... 40 in-lbs
  Stud to cylinder .......... 40 in-lbs
Intake manifold to cylinder .......... 100 in-lbs
Carburetor to intake manifold .......... 68 in-lbs
Carburetor adapter-to-cylinder bolts .......... 85 in-lbs
Air cleaner to carburetor .......... 15 in-lbs
Recoil starter mounting bolts
  8-32 thread bolts .......... 22 in-lbs
  Top mount .......... 50 in-lbs
  Side mount plastic .......... 85 in-lbs
  Side mount metal .......... 60 in-lbs
Electric starter
  L-head light duty engines
    HH60, HH70, V70, VH60, VH70 .......... 155 in-lbs
    All others .......... 65 in-lbs
  OHV light duty engines
    OHV13.5-17 .......... 150 in-lbs
    All others .......... 100 in-lbs
  L-head and OHV heavy duty engines .......... 160 in-lbs
Valve cover bolts
  Hex jam nut type .......... 18 ft-lbs
  Stud lock nut type .......... 120 in-lbs
  Allen screw type .......... 110 in-lbs
  Four screw type .......... 55 in-lbs
Rocker arm nuts
  OHV light duty engines
    OVRM60, OHH55, OHSK55 .......... 40 in-lbs
    OVXL, OHM, OVM, OHSK .......... 45 in-lbs
    OHV11-13, OHV13.5-17 .......... 18 ft-lbs
    Four screw type .......... 55 in-lbs
  OHV heavy duty engines
    Bolts-to-cylinder head .......... 90 in-lbs
    Bolts-to-manufacturing stud .......... 30 in-lbs
Oil drain plug .......... 15 in-lbs

All models have point setting of .020", spark plug gap of .030", air gap .0125", valve seat angle of 46°.
All dimensions are in inches.

| SPECIFICATION | V60, VH60, TVM140, H60, HH60 | V70, VH70,VM70 H70, HH70 | TVM170, HM70 | V80, VM80 , H80, HM80* | VM80 , TVM195, HM80*, HHM80 | VM100, HM100* | TVM220, HM100* |
|---|---|---|---|---|---|---|---|
| HP (Approx.) | 6 | 7 | 7 | 8 | 8 | 10 | 10 |
| Displacement (in³) | 13.53 | 15.04 | 17.17 | 18.65 | 19.43 note A | 20.2 | 21.82 |
| Stroke | 2.5 | 2.532 | 2.532 | 2.532 | 2.532 note B | 2.532 | 2.532 |
| Bore | 2.6250 2.6260 | 2.7500 2.7510 | 2.9375 2.9385 | 3.0620 3.0630 | 3.1250 3.1260 | 3.1870 3.1880 | 3.3120 3.3130 |
| Timing Dim. B.T.D.C. | 0.080 | 0.080 | 0.090 | 0.090 | 0.090 | 0.090 | 0.090 |
| Valve Clearance | .004-.010 Both | .004-.010 Both | .004-.010 Both | .004-.010 Both | .004-.010 Both | .004-.010 Both | .004-.010 Both |
| Valve Seat Width | .042 .052 | .042 .052 | .042 .052 | .042 .052 | .042 .052 | .042 .052 | .042 .052 |
| Valve Guide Oversize Dim. | .3432 .3442 | .3432 .3442 | .3432 .3442 | .3432 .3442 | .3432 .3442 | .3432 .3442 | .3432 .3442 |
| Crankshaft End Play | .005 .027 note F | .005 .027 note F | .005 .027 | .005 .027 | .005 .027 | .005 .027 | .005 .027 |
| Crankpin Journal Dia. | 1.0615 1.0620 | 1.1860 1.1865 | 1.1860 1.1865 | 1.1860 1.1865 | 1.1860 1.1865 | 1.1860 1.1865 | 1.1860 1.1865 |
| Crankshaft Mag. Main Brg. Dia. | .9985 .9990 | .9985 .9990 | .9985 .9990 | .9985 .9990 | .9985 .9990 | .9985 .9990 | .9985 .9990 |
| Crankshaft P.T.O. Main Brg. Dia. | .9985 .9990 | .9985 .9990 | 1.1870 1.1875 | 1.1870 1.1875 | 1.1870 1.1875 | 1.1870 1.1875 | 1.1870 1.1875 |
| Camshaft Bearing | .6230 .6235 | .6230 .6235 | .6230 .6235 | .6230 .6235 | .6230 .6235 | .6230 .6235 | .6230 .6235 |
| Conn. Rod Dia. Crank Brg. | 1.0630 1.0635 | 1.1880 1.1885 | 1.1880 1.1885 | 1.1880 1.1885 | 1.1880 1.1885 | 1.1880 1.1885 | 1.1880 1.1885 |
| Piston Diameter Bottom Of Skirt | 2.6210 2.6215 | 2.7450 2.7455 | 2.9325 2.9335 | 3.0575 3.0585 note C | 3.1195 3.1205 | 3.1815 3.1825 | 3.3090 3.3105 |
| Ring Groove Side Clearance 1st & 2nd Comp. | .002 .004 | .002 .003 | .002 .005 | .002 .005 | .002 .005 | .002 .005 | .002 .005 |
| Ring Groove Side Clearance Bottom Oil | .002 .004 | .001 .003 | .001 .004 | .001 .004 | .001 .004 | .001 .004 | .001 .004 |
| Piston Skirt Clearance | .0035 .0050 note D | .0045 .0060 note E | .004 .006 | .0035 .0055 | .0045 .0065 | .0045 .0065 | .0015 .0040 |
| Ring End Gap | .010 .020 | .010 .020 | .010 .020 | .010 .020 | .010 .020 | .010 .020 | .010 .020 |
| Cylinder Main Brg. | 1.0005 1.0010 | 1.0005 1.0010 | 1.0005 1.0010 | 1.0005 1.0010 | 1.0005 1.0010 | 1.0005 1.0010 | 1.0005 1.0010 |
| Cylinder Cover/Flange Main Bearing Diameter | 1.0005 1.0010 | 1.0005 1.0010 | 1.1890 1.1895 | 1.1890 1.1895 | 1.1890 1.1895 | 1.1890 1.1895 | 1.1890 1.1895 |

* Check to detemine bore size

Notes: (A) VM & HM80 - Displacement 19.41 (B) VM & HM80 - Bore 3-1/8" (C) VM & HM80 Piston Dia. 3.1195/3.1205
(D) VH50, 60 .0015/.0055 (E) VH70 .0038/.0073 (F) VH, HH50-70 Models .003/.031

**Engine specifications for L-head light duty engines with standard point ignition**

*Tecumseh/Craftsman engine specifications courtesy of and with permission of Tecumseh Products Company.*

All models have point setting of .020", spark plug gap of .030", air gap .0125", valve seat angle of 46°.
All dimensions are in inches.

| SPECIFICATION | HSK60 VH60, TVM140, H60, HH60 | V70, VH70,H70, HSK70, HH70, TVM170 (E) MODEL | HM70 (Models ending in C) | HM70 (Models ending in D) | TVM170 Models (F & UP ), HM70 Models (E & up) | HMSK80 VM80*, TVM195 (A-K), HM80** | HM80** TVM195 (L & up), TVXL195 | TVM220 (A-F), HM 100** | HMSK100 TVM220 (G & up), TVXL220, HM100** |
|---|---|---|---|---|---|---|---|---|---|
| HP (Approx.) | 6 | 7 | 7 | 7 | 7 | 8 | 8 | 10 | 10 |
| Displacement (in³) | 13.53 | 15.04 | 17.17 | 17.17 | 19.43 | 19.43 | 19.43 | 21.82 | 21.82 |
| Stoke | 2.500 | 2.532 | 2.532 | 2.532 | 2.532 | 2.532 | 2.532 | 2.532 | 2.532 |
| Bore | 2.625 2.626 | 2.750 2.751 | 2.9375 2.9385 | 2.9375 2.9385 | 3.125 3.126 | 3.125 3.126 | 3.125 3.126 | 3.312 3.313 | 3.312 3.313 |
| Valve Clearance | .004-.010 Both | .004-.010 Both | .004-.010 Both | .004-.010 Both | .004-.010 Both | .004-.010 Both | .004-.010 Both | .004-.010 Both | .004-.010 Both |
| Valve Seat Width | .042 .052 | .042 .052 | .042 .052 | .042 .052 | .042 .052 | .042 .052 | .042 .052 | .042 .052 | .042 .052 |
| Valve Guide Oversize Dim. | .3432 .3442 | .3432 .3442 | .3432 .3442 | .3432 .3442 | .3432 .3442 | .3432 .3442 | .3432 .3442 | .3432 .3442 | .3432 .3442 |
| Crankshaft End Play | .005 .027 Note (A) | .005 .027 Note (A) | .005 .027 | .007 .029 | .007 .029 | .007 .029 | .007 .029 | .007 .029 | .007 .029 Note (G) |
| Crankpin Journal Dia. | 1.0615 1.0620 | 1.1862 1.1865 | 1.1860 1.1865 | 1.3740 1.3745 | 1.3740 1.3745 | 1.1860 1.1865 | 1.3740 1.3745 | 1.1860 1.1865 | 1.3740 1.3745 |
| Crankshaft Mag. Main Brg. Dia. | .9985 .9990 | .9985 .9990 | .9985 .9990 | 1.3745 1.3750 | 1.3745 1.3750 | .9985 .9990 | 1.3745 1.3750 | .9985 .9990 | 1.3745 1.3750 |
| Crankshaft P.T.O. Main Brg. Dia. | .9985 .9990 | .9985 .9990 | 1.1870 1.1875 | 1.3745 1.3750 | 1.3745 1.3750 | 1.1870 1.1875 | 1.3745 1.3750 | 1.1870 1.1875 | 1.3745 1.3750 |
| Camshaft Bearing | .6230 .6235 | .6230 .6235 | .6230 .6235 | .6230 .6235 | .6230 .6235 | .6230 .6235 | .6230 .6235 | .6230 .6235 | .6230 .6235 |
| Conn. Rod Dia. Crank Brg. | 1.0630 1.0635 | 1.0630 1.0635 | 1.1880 1.1885 | 1.3760 1.3765 Note (F) | 1.3760 1.3765 Note (F) | 1.1880 1.1885 | 1.3760 1.3765 Note (F) | 1.1880 1.1885 | 1.3760 1.3765 Note (F) |
| Piston Diameter Bottom Of Skirt | 2.6212 2.6220 Note (D) | 2.6212 2.6220 Note (E) | 2.9325 2.9335 | 2.9325 2.9335 | 3.1195 3.1205 | 3.1195 3.1205 | 3.1195 3.1205 | 3.3090 3.3105 | 3.3098 3.3108 |
| Ring Groove Side Clearance 1st & 2nd Comp. | .002 .005 | .002 .005 | .002 .005 | .002 .005 | .002 .005 | .002 .005 | .002 .005 | .0015 .0035 | .0015 .0035 |
| Ring Groove Side Clearance Bottom Oil | .001 .004 | .001 .004 | .001 .004 | .001 .004 | .001 .004 | .001 .004 | .001 .004 | .001 .004 | .001 .004 |
| Piston Skirt Clearance | .0030 .0048 Note (B) | .0030 .0048 Note (C) | .004 .006 | .004 .006 | .0045 .0065 | .0045 .0065 | .0045 .0065 | .0015 .0040 | .0012 .0032 |
| Ring End Gap | .010 .020 | .010 .020 | .010 .020 | .010 .020 | .010 .020 | .010 .020 | .010 .020 | .010 .020 | .010 .020 |
| Cylinder Main Brg. | 1.0005 1.0010 | 1.0005 1.0010 | 1.0005 1.0010 | 1.3765 1.3770 | 1.3765 1.3770 | 1.0005 1.0010 | 1.3765 1.3770 | 1.0005 1.0010 | 1.3765 1.3770 |
| Cylinder Cover/Flange Main Bearing Diameter | 1.0005 1.0010 | 1.0005 1.0010 | 1.1890 1.1895 | 1.3765 1.3770 | 1.3765 1.3770 | 1.1890 1.1895 | 1.3765 1.3770 | 1.1890 1.1895 | 1.3765 1.3770 |

* Check to detemine bore size
Notes: (A) VH, HH50-70 models .003/.031 (B) VH, HH50-60 .0015/.005 (C) VH, HH70 .0038/.0073
(D) VH, HH50-60 2.6235/2.6205 (E) VH, HH70 2.7462/2.7437 (F) After Serial Number 9274 1.3775/1.3780 (G) TVM 220 Ultra Balance .002/.042
** Check to determine crankshaft bearing diameters

**Engine specifications for L-head light duty engines with solid state and external ignition**

| All dimensions are in inches | OVRM50-60 | OHH50<br>OHSK50 | OHH55<br>OHSK55 | OHM, OHSK, OVM, OVXL, & OHV11-13 | OHV13.5-17 |
|---|---|---|---|---|---|
| Displacement | 10.49<br>Note (A) | 10.49 | 11.9 | 21.82 | 29.9 |
| Stroke (In$^3$) | 1.938 | 1.938 | 1.938 | 2.532 | 3.00 |
| Bore | 2.625<br>2.626<br>Note (B) | 2.625<br>2.626 | 2.795<br>2.796 | 3.312<br>3.313 | 3.562<br>3.563 |
| Ignition Module Air Gap | .0125 | .0125 | .0125 | .0125 | .0125 |
| Spark Plug Gap | .030 | .030 | .030 | .030 | .030 |
| Valve Clearance In.<br>Ex. | .004<br>.004 | .004<br>.004 | .004<br>.004 | .004<br>.004 | .004<br>.004 |
| Valve Seat Angle | 46$^{0}$ | 46$^{0}$ | 46$^{0}$ | 46$^{0}$ | 46$^{0}$ |
| Valve Seat Width | .035<br>.045 | .035<br>.045 | .035<br>.045 | .035<br>.045 | .042<br>.052 |
| Valve Guide Oversize Dimension | INT. .2807 .2817<br>EX. .2787 .2797 | .2807<br>.2817 | .2807<br>.2817 | .3432<br>.3442 | .3432<br>.3442 |
| Crankshaft End Play | .006<br>.027 | .006<br>.027 | .006<br>.027 | .002<br>.042 | .0025<br>.0335 |
| Crankpin Journal Dia. | .9995<br>1.000 | .9995<br>1.000 | .9995<br>1.000 | 1.3740<br>1.3745 | 1.6223<br>1.6228 |
| Crankshaft Dia. Flywheel End Main Brg. | .9985<br>.9990 | .9985<br>.9990 | .9985<br>.9990 | 1.3745<br>1.3750 | 1.6245<br>1.6250 |
| Crankshaft Dia. P.T.O. Main Brg. | 1.0005<br>1.0010 | .9985<br>.9990 | .9985<br>.9990 | 1.3745<br>1.3750 | 1.6245<br>1.6250 |
| Conn. Rod Dia. Crank Brg. | 1.0005<br>1.0010 | 1.0005<br>1.0010 | 1.0005<br>1.0010 | 1.3775<br>1.3780 | 1.6234<br>1.6240 |
| Camshaft Bearing Diameter | .4975<br>.4980 | .4975<br>.4980 | .4975<br>.4980 | .6230<br>.6235 | .6235<br>.6240 |
| Piston Dia. Bottom of Skirt | 2.6204<br>2.6220 | 2.6204<br>2.6220 | 2.7904<br>2.7920 | 3.3095<br>3.3105 | 3.5595<br>3.5605 |
| Ring Groove Side Clearance 1st & 2nd Comp. | .0020<br>.0050 | .0020<br>.0050 | .0020<br>.0050 | .0020<br>.0040 | .0020<br>.0040 |
| Ring Groove Side Clearance Bottom Oil | .0005<br>.0035 | .0005<br>.0035 | .0005<br>.0035 | .0010<br>.0030 | .0009<br>.0029 |
| Piston Skirt to Cylinder Clearance | .0030<br>.0056 | .0030<br>.0056 | .0030<br>.0056 | .0015<br>.0035 | .0015<br>.0030 |
| Ring End Gap | .010<br>.020 | .007<br>.017 | .007<br>.017 | .010<br>.020 | .012<br>.022 |
| Cylinder Main Bearing Diameter | 1.0005<br>1.0010 | 1.0005<br>1.0010 | 1.0005<br>1.0010 | 1.3765<br>1.3770 | 1.6265<br>1.6270 |
| Cylinder Cover / Flange Main Brg. Diameter | 1.0005<br>1.0010 | 1.0005<br>1.0010 | 1.0005<br>1.0010 | 1.3765<br>1.3770 | 1.6265<br>1.6270 |

**Engine specifications for OHV light duty engines**

| Model | | HH80 | HH100 | HH120 |
|---|---|---|---|---|
| Displacement | | 23.7 | 23.7 | 27.66 |
| Stroke | | 2-3/4" | 2-3/4" | 2-7/8" |
| Bore | | 3.3120<br>3.3130 | 3.3120<br>3.3130 | 3.500<br>3.501 |
| Timing Dimension | | TDC-Start<br>.095 Run | TDC-Start<br>.095 Run | TDC-Start<br>.095 Run |
| Point Gap | | .020 | .020 | .020 |
| Spark Plug Gap | | .028<br>.033 | .028<br>.033 | .028<br>.033 |
| Valve Clearance | Intake | .010 | .010 | .010 |
| | Exhaust | .020 | .020 | .020 |
| Valve Seat Angle | | 46° | 46° | 46° |
| Valve Seat Width | | .042<br>.052 | .042<br>.052 | .042<br>.052 |
| Valve Face Angle | | 45° | 45° | 45° |
| Valve Face Width | | .089<br>.099 | .089<br>.099 | .089<br>.099 |
| Valve Lip Width | | .06 | .06 | .06 |
| Valve Spring Free Length | | 1.885 | 1.885 | 1.885 |
| Valve Guides STD Diameter | | .312<br>.313 | .312<br>.313 | .312<br>.313 |
| Valve Guides Over-Size Dimensions | | .343<br>.344 | .343<br>.344 | .343<br>.344 |
| Dia. Crankshaft Conn. Rod Journal | | 1.3750<br>1.3755 | 1.3750<br>1.3755 | 1.3750<br>1.3755 |
| Maximum Conn. Rod Dia. Crank Bearing | | Maximum<br>1.3765 | Maximum<br>1.3765 | Maximum<br>1.3765 |
| Shaft Seat Dia. for Roller Bearings | | 1.1865<br>1.1870 | 1.1865<br>1.1870 | 1.1865<br>1.1870 |
| Crankshaft End Play | | None | None | None |
| Piston Diameter | | 3.3080<br>3.3100 | 3.3080<br>3.3100 | 3.4950<br>3.4970 |
| Piston Pin Diameter | | .6873<br>.6875 | .6873<br>.6875 | .6873<br>.6875 |
| Width Comp. Ring Groove | | .0950<br>.0960 | .0950<br>.0960 | .0950<br>.0960 |
| Width Oil Ring Groove | | .1880<br>.1900 | .1880<br>.1900 | .1880<br>.1890 |
| Side Clearance Ring Groove | | .0020<br>.0035 | .0020<br>.0035 | .0020<br>.0035 |
| Ring End Gap | | .007<br>.020 | .007<br>.020 | .007<br>.020 |
| Top Piston Land Clearance | | .0305<br>.0335 | .030<br>.035 | .031<br>.036 |
| Piston Skirt Clearance | | .002<br>.005 | .002<br>.005 | .003<br>.006 |
| Camshaft Bearing Diameter | | .6235<br>.6240 | .6235<br>.6240 | .6235<br>.6240 |
| Cam Lobe Diameter Nose to Heel | | 1.3045<br>1.3085 | 1.3045<br>1.3085 | 1.3045<br>1.3085 |
| Magneto Air Gap | | .006<br>.010 | .006<br>.010 | .006<br>.010 |

**Engine specifications for L-head heavy duty horizontal shaft engines**

| Model | | VH80 | VH100 |
|---|---|---|---|
| Displacement | | 23.7 | 23.7 |
| Stroke | | 2-3/4″ | 2-3/4″ |
| Bore | | 3.3120 / 3.3130 | 3.3120 / 3.3130 |
| Timing Dimension | | Solid State | Solid State |
| Point Gap | | Solid State | Solid State |
| Spark Plug Gap | | .035 | .035 |
| Valve Clearance | Intake | .010 | .010 |
| | Exhaust | .020 | .020 |
| Valve Seat Angle | | 46° | 46° |
| Valve Seat Width | | .042 / .052 | .042 / .052 |
| Valve Face Angle | | 45° | 45° |
| Valve Face Width | | .089 / .099 | .089 / .099 |
| Valve Lip Width | | .06 | .06 |
| Valve Spring Free Length | | 1.885 | 1.885 |
| Valve Guides STD Diameter | | .312 / .313 | .312 / .313 |
| Valve Guides Over-Size Dimensions | | .344 / .345 | .344 / .345 |
| Diameter Crankshaft Conn. Rod Journal | | 1.3755 / 1.3750 | 1.3755 / 1.3750 |
| Maximum Conn. Rod Dia. Crank Bearing | | 1.3761 | 1.3761 |
| Shaft Seat Dia. for Roller Bearings | | 1.1865 / 1.1870 | 1.1865 / 1.1870 |
| Crankshaft End Play | | None | None |
| Piston Diameter | | 3.308 / 3.310 | 3.308 / 3.310 |
| Piston Pin Diameter | | .6873 / .6875 | .6873 / .6875 |
| Width Comp. Ring Groove | | .0950 / .0960 | .0950 / .0960 |
| Width Oil Ring Groove | | .1800 / .1900 | .1800 / .1900 |
| Side Clearance Ring Groove | | .0025 / .0030 | .0025 / .0030 |
| Ring End Gap | | .010 / .020 | .010 / .020 |
| Top Piston Land Clearance | | .0305 / .0335 | .0305 / .0335 |
| Piston Skirt Clearance | | .003 | .003 |
| Camshaft Bearing Diameter | | .6235 / .6240 | .6235 / .6240 |
| Cam Lobe Diameter Nose to Heel | | 1.3045 / 1.3085 | 1.3045 / 1.3085 |
| Magneto Air Gap | | .006 / .010 | .006 / .010 Solid State |

**Engine specifications for L-head heavy duty vertical shaft engines**

| Description | | Model OH120 | Model OH140 | Model OH150 | Model OH160 | Model OH180 |
|---|---|---|---|---|---|---|
| Displacement | | 21.1 | 23.7 | 27.66 | 27.66 | 30.0 |
| Stroke | | 2.75 | 2.75 | 2.875 | 2.875 | 2.875 |
| Bore | | 3.125<br>3.126 | 3.312<br>3.313 | 3.500<br>3.501 | 3.500<br>3.501 | 3.625<br>3.626 |
| Spark Plug Gap | | .030 | .030 | .030 | .030 | .030 |
| Valve Clearance | Intake | .005 | .005 | .005 | .005 | .005 |
| | Exhaust | .010 | .010 | .010 | .010 | .010 |
| Valve Seat Angle | | 46° | 46° | 46° | 46° | 46° |
| Valve Seat Width | | .042<br>.052 | .042<br>.052 | .042<br>.052 | .042<br>.052 | .042<br>.052 |
| Valve Face Angle | | 45° + 0°15′ | 45° + 0°15′ | 45° + 0°15′ | 45° + 0°15′ | 45° + 0°15′ |
| Valve Face Width | | .094 | .133 | .133 | .133 | .133 |
| Valve Lip Width | | .06 | .06 | .06 | .06 | .06 |
| Valve Spring Free Length | | 1.915 | 1.980 | 1.980 | 1.980 | 1.980 |
| Valve Guides STD Diameter | | .312<br>.313 | .312<br>.313 | .312<br>.313 | .312<br>.313 | .312<br>.313 |
| Dia. Crankshaft Conn. Rod Journal | | 1.3750<br>1.3755 | 1.3750<br>1.3755 | 1.3750<br>1.3755 | 1.3750<br>1.3755 | 1.3750<br>1.3755 |
| Maximum Conn. Rod Dia. Crank Bearing | | 1.3765 | 1.3765 | 1.3765 | 1.3765 | 1.3765 |
| Shaft Seat Dia. for Roller Bearings | | 1.1865<br>1.1870 | 1.1865<br>1.1870 | 1.1865<br>1.1870 | 1.1865<br>1.1870 | 1.1865<br>1.1870 |
| Crankshaft End Play | | *None (.001 to .007 Preload) | *None (.001 to .007 Preload) | *None (.001 to .007 Preload) | *None (.001 to .007 Preload) | *None (.001 to .007 Preload) |
| Piston Skirt Diameter | | 3.121<br>3.123 | 3.3080<br>3.3100 | 3.4950<br>3.4970 | 3.4950<br>3.4970 | 3.620<br>3.622 |
| Piston Pin Diameter | | .6876<br>.6880 | .6876<br>.6880 | .6876<br>.6880 | .6876<br>.6880 | .7810<br>.7812 |
| Width Comp. Ring Groove | | .0950<br>.0960 | .0950<br>.0960 | .0950<br>.0960 | .0950<br>.0960 | .0955<br>.0965 |
| Width Oil Ring Groove | | .1880<br>.1900 | .1880<br>.1900 | .1880<br>.1890 | .1880<br>.1890 | .1880<br>.1895 |
| Side Clearance Ring Groove Compression | | .0015<br>.0035 | .0015<br>.0035 | .0015<br>.0035 | .0015<br>.0035 | .0015<br>.0035 |
| Ring End Gap | | .010<br>.020 | .010<br>.020 | .010<br>.020 | .010<br>.020 | .010<br>.020 |
| Top Piston Land Clearance | | .031<br>.034 | .030<br>.035 | .031<br>.036 | .031<br>.036 | .031<br>.036 |
| Camshaft Bearing Diameter | | .6235<br>.6240 | .6235<br>.6240 | .6235<br>.6240 | .6235<br>.6240 | .6235<br>.6240 |
| Cam Lobe Diameter Nose to Heel | | 1.3117<br>1.3167 | 1.3117<br>1.3167 | 1.3117<br>1.3167 | 1.3117<br>1.3167 | 1.3117<br>1.3167 |

**Engine specifications for OHV heavy duty engines**

# SEARS CRAFTSMAN CROSS REFERENCE CHARTS

| | | | | | | | |
|---|---|---|---|---|---|---|---|
| 143-254012 | LAV 03540810K | 143-255042 | LAV 05062037 | 143-264402 | ECV 100145097A | 143-266162 | VM 080150039C |
| 143-254022 | LAV 03540818K | 143-255052 | LAV 05062039A | 143-264412 | ECV 100145098A | 143-266172 | VM 080150067C |
| 143-254032 | LAV 03540819K | 143-255062 | LAV 05062043 | 143-264422 | LAV 03540917K | 143-266182 | VM 080150062C |
| 143-254042 | LAV 03540820K | 143-255072 | LAV 05062043A | 143-264432 | ECV 100145099A | 143-266202 | VM 080150058C |
| 143-254052 | LAV 03540821K | 143-255082 | LAV 05062015A | 143-264452 | ECV 100145101A | 143-266212 | VM 080150064C |
| 143-254062 | ECV 100145069A | 143-255092 | LAV 05062037A | 143-264462 | ECV 100145102A | 143-266222 | VM 080150016C |
| 143-254072 | LAV 03540811K | 143-255102 | LAV 05062029A | 143-264472 | LAV 03540922K | 143-266232 | VM 080150065C |
| 143-254082 | LAV 03540812K | 143-255112 | LAV 05062039A | 143-264482 | ECV 100145103A | 143-266242 | VM 080150017C |
| 143-254092 | LAV 03540813K | 143-256012 | VM 080150076A | 143-264492 | LAV 03540923K | 143-266252 | V 06070234J |
| 143-254102 | LAV 03540814K | 143-256022 | V 06070291H | 143-264502 | LAV 03540924K | 143-266262 | VM 080150083C |
| 143-254112 | LAV 03540815K | 143-256032 | V 070125202A | 143-264512 | ECV 100145104A | 143-266272 | V 070125202C |
| 143-254122 | LAV 03540816K | 143-256042 | VM 080150080A | 143-264522 | LAV 03540927K | 143-266282 | V 070125206C |
| 143-254142 | ECV 100145062A | 143-256052 | V 06070301H | 143-264542 | LAV 03540930K | 143-266302 | V 070125210C |
| 143-254152 | ECV 100145063A | 143-256062 | VM 080150083A | 143-264562 | ECV 100145106A | 143-266312 | VM 080150096C |
| 143-254162 | ECV 100145064A | 143-256072 | VM 080150017A | 143-264572 | ECV 100145107A | 143-266322 | VM 080150097C |
| 143-254172 | ECV 100145065A | 143-256082 | V 06070303H | 143-264582 | ECV 100145108A | 143-266332 | V 070125211C |
| 143-254182 | ECV 100145066A | 143-256092 | V 06070304H | 143-264592 | ECV 100145109A | 143-266342 | VM 080150098C |
| 143-254192 | ECV 100145067A | 143-256102 | V 070125185A | 143-264602 | ECV 100145110A | 143-266352 | V 070125213C |
| 143-254202 | ECV 100145068A | 143-256112 | VM 080150047A | 143-264612 | ECV 100145111A | 143-266362 | VM 080150066C |
| 143-254212 | LAV 03540829K | 143-256122 | V 06070314H | 143-264622 | ECV 100145112A | 143-266372 | V 06070284J |
| 143-254222 | LAV 03540839K | 143-256132 | V 070125206A | 143-264632 | ECV 100145113A | 143-266382 | V 06070281J |
| 143-254232 | ECV 100145070A | 143-257012 | LAV 04050358D | 143-264642 | ECV 100145114A | 143-266392 | V 06070303J |
| 143-254232 | ECV 100145070A | 143-257022 | LAV 04050358C | 143-264652 | ECV 100145115A | 143-266402 | V 06070304J |
| 143-254242 | ECV 100145071A | 143-257032 | LAV 04050366D | 143-264662 | LAV 03540926K | 143-266412 | V 06070231J |
| 143-254252 | ECV 100145072A | 143-257042 | LAV 04050201D | 143-264672 | ECV 100145116A | 143-266422 | VM 080150063C |
| 143-254262 | ECV 100145073A | 143-257052 | LAV 04050369D | 143-264682 | LAV 03540933K | 143-266432 | V 06070327J |
| 143-254272 | ECV 100145074A | 143-257062 | LAV 04050369E | 143-265012 | LAV 05062015A | 143-266442 | V 06070118J |
| 143-254282 | ECV 100145075A | 143-257072 | LAV 04050366E | 143-265032 | LAV 05062047A | 143-266452 | V 06070117J |
| 143-254292 | ECV 100145076A | 143-264012 | LAV 03540888K | 143-265042 | LAV 05062030A | 143-266462 | VM 100157007 |
| 143-254302 | LAV 03540846K | 143-264022 | LAV 03540889K | 143-265052 | LAV 05062049A | 143-266472 | VM 100157008A |
| 143-254312 | LAV 03540847K | 143-264022 | LAV 03540898K | 143-265062 | LAV 05062039B | 143-266482 | VM 080150100D |
| 143-254322 | ECV 100145077A | 143-264032 | LAV 03540890K | 143-265072 | LAV 05062047B | 143-266492 | VM 100157009 |
| 143-254322 | ECV 100145077A | 143-264042 | LAV 03540891K | 143-265082 | LAV 05062015B | 143-267012 | LAV 04050368E |
| 143-254332 | LAV 03540857K | 143-264052 | ECV 100145088A | 143-265092 | LAV 05062029B | 143-267022 | LAV 04050368F |
| 143-254342 | ECV 100145078A | 143-264062 | ECV 100145089A | 143-265112 | LAV 05062037B | 143-267032 | LAV 04050366F |
| 143-254352 | ECV 100145079A | 143-264072 | ECV 100145090A | 143-265122 | LAV 05062043B | 143-267042 | LAV 04050369F |
| 143-254362 | LAV 03540858K | 143-264082 | ECV 100145091A | 143-265132 | LAV 05062050B | 143-274012 | ECV 100145092B |
| 143-254372 | ECV 100145080A | 143-264092 | LAV 03540893K | 143-265142 | LAV 05062051B | 143-274022 | ECV 10014511 [illegible] |
| 143-254382 | ECV 100145081A | 143-264102 | ECV 100145092A | 143-265152 | LAV 05062052B | 143-274032 | ECV 10014 [illegible] |
| 143-254392 | LAV 03540859K | 143-264232 | LAV 03540896K | 143-265162 | LAV 05062053B | 143-274042 | ECV 1001 [illegible] |
| 143-254402 | ECV 100145084A | 143-264242 | LAV 03540897K | 143-265172 | LAV 05062024B | 143-274052 | ECV 10 [illegible] |
| 143-254412 | ECV 100145083A | 143-264252 | LAV 03540898K | 143-265192 | LAV 05062049B | 143-274062 | ECV [illegible] |
| 143-254422 | ECV 100145082A | 143-264262 | LAV 03540899K | 143-266012 | VM 080150029C | 143-274072 | [illegible] |
| 143-254432 | LAV 03540872K | 143-264272 | LAV 03540900K | 143-266022 | VM 080150076C | 143-274092 | [illegible] |
| 143-254442 | ECV 100145085A | 143-264282 | LAV 03540902K | 143-266032 | V 06070259J | 143-274102 | [illegible] |
| 143-254452 | LAV 03540875K | 143-264292 | LAV 03540906K | 143-266032 | V 06070259J | 143-27411 [illegible] | [illegible] |
| 143-254462 | ECV 100145086A | 143-264302 | LAV 03540905K | 143-266042 | V 070125162C | 143-2741 [illegible] | [illegible] |
| 143-254472 | LAV 03540877K | 143-264312 | LAV 03540911K | 143-266052 | V 070125176C | 143-274 [illegible] | [illegible] |
| 143-254482 | LAV 03540879K | 143-264322 | LAV 03540912K | 143-266062 | V 06070200J | 143-27 [illegible] | [illegible] |
| 143-254492 | ECV 100145087A | 143-264332 | LAV 03540913K | 143-266082 | V 06070301J | 143- [illegible] | [illegible] |
| 143-254502 | LAV 03540881K | 143-264342 | LAV 03540914K | 143-266092 | V 070125197C | 143- [illegible] | [illegible] |
| 143-254512 | LAV 03540882K | 143-264352 | ECV 100145093A | 143-266102 | V 070125185C | 143 [illegible] | [illegible] |
| 143-254522 | LAV 03540883K | 143-264362 | ECV 100145094A | 143-266112 | V 070125201C | 14 [illegible] | [illegible] |
| 143-254532 | LAV 03540884K | 143-264372 | ECV 100145095A | 143-266122 | V 070125083C | 14 [illegible] | [illegible] |
| 143-255012 | LAV 05062027 | 143-264382 | LAV 03540916K | 143-266132 | V 070125174C | 14 [illegible] | [illegible] |
| 143-255022 | LAV 05062029 | 143-264392 | ECV 100145096A | 143-266142 | VM 080150080C | 1 [illegible] | [illegible] |

**Sears Craftsman cross reference chart for L-head light duty engines ( [illegible]**

| | | | | | | | |
|---|---|---|---|---|---|---|---|
| 143-274222 | ECV 100145118A | 143-274792 | LAV 03540774L | 143-276442 | VM 080150063D | 143-284552 | LAV 03541000M |
| 143-274232 | ECV 100145118B | 143-275012 | LAV 05062059B | 143-276452 | VM 080150065D | 143-284562 | LAV 03541001M |
| 143-274242 | ECV 100145098B | 143-275012 | LAV 05062059B | 143-276462 | VM 080150115D | 143-284572 | LAV 03541002M |
| 143-274252 | LAV 03540751L | 143-275022 | LAV 05062060B | 143-276472 | VM 080150116D | 143-284582 | ECV 100145148C |
| 143-274262 | ECV 100145109B | 143-275042 | LAV 05062018B | 143-276482 | VM 100157015A | 143-284592 | LAV 03541003M |
| 143-274272 | LAV 03540948L | 143-275052 | LAV 05062063B | 143-277012 | LAV 04050387D | 143-284602 | ECV 100145149C |
| 143-274282 | LAV 03540949L | 143-275062 | LAV 05062065B | 143-277022 | LAV 04050382D | 143-284612 | ECV 100145150C |
| 143-274292 | LAV 03540950L | 143-275072 | LAV 05062019B | 143-284012 | LAV 03540969M | 143-284622 | ECV 100145131C |
| 143-274302 | LAV 03540951L | 143-275082 | LAV 05062066B | 143-284022 | LAV 03540970M | 143-284632 | LAV 03541006M |
| 143-274312 | LAV 03540952L | 143-276012 | VM 080150058D | 143-284032 | LAV 03540971M | 143-284642 | ECV 100145087C |
| 143-274322 | LAV 03540953L | 143-276022 | VM 100157009A | 143-284042 | ECV 00145140C | 143-284652 | LAV 03540923M |
| 143-274332 | LAV 03540954L | 143-276032 | VM 080150016D | 143-284052 | LAV 03540972M | 143-284662 | LAV 03540938M |
| 143-274342 | ECV 100145119B | 143-276042 | VM 080150067D | 143-284062 | LAV 03540973M | 143-284672 | ECV 100145075C |
| 143-274352 | LAV 03540955L | 143-276052 | VM 100157002A | 143-284072 | ECV 00145141C | 143-284682 | ECV 100145106C |
| 143-274362 | ECV 100145120B | 143-276062 | VM 080150102D | 143-284082 | LAV 03540974M | 143-284692 | ECV 100145110C |
| 143-274372 | LAV 03540956L | 143-276072 | VM 080150103D | 143-284092 | LAV 03540975M | 143-284702 | ECV 100145116C |
| 143-274382 | ECV 100145121B | 143-276082 | VM 080150104D | 143-284102 | ECV 00145142C | 143-284712 | LAV 03540926M |
| 143-274392 | ECV 100145122B | 143-276092 | V 070125220C | 143-284112 | LAV 03540977M | 143-284722 | LAV 03540917M |
| 143-274402 | ECV 100145123B | 143-276092 | V 070125220C | 143-284142 | LAV 03540980M | 143-284732 | LAV 03540859M |
| 143-274412 | ECV 100145124B | 143-276102 | V 070125221C | 143-284152 | LAV 03540981M | 143-284742 | ECV 100145151C |
| 143-274422 | ECV 100145087B | 143-276102 | V 070125221C | 143-284162 | LAV 03540982M | 143-284752 | ECV 100145108C |
| 143-274432 | ECV 100145095B | 143-276112 | V 070125222C | 143-284182 | LAV 03540986M | 143-284762 | LAV 03540463M |
| 143-274442 | ECV 100145106B | 143-276122 | VM 080150106D | 143-284212 | ECV 00145143C | 143-284772 | ECV 100145115C |
| 143-274452 | ECV 100145111B | 143-276122 | VM 080150106D | 143-284222 | LAV 03030538M | 143-284782 | ECV 100145107C |
| 143-274462 | ECV 100145116B | 143-276132 | VM 100157010A | 143-284232 | TVS 09043001 | 143-285012 | LAV 05062050C |
| 143-274472 | LAV 03540957L | 143-276132 | VM 100157010A | 143-284242 | LAV 03540990M | 143-285022 | LAV 05062024C |
| 143-274482 | ECV 100145114B | 143-276142 | VM 080150017D | 143-284252 | LAV 03540991M | 143-285032 | LAV 05062066C |
| 143-274492 | LAV 03540858L | 143-276152 | VM 080150069D | 143-284262 | ECV 00145144C | 143-285042 | LAV 05062049C |
| 143-274502 | ECV 100145125B | 143-276162 | VM 080150098D | 143-284272 | TVS 09043002 | 143-285052 | LAV 05062067C |
| 143-274512 | ECV 100145126B | 143-276172 | V 070125224C | 143-284282 | LAV 03540992M | 143-285062 | LAV 05062065C |
| 143-274522 | ECV 100145127B | 143-276172 | V 070125224C | 143-284292 | LAV 03540993M | 143-285072 | LAV 05062037C |
| 143-274542 | ECV 100145075B | 143-276182 | V 06070223J | 143-284302 | LAV 03540994M | 143-285082 | LAV 05062063C |
| 143-274552 | LAV 03540961L | 143-276192 | V 070125225C | 143-284312 | LAV 03540882M | 143-285092 | LAV 05062068C |
| 143-274562 | ECV 100145129B | 143-276202 | V 06070333J | 143-284322 | LAV 03540858M | 143-285102 | LAV 05062043C |
| 143-274572 | ECV 100145130B | 143-276212 | V 070125226C | 143-284332 | ECV 00145058C | 143-286012 | V 05060233J |
| 143-274582 | ECV 100145131B | 143-276222 | V 070125227C | 143-284342 | ECV 00145093C | 143-286022 | V 06070352J |
| 143-274592 | LAV 03540962L | 143-276232 | VM 080150107D | 143-284352 | ECV 00145109C | 143-286032 | V 070125241C |
| [illegible]3-274602 | ECV 100145110B | 143-276242 | VM 080150080D | 143-284362 | ECV 100145111C | 143-286042 | VM 100157016A |
| [illegible]4612 | ECV 100145115B | 143-276252 | V 06070339J | 143-284372 | ECV 00145085C | 143-286052 | VM 100157018A |
| [illegible]22 | ECV 100145108B | 143-276262 | VM 080150110D | 143-284382 | ECV 100145112C | 143-286062 | V 070125243C |
| [illegible] | ECV 100145113B | 143-276272 | VM 080150066D | 143-284392 | LAV 03540751M | 143-286072 | VM 080150080E |
| [illegible] | LAV 03540626L | 143-276282 | V 070125232C | 143-284402 | LAV 03540995M | 143-286082 | VM 080150098E |
| [illegible] | [illegible]V 100145132B | 143-276292 | VM 080150039D | 143-284412 | LAV 03540996M | 143-286092 | VM 080150039E |
| [illegible] | [illegible]3540967L | 143-276302 | VM 080150064D | 143-284422 | LAV 03540962M | 143-286102 | VM 100157019A |
| [illegible] | [illegible]0145133B | 143-276322 | V 070125233C | 143-284432 | ECV 00145135C | 143-286112 | VM 080150117E |
| [illegible] | [illegible]0882L | 143-276332 | V 070125234C | 143-284442 | LAV 03030539M | 143-286122 | VM 080150118E |
| [illegible] | [illegible]45062B | 143-276342 | V 070125235C | 143-284452 | ECV 00145139C | 143-286132 | V 070125245C |
| [illegible] | [illegible]26L | 143-276352 | VM 080150114D | 143-284462 | ECV 00145145C | 143-286142 | VM 080150119E |
| [illegible] | [illegible]134B | 143-276362 | VM 100157013A | 143-284472 | ECV 00145134C | 143-286152 | VM 080150115E |
| [illegible] | [illegible]135B | 143-276372 | V 070125236C | 143-284482 | LAV 03540961M | 143-286162 | VM 080150116E |
| [illegible] | [illegible]6B | 143-276382 | V 070125237C | 143-284492 | ECV 00145126C | 143-286172 | VM 080150016E |
| [illegible] | [illegible]7B | 143-276392 | V 070125238C | 143-284502 | ECV 00145146D | 143-286182 | VM 080150121E |
| [illegible] | [illegible]B | 143-276402 | VM 100157014A | 143-284512 | LAV 03540997M | 143-286192 | VM 100157024A |
| [illegible] | [illegible]B | 143-276412 | V 06070351J | 143-284522 | LAV 03540998M | 143-286202 | VM 080150122E |
| [illegible] | [illegible] | 143-276422 | V 070125240C | 143-284532 | ECV 00145147C | 143-286212 | VM 100157025A |
| [illegible] | [illegible] | 143-276432 | VM 080150062D | 143-284542 | LAV 03540999M | 143-286222 | V 070125246C |

[illegible] Craftsman cross reference chart for L-head light duty engines (2 of 11)

| | |
|---|---|
| 143-286232 | VM 100157026A |
| 143-286242 | VM 100157027A |
| 143-286252 | V 070125247C |
| 143-286262 | VM 100157028A |
| 143-286272 | VM 100157029A |
| 143-286282 | V 070125248C |
| 143-286292 | VM 100157030A |
| 143-286312 | VM 080150063E |
| 143-286322 | V 070125250C |
| 143-286332 | V 070125251C |
| 143-286342 | V 06070357J |
| 143-286352 | VM 080150114E |
| 143-286362 | VM 100157018B |
| 143-287012 | LAV 04050389E |
| 143-287022 | LAV 04050391D |
| 143-287032 | LAV 04050392E |
| 143-293012 | TVS 07533002A |
| 143-294012 | TVS 09043020A |
| 143-294022 | TVS 09043021A |
| 143-294032 | TVS 09043022A |
| 143-294042 | TVS 09043023A |
| 143-294052 | TVS 09043024A |
| 143-294062 | TVS 09043025A |
| 143-294072 | TVS 09043026A |
| 143-294092 | TVS 09043028A |
| 143-294102 | TVS 09043029A |
| 143-294112 | TVS 09043030A |
| 143-294122 | TVS 09043031A |
| 143-294132 | TVS 09043032A |
| 143-294142 | ECV 00145155D |
| 143-294152 | ECV 00145156D |
| 143-294162 | ECV 00145157D |
| 143-294172 | ECV 00145158D |
| 143-294182 | TVS 09043033A |
| 143-294192 | TVS 09043017A |
| 143-294202 | TVS 09043018A |
| 143-294212 | TVS 09043019A |
| 143-294222 | ECV 00145058D |
| 143-294232 | ECV 00145085D |
| 143-294242 | TVS 09043036A |
| 143-294242 | TVS 09043036A |
| 143-294252 | TVS 09043037A |
| 143-294252 | TVS 09043037A |
| 143-294262 | TVS 09043038A |
| 143-294262 | TVS 09043038A |
| 143-294272 | TVS 09043039A |
| 143-294272 | TVS 09043039A |
| 143-294282 | TVS 09043040A |
| 143-294282 | TVS 09043040A |
| 143-294292 | TVS 09043041A |
| 143-294292 | TVS 09043041A |
| 143-294302 | TVS 09043042A |
| 143-294302 | TVS 09043042A |
| 143-294312 | TVS 09043043A |
| 143-294312 | TVS 09043043A |
| 143-294322 | TVS 09043044A |

| | |
|---|---|
| 143-294322 | TVS 09043044A |
| 143-294332 | ECV 100145159D |
| 143-294342 | TVS 09043045A |
| 143-294352 | ECV 100145110D |
| 143-294362 | ECV 100145112D |
| 143-294372 | ECV 100145116D |
| 143-294382 | ECV 100145109D |
| 143-294392 | ECV 100145106D |
| 143-294402 | ECV 100145111D |
| 143-294412 | ECV 100145108D |
| 143-294422 | ECV 100145151D |
| 143-294432 | TVS 09043055A |
| 143-294442 | TVS 09043057A |
| 143-294452 | TVS 09043058A |
| 143-294462 | TVS 09043059A |
| 143-294472 | ECV 100145152D |
| 143-294482 | ECV 100145148D |
| 143-294492 | TVS 09043061A |
| 143-294502 | TVS 09043065A |
| 143-294512 | TVS 09043066A |
| 143-294522 | TVS 09043067A |
| 143-294532 | TVS 09043068A |
| 143-294542 | ECV 100145075D |
| 143-294552 | TVS 10553012A |
| 143-294562 | TVS 10553013A |
| 143-294572 | ECV 100145115D |
| 143-294582 | ECV 100145139D |
| 143-294592 | ECV 100145147D |
| 143-294602 | TVS 09043071A |
| 143-294612 | TVS 09043072A |
| 143-294622 | TVS 10553017A |
| 143-294632 | TVS 10553018A |
| 143-294642 | TVS 10553016A |
| 143-294652 | TVS 09043073A |
| 143-294662 | ECV 100145131D |
| 143-294672 | ECV 100145107D |
| 143-294682 | ECV 100145160D |
| 143-294692 | TVS 09043088A |
| 143-294702 | TVS 10553025A |
| 143-294712 | TVS 09043092A |
| 143-294722 | ECV 100145161D |
| 143-294732 | ECV 100145149D |
| 143-294742 | ECV 100145168D |
| 143-295012 | LAV 05062071C |
| 143-295022 | LAV 05062072C |
| 143-295032 | LAV 05062053C |
| 143-295042 | ECV 120152023A |
| 143-296012 | VM 080150080F |
| 143-296022 | VM 100157028B |
| 143-296032 | VM 080150126F |
| 143-296042 | VM 100157034B |
| 143-296052 | VM 080150115F |
| 143-296062 | VM 080150116F |
| 143-296072 | VM 100157016B |
| 143-296082 | VM 080150017F |
| 143-296092 | VM 080150069F |

| | |
|---|---|
| 143-296102 | VM 100157014B |
| 143-296112 | VM 100157025B |
| 143-296122 | VM 100157026B |
| 143-296132 | VM 100157027B |
| 143-296142 | VM 080150122F |
| 143-296152 | VM 080150118F |
| 143-296162 | VM 080150064F |
| 143-296172 | VM 080150065F |
| 143-296182 | VM 080150039F |
| 143-296192 | VM 080150114F |
| 143-296202 | VM 100157035B |
| 143-296212 | VM 100157030B |
| 143-296222 | VM 080150121F |
| 143-296232 | VM 100157024B |
| 143-296242 | VM 080150016F |
| 143-296252 | VM 080150062F |
| 143-296262 | VM 080150071F |
| 143-297012 | TVS 10553006A |
| 143-304012 | TVS 09043094A |
| 143-304032 | ECV 100145163D |
| 143-304042 | ECV 100145164D |
| 143-304052 | TVS 09043095A |
| 143-304072 | ECV 100145166D |
| 143-304092 | TVS 09043096A |
| 143-304102 | ECV 100145169D |
| 143-304112 | ECV 100145170D |
| 143-304122 | ECV 100145171D |
| 143-304132 | ECV 100145174D |
| 143-304142 | ECV 100145172D |
| 143-304152 | ECV 100145173D |
| 143-304162 | ECV 100145175D |
| 143-304172 | ECV 100145176D |
| 143-304182 | ECV 100145177D |
| 143-304192 | ECV 100145178D |
| 143-304202 | TVS 09043107A |
| 143-304212 | TVS 09043108A |
| 143-304222 | TVS 09043109A |
| 143-304232 | TVS 09043110A |
| 143-304242 | TVS 09043111A |
| 143-304252 | TVS 09043112A |
| 143-304262 | TVS 09043113A |
| 143-304272 | TVS 09043115A |
| 143-304282 | ECV 100145179D |
| 143-304292 | ECV 100145180D |
| 143-304302 | TVS 09043116A |
| 143-304312 | ECV 100145180D |
| 143-304322 | ECV 100145182D |
| 143-304332 | ECV 100145183D |
| 143-304342 | TVS 09043121A |
| 143-304352 | ECV 100145137D |
| 143-304362 | LAV 03540906M |
| 143-305012 | ECV 120152024A |
| 143-305022 | ECV 120152025A |
| 143-305032 | ECV 120152026A |
| 143-305042 | LAV 05062073C |
| 143-305052 | ECV 120152027A |

| | |
|---|---|
| 143-305062 | LAV 05062074C |
| 143-306012 | VM 080150134F |
| 143-306022 | VM 100157042B |
| 143-306032 | VM 100157015B |
| 143-306042 | VM 100157047B |
| 143-313012 | TVS 07533010B |
| 143-313022 | TVS 07533012B |
| 143-314012 | ECV 100145184E |
| 143-314022 | ECV 100145185E |
| 143-314032 | TVS 09043020B |
| 143-314042 | TVS 09043022B |
| 143-314052 | TVS 09043023B |
| 143-314062 | TVS 09043025B |
| 143-314072 | TVS 09043032B |
| 143-314082 | TVS 09043033B |
| 143-314092 | TVS 09043094B |
| 143-314102 | TVS 09043095B |
| 143-314112 | TVS 09043096B |
| 143-314122 | ECV 100145163E |
| 143-314132 | ECV 100145164E |
| 143-314142 | ECV 100145169E |
| 143-314152 | ECV 100145170E |
| 143-314162 | ECV 100145180E |
| 143-314172 | ECV 100145171E |
| 143-314182 | TVS 09043121B |
| 143-314192 | ECV 100145186E |
| 143-314202 | ECV 100145139E |
| 143-314212 | ECV 100145147E |
| 143-314222 | ECV 100145148E |
| 143-314232 | ECV 100145173E |
| 143-314242 | ECV 100145174E |
| 143-314252 | ECV 100145182E |
| 143-314262 | TVS 09043059B |
| 143-314272 | TVS 09043061B |
| 143-314282 | TVS 09043107B |
| 143-314292 | TVS 09043019B |
| 143-314302 | TVS 09043134B |
| 143-314312 | ECV 100145179E |
| 143-314322 | TVS 09043040B |
| 143-314332 | TVS 09043112B |
| 143-314342 | TVS 09043111B |
| 143-314362 | TVS 09043145B |
| 143-314372 | ECV 100145187E |
| 143-314382 | TVS 09043044B |
| 143-314392 | ECV 100145188E |
| 143-314402 | TVS 09043068B |
| 143-314412 | TVS 09043108B |
| 143-314422 | ECV 100145172E |
| 143-314432 | LAV 03540917N |
| 143-314442 | ECV 100145183E |
| 143-314452 | ECV 100145177E |
| 143-314462 | ECV 100145176E |
| 143-314472 | ECV 100145181E |
| 143-314482 | TVS 09043150B |
| 143-314502 | ECV 100145190E |
| 143-314512 | ECV 100145191E |

**Sears Craftsman cross reference chart for L-head light duty engines (3 of 11)**

| 143-314522 | ECV 100145192E |
|---|---|
| 143-314532 | LAV 03541003N |
| 143-314542 | TVS 09043066B |
| 143-314552 | TVS 09043088B |
| 143-314562 | TVS 09043116B |
| 143-314572 | TVS 09043067B |
| 143-314582 | ECV 100145168E |
| 143-314592 | ECV 100145131E |
| 143-314602 | ECV 100145085E |
| 143-314612 | ECV 100145110E |
| 143-314622 | ECV 100145115E |
| 143-314632 | ECV 100145106E |
| 143-314642 | ECV 100145111E |
| 143-314652 | ECV 100145109E |
| 143-314662 | ECV 100145108E |
| 143-314672 | ECV 100145107E |
| 143-314682 | ECV 100145112E |
| 143-314692 | ECV 100145116E |
| 143-314702 | LAV 03540906N |
| 143-314722 | TVS 09043139B |
| 143-314732 | TVS 09043140B |
| 143-314742 | TVS 09043141B |
| 143-314752 | TVS 09043142B |
| 143-314762 | TVS 09043143B |
| 143-314772 | TVS 09043144B |
| 143-314782 | ECV 100145166E |
| 143-314792 | TVS 09043102A |
| 143-314802 | ECV 100145178E |
| 143-314812 | ECV 100145137E |
| 143-315012 | ECV 100152024B |
| 143-315022 | LAV 05062071D |
| 143-315032 | TVS 10553018B |
| 143-315042 | TVS 10553025B |
| 143-315052 | ECV 120152027B |
| 143-315062 | LAV 05062065D |
| 143-315072 | TVS 10553006B |
| 143-315082 | ECV 120152026B |
| 143-315092 | LAV 05062037D |
| 143-315102 | LAV 05062053D |
| 143-315112 | LAV 05062050D |
| 143-315122 | LAV 05062067D |
| 143-315132 | ECV 120152025B |
| 143-316022 | VM 080150063F |
| 143-316032 | V 070125255C |
| 143-316042 | VM 070127008B |
| 143-316052 | VM 080150116G |
| 143-316062 | VM 080150134G |
| 143-316082 | VM 080150121G |
| 143-316092 | VM 100157034C |
| 143-316102 | VM 100150080G |
| 143-316112 | VM 080150114G |
| 143-316122 | VM 100157027C |
| 143-316132 | VM 100157035C |
| 143-316142 | VM 100157028C |
| 143-316152 | VM 080150121G |
| 143-316162 | VM 100157026C |

| 143-316172 | VM 070127009B |
|---|---|
| 143-316182 | VM 080150071G |
| 143-316192 | VM 100157047C |
| 143-316202 | VM 100157052C |
| 143-316222 | VM 080150016G |
| 143-316232 | VM 080150017G |
| 143-316242 | VM 080150065G |
| 143-316252 | VM 080150039G |
| 143-316262 | VM 100157014C |
| 143-316272 | VM 100157025C |
| 143-316282 | VM 080150122G |
| 143-316292 | VM 080150119G |
| 143-316302 | VM 100157018C |
| 143-316312 | VM 080150063G |
| 143-321012 | TVS 07533010C |
| 143-321022 | TVS 07533012C |
| 143-324012 | ECV 100145193E |
| 143-324022 | ECV 100145194E |
| 143-324042 | ECV 100145195E |
| 143-324052 | TVS 09043020C |
| 143-324062 | ECV 100145196E |
| 143-324072 | ECV 100145200E |
| 143-324082 | ECV 100145197E |
| 143-324102 | ECV 100145199E |
| 143-324112 | TVS 09043116C |
| 143-324122 | TVS 09043067C |
| 143-324132 | ECV 100145201E |
| 143-324142 | TVS 09043139C |
| 143-324152 | TVS 09043121C |
| 143-324162 | TVS 09043025C |
| 143-324172 | TVS 09043142C |
| 143-324182 | TVXL10554009A |
| 143-324192 | TVS 09043144C |
| 143-324202 | ECV 100145202E |
| 143-324212 | ECV 100145203E |
| 143-324222 | ECV 100145204E |
| 143-324232 | ECV 100145205E |
| 143-326012 | TVM 195150016H |
| 143-326022 | TVM 195150062H |
| 143-326032 | TVM 195150063H |
| 143-326042 | TVM 195150114H |
| 143-326052 | TVM 195150122H |
| 143-326062 | TVM 195150017H |
| 143-326072 | TVM 195150065H |
| 143-326082 | TVM 195150071H |
| 143-326092 | TVM 195150039H |
| 143-326102 | TVM 195150080H |
| 143-326112 | TVM 195150064H |
| 143-326122 | TVM 220157028D |
| 143-326132 | TVM 220157035D |
| 143-326142 | TVM 220157014D |
| 143-326152 | TVM 220157026D |
| 143-326162 | TVM 220157047D |
| 143-326172 | TVM 220157027D |
| 143-326182 | TVM 220157025D |
| 143-326192 | V 070125250D |

| 143-326202 | V 070125232D |
|---|---|
| 143-326212 | V 070125247D |
| 143-326222 | V 070125234D |
| 143-326232 | V 070125237D |
| 143-326242 | V 070125185D |
| 143-326252 | V 070125236D |
| 143-326262 | V 070125251D |
| 143-326272 | V 070125246D |
| 143-326282 | TVM 14070281K |
| 143-326292 | TVM 14070259K |
| 143-326312 | V 070125255D |
| 143-326322 | TVM 170127008C |
| 143-326332 | TVM 195150116H |
| 143-326342 | TVM 195150134H |
| 143-326372 | TVM 170127009C |
| 143-331012 | TVS 07533025D |
| 143-331022 | TVS 07533012D |
| 143-334022 | TVS 09043173D |
| 143-334032 | TVS 09043174D |
| 143-334042 | ECV 100145207F |
| 143-334052 | TVXL10554012B |
| 143-334062 | TVS 09043175D |
| 143-334072 | TVS 09043176D |
| 143-334082 | ECV 100145208E |
| 143-334102 | ECV 100145210F |
| 143-334112 | TVS 09043177D |
| 143-334122 | TVS 09043178D |
| 143-334132 | ECV 100145211F |
| 143-334142 | TVS 09043068D |
| 143-334152 | TVS 09043019D |
| 143-334162 | TVS 09043145D |
| 143-334172 | ECV 100145200F |
| 143-334182 | ECV 100145213F |
| 143-334192 | LAV 03540917P |
| 143-334202 | TVS 09043116D |
| 143-334212 | ECV 100145214F |
| 143-334222 | ECV 100145215F |
| 143-334232 | ECV 100145217F |
| 143-334242 | ECV 100145216F |
| 143-334252 | ECV 100145218F |
| 143-334262 | TVS 09043139D |
| 143-334272 | TVS 09043140D |
| 143-334282 | TVS 09043141D |
| 143-334292 | TVS 09043142D |
| 143-334302 | TVS 09043143D |
| 143-334312 | TVS 09043144D |
| 143-334322 | ECV 100145219F |
| 143-334332 | TVS 09043201D |
| 143-334342 | ECV 100145202F |
| 143-334352 | TVS 09043121D |
| 143-334362 | TVS 09043215D |
| 143-334372 | TVS 09043220D |
| 143-334382 | TVS 09043226D |
| 143-335012 | ECV 120152028C |
| 143-335022 | ECV 120152029C |
| 143-335032 | LAV 05062037E |

| 143-335042 | LAV 05062067E |
|---|---|
| 143-335052 | TVS 12063209A |
| 143-335062 | LAV 05062065E |
| 143-335072 | TVS 12063210A |
| 143-336012 | TVM 14070371K |
| 143-336022 | TVM 220157058D |
| 143-336032 | TVM 220157062D |
| 143-336042 | TVM 220157069D |
| 143-341012 | TVS 07533039D |
| 143-344022 | TVS 09043213D |
| 143-344032 | TVS 09043214D |
| 143-344042 | TVS 09043221D |
| 143-344052 | ECV 100145221F |
| 143-344062 | ECV 100145220F |
| 143-344072 | TVS 09043219D |
| 143-344082 | ECV 100145222F |
| 143-344092 | ECV 100145223F |
| 143-344102 | TVS 09043222D |
| 143-344112 | TVXL10554017B |
| 143-344122 | ECV 100145224F |
| 143-344132 | ECV 100145225F |
| 143-344142 | TVS 09043230D |
| 143-344152 | ECV 100145226F |
| 143-344162 | TVS 09043250D |
| 143-344172 | ECV 100145227F |
| 143-344182 | TVS 09043252D |
| 143-344192 | TVS 09043253D |
| 143-344202 | TVS 09043254D |
| 143-344212 | TVS 09043255D |
| 143-344222 | TVS 09043256D |
| 143-344232 | ECV 100145228F |
| 143-344242 | ECV 100145229F |
| 143-344252 | ECV 100145230F |
| 143-344262 | ECV 100145231F |
| 143-344272 | ECV 100145232F |
| 143-344282 | ECV 100145233F |
| 143-344292 | ECV 100145234F |
| 143-344302 | ECV 100145235F |
| 143-344312 | ECV 100145236F |
| 143-344322 | ECV 100145237F |
| 143-344332 | ECV 100145238F |
| 143-344342 | ECV 100145239F |
| 143-344352 | ECV 100145240F |
| 143-344362 | ECV 100145241F |
| 143-344372 | ECV 100145242F |
| 143-344382 | ECV 100145243F |
| 143-344392 | ECV 100145244F |
| 143-344402 | TVXL10554020B |
| 143-344412 | TVXL10554021B |
| 143-344422 | TVS 09043268D |
| 143-344432 | TVS 09043269D |
| 143-344442 | TVS 10553059D |
| 143-344452 | ECV 100145245F |
| 143-344462 | TVS 10553067D |
| 143-344472 | ECV 100145246F |
| 143-345012 | ECV 20152031C |

**Sears Craftsman cross reference chart for L-head light duty engines (4 of 11)**

| | |
|---|---|
| 143-345022 | ECV 120152030C |
| 143-345032 | TVS 12063211A |
| 143-345042 | LAV 05062082E |
| 143-345052 | ECV 120152032C |
| 143-345062 | ECV 120152034C |
| 143-346012 | TVM220157058E |
| 143-346022 | TVM 220157062E |
| 143-346032 | TVM 170127008D |
| 143-346042 | TVM 195150116J |
| 143-346052 | TVM 195150134J |
| 143-346062 | TVM 220157069E |
| 143-346072 | TVM 220157035E |
| 143-346082 | TVM 170127009D |
| 143-346092 | TVM 195150016J |
| 143-346102 | TVM 195150114J |
| 143-346112 | TVM 195150122J |
| 143-346122 | TVM 195150039J |
| 143-346132 | TVM 195150080J |
| 143-346142 | TVM 220157028E |
| 143-346152 | TVM 220157014E |
| 143-346162 | TVM 220157026E |
| 143-346172 | TVM 220157047E |
| 143-346182 | TVM 220157027E |
| 143-346192 | TVM 220157025E |
| 143-346202 | TVM 12560249K |
| 143-351012 | TVS 07533046D |
| 143-351022 | TVS 07533052D |
| 143-354012 | TVS 09043290D |
| 143-354022 | ECV 100145248F |
| 143-354032 | ECV 100145249F |
| 143-354052 | ECV 100145251F |
| 143-354062 | TVS 09043291D |
| 143-354082 | ECV 100145253F |
| 143-354092 | TVS 09043292D |
| 143-354102 | TVS 09043293D |
| 143-354112 | ECV 100145254F |
| 143-354122 | TVS 09043294D |
| 143-354132 | TVXL10554024B |
| 143-354142 | TVS 09043295D |
| 143-354152 | ECV 100145255F |
| 143-354162 | TVS 09043289D |
| 143-354172 | TVS 09043288D |
| 143-354182 | TVS 09043296D |
| 143-354192 | TVS 09043297D |
| 143-354202 | TVS 09043298D |
| 143-354212 | TVS 09043299D |
| 143-354222 | ECV 100145256F |
| 143-354232 | TVS 09043300D |
| 143-354242 | ECV 100145257F |
| 143-354262 | ECV 100145259F |
| 143-354272 | ECV 100145260F |
| 143-354282 | LAV 03540917R |
| 143-354292 | TVS 09043307D |
| 143-354302 | ECV 100145262F |
| 143-354312 | TVS 09043312D |
| 143-354322 | TVS 09043315D |

| | |
|---|---|
| 143-354332 | TVS 09043316D |
| 143-354342 | TVS 09043321D |
| 143-354352 | TVS 09043322D |
| 143-354362 | ECV 100145263F |
| 143-354372 | ECV 100145264F |
| 143-354382 | ECV 100145265F |
| 143-354392 | ECV 100145266F |
| 143-354402 | ECV 100145270F |
| 143-354412 | ECV 100145271F |
| 143-354422 | ECV 100145267F |
| 143-354432 | ECV 100145268F |
| 143-354442 | ECV 100145261F |
| 143-354452 | ECV 100145269F |
| 143-354462 | ECV 100145273F |
| 143-354482 | TVS 10553077D |
| 143-354492 | TVS 10553083D |
| 143-354502 | TVS 10553084D |
| 143-355012 | ECV 100145033C |
| 143-355022 | ECV 120152035C |
| 143-355032 | LAV 05062050F |
| 143-356012 | TVM 220157081F |
| 143-356022 | TVM 12560249L |
| 143-356032 | TVM 195150134K |
| 143-356042 | TVM 220157069F |
| 143-356052 | TVM 195150151K |
| 143-356062 | TVM 12560251L |
| 143-356072 | TVM 195150152K |
| 143-356082 | TVM 220157083F |
| 143-356092 | TVM 220157084F |
| 143-356102 | TVM 170127013E |
| 143-356122 | TVM 195150154K |
| 143-356132 | TVM 195150155K |
| 143-356142 | TVM 195150156K |
| 143-356152 | TVM 195150157K |
| 143-356162 | TVM 220157085F |
| 143-356172 | TVM 220157086F |
| 143-356182 | TVM 220157087F |
| 143-356192 | TVM 220157088F |
| 143-356202 | TVM 220157089F |
| 143-356212 | TVM 220157090F |
| 143-356222 | TVM 220157091F |
| 143-356232 | TVM 220157093F |
| 143-356252 | TVM 220157097F |
| 143-356362 | TVM 12560251L |
| 143-361012 | TVS 07533054D |
| 143-364012 | TVS 09043333D |
| 143-364022 | ECV 100145274F |
| 143-364032 | ECV 100145275F |
| 143-364042 | ECV 100145276F |
| 143-364052 | ECV 100145277F |
| 143-364062 | ECV 100145278F |
| 143-364072 | ECV 100145279F |
| 143-364082 | TVS 09043334D |
| 143-364092 | ECV 100145280F |
| 143-364102 | TVS 09043335D |
| 143-364112 | TVS 09043336D |

| | |
|---|---|
| 143-364122 | TVS 09043337D |
| 143-364132 | TVS 09043338D |
| 143-364142 | TVS 09043339D |
| 143-364152 | TVXL10554029C |
| 143-364162 | ECV 100145281F |
| 143-364172 | ECV 100145282F |
| 143-364182 | ECV 100145283F |
| 143-364192 | ECV 100145284F |
| 143-364202 | TVS 09043341D |
| 143-364212 | ECV 100145285F |
| 143-364222 | TVS 09043342D |
| 143-364232 | ECV 100145286F |
| 143-364242 | ECV 100145287F |
| 143-364252 | ECV 100145288F |
| 143-364262 | TVS 10553087D |
| 143-364272 | ECV 100145290F |
| 143-364282 | ECV 100145289F |
| 143-364292 | ECV 100145291F |
| 143-364302 | ECV 100145292F |
| 143-364312 | ECV 100145293F |
| 143-364322 | ECV 100145294F |
| 143-364332 | ECV 100145295F |
| 143-364342 | ECV 100145296F |
| 143-364352 | TVS 09043346D |
| 143-364362 | TVS 09043347D |
| 143-364372 | TVS 09043348D |
| 143-364382 | ECV 100145297F |
| 143-364392 | TVS 09043351D |
| 143-364402 | TVS 10553090D |
| 143-365012 | ECV 120152037C |
| 143-365022 | ECV 120152038C |
| 143-366022 | TVM 195150134L |
| 143-366032 | TVM 220157069G |
| 143-366042 | TVM 195150151L |
| 143-366052 | TVM 220157093G |
| 143-366062 | TVM 220157081G |
| 143-366082 | TVM 12560252L |
| 143-366102 | TVM 195150152L |
| 143-366112 | TVM 220157083G |
| 143-366122 | TVM 220157084G |
| 143-366132 | TVM 220157097G |
| 143-366152 | TVM 195150163L |
| 143-366172 | TVM 220157108G |
| 143-366182 | TVM 12560254L |
| 143-366192 | TVM 220157106G |
| 143-366222 | TVM 220157110G |
| 143-371012 | TVS 07533056E |
| 143-371022 | TVS 07533057E |
| 143-371032 | TVS 07533059E |
| 143-374012 | TVS 09043352E |
| 143-374022 | TVS 09043353E |
| 143-374032 | TVS 09043354E |
| 143-374052 | TVS 09043356E |
| 143-374062 | TVS 09043357E |
| 143-374072 | TVS 09043358E |
| 143-374082 | TVS 09043359E |

| | |
|---|---|
| 143-374092 | ECV 100145298F |
| 143-374102 | ECV 100145299F |
| 143-374112 | ECV 100145300F |
| 143-374122 | ECV 100145301F |
| 143-374132 | ECV 100145302F |
| 143-374142 | ECV 100145303F |
| 143-374152 | ECV 100145304F |
| 143-374162 | ECV 100145305F |
| 143-374172 | ECV 100145306F |
| 143-374182 | ECV 100145307F |
| 143-374192 | ECV 100145308F |
| 143-374202 | ECV 100145309F |
| 143-374212 | TVS 09043360E |
| 143-374222 | TVS 09043361E |
| 143-374232 | TVS 09043362E |
| 143-374282 | TVS 09043601E |
| 143-374292 | TVS 10553601E |
| 143-374302 | TVS 09043371E |
| 143-374312 | TVS 10553101E |
| 143-374322 | TVS 09043342E |
| 143-374332 | TVS 09043375E |
| 143-374342 | ECV 100145310F |
| 143-374362 | TVS 09043307E |
| 143-374372 | TVS 10553602E |
| 143-374382 | TVS 09043215E |
| 143-374402 | ECV 100145312F |
| 143-374412 | ECV 100145311F |
| 143-374422 | TVS 10553102E |
| 143-374432 | TVS 09043389E |
| 143-374452 | ECV 100145320F |
| 143-375012 | ECV 120152039C |
| 143-375022 | ECV 120152040C |
| 143-375032 | LAV 05062082F |
| 143-375042 | LAV 05062037F |
| 143-375052 | ECV 120152043C |
| 143-376022 | TVM 220157106H |
| 143-376042 | TVM 195150164M |
| 143-376052 | TVM 220157115H |
| 143-376062 | TVM 195150151M |
| 143-376092 | TVM 220157083H |
| 143-381012 | TVS 07533061F |
| 143-381022 | TVS 07533059F |
| 143-384012 | TVS 09043379F |
| 143-384022 | TVS 09043380F |
| 143-384032 | TVS 09043381F |
| 143-384042 | TVS 09043382F |
| 143-384052 | TVS 09043383F |
| 143-384062 | TVS 09043384F |
| 143-384072 | TVS 09043385F |
| 143-384082 | TVS 09043386F |
| 143-384092 | ECV 100145313G |
| 143-384102 | ECV 100145314G |
| 143-384112 | ECV 100145315G |
| 143-384122 | ECV 100145316G |
| 143-384172 | ECV 100145317G |
| 143-384202 | ECV 100145318G |

**Sears Craftsman cross reference chart for L-head light duty engines (5 of 11)**

| 143-384212 | ECV 100145319G |
|---|---|
| 143-384222 | ECV 100145258G |
| 143-384232 | ECV 100145295G |
| 143-384242 | ECV 100145296G |
| 143-384252 | ECV 100145286G |
| 143-384262 | ECV 100145287G |
| 143-384272 | TVS 09043342F |
| 143-384282 | TVS 09043347F |
| 143-384292 | TVS 09043346F |
| 143-384302 | TVS 09043215F |
| 143-384312 | TVS 09043396F |
| 143-384322 | ECV 100145321G |
| 143-384332 | ECV 100145322G |
| 143-384342 | TVS 09043348F |
| 143-384352 | ECV 100145285G |
| 143-384362 | ECV 100145294G |
| 143-384372 | ECV 100145293G |
| 143-384382 | TVS 09043402F |
| 143-384392 | TVS 09043403F |
| 143-384402 | TVS 10553107F |
| 143-384412 | TVS 10553602F |
| 143-384422 | TVS 10553607F |
| 143-384432 | TVS 10044604B |
| 143-384442 | TVS 09043405F |
| 143-384452 | TVS 09043375F |
| 143-384462 | ECV 100145273G |
| 143-384472 | ECV 100145291G |
| 143-384482 | ECV 100145292G |
| 143-384492 | ECV 100145266G |
| 143-384502 | ECV 100145290G |
| 143-384512 | ECV 100145288G |
| 143-384522 | ECV 100145297G |
| 143-384532 | ECV 100145289G |
| 143-384542 | ECV 100145310G |
| 143-384552 | TVS 09043389F |
| 143-384562 | ECV 100145320G |
| 143-384572 | TVS 09043415F |
| 143-385012 | ECV 120152041D |
| 143-385022 | ECV 120152042D |
| 143-385032 | ECV 120152036D |
| 143-385042 | LAV 05062037G |
| 143-385052 | LAV 05062082G |
| 143-386022 | TVM 220157120J |
| 143-386042 | TVM 220157122J |
| 143-386052 | TVM 195150152N |
| 143-386062 | TVM 220157083J |
| 143-386072 | TVM 220157084J |
| 143-386082 | TVM 220157097J |
| 143-386122 | TVM 195150151N |
| 143-386132 | TVM 195150164N |
| 143-386142 | TVM 220157115J |
| 143-386172 | TVM 220157126H |
| 143-386182 | TVM 220157128J |
| 143-391012 | TVS 07533066F |
| 143-391022 | TVS 07533067F |
| 143-394012 | ECV 100145323G |

| 143-394022 | TVS 09043420F |
|---|---|
| 143-394032 | TVS 09043422F |
| 143-394042 | TVS 09043423F |
| 143-394052 | TVS 09043424F |
| 143-394062 | TVS 09043425F |
| 143-394072 | TVS 09043426F |
| 143-394082 | ECV 100145324G |
| 143-394122 | TVS 09043438F |
| 143-394132 | TVS 09043421F |
| 143-394142 | TVS 09043428F |
| 143-394152 | TVS 09043443F |
| 143-394162 | ECV 100145333G |
| 143-394172 | ECV 100145327G |
| 143-394182 | ECV 100145328G |
| 143-394222 | ECV 100145326G |
| 143-394232 | ECV 100145325G |
| 143-394242 | TVS 09043451F |
| 143-394252 | ECV 100145330G |
| 143-394262 | ECV 100145332G |
| 143-394272 | ECV 100145331G |
| 143-394282 | ECV 100145329G |
| 143-394302 | TVS 09043454F |
| 143-394312 | TVS 09043455F |
| 143-394322 | TVS 09043456F |
| 143-394332 | TVS 09043457F |
| 143-394342 | ECV 100145334G |
| 143-394352 | ECV 100145335G |
| 143-394362 | ECV 100145336G |
| 143-394372 | ECV 100145337G |
| 143-394382 | ECV 100145338G |
| 143-394392 | ECV 100145339G |
| 143-394402 | ECV 100145340G |
| 143-394412 | ECV 100145341G |
| 143-394422 | ECV 100145342G |
| 143-394432 | ECV 100145343G |
| 143-394442 | ECV100145344G |
| 143-394452 | ECV 100145345G |
| 143-394462 | ECV 100145346G |
| 143-394472 | ECV 100145347G |
| 143-394482 | ECV 100145348G |
| 143-394492 | TVS 09043458F |
| 143-394502 | LAV 03540917S |
| 143-394512 | ECV 100145349G |
| 143-394522 | TVS 10044605B |
| 143-395012 | ECV 120152044D |
| 143-395022 | ECV 120152045D |
| 143-396022 | TVXL220157213 |
| 143-396042 | TVXL220157206 |
| 143-396052 | TVXL220157205 |
| 143-396082 | TVXL220157215 |
| 143-396102 | TVM 12560258L |
| 143-396122 | TVXL220157220 |
| 143-401012 | TVS 07533070F |
| 143-404022 | TVS 09043490F |
| 143-404032 | TVS 09043491F |
| 143-404042 | TVS 10553123G |

| 143-404052 | TVS 10553124G |
|---|---|
| 143-404062 | TVS 10553125G |
| 143-404072 | TVS 10553126G |
| 143-404082 | TVS 10553901G |
| 143-404092 | TVS 10553902G |
| 143-404122 | TVS 12063114F |
| 143-404132 | TVS 10553130G |
| 143-404142 | TVS 10553903G |
| 143-404152 | TVS 12063115F |
| 143-404162 | TVS 10553132G |
| 143-404172 | TVS 10553131G |
| 143-404182 | TVS 12063901F |
| 143-404202 | TVS 10553136G |
| 143-404222 | TVS 10553137G |
| 143-404232 | TVS 10553138G |
| 143-404242 | TVS 10553133G |
| 143-404252 | TVS 10553134G |
| 143-404282 | TVS 10553139G |
| 143-404292 | TVS 12063117F |
| 143-404312 | TVS 10553140G |
| 143-404322 | TVS 10553153G |
| 143-404332 | TVS 10553904G |
| 143-404342 | TVS 09043498F |
| 143-404352 | TVS 09043499F |
| 143-404362 | TVS 10553143G |
| 143-404372 | TVS 10553905G |
| 143-404382 | TVS 10553906G |
| 143-404392 | TVS 10553907G |
| 143-404402 | TVS 12063902F |
| 143-404412 | TVS 10553147G |
| 143-404422 | TVS 10553148G |
| 143-404432 | TVS 10553149G |
| 143-404442 | TVS 10553150G |
| 143-404452 | TVS 10553151G |
| 143-404462 | TVS 10553152G |
| 143-404472 | TVS 12063120F |
| 143-404482 | TVS 12063903F |
| 143-404502 | TVS 09043504F |
| 143-404532 | TVS 09043497F |
| 143-406022 | TVXL220157205A |
| 143-406032 | TVXL220157215A |
| 143-406042 | TVXL220157220A |
| 143-406082 | TVM 12560261L |
| 143-406092 | TVXL195150233 |
| 143-406102 | TVXL220157230A |
| 143-406122 | TVXL220157206A |
| 143-406172 | TVXL195150238 |
| 143-414012 | TVS 09043512G |
| 143-414022 | TVS 10553162H |
| 143-414032 | TVS 09043513G |
| 143-414042 | TVS 09043514G |
| 143-414052 | TVS 09043515G |
| 143-414062 | TVS 10553163H |
| 143-414072 | TVS 10553165H |
| 143-414082 | TVS 09043497G |
| 143-414092 | ECV 100145334H |

| 143-414102 | ECV 100145335H |
|---|---|
| 143-414112 | ECV 100145339H |
| 143-414122 | ECV 100145340H |
| 143-414132 | ECV 100145341H |
| 143-414142 | ECV 100145342H |
| 143-414152 | ECV 100145344H |
| 143-414162 | ECV 100145345H |
| 143-414172 | TVS 09043298G |
| 143-414182 | TVS 09043299G |
| 143-414192 | ECV 100145337H |
| 143-414202 | ECV 100145338H |
| 143-414212 | TVS 09043389G |
| 143-414222 | TVS 10553167H |
| 143-414222 | TVS 10553167H |
| 143-414232 | TVS 09043526G |
| 143-414242 | TVS 09043375G |
| 143-414252 | TVS 09043215G |
| 143-414262 | ECV 100145346H |
| 143-414272 | ECV 100145347H |
| 143-414282 | TVS 09043528G |
| 143-414292 | TVS 10553153H |
| 143-414292 | TVS 10553153H |
| 143-414302 | TVS 12063124G |
| 143-414312 | TVS 10553130H |
| 143-414322 | TVS 10553901H |
| 143-414332 | TVS 09043504G |
| 143-414342 | TVS 10553903H |
| 143-414352 | TVS 12063901G |
| 143-414362 | TVS 10553911H |
| 143-414372 | TVS 10553169H |
| 143-414382 | TVS 10553151H |
| 143-414392 | TVS 12063903G |
| 143-414402 | TVS 10553902H |
| 143-414412 | TVS 10553168H |
| 143-414422 | TVS 12063115G |
| 143-414442 | TVS 10553132H |
| 143-414452 | TVS 10553152H |
| 143-414462 | TVS 10553137H |
| 143-414472 | TVS 12063117G |
| 143-414482 | TVS 10553139H |
| 143-414492 | TVS 12063120G |
| 143-414502 | TVS 09043534G |
| 143-414512 | TVS 09043535G |
| 143-414522 | TVS 09043901G |
| 143-414532 | TVS 09043533G |
| 143-414542 | TVS 10553907H |
| 143-414552 | TVS 12063906G |
| 143-414562 | TVS 10553170H |
| 143-414572 | TVS 12063902G |
| 143-414582 | TVS 10553910H |
| 143-414592 | TVS 10553912H |
| 143-414602 | TVS 10553913H |
| 143-414612 | TVS 09043537G |
| 143-414622 | TVS 12063127G |
| 143-414632 | TVS 10553914H |
| 143-414642 | TVS 12063907G |

**Sears Craftsman cross reference chart for L-head light duty engines (6 of 11)**

| 143-414652 | TVS 10553175H |
|---|---|
| 143-414662 | TVS 10553176H |
| 143-414672 | TVS 10553177H |
| 143-414682 | ECV 00145349H |
| 143-414692 | TVS 10044022C |
| 143-416022 | TVXL195150239 |
| 143-416032 | TVXL220157240A |
| 143-416052 | TVM 12560254M |
| 143-416062 | TVM 12560265M |
| 143-416072 | TVXL220157241A |
| 143-424012 | TVS 09043504H |
| 143-424022 | TVS 10556001 |
| 143-424032 | TVS 09043497H |
| 143-424042 | TVS 10553153J |
| 143-424052 | TVS 09043526H |
| 143-424062 | TVS 12063129H |
| 143-424072 | TVS 10044026D |
| 143-424082 | TVS 10556904 |
| 143-424102 | TVS 12063910H |
| 143-424112 | TVS 10044029D |
| 143-424122 | TVS 10044030D |
| 143-424132 | TVS 10044031D |
| 143-424142 | TVS 10556905 |
| 143-424152 | TVS 12063911H |
| 143-424162 | TVS 10556906 |
| 143-424172 | TVS 12063130H |
| 143-424182 | TVS 10044032D |
| 143-424192 | TVS 10044033D |
| 143-424202 | TVS 09043215H |
| 143-424212 | TVS 09043514H |
| 143-424222 | TVS 09043513H |
| 143-424232 | TVS 09043375H |
| 143-424242 | TVS 09043553H |
| 143-424252 | TVS 09043528H |
| 143-424262 | TVS 10553163J |
| 143-424272 | TVS 10553165J |
| 143-424282 | TVS 10553912J |
| 143-424292 | TVS 10553913J |
| 143-424302 | TVS 10553179J |
| 143-424312 | TVS 10556005 |
| 143-424322 | TVS 10556006 |
| 143-424332 | TVS 12063134H |
| 143-424342 | TVS 12063135H |
| 143-424352 | TVS 10553180J |
| 143-424362 | TVS 09043555H |
| 143-424372 | TVS 09043556H |
| 143-424382 | TVS 10556007 |
| 143-424392 | TVS 10556907 |
| 143-424402 | TVS 12063902H |
| 143-424412 | TVS 09043558H |
| 143-424422 | TVS 10044036D |
| 143-424432 | TVS 09043512H |
| 143-424442 | TVS 10553162J |
| 143-424452 | TVS 10044037D |
| 143-424462 | TVS 10044038D |
| 143-424472 | TVS 09043515H |

| 143-424482 | TVS 12063137H |
|---|---|
| 143-424492 | TVS 10550920J |
| 143-424502 | TVS 12063915H |
| 143-424512 | TVS 09043298H |
| 143-424522 | TVS 10044042D |
| 143-424532 | TVS 10044043D |
| 143-424542 | TVS 10044045D |
| 143-424552 | TVS 10044046D |
| 143-424562 | TVS 09043389H |
| 143-424572 | TVS 09043299H |
| 143-424582 | TVS 12063916H |
| 143-426012 | TVM 12560261M |
| 143-426022 | TVXL195150246A |
| 143-426032 | TVXL195150238A |
| 143-426042 | TVXL220157205B |
| 143-426052 | TVXL220157206B |
| 143-426062 | TVXL220157220B |
| 143-426072 | TVXL220157245B |
| 143-426132 | TVXL220157215B |
| 143-434012 | TVS 09043504J |
| 143-434022 | TVS 09043526J |
| 143-434032 | TVS 11561902 |
| 143-434042 | TVS 11556007A |
| 143-434052 | TVS 11556012A |
| 143-434062 | TVS 11557902A |
| 143-434072 | TVS 09043572J |
| 143-434082 | TVS 11556011A |
| 143-434092 | TVS 09046005 |
| 143-434102 | TVS 11561002 |
| 143-434122 | TVS 11556010A |
| 143-434132 | TVS 11561901 |
| 143-434142 | TVS 09043497J |
| 143-434152 | TVS 11556906A |
| 143-434162 | TVS 11556001A |
| 143-434182 | TVS 11556017A |
| 143-434192 | TVS 09046003 |
| 143-434202 | TVS 11557012A |
| 143-434212 | TVS 09043576J |
| 143-434222 | TVS 09046012 |
| 143-434232 | TVS 11556016A |
| 143-434242 | TVS 09046013 |
| 143-434262 | TVS 09046007 |
| 143-434272 | TVS 09046015 |
| 143-434282 | TVS 12063917J |
| 143-434292 | TVS 09046017 |
| 143-434302 | TVS 09046018 |
| 143-434312 | TVS 09046019 |
| 143-434332 | TVS 12063918J |
| 143-434342 | TVS 10044037E |
| 143-434352 | TVS 09043375J |
| 143-434362 | TVS 10044033E |
| 143-434372 | TVS 09043513J |
| 143-434382 | TVS 10044031E |
| 143-434392 | TVS 09043515J |
| 143-434402 | TVS 09043553J |
| 143-434412 | TVS 09043298J |

| 143-434422 | TVS 10044043E |
|---|---|
| 143-434432 | TVS 09043215J |
| 143-434442 | TVS 10044030E |
| 143-434452 | TVS 10044038E |
| 143-434462 | TVS 10044032E |
| 143-434472 | TVS 10044036E |
| 143-434482 | TVS 09043528J |
| 143-434492 | TVS 10553913K |
| 143-434502 | TVS 10553163K |
| 143-434512 | TVS 11561016 |
| 143-434522 | TVS 11561906 |
| 143-434532 | TVS 09043514J |
| 143-434542 | TVS 10044029E |
| 143-434552 | TVS 10044045E |
| 143-434562 | TVS 09043299J |
| 143-434572 | TVS 09043512J |
| 143-434582 | TVS 10044048E |
| 143-434592 | TVS 11556911A |
| 143-434602 | TVS 11556031A |
| 143-436012 | TVXL220157245C |
| 143-436052 | TVM 12560267N |
| 143-436062 | TVXL195150246B |
| 143-436072 | TVXL220157220C |
| 143-436082 | TVXL220157215C |
| 143-436112 | TVXL220157206C |
| 143-436122 | TVXL220157205C |
| 143-436162 | TVM 12560254N |
| 143-436172 | TVXL195150238B |
| 143-651012 | H 03035310J |
| 143-651022 | H 03035312J |
| 143-651032 | H 03035311J |
| 143-651042 | H 03035309J |
| 143-651052 | H 03035311K |
| 143-651062 | H 03035310K |
| 143-651072 | H 03035312K |
| 143-654012 | H 03545421H |
| 143-654022 | H 03545424J |
| 143-654032 | H 03545424K |
| 143-654042 | H 03545427J |
| 143-654052 | H 03545428J |
| 143-654062 | H 03545441H |
| 143-654072 | H 03545448K |
| 143-654082 | H 03545428K |
| 143-654092 | H 03545441J |
| 143-654102 | H 03545285K |
| 143-654112 | H 03545379K |
| 143-654122 | H 03545328K |
| 143-654132 | H 03545371K |
| 143-654142 | H 03545133K |
| 143-654152 | H 03545392K |
| 143-654162 | H 03545427K |
| 143-654172 | H 03545460K |
| 143-654182 | H 03545463K |
| 143-654192 | H 03545462K |
| 143-654202 | H 03545464K |
| 143-654212 | H 03545391K |

| 143-654222 | H 03545205H |
|---|---|
| 143-654232 | H 03545471K |
| 143-654242 | H 03545284K |
| 143-654252 | H 03545471L |
| 143-654262 | H 03545371L |
| 143-654272 | H 03545428L |
| 143-654282 | H 03545133L |
| 143-654292 | H 03545379L |
| 143-654302 | H 03545474K |
| 143-654312 | H 03545476L |
| 143-654322 | H 03545328L |
| 143-655012 | HS 05067094A |
| 143-655022 | HS 05067105A |
| 143-655032 | HS 05067110A |
| 143-656012 | H 06075394J |
| 143-656022 | H 06075398J |
| 143-656032 | H 06075403K |
| 143-656042 | H 06075404J |
| 143-656052 | H 06075407J |
| 143-656062 | H 070130172B |
| 143-656072 | HM 080155106A |
| 143-656072 | HM 080155106A |
| 143-656072 | HM 080155106A |
| 143-656082 | HM 080155109A |
| 143-656092 | H 06075411J |
| 143-656102 | H 070130193B |
| 143-656122 | H 070130196B |
| 143-656132 | H 070130197B |
| 143-656152 | H 070130200B |
| 143-656172 | H 06075417J |
| 143-656182 | H 06075416J |
| 143-656192 | H 070130069B |
| 143-656202 | H 06075420J |
| 143-656232 | H 070130202B |
| 143-656242 | HM 080155032B |
| 143-656252 | H 06075426J |
| 143-656262 | H 070130203C |
| 143-656272 | H 070130205C |
| 143-656282 | HM 080155129B |
| 143-657012 | HS 04055212E |
| 143-657022 | HS 04055477E |
| 143-657032 | HS 04055482E |
| 143-657042 | HS 04055484D |
| 143-657052 | HS 04055212F |
| 143-661012 | H 03035310L |
| 143-661022 | H 03035311L |
| 143-661032 | H 03035312L |
| 143-661042 | H 03035310M |
| 143-661052 | H 03035311M |
| 143-661062 | H 03035312M |
| 143-664012 | H 03545424L |
| 143-664022 | H 03545372L |
| 143-664032 | H 03545462L |
| 143-664042 | H 03545474L |
| 143-664052 | H 03545392L |
| 143-664062 | H 03545448L |

**Sears Craftsman cross reference chart for L-head light duty engines (7 of 11)**

| 143-664072 | H 03545441K |
|---|---|
| 143-664082 | H 03545427L |
| 143-664092 | H 03545462L |
| 143-664102 | H 03545284L |
| 143-664112 | H 03545392M |
| 143-664122 | H 03545441M |
| 143-664132 | H 03545474M |
| 143-664142 | H 03545476M |
| 143-664152 | H 03545328M |
| 143-664162 | H 03545284M |
| 143-664172 | H 03545133M |
| 143-664182 | H 03545379M |
| 143-664192 | H 03545471M |
| 143-664202 | H 03545462M |
| 143-664212 | H 03545448M |
| 143-664222 | H 03545428M |
| 143-664232 | H 03545427M |
| 143-664242 | H 03545424M |
| 143-664252 | H 03545372M |
| 143-664262 | H 03545371M |
| 143-664272 | H 03545235M |
| 143-664282 | H 03545464M |
| 143-664292 | H 03545391M |
| 143-664302 | H 03545463M |
| 143-664312 | H 03545486M |
| 143-664332 | H 03545487K |
| 143-665012 | HS 05067037B |
| 143-665022 | HS 05067062B |
| 143-665032 | HS 05067062C |
| 143-665042 | HS 05067037C |
| 143-665052 | HS 05067128C |
| 143-665062 | HS 05067132B |
| 143-665072 | HS 05067135C |
| 143-665082 | HS 05067142B |
| 143-666012 | H 070130097C |
| 143-666022 | H 070130206C |
| 143-666032 | HM 080155131B |
| 143-666042 | H 070130207C |
| 143-666052 | H 070130193C |
| 143-666062 | H 070130197C |
| 143-666072 | H 070130202C |
| 143-666082 | HM 080155121C |
| 143-666092 | HM 080155122C |
| 143-666102 | H 06075420K |
| 143-666112 | H 06075426K |
| 143-666122 | H 06075411K |
| 143-666132 | H 06075404K |
| 143-666142 | H 06075398K |
| 143-666152 | HM 080155106C |
| 143-666162 | HM 080155109C |
| 143-666172 | H 06075403L |
| 143-666182 | H 06075437K |
| 143-666192 | H 06075438K |
| 143-666192 | H 06075438K |
| 143-666202 | H 06075439K |
| 143-666222 | H 070130200C |

| 143-666232 | HM 080155083C |
|---|---|
| 143-666242 | HH 060105096F |
| 143-666252 | H 070130172C |
| 143-666272 | H 06075416K |
| 143-666282 | H 070130211C |
| 143-666292 | H 06075442M |
| 143-666302 | H 070130212D |
| 143-666312 | H 070130213D |
| 143-666322 | HM 080155137C |
| 143-666332 | HM 100159008B |
| 143-666342 | H 070130006C |
| 143-666352 | HM 080155129C |
| 143-666362 | HM 100159011B |
| 143-666372 | H 06075445K |
| 143-666382 | H 070130205D |
| 143-667012 | HS 04055477F |
| 143-667022 | HS 04055486D |
| 143-667032 | HS 05067117A |
| 143-667042 | HS 04055482F |
| 143-667052 | HS 04055482G |
| 143-667062 | HS 04055477G |
| 143-667072 | HS 04055212G |
| 143-667082 | HS 04055495G |
| 143-674012 | H 03545490M |
| 143-674032 | H 03545285M |
| 143-674042 | H 03545205K |
| 143-674052 | H 03545496K |
| 143-674062 | H 03545460M |
| 143-675012 | HS 05067146C |
| 143-675022 | HS 05067149C |
| 143-675032 | H 05065398L |
| 143-675042 | HS 05067152B |
| 143-675052 | HS 05067117B |
| 143-675062 | H 05065406K |
| 143-676012 | HM 080155121D |
| 143-676022 | HM 080155122D |
| 143-676032 | H 070130219C |
| 143-676042 | HM 080155145D |
| 143-676052 | HM 080155146D |
| 143-676062 | HM 100159014A |
| 143-676072 | HM 100159015A |
| 143-676082 | HM 100159016A |
| 143-676092 | HM 100159017A |
| 143-676102 | H 070130221C |
| 143-676112 | H 06075452K |
| 143-676112 | H 06075452K |
| 143-676122 | H 070130211D |
| 143-676132 | H 06075403M |
| 143-676142 | HM 080155151D |
| 143-676152 | HM 100159019A |
| 143-676162 | HM 100159020A |
| 143-676172 | H 070130172D |
| 143-676182 | HM 080155083D |
| 143-676192 | H 070130224C |
| 143-676202 | HM 080155131C |
| 143-676212 | HM 100159011C |

| 143-676222 | HM 080155129D |
|---|---|
| 143-676232 | H 06075456M |
| 143-676242 | H 06075457M |
| 143-676252 | HM 080155137D |
| 143-676262 | HM 100159008C |
| 143-677012 | HS 04055469D |
| 143-677022 | HS 04055363G |
| 143-684012 | H 03545518M |
| 143-685012 | HS 05067117B |
| 143-685022 | HS 05067163C |
| 143-685032 | HS 05067177C |
| 143-686012 | HM 070132007A |
| 143-686022 | HM 070132008A |
| 143-686032 | HM 080155122E |
| 143-686042 | HM 080155121E |
| 143-686052 | HM 080155146E |
| 143-686062 | H 070130206D |
| 143-686072 | H 05065413K |
| 143-686082 | HM 080155164E |
| 143-686092 | HM 080155170E |
| 143-686102 | HM 100159034C |
| 143-686112 | H 05065423K |
| 143-686112 | H 05065423K |
| 143-686122 | H 06075461M |
| 143-686122 | H 06075461M |
| 143-686132 | H 070130232D |
| 143-686132 | H 070130232D |
| 143-686142 | HM 080155171E |
| 143-686142 | HM 080155171E |
| 143-686152 | HM 100159036C |
| 143-686152 | HM 100159036C |
| 143-686162 | H 06075462M |
| 143-686172 | HM 100159040C |
| 143-686182 | H 06075464K |
| 143-687012 | HS 04055502G |
| 143-687022 | HS 05067169B |
| 143-687032 | HS 05067170B |
| 143-687032 | HS 05067170B |
| 143-687042 | HS 04055514G |
| 143-694012 | H 03545528L |
| 143-694022 | H 03545496L |
| 143-694032 | H 03545205L |
| 143-696012 | H 06075465K |
| 143-696032 | HM 080155145E |
| 143-696042 | H 06075461N |
| 143-696052 | H 070130232E |
| 143-696062 | HM 080155171F |
| 143-696072 | HM 100159036D |
| 143-696082 | HS 05067181C |
| 143-696092 | HM 080155170F |
| 143-696102 | H 070130205E |
| 143-696112 | HM 100159034D |
| 143-696122 | H 06075462N |
| 143-696132 | HM 080155131F |
| 143-696142 | H 05065413L |
| 143-696152 | H 06075403N |

| 143-697012 | HS 05067178B |
|---|---|
| 143-697022 | HS 05067170C |
| 143-697032 | HS 04055519F |
| 143-697042 | HS 05067117C |
| 143-697052 | HS 05067178C |
| 143-701012 | H 03035322M |
| 143-701022 | H 03035325M |
| 143-704022 | H 03545542M |
| 143-704032 | H 03545543L |
| 143-704042 | H 03545544M |
| 143-704062 | H 03545554M |
| 143-706012 | H 06075467K |
| 143-706022 | H 06075468K |
| 143-706032 | HM 080155190E |
| 143-706042 | HM 080155189E |
| 143-706052 | HM 080155189E |
| 143-706062 | H 06075469K |
| 143-706072 | HM 080155194E |
| 143-706082 | HM 080155193E |
| 143-706092 | H 070130172E |
| 143-706102 | H 06075470P |
| 143-706112 | H 070130240F |
| 143-706122 | HM 080155195G |
| 143-706132 | HM 100159055E |
| 143-706142 | H 06075471P |
| 143-706152 | HM 100159034E |
| 143-706162 | HM 080155170G |
| 143-706172 | H 070130205F |
| 143-706182 | H 070130172F |
| 143-706212 | H 05065447M |
| 143-706222 | HM 080155204G |
| 143-706232 | HM 100159062E |
| 143-707012 | HS 05067188C |
| 143-707042 | HS 04055524G |
| 143-707052 | HS 05067190C |
| 143-707062 | HS 04055525F |
| 143-707072 | HS 05067191C |
| 143-707082 | HS 05067192C |
| 143-707092 | HS 05067193D |
| 143-707102 | HS 04055526G |
| 143-707112 | HS 05067195D |
| 143-707122 | HS 04055534G |
| 143-707132 | HS 05067200D |
| 143-711012 | H 03035326N |
| 143-711022 | H 03035327N |
| 143-711032 | H 03035312N |
| 143-711042 | H 03035325N |
| 143-711052 | H 03035332N |
| 143-714012 | H 03545372N |
| 143-714022 | H 03545460N |
| 143-714032 | H 03545542N |
| 143-714042 | H 03545462N |
| 143-714052 | H 03545427N |
| 143-714062 | H 03545371N |
| 143-714072 | H 03545474N |
| 143-714082 | H 03545543M |

**Sears Craftsman cross reference chart for L-head light duty engines (8 of 11)**

| | |
|---|---|
| 143-714092 | H 03545285N |
| 143-714102 | H 03545133N |
| 143-714112 | H 03545379N |
| 143-714122 | H 03545328N |
| 143-714132 | H 03545544N |
| 143-716012 | HM 070132014B |
| 143-716022 | HM 080155208F |
| 143-716032 | H 06075465L |
| 143-716042 | H 06075445L |
| 143-716052 | HM 070132007B |
| 143-716062 | HM 070132008B |
| 143-716072 | HM 080155189F |
| 143-716082 | HM 080155190F |
| 143-716092 | HM 100159014B |
| 143-716102 | HM 100159015B |
| 143-716112 | H 070130221D |
| 143-716122 | H 070130207D |
| 143-716132 | H 070130193D |
| 143-716142 | H 070130197D |
| 143-716152 | HM 080155194F |
| 143-716162 | H 05065398M |
| 143-716172 | HM 080155122F |
| 143-716182 | HM 080155145F |
| 143-716192 | HM 080155146F |
| 143-716202 | H 070130232F |
| 143-716212 | HM 100159066 |
| 143-716222 | HM 080155211F |
| 143-716232 | H 06075469L |
| 143-716242 | H 05065413M |
| 143-716252 | H 070130224D |
| 143-716262 | HM 100159020B |
| 143-716272 | H 06075416L |
| 143-716282 | H 06075439L |
| 143-716292 | H 06075437L |
| 143-716302 | HH 060105096G |
| 143-716312 | H 070130006D |
| 143-716322 | H 070130196D |
| 143-716332 | HM 100159019B |
| 143-716342 | HM 070132015B |
| 143-716352 | H 06075438L |
| 143-716362 | H 070130205G |
| 143-716372 | HM 100159034F |
| 143-716382 | HM 100159055F |
| 143-716392 | HM 080155204H |
| 143-716402 | H 06075479R |
| 143-716412 | H 06075404L |
| 143-716422 | H 05065447N |
| 143-716432 | HM 100159062F |
| 143-717012 | HS 04055482H |
| 143-717022 | HS 04055502H |
| 143-717032 | HS 04055524H |
| 143-717042 | HS 05067190D |
| 143-717052 | HS 04055363H |
| 143-717062 | HS 05067128D |
| 143-717072 | HS 05067163D |
| 143-717082 | HS 05067192D |
| 143-717092 | HS 05067 95E |
| 143-717102 | HS 04055537H |
| 143-717112 | HS 05067206E |
| 143-721012 | H 03035334N |
| 143-721022 | H 03035333N |
| 143-721032 | H 03035335N |
| 143-724012 | H 03545568N |
| 143-724022 | H 03545569N |
| 143-724032 | H 03545571N |
| 143-724042 | H 03545474N |
| 143-724052 | HS 04055526H |
| 143-725012 | HS 05067210E |
| 143-726012 | H 06075480M |
| 143-726022 | HM 100159019C |
| 143-726032 | HM 080155228F |
| 143-726042 | H 070130207E |
| 143-726052 | H 070130224E |
| 143-726082 | H 06075445M |
| 143-726092 | H 05065461P |
| 143-726102 | HM 100159066C |
| 143-726112 | H 06075483S |
| 143-726132 | H 06075416M |
| 143-726142 | HH 060105096H |
| 143-726152 | HM 080155229H |
| 143-726182 | H 070130193E |
| 143-726192 | H 070130197E |
| 143-726202 | H 070130006E |
| 143-726212 | H 070130196E |
| 143-726222 | H 06075465M |
| 143-726232 | HM 100159014C |
| 143-726242 | H 06075469M |
| 143-726252 | H 070130206G |
| 143-726262 | H 06075439M |
| 143-726272 | HM 080155231F |
| 143-726282 | H 070130205H |
| 143-726292 | HM 100159034G |
| 143-726302 | H 05065413P |
| 143-726312 | HM 100159072G |
| 143-726322 | HM 100159020C |
| 143-731012 | H 03035337N |
| 143-734012 | H 03545573N |
| 143-734022 | H 03545575N |
| 143-734032 | H 03545576N |
| 143-734042 | HS 04055542H |
| 143-735012 | HS 05067220E |
| 143-735022 | HS 05067224D |
| 143-736032 | H 06075484M |
| 143-736042 | HM 080155235F |
| 143-736052 | H 06075486M |
| 143-736062 | H 06075487M |
| 143-736072 | H 05065447P |
| 143-736082 | HM 080155238F |
| 143-736092 | H 070130252H |
| 143-736102 | HM 100159079H |
| 143-736112 | H 05065473P |
| 143-736122 | HM 080155246J |
| 143-736132 | HM 080155247J |
| 143-736142 | HM 100159086H |
| 143-741012 | H 03035327P |
| 143-741022 | H 03035333P |
| 143-741032 | H 03035337P |
| 143-741042 | H 03035342P |
| 143-741052 | H 03035350P |
| 143-741062 | H 03035351P |
| 143-741072 | H 03035354P |
| 143-741082 | H 03035362P |
| 143-741092 | H 03035363P |
| 143-742032 | HM 080155256F |
| 143-742042 | H 05065479N |
| 143-742052 | H 05065480N |
| 143-744012 | H 03545581P |
| 143-744022 | H 03545575P |
| 143-744032 | H 03545569P |
| 143-744052 | H 03545474P |
| 143-744062 | H 03545576P |
| 143-744072 | H 03545554N |
| 143-744082 | H 03545587P |
| 143-744092 | HS 05067238E |
| 143-744102 | HS 04055546H |
| 143-744112 | HS 05067247E |
| 143-744122 | H 03545592P |
| 143-746012 | H 06075489M |
| 143-746022 | HM 080155250F |
| 143-746062 | H 070130256J |
| 143-746072 | HM 100159095J |
| 143-746082 | HM 080155279J |
| 143-746092 | HM 080155280J |
| 143-746102 | HM 100159101J |
| 143-751012 | H 03035342R |
| 143-751022 | H 03035333R |
| 143-751032 | H 03035362R |
| 143-751042 | H 03035363R |
| 143-751052 | H 03035350R |
| 143-751062 | H 03035351R |
| 143-754012 | H 03545581R |
| 143-754022 | H 03545379R |
| 143-754032 | H 03545575R |
| 143-754042 | H 03545592R |
| 143-754052 | H 03545576R |
| 143-754062 | HS 05067224E |
| 143-754072 | H 03545595R |
| 143-754082 | HS 05067163E |
| 143-754092 | HS 05067192E |
| 143-754102 | H 03545587R |
| 143-754112 | HS 05067238F |
| 143-754122 | HS 04055546J |
| 143-754132 | HS 05067247F |
| 143-754142 | HS 05067200F |
| 143-754152 | H 03545554R |
| 143-756012 | H 06075487N |
| 143-756022 | H 06075489N |
| 143-756042 | H 070130207F |
| 143-756052 | H 070130006F |
| 143-756062 | HM 100159020D |
| 143-756072 | H 06075469N |
| 143-756082 | HM 080155228G |
| 143-756092 | H 06075465N |
| 143-756102 | HM 080155146G |
| 143-756112 | HM 100159014D |
| 143-756122 | HM 080155256G |
| 143-756132 | H 070130224F |
| 143-756142 | H 05065479P |
| 143-756152 | HM 100159019D |
| 143-756162 | H 070130256K |
| 143-756172 | HM 100159095K |
| 143-756182 | HM 080155279K |
| 143-756192 | HM 080155280K |
| 143-756202 | HM 100159101K |
| 143-756212 | HM 080155299G |
| 143-756222 | H 070130260K |
| 143-756232 | HM 100159111K |
| 143-764012 | HS 05067178F |
| 143-764022 | HS 05067265F |
| 143-764032 | HS 04055556J |
| 143-764042 | HS 05067268F |
| 143-764052 | H 03545604R |
| 143-764062 | HS 05067273F |
| 143-764072 | HS 05067274F |
| 143-766012 | HM 080155302H |
| 143-766072 | HM 080155306H |
| 143-766082 | HM 080155299H |
| 143-766092 | HM 100159111L |
| 143-766102 | HM 080155308L |
| 143-766112 | HM 080155309L |
| 143-766122 | HM 100159115L |
| 143-766132 | H 070130263K |
| 143-766142 | HM 100159125L |
| 143-766152 | HM 080155321L |
| 143-774012 | H 03545605R |
| 143-774022 | H 03035342S |
| 143-774032 | H 03545581S |
| 143-774052 | H 03035363S |
| 143-774102 | H 03035374R |
| 143-774112 | HS 04055534J |
| 143-774122 | H 03545612R |
| 143-774132 | HS 05067280E |
| 143-776012 | HM 080155299J |
| 143-776022 | H 070130264F |
| 143-776042 | HM 080155327J |
| 143-776052 | HM 100159134F |
| 143-776062 | HM 100159135F |
| 143-776072 | H 070130267L |
| 143-784012 | HS 04055556K |
| 143-784022 | HS 05067268G |
| 143-784032 | HS 05067274G |
| 143-784042 | HS 04055562K |
| 143-784052 | H 03035351S |
| 143-784062 | H 03035382S |

**Sears Craftsman cross reference chart for L-head light duty engines (9 of 11)**

| | |
|---|---|
| 143-784072 | HS 05067192F |
| 143-784082 | H 03035333S |
| 143-784092 | H 03035362S |
| 143-784102 | H 03035374S |
| 143-784102 | H 03035374S |
| 143-784112 | H 03545592S |
| 143-784112 | H 03545592S |
| 143-784122 | H 03545612S |
| 143-784132 | HS 05067280F |
| 143-784142 | H 03545379S |
| 143-784152 | HS 05067163F |
| 143-784162 | H 03035350S |
| 143-784172 | H 03035393S |
| 143-784182 | H 03545595S |
| 143-784192 | H 03545554T |
| 143-784202 | HS 04055534L |
| 143-786012 | HM 080155308M |
| 143-786022 | HM 080155309M |
| 143-786032 | HM 080155321M |
| 143-786042 | HM 100159115M |
| 143-786052 | HM 080155338M |
| 143-786062 | HM 100159140M |
| 143-786072 | HM 100159141M |
| 143-786082 | H 06075507P |
| 143-786092 | HM 080155340L |
| 143-786102 | H 070130268G |
| 143-786112 | HM 080155346K |
| 143-786122 | HM 100159135G |
| 143-786132 | H 070130006G |
| 143-786142 | H 070130269G |
| 143-786152 | H 070130264G |
| 143-786162 | H 06075469P |
| 143-786172 | HM 080155327K |
| 143-786182 | HM 100159111N |
| 143-786192 | HM 100159134G |
| 143-786202 | HM 100159158G |
| 143-794012 | HS 04055556L |
| 143-794022 | HS 04055562L |
| 143-794032 | HS 05067274H |
| 143-794042 | HS 05067268H |
| 143-794052 | HS 04055572L |
| 143-794053 | HS 05067291H |
| 143-794072 | HS 04055573L |
| 143-794082 | HS 05067300H |
| 143-796012 | HM 080155308N |
| 143-796022 | HM 080155309N |
| 143-796032 | HM 080155321N |
| 143-796042 | HM 080155338N |
| 143-796052 | HM 100159115N |
| 143-796062 | HM 100159140N |
| 143-796072 | HM 100159141N |
| 143-796082 | HM 080155365N |
| 143-796092 | HM 080155366N |
| 143-796102 | HM 100159162N |
| 143-796122 | H 070130224G |
| 143-796132 | HM 080155370N |
| 143-796142 | H 070130267M |
| 143-796152 | HM 080155346L |
| 143-796162 | HM 100159169H |
| 143-796172 | HM 100159135H |
| 143-796182 | H 070130006H |
| 143-796192 | HM 080155327L |
| 143-796202 | HM 080155384L |
| 143-804012 | HS 05067300J |
| 143-804022 | HS 05067268J |
| 143-804032 | HS 04055573M |
| 143-804042 | HS 04055556M |
| 143-804052 | HS 04055585M |
| 143-804062 | HS 04055586M |
| 143-804072 | HS 05067309J |
| 143-804082 | H 03035419S |
| 143-804092 | H 03035420S |
| 143-804102 | H 03545629S |
| 143-804112 | H 03035424S |
| 143-806012 | HM 080155370P |
| 143-806022 | HM 080155309P |
| 143-806032 | HM 080155389P |
| 143-806042 | HM 100159115P |
| 143-806052 | HM 100159180P |
| 143-806072 | HM 080155308P |
| 143-806082 | HM 080155394P |
| 143-806092 | HM 100159183P |
| 143-806092 | HM 100159183P |
| 143-806102 | HM 080155400L |
| 143-806112 | H 070130264H |
| 143-806122 | H 070130267N |
| 143-806132 | HM 100159192J |
| 143-806142 | H 070130268H |
| 143-806152 | HM 080155411L |
| 143-806162 | HM 100159135J |
| 143-806172 | HM 100159169J |
| 143-806182 | HM 080155424L |
| 143-814012 | HS 04055586N |
| 143-814022 | HS 05067309K |
| 143-814032 | H 03035426S |
| 143-814042 | H 03035427S |
| 143-814052 | H 03545631S |
| 143-814062 | H 03545633S |
| 143-814072 | H 03035431S |
| 143-816012 | HM 100159183R |
| 143-816022 | HM 080155416P |
| 143-816032 | HM 100159199R |
| 143-816052 | HM 080155433L |
| 143-816062 | H 06075469S |
| 143-816072 | H 070130264J |
| 143-824012 | H 03035427T |
| 143-824022 | H 03035426T |
| 143-824022 | H 03035426T |
| 143-824032 | HS 05067330K |
| 143-824042 | H 03035431T |
| 143-826012 | HM 080155445P |
| 143-826022 | HM 100159209R |
| 143-826032 | HM 080155433M |
| 143-826042 | HM 080155454M |
| 143-826052 | HM 100159135K |
| 143-826062 | H 06075537S |
| 143-826072 | HM 080155424M |
| 143-826082 | HM 100159169K |
| 143-826092 | H 06075538S |
| 143-826102 | HM 080155462M |
| 143-826112 | H 06075539S |
| 143-826122 | HM 080155400M |
| 143-834012 | HSSK05067338L |
| 143-834022 | H 03035426U |
| 143-834032 | HS 05067192G |
| 143-834042 | H 03545595T |
| 143-834052 | HS 05067163G |
| 143-836012 | HMSK080155478R |
| 143-836022 | HMSK080155416R |
| 143-836032 | HMSK100159199S |
| 143-836042 | HMSK100159244S |
| 143-836082 | HM 080155487M |
| 143-836092 | H 06075537T |
| 143-836102 | H 06075538T |
| 143-836112 | H 06075469T |
| 143-836122 | H 06075539T |
| 143-836132 | H 06075554T |
| 143-941000 | TVXL220157245D |
| 143-941001 | HMSK100159244T |
| 143-941002 | TVXL220157205D |
| 143-941003 | HMSK100159261T |
| 143-941004 | TVXL220157215D |
| 143-941005 | HM 100159262K |
| 143-941006 | TVXL220157206D |
| 143-941008 | TVXL220157220D |
| 143-941009 | HM 100159262L |
| 143-943001 | H 03035426V |
| 143-943003 | H 03035431V |
| 143-943005 | H 03035450V |
| 143-943009 | H 03035453V |
| 143-943501 | H 03545655V |
| 143-943502 | TVS 09043515K |
| 143-943503 | H 03545654V |
| 143-943504 | TVS 09043513K |
| 143-943505 | H 03545595U |
| 143-943506 | TVS 09043215K |
| 143-943507 | H 03545657U |
| 143-943508 | TVS 09043572K |
| 143-943510 | TVS 09043298K |
| 143-943512 | TVS 09043375K |
| 143-943514 | TVS 09043576K |
| 143-943526 | TVS 09043512K |
| 143-943528 | TVS 09043299K |
| 143-943530 | TVS 09043514K |
| 143-943800 | TVS 10044048F |
| 143-943802 | TVS 10044036F |
| 143-943804 | TVS 09046030A |
| 143-943806 | TVS 09046031A |
| 143-943808 | TVS 09046032A |
| 143-943810 | TVS 10044043F |
| 143-943812 | TVS 09046035A |
| 143-943814 | TVS 09046036A |
| 143-943816 | TVS 09046037A |
| 143-943818 | TVS 10044038F |
| 143-943820 | TVS 10044029F |
| 143-943830 | TVS 10044046F |
| 143-943832 | TVS 10044030F |
| 143-943834 | TVS 10044031F |
| 143-943838 | TVS 10044033F |
| 143-943842 | TVS 10044045F |
| 143-943844 | TVS 10044032F |
| 143-944000 | TVS 10553163L |
| 143-944002 | TVS 11556032B |
| 143-944004 | TVS 11556031B |
| 143-944006 | TVS 11556033B |
| 143-944008 | TVS 10553912L |
| 143-944010 | TVS 11556012B |
| 143-944012 | TVS 11556010B |
| 143-944014 | TVS 11556036B |
| 143-944016 | TVS 11556037B |
| 143-944018 | TVS 11556016B |
| 143-944022 | TVS 11556043B |
| 143-944024 | TVS 11556042B |
| 143-944026 | TVS 10553913L |
| 143-944500 | TVS 11557020B |
| 143-944502 | TVS 11557023B |
| 143-945000 | TVM 12560254P |
| 143-945001 | HSSK05067338M |
| 143-945002 | TVS 11561901A |
| 143-945003 | HS 05067163H |
| 143-945004 | TVS 11561021A |
| 143-945006 | TVS 11561022A |
| 143-945010 | TVS 11561906A |
| 143-945012 | TVS 11561016A |
| 143-945014 | TVS 11561002A |
| 143-945016 | TVS 11561024A |
| 143-945300 | TVS 12063918K |
| 143-945302 | TVS 12063919K |
| 143-945502 | TVS 12063921K |
| 143-945504 | TVS 12063920K |
| 143-945506 | TVS 12063922K |
| 143-945508 | TVS 12063923K |
| 143-946001 | H 06075539U |
| 143-946003 | H 06075469U |
| 143-946005 | H 06075537U |
| 143-946007 | H 06075554U |
| 143-948000 | TVXL195150238C |
| 143-948001 | HMSK080155478S |
| 143-948003 | HMSK080155502S |
| 143-948005 | HM 080155487N |
| 143-948007 | HM 080155433N |
| 143-948009 | HM 080155424N |
| 143-955001 | HSSK05067338N |
| 143-744042 | H 03545379P |

**Sears Craftsman cross reference chart for L-head light duty engines (10 of 11)**

| 143-955001 | HSSK05067338N |
|---|---|
| 143-945300 | TVS 12063918K |
| 143-948001 | HMSK080155478S |
| 143-945502 | TVS 12063921K |
| 143-958001 | HMSK080155535S |
| 143-951001 | HMSK100159282T |
| 143-945016 | TVS 11561024A |
| 143-943009 | H 03035453V |
| 143-944506 | TVS 11557030B |
| 143-943532 | TVS 09043700K |
| 143-944028 | TVS 11556044B |
| 143-944032 | TVS 11556912B |
| 143-943509 | H 03545661V |
| 143-941001 | HMSK100159244T |
| 143-943508 | TVS 09043572K |
| 143-944034 | TVS 11556047B |
| 143-944036 | TVS 11556048B |
| 143-944504 | TVS 11557028B |
| 143-945018 | TVS 11561026A |
| 143-945020 | TVS 11561027A |
| 143-945510 | TVS 12063924K |
| 143-944030 | TVS 11556046B |
| 143-951000 | TVM220157245E |
| 143-951003 | HM 100159262M |
| 143-954000 | TVS 11556033C |
| 143-954002 | TVS 11556036C |
| 143-953800 | TVS 09046036B |
| 143-954500 | TVS 11557031C |
| 143-955500 | TVS 12063920L |
| 143-953500 | TVS 09043576L |
| 143-954502 | TVS 11557023C |
| 143-954504 | TVS 11557032C |
| 143-954506 | TVS 11557030C |
| 143-955010 | TVS 11561021B |
| 143-955002 | TVS 11561024B |
| 143-953003 | H 03035453W |
| 143-955004 | TVS 11561032B |
| 143-951002 | TVM 220157206E |
| 143-951004 | TVM 220157215E |
| 143-958000 | TVM 195150238D |
| 143-951006 | TVM 220157205E |
| 143-951008 | TVM 220157220E |
| 143-955000 | TVS 11561907B |
| 143-953005 | H 03035450W |
| 143-953503 | H 03545654W |
| 143-953505 | H 03545655W |
| 143-953504 | TVS 09043215L |
| 143-953808 | TVS 10044037G |
| 143-953812 | TVS 10044038G |
| 143-953810 | TVS 10044046G |
| 143-955006 | TVS 11561906B |
| 143-953001 | H 03035431W |

**Sears Craftsman cross reference chart for L-head light duty engines (11 of 11)**

| OVERHEAD VALVE SEARS CRAFTSMAN CROSS REFFERENCE CHART | | | | | | | | |
|---|---|---|---|---|---|---|---|---|
| 143.366012 | OVM120 | 200004 | 143.396032 | OVXL120 | 202036C | 143.436092 | OHV125 | 203015A |
| 143.366072 | OVM120 | 200006 | 143.396062 | OVXL120 | 202034C | 143.436102 | OHV125 | 203018A |
| 143.366202 | OVM120 | 200012 | 143.396072 | OVXL120 | 202035C | 143.436132 | OHV125 | 203020A |
| 143.366212 | OVM120 | 200014A | 143.396092 | OVXL120 | 200023C | 143.436142 | OHV12 | 202715A |
| 143.366232 | OVM120 | 200015A | 143.396112 | OVXL120 | 202039C | 143.436152 | OHV125 | 203023A |
| 143.376012 | OVM120 | 200017A | 143.396132 | OVXL120 | 202040C | 143.796112 | OHSK120 | 222005 |
| 143.376032 | OVM120 | 200012A | 143.404112 | OVRM50 | 52901 | 143.806062 | OHSK120 | 222005A |
| 143.376072 | OVM120 | 200006A | 143.406012 | OVXL120 | 202040D | 143.816042 | OHSK120 | 222007B |
| 143.376082 | OVM120 | 200018A | 143.406052 | OVXL120 | 202034D | 143.836052 | OHSK120 | 222015C |
| 143.384132 | OVRM40 | 42002 | 143.406062 | OVXL120 | 202035D | 143.941200 | OHV125 | 203023B |
| 143.384142 | OVRM40 | 42003 | 143.406072 | OVXL120 | 202047D | 143.941201 | OHSK120 | 222015D |
| 143.384152 | OVRM40 | 42001 | 143.406132 | OVXL120 | 202039D | 143.941202 | OHV12 | 202720B |
| 143.384162 | OVRM40 | 42004 | 143.406142 | OVXL120 | 202049D | 143.951600 | OHV165 | 204402B |
| 143.384182 | OVRM40 | 42005 | 143.406152 | OVXL120 | 202050D | 143.951602 | OHV165 | 204403B |
| 143.384192 | OVRM40 | 42006 | 143.406162 | OVXL120 | 202051D | 143.955003 | OHH50 | 68001A |
| 143.386012 | OVM120 | 200020B | 143.414432 | OVRM50 | 52901A | 143.955007 | OHH50 | 68004A |
| 143.386032 | OVM120 | 200006B | 143.416012 | OVXL120 | 202056D | 143.961201 | OHSK120 | 222026E |
| 143.386092 | OVM120 | 200014B | 143.416042 | OVXL125 | 202401 | 143.965007 | OHH50 | 68001B |
| 143.386102 | OVM120 | 200018B | 143.416082 | OVXL125 | 202403 | 143.965009 | OHH50 | 68036B |
| 143.386152 | OVM120 | 200015B | 143.426092 | OVXL120 | 202063E | 143.965011 | OHH50 | 68044B |
| 143.386162 | OVM120 | 200022A | 143.426112 | OVXL120 | 202062E | 143.965015 | OHH50 | 68060B |
| 143.394092 | OVRM40 | 42008 | 143.426122 | OVXL125 | 202064E | 143.965017 | OHH50 | 68060C |
| 143.394102 | OVRM40 | 42010 | 143.426142 | OVXL120 | 202034E | 143.965019 | OHH50 | 68044C |
| 143.394112 | OVRM40 | 42011 | 143.426152 | OVXL120 | 202039E | 143.965021 | OHH50 | 68036C |
| 143.394192 | OVRM40 | 42007 | 143.426172 | OVXL125 | 202067E | 143.965023 | OHH50 | 68004C |
| 143.394202 | OVRM40 | 42009 | 143.436022 | OHV12 | 202709A | 143.965501 | OHH50 | 69020A |
| 143.394212 | OVRM40 | 42012 | 143.436032 | OHV125 | 203013A | 143.965503 | OHH50 | 69020B |
| 143.396012 | OVXL120 | 202031C | 143.436042 | OHV125 | 203012A | 143.971201 | OHSK120 | 222026F |

**Sears Craftsman cross reference chart for OHV light duty engines**

| Craftsman Engine Model | See Model |
|---|---|
| 143.558012 | HH80 |
| 143.558022 | HH80 |
| 143.558032 | HH80 |
| 143.558052 | HH80 |
| 143.559012 | HH100 |
| 143.559022 | HH100 |
| 143.559032 | HH100 |
| 143.559042 | HH100 |
| 143.562012 | HH120 |
| 143.562022 | HH120 |
| 143.562032 | HH120 |
| 143.568032 | HH80 |
| 143.569022 | HH100 |
| 143.569032 | HH100 |
| 143.569042 | HH100 |
| 143.569052 | HH100 |
| 143.569082 | HH100 |
| 143.572012 | HH120 |
| 143.572022 | HH120 |
| 143.572032 | HH120 |
| 143.572042 | HH120 |
| 143.572052 | HH120 |
| 143.572062 | HH120 |
| 143.572092 | HH120 |
| 143.572102 | HH120 |
| 143.578012 | HH80 |
| 143.578022 | HH80 |
| 143.578052 | HH80 |
| 143.578062 | HH80 |
| 143.578072 | HH80 |
| 143.579012 | HH100 |
| 143.579022 | HH100 |
| 143.579032 | HH100 |
| 143.579042 | HH100 |
| 143.579052 | HH100 |
| 143.579062 | HH100 |
| 143.579072 | HH100 |
| 143.579082 | HH100 |
| 143.579092 | HH100 |
| 143.579102 | HH100 |
| 143.579112 | HH100 |
| 143.579132 | HH100 |
| 143.582012 | HH120 |
| 143.582022 | HH120 |
| 143.582032 | HH120 |
| 143.582042 | HH120 |
| 143.582052 | HH120 |
| 143.582062 | HH120 |
| 143.582072 | HH120 |
| 143.582082 | HH120 |
| 143.582092 | HH120 |
| 143.582102 | HH120 |
| 143.582112 | HH120 |
| 143.582122 | HH120 |
| 143.582132 | HH120 |
| 143.582142 | HH120 |
| 143.582172 | HH120 |
| 143.588012 | HH80 |
| 143.588032 | HH80 |
| 143.589012 | HH100 |
| 143.589022 | HH100 |
| 143.589032 | HH100 |
| 143.589042 | HH100 |
| 143.589052 | HH100 |
| 143.589072 | HH100 |
| 143.592012 | HH120 |
| 143.592022 | HH120 |
| 143.592032 | HH120 |
| 143.592052 | HH120 |
| 143.592062 | HH120 |
| 143.592072 | HH120 |
| 143.592082 | HH120 |
| 143.598012 | HH80 |
| 143.599012 | HH100 |
| 143.599022 | HH100 |
| 143.599042 | HH100 |
| 143.599052 | HH100 |
| 143.599062 | HH100 |
| 143.602012 | HH120 |
| 143.602022 | HH120 |
| 143.602032 | HH120 |
| 143.602052 | HH120 |
| 143.602062 | HH120 |
| 143.602072 | HH120 |
| 143.602082 | HH120 |
| 143.602092 | HH120 |
| 143.602102 | HH120 |
| 143.602112 | HH120 |
| 143.602122 | HH120 |
| 143.608012 | HH80 |
| 143.608022 | HH80 |
| 143.608032 | HH80 |
| 143.609022 | HH100 |
| 143.609032 | HH100 |
| 143.609042 | HH100 |
| 143.609052 | HH100 |
| 143.609072 | HH100 |
| 143.612012 | HH120 |
| 143.612022 | HH120 |
| 143.619012 | HH100 |
| 143.622012 | HH120 |
| 143.622022 | HH120 |
| 143.622032 | HH120 |
| 143.622042 | HH120 |
| 143.622052 | HH120 |
| 143.622062 | HH120 |
| 143.622072 | HH120 |
| 143.622082 | HH120 |
| 143.622092 | HH120 |
| 143.622102 | HH120 |
| 143.628012 | HH80 |
| 143.628022 | HH80 |
| 143.629012 | HH100 |
| 143.629022 | HH100 |
| 143.629032 | HH100 |
| 143.629042 | HH100 |
| 143.629052 | HH100 |
| 143.629062 | HH100 |
| 143.629072 | HH100 |
| 143.632022 | HH120 |
| 143.632032 | HH120 |
| 143.632042 | HH120 |
| 143.639012 | HH100 |
| 143.642012 | HH120 |
| 143.642022 | HH120 |
| 143.642032 | HH120 |
| 143.642042 | HH120 |
| 143.649012 | HH100 |
| 143.649022 | HH100 |
| 143.652012 | HH120 |
| 143.652022 | HH120 |
| 143.652032 | HH120 |
| 143.652042 | HH120 |
| 143.652052 | HH120 |
| 143.652062 | HH120 |
| 143.652072 | HH120 |
| 143.659012 | HH100 |
| 143.659022 | HH100 |
| 143.659032 | HH100 |
| 143.662012 | HH120 |
| 143.669012 | HH100 |
| 143.672012 | HH120 |
| 143.672022 | HH120 |
| 143.672032 | HH120 |
| 143.672042 | HH120 |
| 143.672052 | HH120 |
| 143.672062 | HH120 |
| 143.672072 | HH120 |
| 143.676082 | HH100 |
| 143.679012 | HH100 |
| 143.679022 | HH100 |
| 143.679032 | HH100 |
| 143.712012 | HH120 |
| 143.712022 | HH120 |
| 143.712032 | HH120 |
| 143.712042 | HH120 |
| 143.760012 | HH120 |
| 143.780012 | HH120 |

**Sears Craftsman cross reference chart for L-head heavy duty engines**

| CROSS REFERENCE LIST | | CROSS REFERENCE LIST | |
|---|---|---|---|
| Craftsman Engine Model | See Model | Craftsman Engine Model | See Model |
| 143.626272 | OH160 | 143.670052 | OH160 |
| 143.630012 | OH160 | 143.670062 | OH160 |
| 143.630022 | OH160 | 143.670072 | OH160 |
| 143.640012 | OH160 | 143.670082 | OH140 |
| 143.640022 | OH160 | 143.670092 | OH140 |
| 143.640032 | OH160 | 143.680012 | OH140 |
| 143.640042 | OH160 | 143.680022 | OH140 |
| 143.640052 | OH160 | 143.680032 | OH160 |
| 143.650012 | OH160 | 143.700012 | OH180 |
| 143.650022 | OH160 | 143.710012 | OH160 |
| 143.650032 | OH160 | 143.710022 | OH140 |
| 143.660012 | OH160 | 143.730012 | OH160 |
| 143.660022 | OH160 | 143.730022 | OH140 |
| 143.670012 | OH140 | 143.740012 | OH140 |
| 143.670032 | OH160 | 143.740022 | OH160 |
| 143.670042 | OH140 | 143.770022 | OH180 |

Sears Craftsman cross reference chart for OHV heavy duty engines

# 8 Kohler engines

## Contents

**1.1 Location of engine identification decal (arrow) on most Kohler models**

## 1 Engine identification numbers/models covered

1 Kohler engines covered by this manual include the K-series and Magnum-series single-cylinder and twin-cylinder L-head models, and the Command-series Overhead Valve (OHV) single-cylinder and twin-cylinder models. The engine designation system used by Kohler consists of a model number, spec number, and serial number. On most models the model number and spec number are on the main ID decal, while the engine serial number is on a second decal, usually right below the main decal and often including a bar-code strip. On Command-series engines, all three identifying numbers are on one decal. On most engines, the decal is located on the side of the flywheel fan housing **(see illustration 1.1)**, but on K-series twin cylinder models, the decal is on top of the housing, near the ignition coil. Knowing the three numbers is important whenever corresponding with a small engine repair shop, a dealer or the manufacturer, or when ordering parts.

2 The model number can be used to determine the major features of the engine by comparing each digit **(see illustration 1.2)**. The letters/digits in the model number can be explained generally as follows:

**K-series and Magnum**

Single-cylinder

| | |
|---|---|
| K141 | 6 HP |
| K161, M8 | 8 HP |
| K241 M10 | 10 HP |
| K301, M12 | 12 HP |
| K321, M14 | 14 HP |
| K341, M16 | 16 HP |

Twin-cylinder

| | |
|---|---|
| MV16 | 16 HP |
| KT-17 | 17 HP |
| K482, M18, MV18 | 18 HP |
| KT-19 | 19 HP |
| K532, M20, MV20 | 20 HP |

**Note:** *K indicates K-series, M indicates Magnum-series, T indicates twin-cylinder and V indicates vertical crankshaft.*

**Command-series Overhead Valve (OHV) engines**

Single-cylinder

| | |
|---|---|
| CH11, CV11 | 11 HP |
| CH125, CV125 | 12.5 HP |
| CH14, CV14 | 14 HP |
| CV15 | 15 HP |

Twin-cylinder

| | |
|---|---|
| CH18, CV18 | 18 HP |
| CH20, CV20 | 20 HP |

**Note:** *On Command-series engines, the letter C indicates Command-series, followed by a V (vertical crankshaft) or H (horizontal crankshaft).*

3 The model number will usually include a letter after the horsepower designation to indicate a version code. Version codes include:

| | |
|---|---|
| A (K-series) | Special oil pan |
| C (K-series, Magnum) | Clutch model |
| EP (K-series, Magnum) | Electric-plant application |
| G (K-series, Magnum) | Generator application (tapered crankshaft) |
| P (K-series, Magnum) | Pump application (threaded crankshaft) |
| Q (K-series) | Quiet model |
| R (K-series, Magnum) | Gear reduction unit |
| S (all models) | Electric start |
| ST (all models) | Electric, with retractable starter backup |
| T (all models) | Retractable starter |

4 The spec number contains some of the same information as the model number, but adds a "variation" code that is only important to a small engine repair shop, dealer or the manufacturer when ordering parts for models which may have gone through changes or improvements.

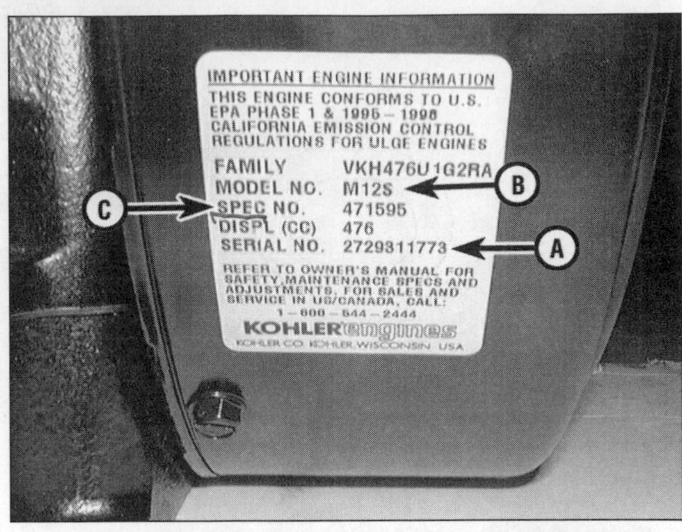


**1.2 Identification of Kohler engines: (A) serial number; (B) engine model number; (C) spec number**

5 The serial number of the engine is also important when ordering parts or service, further defining the date and production run of an engine. The first two digits (of late-model, 10-digit numbers) represents the year manufacture, with 15 indicating 1985, 16 indicating 1986 and so on, with each succeeding year being one number higher. For instance, 25 represents 1995.

# 2 Recoil starter service

## Rope replacement

1 Starter-related problems will require the starter to be removed from the engine and visually inspected. Rope replacement should be done with the correct size and specified length rope to ensure easy starting and pulling. **Warning:** *Retractable (rope) starters incorporate a powerful spring that is always under some tension. Take all safety precautions when working on rope starters, including wearing a safety face shield (not just eye protection).*

2 Remove the starter assembly mounting bolts and separate the starter assembly from the engine **(see illustration 2.2)**.

3 If the rope breaks, the starter doesn't have to be disassembled to replace it, but it may be a good idea to take the opportunity to do a thorough cleaning job and check the spring and dog mechanism. There are two approaches you may be faced with when replacing the rope on a recoil starter. The method you use will depend on the starter type.

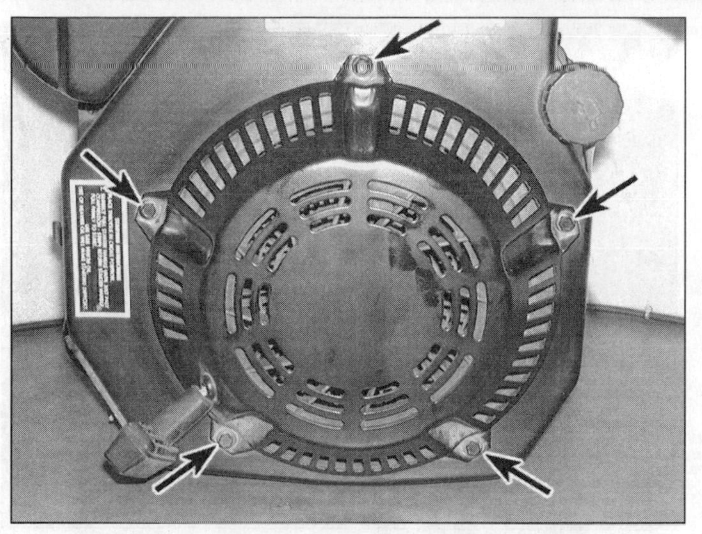

2.2 Recoil starter mounting screws - typical

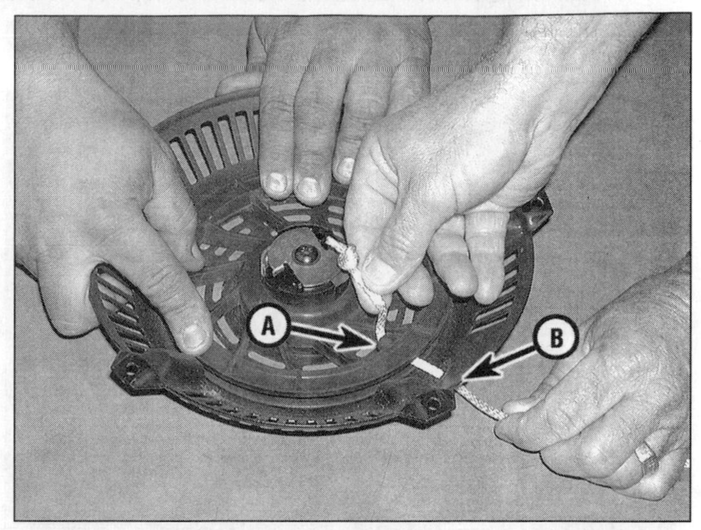


2.12 Have an assistant hold the pulley while it is tensioned, then feed the new rope through the holes in the pulley (A) and the housing (B)

## Early models with Fairbanks-Morse starter

4 Fairbanks-Morse starters on older engines have a cast-aluminum housing. Pull the rope out about one foot and secure the pulley to the housing with a C-clamp. Remove the old rope, starting from the clip end at the hole in the pulley.

5 Thread the new rope, unclipped end first, into the hole in the pulley, around the pulley several times until there is one foot left not wound on.

6 Feed the last foot through the grommet in the cast housing, thread it through the starting handle and double-knot it. Release the C-clamp carefully, and make sure the starter operates before reinstalling it.

## Later starters with stamped housings

7 Remove the screws and remove the starter from the engine.

8 Pull the starter rope out about one foot and tie a slip-knot in it, or clamp it.

9 While holding the rope pulley firmly, release the knot and pull the rope to the inside of the starter. **Note:** *It's helpful to have a helper for this, one person to hold the pulley with two hands, and one person to feed the rope in or out.*

10 Carefully and slowly allow the pulley to rotate with its spring action until all the spring tension has been released. Remove the old rope.

11 With a single knot in one end of the new rope, turn the pulley approximately six turns counterclockwise until it is fully-tensioned. Hold it there.

12 Allow the pulley to turn clockwise only until the pulley's rope hole is in alignment with the rope hole in the housing **(see illustration 2.12)**. Feed the rope through the pulley hole and out through the housing until the knot is snug against the pulley.

13 Clamp the handle end of the rope about a foot from the end (or tie a slip-knot in it), then insert the rope into the handle, tie a knot behind the rope-retaining bar and push the bar back into the handle.

14 Remove the clamp or slip-knot and slowly release the pulley until the rope is wound all the way in. Pull on the rope handle to make sure it operates smoothly all the way in and out.

# Recoil assembly

15 All recoil starters are spring loaded to retract the dog(s) or starter pawls when the engine speed exceeds the turning speed of the starter. The starter dogs engage the drive cup which is attached to the flywheel. If pulling the rope turns the starter but not the engine, you must disassemble the starter. Unless the main spring is broken, most starter problems can be cured with a kit available from a parts distributor. Included are new pawls and springs, pawl retainer, brake spring and other hardware.

## Disassembly

16 Refer to Steps 7 through 10 to release the tension on the pulley and remove the rope. **Warning:** *Wear face protection whenever working on a retractable rope starter, and always have the spring tension fully released before disassembly.*

17 Remove the center screw and brake spring retainer to expose the pawls and their springs **(see illustrations 2.17a and 2.17b)**.

**2.17a Remove the screw and washer (A), then pull up the pawl retainer (B)**

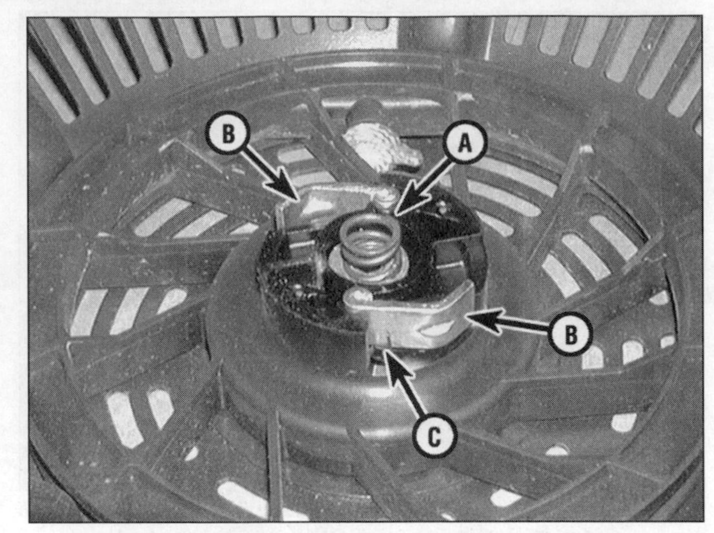


**2.17b Recoil components**

*A Brake spring*
*B Pawls*
*C Pawl springs*

18 Taking note of exactly how the pawls and their springs are installed, remove them and the brake spring and washer.

19 Disengage the main spring by turning the pulley two full turns clockwise. Rotate the pulley back and forth a little while raising it out of the housing. If it doesn't release from the housing, the spring is not disengaged, try again.

20 If the spring is to be replaced, remove the spring/keeper assembly from the pulley as a unit **(see illustration 2.20)**.

## Inspection

21 Examine all starter components at the same time, not just the part suspected of being at fault.

22 Examine the rope for signs of fraying or cuts, the pawls for wear, and the Phillips-drive portion of the center screw. If the drive portion of the screw is a sloppy fit with the screwdriver, the screw must be replaced, since it plays an important role.

23 The manufacturer's starter repair kit will contain all of the mechanical pieces needed to rebuilt the starter, less the rope and the main spring. If the spring is to be replaced, do not remove it from its keeper assembly, a new spring will come in a new keeper.

## Reassembly

24 Lubricate the spring with multi-purpose grease and reinstall the keeper assembly to the pulley with the spring-side towards the pulley **(see illustration 2.20)**.

25 Install the pulley into the housing.

26 Install the pawl springs and pawls just as they were before disassembly **(see illustration 2.17b)**.

27 Place the flat washer in the recess at the center of the starter and install the brake spring over it. The brake spring should be lightly lubed with multi-purpose grease, but do not get any grease in the threads of the starter shaft, or the center screw may not stay tight when installed.

28 Push the pawl retainer over the brake spring and install the center screw and its washer, using a small amount of thread-locking compound on the screw threads. Tighten the screw to the torque listed in this Chapter's Specifications.

29 Install the rope as outlined in Steps 11 through 14.

## Installation

30 Reinstall the starter on the engine, but do not fully tighten the mounting screws. Pull the rope handle out until you feel the pawls engage the cup on the flywheel. This will center the starter assembly on the engine. Hold the rope like this while final-tightening the starter mounting screws.

**2.20 With the spring tension relieved, lift out the spring and keeper housing (arrow) as an assembly**

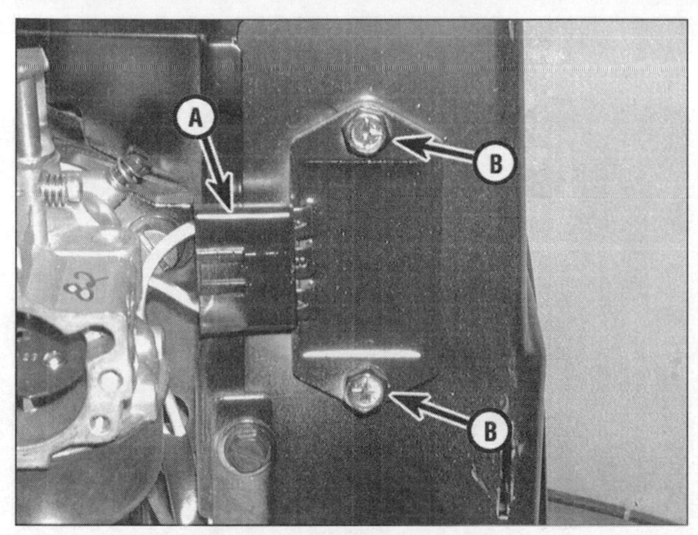

3.5 Regulator electrical connector (A) and mounting screws (B) - typical

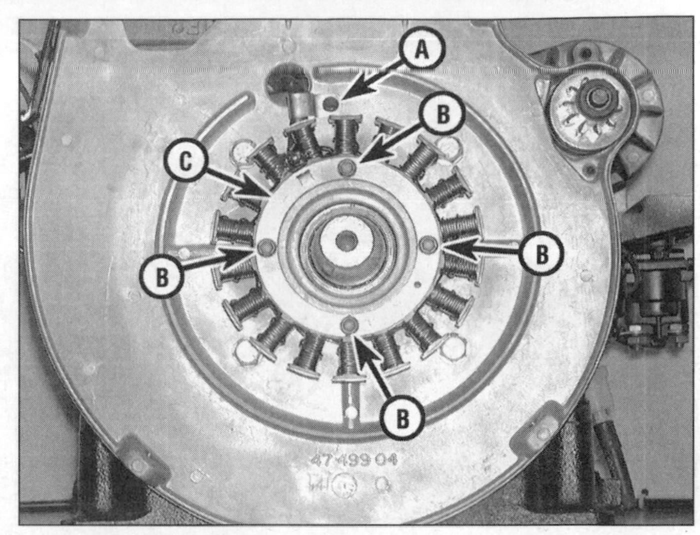

3.9 With the wiring clip (A) disconnected, remove the screws (B) and take off the stator (C)

# 3 Charging and electric starting systems

## Charging system

1 The electrical system consists of three main components - the battery, the starting circuit and charging circuit. The battery functions within both the starting circuit and the charging circuit. Be sure the battery is fully-charged, checked (see Chapter 4) and replaced if necessary, before attempting to diagnose any charging circuit or starting circuit malfunction. **Caution:** *Do not attempt to jump-start your equipment with another battery because the onboard battery is weak. The increased current can damage your starting system. Charge the onboard battery and start the engine without using a jumper battery.*

2 The charging system consists of alternator stator (also called charge coils), a voltage regulator/rectifier, an ignition switch, flywheel magnets and the battery. Some early models with alternators will only possess some of these particular components. The charging system works independently of the starting circuit and other control circuits that regulate lighting, accessories, etc. The engine must be rotating at speed to produce electrical current flow. Most small engine alternators produce AC current when flywheel magnets rotate over the alternator coils (stator) and cut the magnetic field thereby generating a fluctuating AC current. From there, it is necessary to convert the alternating waveform from AC to DC voltage to charge the battery. A rectifier or diode uses only one half of the AC voltage signal thereby allowing the conversion process from alternating to direct current. **Note:** *For troubleshooting of starting and charging systems, refer to Chapter 3.*

3 Some Kohler engines are equipped with auxiliary lighting capability, but without a battery or charging system. In these models, a stator inside the flywheel produces enough voltage for the lighting system only, and the lights can only be used with the engine running.

## Removal and installation

### Regulator

4 Disconnect the negative battery cable.

5 To remove the regulator, mark and disconnect the three electrical connectors from the regulator **(see illustration 3.5)**. **Note:** *Mark the three connectors with masking tape and a pen before disassembly to ensure making the right connections later. On some engines, all three wires are in a single connector that snaps off the regulator/rectifier.*

6 Remove the bolts holding the regulator to the engine shroud or equipment and remove the regulator **(see illustration 3.5)**.

### Stator

7 To remove the stator, refer to Section 8 and remove the flywheel.

8 Disconnect the two stator leads from the regulator, and remove the wiring clamp near where the wires leave the stator.

9 Remove the screws and remove the stator from the engine bearing plate **(see illustration 3.9)**.

10 Installation (regulator and stator) is the reverse of the removal procedure.

## Starting system

11 The starting system consists of a 12 volt battery, battery cables, a starter and ignition switch, safety switches

**3.12 Typical Kohler starter solenoid**

A *Post for battery positive cable and electrical supply to ignition switch*
B *Post for cable to starter motor*
C *Terminal for wire from ignition switch*
D *Mounting bolt*

**3.15 On Command-series OHV engines with Nippondenso starters, remove the two bolts at the rear (A) - (B) is the integral solenoid, mounted by two screws (C)**

and a starter solenoid. The battery functions within both the starting circuit and the charging circuit. The starting system consists of a 12 volt battery, battery cables, a starter and ignition switch, safety switches and a starter solenoid. Be sure the battery is checked and replaced, if necessary, before attempting to diagnose any charging circuit or starting circuit malfunction. **Note:** *For troubleshooting of starting and charging systems, refer to Chapter 3.*

## Removal and installation

### Solenoid

12 The solenoid mounts with only one screw, and is easily removed or installed **(see illustration 3.12)**.

13 Disconnect the negative battery cable, then remove the cable from the battery to the starter solenoid.

14 Remove any other connectors from the solenoid, then remove the mounting screw.

15 On engines with Nippondenso starters, the starter solenoid is mounted on the starter motor itself **(see illustration 3.15)**.

### Starter motor

16 There are several different makes of starter motors used on the various Kohler engine models, but all bolt to the engine with two bolts. The two long starter throughbolts are the mounting bolts on all starters except the Nippondenso and some late-model Eaton starters, which have two separate mounting bolts at the rear **(see illustration 3.15)**.

17 Disconnect the negative battery cable at the battery, and the positive cable from the starter motor (the cable from the solenoid).

18 On all but the Nippondenso and late-Eaton starters, use duct tape or masking tape to hold the commutator end cap in place while you remove the two long mounting bolts **(see illustration 3.18)**. **Note:** *On some models there is a small brace attached between the commutator end of the starter and the engine. Remove this brace. On some models, the oil dipstick tube must be removed to allow full removal of the starter motor.*

19 As soon as the starter is removed from the engine, install two nuts (finger-tight) on the drive-side of the throughbolts to keep them in place. Make matching marks on the case and both ends before disassembling the starter **(see illustration 3.19)**.

20 Installation is the reverse of the removal procedure.

**3.18 Apply masking tape to the commutator end of the starter to hold it in place while the two long throughbolts (A) are removed; also remove the brace (B) - typical starter**

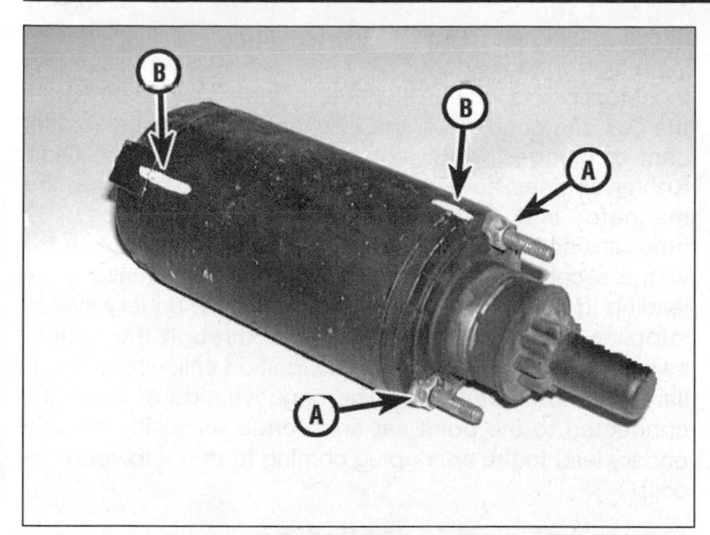


3.19 **Install two nuts (A) to the throughbolts to hold the starter together while it is removed from the engine - make matchmarks (B) to ensure proper reassembly**

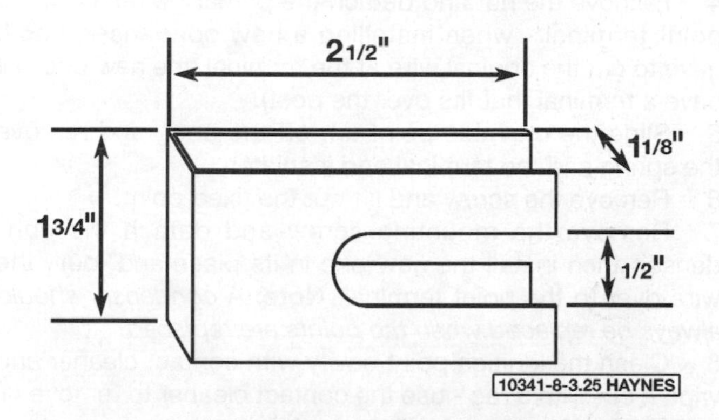


3.25 **Make a tool like this from sheetmetal to ease installation of the end cap with brushes in place**

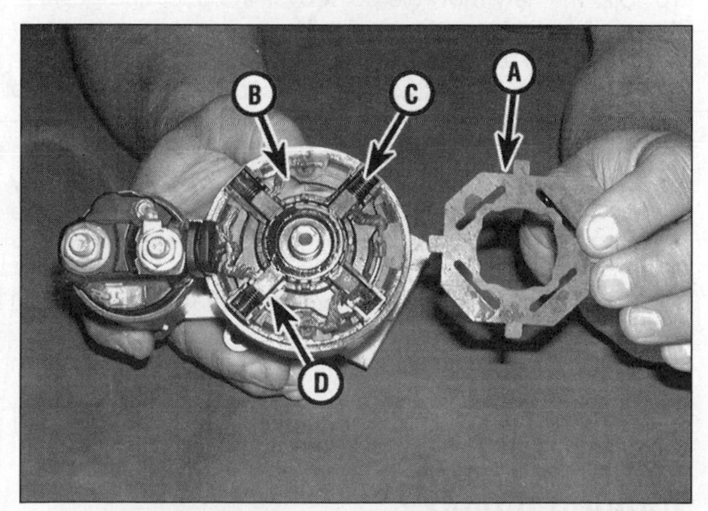


3.26 **Nippondenso starter with end cap and insulator (A) removed - (B) is the brush holder, (C) are the brush springs, and (D) are the brushes**

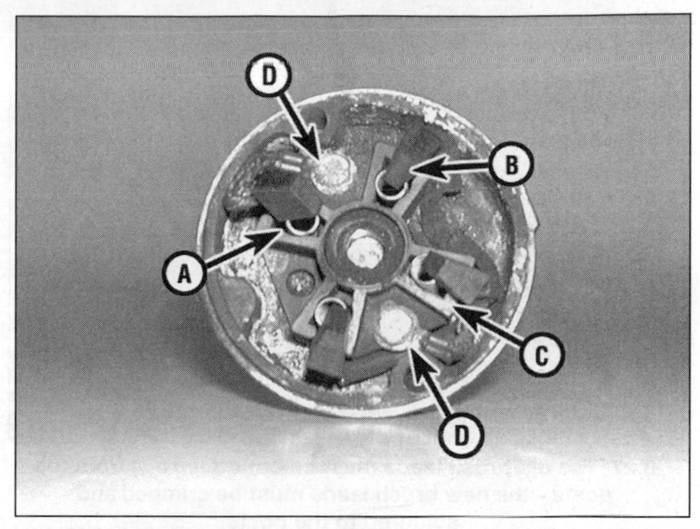


3.24 **Typical starter brushes (all but Eaton and Nippondenso)**

*A Brush springs*
*B Brushes*
*C Brush holder*
*D Self-tapping screws*

## Brush replacement

21 There are two problem areas concerning starter motors. If the motor has been abused (cranked regularly for more than 15 seconds at a time without a cooling off period) or has been in service for many years, the armature or field coils may be damaged. Examination of the starter with the end cap off will indicate it the commutator is worn or damaged. These problems are generally not user-serviced, and a rebuilt or new replacement starter is in order.

22 More common is the general wear of the brushes that ride on the commutator, and this is a normal replacement procedure that can be handled by the home mechanic.

23 The commutator end is the opposite end from the starter drive. Remove the two long throughbolts and remove the commutator end cap.

24 The brushes are contained within the end cap **(see illustration 3.24)**. On Eaton starters, the brushes can't be removed from the end cap, a replacement end cap is available with the new brushes in it. On all other models, the brushes are removed by taking out the self-tapping screws. A new brush kit will includes the brushes and brush springs. **Note:** *All four brushes must be replaced at the same time.*

25 When all the brushes and springs are back in place in the commutator end cap, they must be kept in position while the end cap is slipped over the starter shaft. It is helpful to make a small brush-holder tool from sheetmetal to ease the installation **(see illustration 3.25)**. Slip the tool over the commutator end cap and install the end cap on the starter, slipping the tool out once the cap is started into position, then insert the throughbolts, align the case markings and tighten the bolts with nuts on the drive end.

26 On Nippondenso model starters, the end cap does not contain the brushes, but is really a cover **(see illustration 3.26)**. Remove the throughbolts and cover, then take out the insulator.

**3.27 The old brush leads must be cut off (arrow) from the posts - the new brush leads must be crimped and soldered to the posts**

27 The Nippondenso brushes should be pulled from the brush holder, then remove the brush holder. The brushes are attached to the starter frame and must be cut off with side-cutter pliers. The replacement brushes have ends that must be crimped to the post and then soldered **(see illustration 3.27)**. **Note:** *Clean the posts of any burrs or dirt before soldering the new brush leads on.*

28 With the brushes attached, reinstall the brush holder, align the brushes, then insert the springs. Place the insulator with the same side toward the brushes as it was before, install the end cap and the throughbolts.

29 Starter installation is the reverse of the removal procedure. On models where the throughbolts are also the mounting bolts, remove the nuts you placed on the drive end of the throughbolts before installing the starter.

# 4 Ignition systems

## Breaker points ignition systems

1 The point ignition system, installed on early models and smaller horsepower engines, is also called the "flywheel magneto" ignition system. This mechanical ignition system relies on the opening and the closing of the points to trigger the secondary voltage to the spark plug. The breaker points and the condenser are located under a cover on the governor and operated from a cam on the governor driveshaft. As the flywheel rotates, the magnet mounted on the flywheel passes the coil mounted on the engine case. The magnetic field from the flywheel magnet (magneto) causes a current to generate in the ignition coil primary circuit (winding). As the flywheel continues to rotate, it passes the last pole in the lamination stack on the coil. This produces a large change in the magnetic field ultimately producing high current in the primary circuit. This current change in the primary circuit builds until the points open and the secondary circuit is induced to flow to the spark plug and fire, releasing a high voltage spark.

2 Most points ignition systems include the ignition points, the coil, the condenser, the flywheel magneto, the ignition cam or plunger, the stator plate, and dust cover. Older Kohler engines have used two types of points ignitions, magneto and battery. In the magneto type, the magneto/coil sits above the stator (behind the flywheel), with a secondary lead to the spark plug and smaller wires leading to the point set and condenser. The battery system supplies 12 volts from the battery, through the ignition switch, to the positive side of the ignition coil, which is similar to an automotive coil. The negative side of the coil is connected to the point set and condenser, with the secondary lead to the spark plug coming from the tower of the coil.

## Removal and installation

### Points

3 Release the retainer clip and lift off the point cover and gasket.

4 Remove the nut and detach the primary wires from the point terminal - when installing a new condenser, you'll have to cut the original wire at the terminal (the new one will have a terminal that fits over the post).

5 Slide the movable point up, off the post, and remove the spring and the terminal and insulator.

6 Remove the screw and lift out the fixed point.

7 Remove the mounting screw and detach the condenser, then install the new one in its place and route the wire over to the point terminal. **Note:** *A condenser should always be replaced when the points are replaced.*

8 Clean the ignition point cavity with contact cleaner and wipe it out with a rag - use the contact cleaner to remove oil and dirt from the new ignition point contact faces as well.

9 Install the new fixed point - leave the screw loose enough to allow movement of the plate.

10 Slip the new movable point over the post and position the insulator in the cutout - slip the primary and condenser wires onto the terminal and install the nut.

11 Turn the crankshaft very slowly until the cam opens the movable point as far as possible.

12 Insert a clean feeler gauge - 0.020-inch thick - between the contact points and move the fixed point very carefully with a screwdriver until the gap between the points is the same thickness as the feeler gauge (be careful not to change the position of the movable point as this is done). Point gap can be from 0.017-inch to 0.023-inch, but 0.020 is the best point to adjust them to.

13 Turn the crankshaft and make sure the movable arm opens and closes.

14 Install the gasket and point cover and snap the retainer clip into place.

### Ignition timing

15 All points-type systems require that the engine be timed, either manually or with an electronic timing light.

16 In the bearing plate or on the side of the blower housing, there is a timing inspection hole, and on the outer edge of the flywheel are two marks, a "T" and an "S" (sometimes an "SP"). The T indicates Top Dead Center (TDC), while the S is the mark used for ignition timing.

## Static timing

17 Disconnect and ground the spark plug lead and remove the cover over the ignition points. Looking from the flywheel end of the engine, rotate the engine slowly while observing the points. The S mark should be observable through the timing hole just as the points are opening. This is the exact point in engine revolution at which to set your point gap. You may need a small penlight to see the marks inside the housing.

## Timing light

18 A standard automotive timing light may be used to check your engine's timing dynamically, that is with the engine running, a somewhat more accurate method than the static method.

19 The timing light's clip-type lead simply clamps over the secondary plug lead (with the lead still connected to the plug), while the two alligator clips from the timing light are connected to the positive and negative posts of a 12-volt battery. If your engine/equipment doesn't have battery, use any 12-volt automotive battery. You will have to physically move the battery close enough to the Kohler engine for the timing light leads to reach.

20 Start and warm up the engine, then aim the timing light through the sight hole. At idle speed, you should see the timing marks aligned inside. If not. stop the engine and adjust the breaker plate by loosening the mounting screws and moving the plate slightly. Tighten the screws and recheck the engine timing.

# Electronic ignition systems

21 The electronic ignition system, also called "breakerless" ignition, incorporates solid state electronic components. There are basically two different types of breakerless systems on Kohler engines; the first generation systems used a separate coil (similar to the coil on battery/points systems), while the bulk of modern Kohler engines use a "electronic magneto" ignition. The latter type is found on all models, K-series, Magnum-series and Command-series, and incorporates everything in one magneto/trigger/module unit **(see illustration 4.21)**.

22 This electronic ignition system is more durable than the old points system that commonly fail due to burning, oxidation and mechanical wear, and there is no timing to set. The electronic-magneto system operates similar to a points system but the method used to "break" the electronic signal differs. Flywheel rotation generates electricity in the primary circuit of the ignition coil assembly when the magnets rotate past the charge coil. As the flywheel rotates, it passes a trigger coil where a low voltage signal is produced. The low voltage signal allows the transistor to close allowing the energy stored in the capacitor to fire the secondary circuit or spark plug. The transistor acts as a primary circuit switching device in the same manner as the points in a mechanical (points) ignition system. Spark is produced whenever the engine is rotated, and a tab on the module is connected to either a kill switch or the OFF side of an ignition switch, to ground the module so the engine will stop.

**4.21 Typical Kohler electronic magneto ignition**

A *Secondary lead to spark plug*
B *Air baffle*
C *Lead to kill switch*
D *Ignition module*
E *Module mounting screws*

## Removal and installation

23 Disconnect the spark plug lead from the plug. On models with battery systems, disconnect the negative cable from the battery. **Caution:** *Never connect battery voltage to any wire or component of the electronic-magneto ignition system, or the module will be damaged.*

24 Remove the blower housing and debris screen.

25 At the module, pull the connector from the kill lead tab on the module. Remove the two module mounting screws and remove the module and air baffle.

26 Installation is the reverse of the removal procedure.

27 When installing the module, the gap between the flywheel magnet and the module must be set.

28 Turn the flywheel so that the magnet is directly opposite the module. Loosen the mounting screws slightly and insert a clean (non-magnetic brass) feeler gauge of the correct thickness (0.014-inch for Magnum-series engines, 0.010-inch for Command-series). The magnet will pull the module down to squeeze the feeler gauge between module and magnet. Tighten the module mounting screws.

29 Rotate the flywheel a few revolutions to make sure that the magnet never touches the module's legs. Use feeler gauges to check that both legs of the module are the same distance from the magnet. Correct gap for Magnum-series single-cylinder engine is 0.012-inch to 0.016-inch, and 0.008-inch to 0.012-inch for Magnum-series twin-cylinder and all Command-series engines.

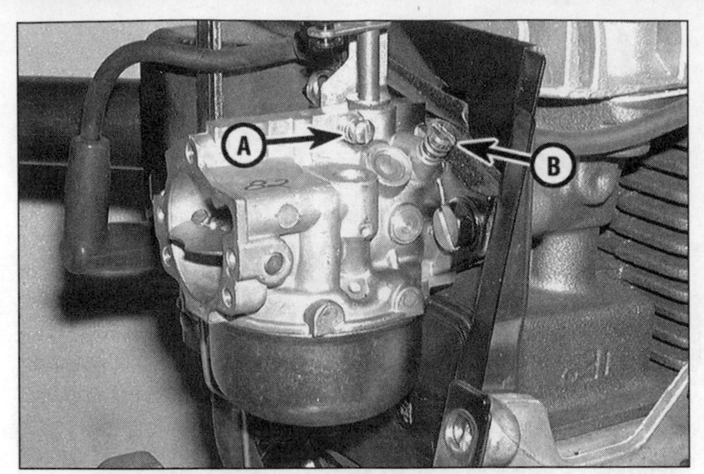

5.2a Typical two-screw Kohler carburetor - (A) is the idle speed screw, (B) is the idle mixture screw

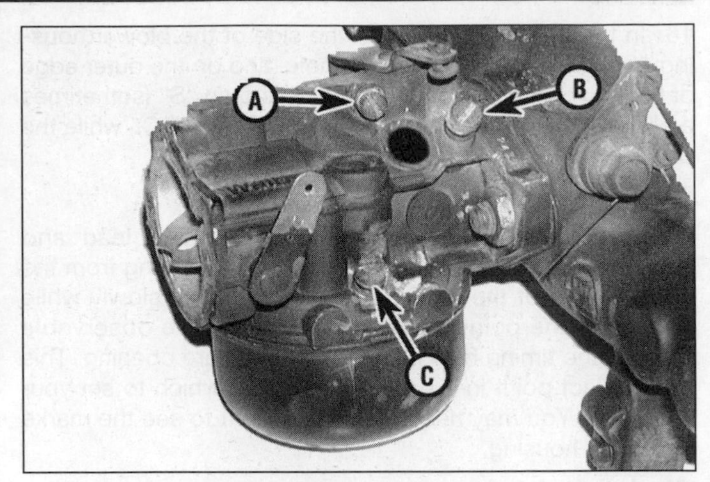

5.2b Typical three-screw Kohler carburetor - (A) is the idle speed screw, (B) is the idle mixture screw, (C) is the high-speed mixture screw

# 5 Carburetor disassembly and reassembly

## General information

1 The following procedures describe how to disassemble and reassemble the carburetor so new parts can be installed. Read the sections in Chapter 5 on carburetor removal and overhaul before doing anything else. In some cases it may be more economical (and much easier) to install a new carburetor rather than attempt to repair the original. Check with a parts distributor to see if parts are readily available and compare the cost of new parts to the price of a complete ready-to-install (standard service) carburetor before deciding how to proceed.

2 Kohler engines have two main types of carburetors, those with two adjusting screws and those with three **(see illustrations 5.2a and 5.2b)**. The former models have adjustable screws for idle speed and idle mixture, while three-screw models also have an adjustable main jet, for controlling the mixture at higher engine speeds. Most carburetors are made by Kohler, but some models have carburetors made by other companies. The other carburetors are built and can be overhauled much like the Kohler carburetors.

## Disassembly

3 The carburetor should be removed from the engine, and notes or sketches made of the linkages for aid in installation. While counting the number of turns, carefully screw the idle mixture adjusting screw in until it bottoms, then remove it along with the spring **(see illustrations 5.2a and 5.2b)**. **Caution:** *Never bottom any carburetor screw with any more than very light force. The needle tips are precisely machined and can easily be damaged, causing irregular engine operation.*

4 Remove the high speed (main) adjustment screw and spring from the fitting in the same manner **(see illustrations 5.2a and 5.2b)**. Counting and recording the number of turns required to bottom the screws will enable you to return them to their original positions and minimize the amount of adjustment required after reassembly.

5 Detach the float bowl and gasket from the carburetor body **(see illustration 5.5)**. On some carburetors the float bowl is held in place with a bolt, while on others, it's held in place with the high speed mixture adjusting screw fitting. Be sure to note the locations of any gaskets/washers used.

6 If you're working on a Walbro carburetor (K-series engine), note how the float spring is positioned before removing the float - it may be a good idea to draw a simple sketch to refer to during reassembly. Push the float pivot pin out of the carburetor body (you may have to use a small punch to do this) **(see illustration 5.6)**.

7 Remove the float assembly and the inlet needle valve. Note how the inlet needle valve is attached to the float - the retainer clip, if used, must be positioned the same way during reassembly.

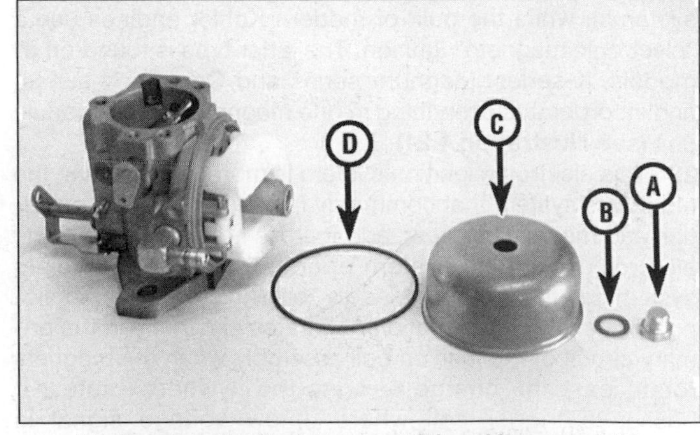

5.5 Remove the float bowl nut (A) and its gasket (B), bowl (C) and bowl gasket (D)

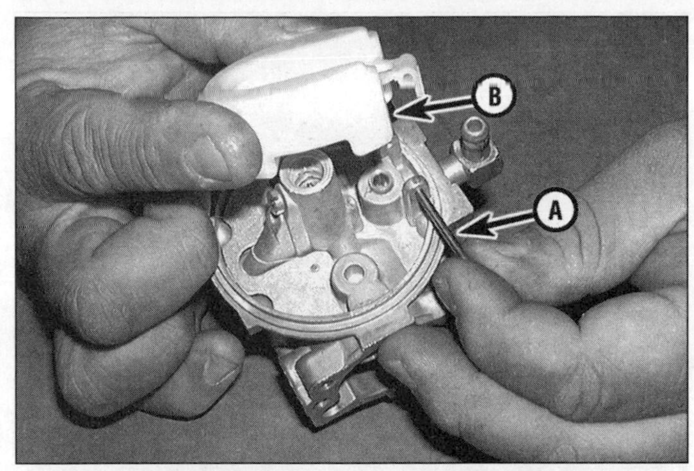


5.6 **Remove the pivot pin (A) and lift straight up on the float to remove it, along with the inlet needle valve (B)**

5.8 **The needle seat (arrow) must be pulled out and a new one pressed in**

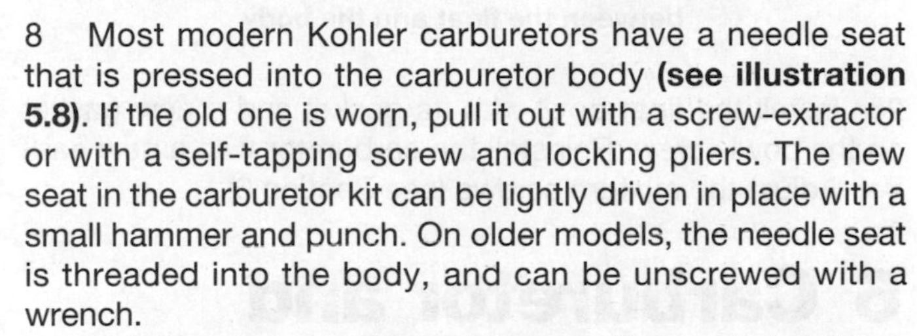

8 Most modern Kohler carburetors have a needle seat that is pressed into the carburetor body **(see illustration 5.8)**. If the old one is worn, pull it out with a screw-extractor or with a self-tapping screw and locking pliers. The new seat in the carburetor kit can be lightly driven in place with a small hammer and punch. On older models, the needle seat is threaded into the body, and can be unscrewed with a wrench.

9 This is as far as most carburetors will need to be disassembled for overhaul, and a standard Kohler carburetor overhaul kit will contain all the parts necessary to restore like-new operation. Inspect the throttle shaft and choke shaft for wear. If the shafts are not loose in the carburetor body, leave them in place. If they are loose, you will need a shaft replacement kit, which includes new throttle and choke shafts, bushings and tools required to install the new bushings.

10 To remove the throttle or choke shafts, make a paint mark on the plate and shaft for alignment purposes in reassembly. Remove the small screws holding the choke or throttle plate to its shaft. **Caution:** *Go slowly and carefully to avoid stripping the tiny screws. Make sure you have the correct-size screwdriver tip that fits snugly*. Slide the shaft out **(see illustration 5.10)**.

11 Follow the instructions included with the shaft/bushing kit. The carburetor body must be held in a vise under a drill press and the shaft bores drilled out precisely. The new bushings should be coated with thread-locking compound and pressed into the carburetor body with a vise, while aligned with long screws and a tool included in the kit. Check the alignment with the old throttle shaft in place while the thread-locking compound sets up. Upon reassembly, install the new shaft(s) and check for smooth operation.

12 To do a thorough cleaning job, remove any Welsh plugs from the carburetor body. Drill a small hole in the center of the Welsh plug and pry it out with a punch or thread a sheet metal screw into the hole, grasp the screw head with a pliers and pull the plug out **(see illustration 5.12)**.

13 Refer to Chapter 5 and follow the cleaning/inspection procedures outlined under Carburetor overhaul. **Note:** *Don't soak plastic or rubber parts in carburetor cleaner. If the float is made of cork, don't puncture it or soak it in carburetor cleaner.*

5.10 **Remove the choke plate screws (arrows) to remove the choke shaft (A)**

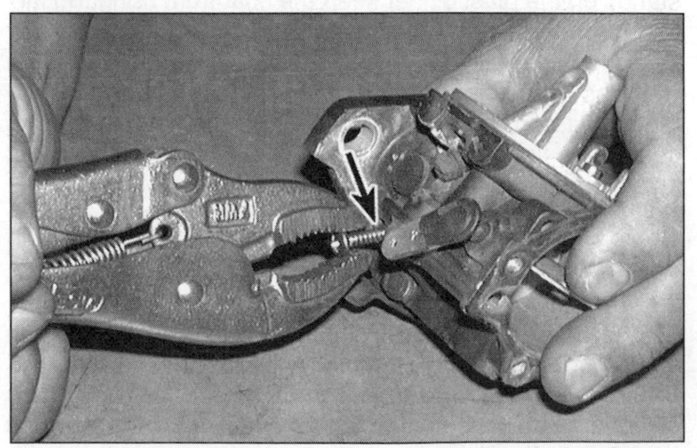

5.12 **To remove a Welsh plug, drill a 1/8-inch hole in it, thread in a self-tapping screw (arrow), then pull it out with locking pliers - don't damage the bore in the process**

## Reassembly

14 Check each mixture adjusting screw tip for damage and distortion. The small taper should be smooth and straight.

15 If the throttle or choke shaft binds after the new parts are reinstalled, loosen the screw and reposition the plate on the shaft. If dust seals are used on the shafts, they should be positioned next to the carburetor body.

16 Check the inlet needle valve and seat. Look for nicks and a pronounced groove or ridge on the tapered end of the valve. If there is one, a new needle and seat should be used when the carburetor is reassembled. They are normally installed as a matched set. Be sure to install a new gasket as well.

17 Check the float pivot pin and the bores in the carburetor casting, the float hinge bearing surfaces and the inlet needle tab for wear - if wear has occurred, excessive amounts of fuel will enter the float bowl and flooding will result.

18 Shake the float to see if there's gasoline in it. If there is, install a new one.

19 Once the carburetor parts have been cleaned thoroughly and inspected, reassemble it by reversing the above procedure. Note the following important points:

a) *Make sure all fuel and air passages in the carburetor body, main nozzle, inlet needle seat and float bowl mounting bolt are clean and clear.*
b) *Be sure to use new gaskets, seals and O-rings. Whenever an O-ring or seal is installed, lubricate it with a small amount of grease or oil.*
c) Don't overtighten small fasteners or they may break off.

20 When installing new Welsh plugs, apply a small amount of non-hardening sealant to the outside edge and seat the plug in the bore with a 1/4-inch or larger diameter pin punch and a hammer. Be careful not to collapse the plug - flatten it just enough to secure it in the opening.

21 When the inlet needle valve assembly is installed, be sure the retaining clip is attached to the float tab, and that a new baffle gasket is in place before installing the float. Most modern Kohler carburetors have no float adjustment, small stands on the plastic float keep the float from rising too far.

22 On models with adjustable floats, invert the carburetor and check the float level **(see illustration 5.22)**. There are two tabs on the hinge end of the float, one rides on the needle/seat assembly and the other is on the outside and fits against the float pin to limit float drop. If the float drop is not as specified, gently bend the float tab with needlenose pliers. Go carefully, only small adjustments are necessary, and do not put pressure on the needle/seat while bending the tab.

23 Turn the carburetor right-side up and check for proper float drop **(see illustration 5.23)**. Adjust the exterior float tab to control the float drop.

24 At the hinge side of the float, check for 0.010-inch clearance between the float and the float hinge tower when the carburetor is inverted. If the feeler gauge won't fit, remove the float and file the towers. After filing, thoroughly clean the carburetor of any filings.

**5.22 With the carburetor inverted, the float level can be checked by inserting a 11/32-inch drill bit between the float and the body**

25 Install the float bowl with its gasket and a new gasket on the bowl screw. Reinstall the carburetor with a new gasket. Adjust the mixture screws (see Section 6).

# 6 Carburetor and governor adjustment

## Carburetor

1 When making carburetor adjustments, the air cleaner must be in place and the fuel tank should be at least half full.

2 If not already done, turn the idle and high speed FUEL/AIR MIXTURE adjusting screws clockwise until they seat lightly, then back them out the same number of turns as you recorded during disassembly. A chart of suggested initial settings is provided here as a good starting point before making running adjustments **(see illustration 6.2)**.

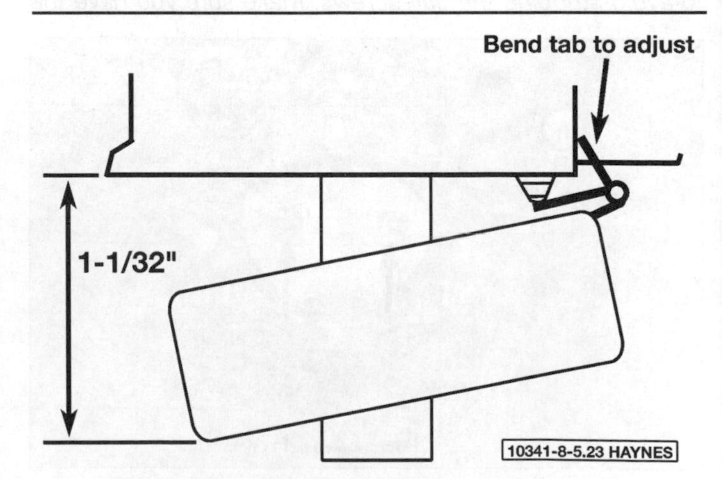


**5.23 Adjust the exterior float tab (arrow) to adjust for proper float drop**

| ENGINE MODEL | MAIN MIXTURE SCREW | IDLE MIXTURE SCREW |
|---|---|---|
| K141, K161 w/o Walbro carb | 3 turns | 1-1/2 turns |
| K141, K161 with Walbro carb | 3/4 turn | 2-1/2 turns |
| K241, K301 w/o Walbro carb | 2 turns | 2-1/2 turns |
| K241, K301 with Walbro carb | 1-1/8 turns | 1-3/4 turns |
| K321 | 3-1/4 turns | 2-1/2 turns |
| K341 w/o Walbro carb | 3-1/2 turns | 2-1/2 turns |
| K341with Walbro carb | 1-1/4 turns | 1-1/4 turns |
| KT17, KT19 w/o Walbro carb | 2-1/2 turns | 1 turn |
| KT17, KT19 with Walbro carb | 1 turn | 1-1/4 turns |
| K482, K532 | 3 turns | 1-1/4 turns |
| M8 | 2 turns | 1-1/4 turns |
| M10, M12 | 1-1/2 turns | 2-1/2 turns |
| M14, M16 | 2-1/2 turns | 2-1/2 turns |
| MV16, MV18 | 1-1/4 turns | 1-1/4 turns |
| MV20 | 1-1/4 turns | 1 turn |
| M18, M20 | 2-1/2 turns | 1 turn |
| CH11, CH12.5 | 11/2 turns | 1-1/4 turns |
| OHV CH14 | 1-1/4 turns | 1-34 turns |
| CV11, CV12.5, CV14, CV15, CV18, CV20 | N/A | 1 turn |
| CH18, CH20 | N/A | 1-1/2turns |

**6.2 Initial carburetor mixture screw adjustments**

3 Start the engine and allow it to reach operating temperature before making the final adjustments.
4 With all carburetor mixture adjustments, the ideal setting point is halfway between the rich and lean positions.
5 Start with the main or high-speed fuel mixture (on models with adjustable main jets). The engine should be run at full throttle, and preferably under some load, such as mowing, pumping, snow-blowing, etc. In small increments, turn the main fuel screw out until the engines runs slower, this is the rich position. Adjust the screw inward until the engine speed increases again (to where it started) and then begins to decrease. This is the lean position. Set the screw to halfway between the rich and lean positions.
6 Place the throttle control lever in the SLOW or IDLE position, then adjust the idle mixture screw in the same manner as in Step 5, except not under load. Leave the idle mixture screw halfway between the rich and lean positions.
9 If the engine smokes under load, this indicates the high-speed mixture is too rich, and if it backfires under load, this indicates the high-speed mixture is too lean. If the engine was adjusted in warm weather, but stops frequently in cold weather, try adjusting the main fuel mixture slightly richer.
10 Move the speed control lever from SLOW to FAST - the engine should accelerate without hesitating or sputtering.
11 If the engine dies, it's too lean - turn the adjusting screws out in small increments.
12 If the engine sputters and runs rough before picking up the load, it's too rich - turn the adjusting screws in slightly.
13 If the adjustments are "touchy", check the float level and make sure it isn't sticking.

# Governor

14 It is best not to change the relationship of any linkages or other components in the governor system, unless a worn or broken component is to be replaced. If the engine is being overhauled, do not change any adjustments. If the governor was working suitably before the overhaul, and none of the connections are altered, it will continue to work properly. While governor adjustment should be made with a small-engine tachometer to avoid setting the engine to a higher rpm than it is designed for, some basic adjustments can be made by the home mechanic.
15 On most engines, a governor shaft sticks through the side of the crankcase, while the governor mechanism is protected inside the crankcase. An arm attaches to this shaft with a clamp-bolt/nut. The basic initial adjustment on governor linkage begins with loosening the nut/bolt where the arm attaches to the shaft **(see illustration 6.15)**. Pull the arm back away from the carburetor to its limit and hold it there. Grip the end of the shaft that sticks out beyond the arm with pliers and turn it counterclockwise to its limit. Tighten the bolt/nut on the arm while the arm and shaft are in this relationship.
16 Sensitivity is the other common adjustment for governors. If the engine seems like it is hunting for consistent speed as the load changes, the sensitivity is too sensitive and must be lowered. If the engine speed drops too much when the engine encounters a load, the governor adjustment isn't sensitive enough.
17 The governor sensitivity is adjusted by moving the position of the governor spring, where it hooks into the lower portion of the governor arm **(see illustration 6.17)**. A series of holes are provided. To increase the sensitivity, raise the spring into the next higher hole in the arm. To decrease sensitivity, hook the spring into a hole lower than its original position. The center of this arm is the original position, but varying equipment and loads are accommodated by the extra adjustment holes.

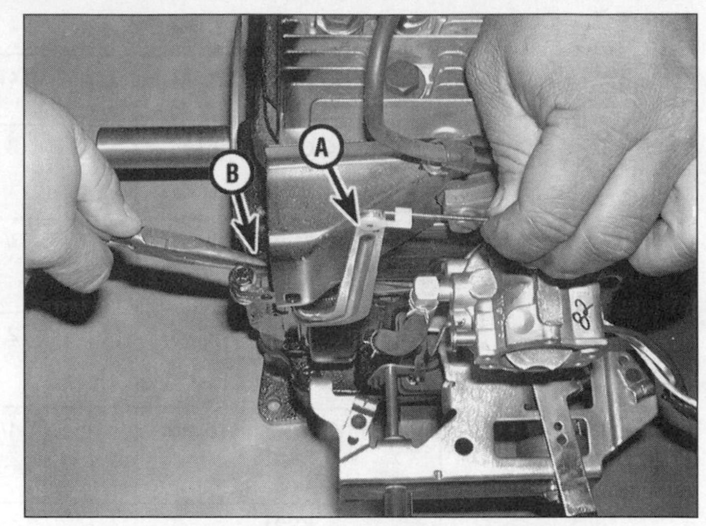

**6.15 Pull back the governor arm (A) while turning the governor shaft end (B) counterclockwise**

**6.17 The governor sensitivity is controlled by changing the position of the governor spring end in the holes in the governor arm (arrow)**

# 7 Valve clearance - check and adjustment

1 Correct valve tappet clearance is essential for efficient fuel use, easy starting, maximum power output, prevention of overheating and smooth engine operation. It also ensures the valves will last as long as possible.

2 When the valve is closed, clearance should exist between the end of the stem and the tappet. The clearance is very small - measured in thousandths of an inch - but it's very important. The recommended valve clearance is listed in the Specifications at the end of this Chapter. Note that intake and exhaust valves often require different clearance values. **Note:** *The engine must be cold when the valve clearance is checked.* A feeler gauge with a blade thickness equal to the valve clearance will be needed for this procedure.

3 On some engines, if the clearance is too small, the valves will have to be removed and the stem ends ground down carefully and lapped to provide more clearance (this is a major job, covered in the overhaul and repair procedures. If the clearance is too great, new valves will have to be installed (again, a major repair procedure). On many other engines, the tappets are adjustable, and with two wrenches, the clearance can be set to Specifications. **Note:** *Kohler OHV engines feature hydraulic lifters, so valve adjustment is not necessary.*

4 Disconnect the wire from the spark plug and ground it on the engine.

5 Remove the bolts and detach the tappet cover plate or the crankcase breather assembly **(see Section 8)**. **Note:** *On some models the crankcase breather is behind the carburetor, so the carburetor will have to be removed first.*

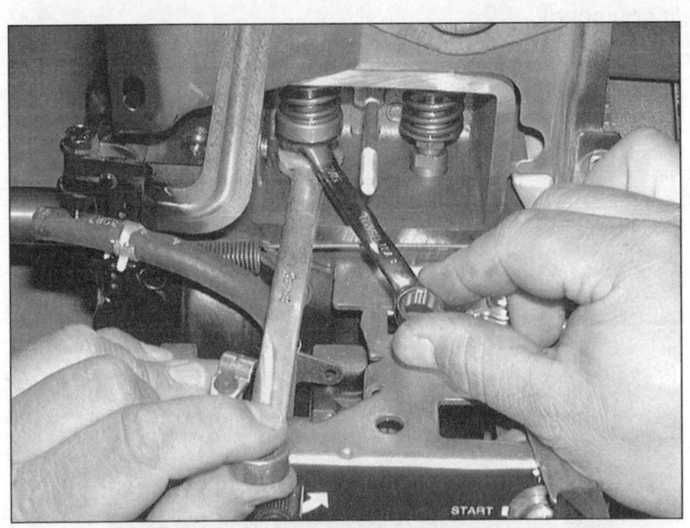


**7.9 Tappets on some models are adjusted by holding the body of the tappet with one wrench, while using another wrench to turn the tip of the tappet in or out**

6 Turn the crankshaft by hand and watch the valves to see if they stick in the guide(s).

7 Turn the crankshaft until the intake valve is wide open, then turn it an additional 360-degrees (one complete turn). This will ensure the valves are completely closed for the clearance check. **Note:** *On twin-cylinder engines, turn the crankshaft until the no. 1 piston is at TDC (the no. 1 cylinder is closest to the flywheel), with both valves closed. Check the valve clearance on cylinder no.1, then rotate the crankshaft 360 degrees and check the valves for cylinder no. 2.*

8 Select a feeler gauge thickness equal to the specified valve clearance and slip it between the valve stem end and the tappet **(see Chapter 4)**.

9 If the feeler gauge can be moved back-and-forth with a slight drag, the clearance is correct. If it's loose, the clearance is excessive; if it's tight (watch the valve to see if it's forced open slightly), the clearance is inadequate. On Magnum engines, and other models with adjustable tappets, use two wrenches on the tappet to either extend or retract it to achieve the proper valve clearance **(see illustration 7.9)**.

10 If the clearance is incorrect, refer to Chapter 5 for valve service procedures.

11 Reinstall the crankcase breather or tappet cover plate, using a new gasket if necessary.

# 8 K-series and Magnum engines

## Single-cylinder engines

The models covered by this Section include the following:

**K-series and Magnum L-head single-cylinder engines**

| | |
|---|---|
| K141 | 6 HP |
| K161, M8 | 8 HP |
| K241, M10 | 10 HP |
| K301, M12 | 12 HP |
| K321, M14 | 14 HP |
| K341, M16 | 16 HP |

### Disassembly

The engine components should be removed in the following general order:

*Oil filter*
*Engine cover*
*Air filter*
*Fuel tank*
*Cooling shroud/recoil starter*
*Carburetor/intake manifold*
*Muffler*
*Electrical components*
*Fuel pump*
*Cylinder head*

**8.1 Begin disassembly by removing the exterior components**

*A Air cleaner* *C Muffler*
*B Oil drain plug* *D Dipstick tube*

*Flywheel*
*Flywheel brake components (if equipped)*
*Ignition components*
*Crankcase breather assembly*
*Oil pan*
*Piston/connecting rod assembly*
*Crankshaft and bearing plate*
*Camshaft*
*Tappets*
*Valves*
*Governor components*

1 For fuel tank, carburetor/intake manifold, muffler and oil draining procedures, refer to Chapters 4 and 5 as necessary **(see illustration 8.1)**. For the recoil assembly removal (if equipped), refer to Section 2.

**8.3 Fuel tank mounting bolts (arrows)**

2 Disconnect the ignition "kill wire" from the ignition module and remove the mounting bolt. The engine should be removed from the equipment, with the battery (on models so equipped) disconnected.

3 Drain and remove the fuel tank **(see illustration 8.3)**. On some models, there may be a fuel shutoff valve at the bottom of the tank. If so, simply shut off the fuel valve before removing the tank, there is no need to drain this type. On many models, the starter solenoid is attached to the bottom of the fuel tank and just be removed to access the fuel tank mounting bolts.

4 Look at the arrangement of throttle, choke and governor linkages on the carburetor before disconnecting anything **(see illustration 8.4)**. If necessary, make a sketch to aid in reassembly.

5 Disconnect the fuel line to the carburetor and remove the carburetor.

6 Refer to Section 3 and remove the starter and solenoid (if equipped).

**8.4 Disconnect the choke (A), throttle linkage (B) and governor linkage (C), before removing the carburetor (arrows indicate carburetor mounting bolts)**

**8.7a Remove the breather assembly (A), the camshaft gear cover (B), and the fuel pump (C) - (D) is a control panel containing choke and ignition switch (not on all models)**

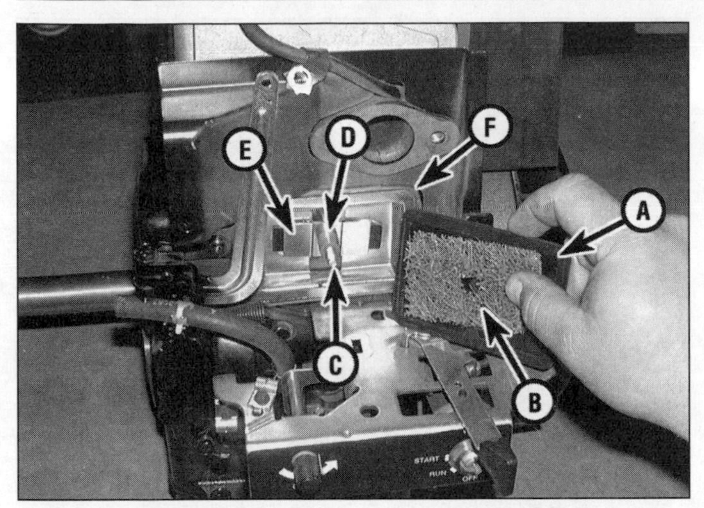


**8.7b Crankcase breather components**

| | | | |
|---|---|---|---|
| *A* | *Cover and gasket* | *D* | *Seal* |
| *B* | *Filter* | *E* | *Reed stop* |
| *C* | *Stud* | *F* | *Breather plate* |

7 With the carburetor off, there is better access to remove the crankcase breather assembly, camshaft gear cover and fuel pump **(see illustration 8.7a)**. Once the breather cover is off, remove the mounting stud, reed, gasket and breather plate **(see illustration 8.7b)**.

8 Remove the bolts holding the cooling shrouds and blower housing **(see illustration 8.8)**. Disconnect the electrical connector from the regulator/rectifier on the blower housing. The regulator may be left mounted on the blower housing, but take care when setting the housing aside that the regulator isn't damaged. On the carburetor side of the engine, remove the screw holding the spark plug wire to the cooling shroud before removing the shroud. **Note:** *On K-series engines, there may be an ignition coil mounted on the blower housing. Mark and disconnect the wires connected to it and leave it on the housing.*

9 Unbolt the ignition module and its air baffle from the engine, mark and disconnect any electrical connectors, and

**8.8 Remove the bolts/screws and take off the blower housing and shrouds - arrows indicate screws on carburetor side, two others are on the opposite side**

remove the module with the secondary spark plug lead **(see illustration 4.21)**.

10 Remove the cylinder head bolts, keeping note of the length of the bolts and where they go. On some models, a fuel tank upper bracket is under two of the head bolts, remove it with the bolts. Remove the cylinder head.

11 Remove the plastic or metal debris screen from the fan. Hold the flywheel using a flywheel-strap tool and remove the center bolt with a breaker bar and socket **(see illustration 8.11)**. On models with a rope starter, remove the bolt, then take off the washer and starter drive cup. After the nut is removed, the starter cup and debris screen can be detached. **Caution:** *The bolt is under considerable torque. Make sure your socket is squarely on the bolt head - slipping off under pressure could cause some injury.*

12 The flywheel can be removed with a standard automotive puller **(see illustration 8.12)**. Never pry behind the flywheel to remove it.

13 Place the flywheel upside-down on a wooden surface

**8.11 Hold the flywheel with a strap wrench while removing the flywheel center bolt**

**8.12 Use an automotive puller with two bolts into threaded holes in the flywheel to remove the flywheel**

8.14 Remove the fan-to-flywheel bolts (arrows)

8.17 Oil pan bolts (arrows indicate three that can be seen here, there are four)

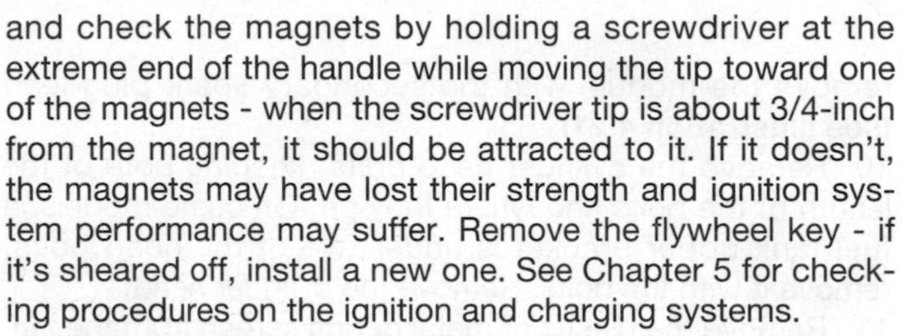

and check the magnets by holding a screwdriver at the extreme end of the handle while moving the tip toward one of the magnets - when the screwdriver tip is about 3/4-inch from the magnet, it should be attracted to it. If it doesn't, the magnets may have lost their strength and ignition system performance may suffer. Remove the flywheel key - if it's sheared off, install a new one. See Chapter 5 for checking procedures on the ignition and charging systems.

14 If the fan is damaged, remove the bolts securing it to the flywheel and install a new one **(see illustration 8.14)**. **Note:** *Don't loose the spacers that go with each fan-to-flywheel bolt.*

15 Refer to Section 3 and remove the stator and its wiring harness from the engine's bearing plate.

16 Rotate the engine until the piston is at the top and both valves are closed. Use a valve spring compressor to compress the spring just enough to remove the keepers, then slowly release to tool to remove the valve spring retainers and springs (see Chapter 5). **Note:** *There are upper retainers (case side) and lower retainers (tappet end).*

17 With the engine still at TDC, remove the bolts holding the oil pan to the case **(see illustration 8.17)**. Remove the oil pan and its gasket.

18 Turn the engine over on its side on the bench to work on the bottom end. Apply match marks to the connecting rod and cap. Remove the two connecting rod nuts and the rod cap with oil dipper **(see illustration 8.18)**.

19 Check the top of the cylinder with your finger for a ridge, and use fine emery paper to remove it if present. Using a wooden hammer handle, push the rod and piston up and out of the engine from the bottom.

20 Refer to Chapter 5 for piston/rod inspection and measuring procedures. If it is determined that the piston is worn beyond Specifications or otherwise damaged, use snap-ring pliers to remove the piston pin snap-rings, then slide the piston pin out and separate the piston from the rod (see Chapter 5). A new piston generally comes with a new pin and snap-rings.

21 Remove the bolts holding the bearing plate, then use a

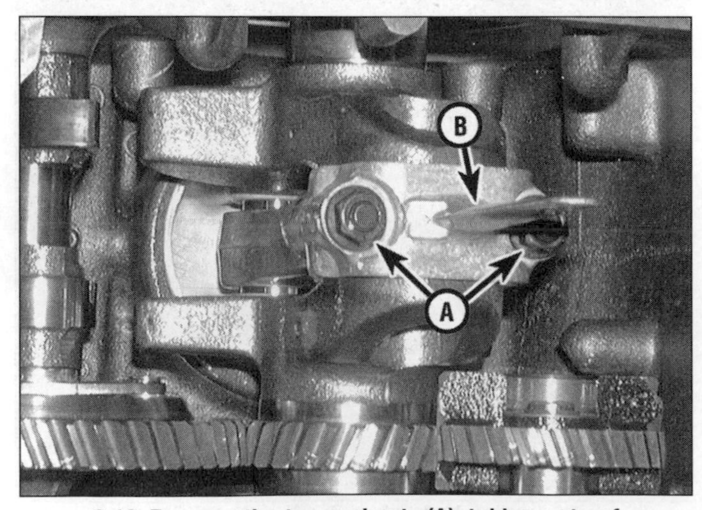


8.18 Remove the two rod nuts (A), taking note of how the oil dipper (B) faces

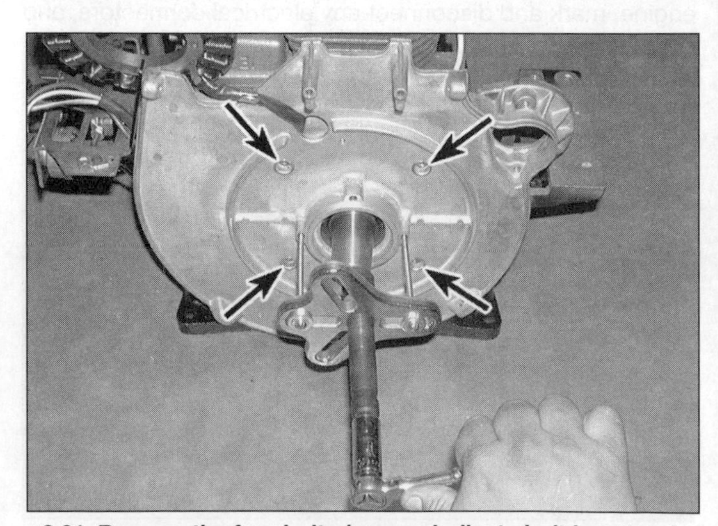

8.21 Remove the four bolts (arrows indicate bolt locations), then use a puller to remove the bearing plate

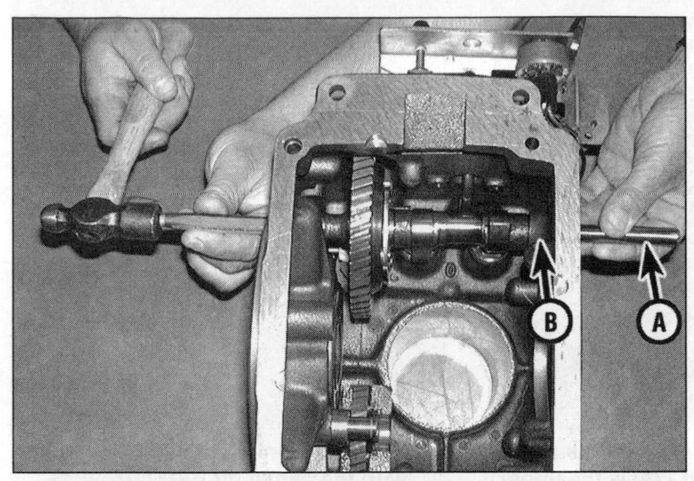


8.23 Drive the camshaft pin (A) out of the case with a hammer and long punch - keep track of the endplay shim stack (B) when removing the camshaft from the crankcase

8.24a Mark, then remove, the two tappets - I is intake, E is exhaust

two-bolt puller to remove the plate **(see illustration 8.21)**. Do not use a prybar behind the plate to remove it.

22 Slide the crankshaft out of the crankcase and set it aside.

23 The camshaft is retained in the crankcase by a long pin inside, which is covered by a cup plug pressed into the flywheel side of the crankcase. Use a long punch , from the PTO side of the crankcase, to drive the camshaft pin (and its cup plug) out the other side **(see illustration 8.23)**. Then remove the camshaft from the crankcase, with the camshaft endplay shims, from the flywheel side of the case.

24 After the camshaft is removed, mark the tappets (intake is the closest one to the flywheel side) and store them in marked containers so they can be returned to their original locations **(see illustration 8.24a)**. **Note:** *Some engines are equipped with ACR (Automatic compression Release) which is a mechanism that reduces compression during low engine rpm, to make starting easier.* Normal compression is resumed once the engine starts. The mechanism consists of a pair of flyweights and a spring **(see illustration 8.24b)**. Inspect these components during disassembly.

25 Do not separate the lever from the governor shaft unless new parts are needed.

26 All Magnum engines and most late model K-series engines have two balance gears. These mesh with the crankshaft gear and increase engine smoothness. To remove them, use snap-ring pliers to remove the retaining rings **(see illustration 8.26)**. When the retaining rings and upper shims are removed, the balance gears can be removed. In some models, needle bearings inside the bore of the balance gears are caged, but on models where they are not, the needle bearings will fall out into the case when the gears are pulled out. Use a magnet to pick them all up. Count to make sure you have them all, there are 27 to each of the two balance gears. Keep track of the location of all shims (above and below the gears) and spacers, and mark the components for proper reassembly. Keep the bearings, shims and spacers in a zip-lock plastic bag, marked for either the upper or lower balance gear.

8.24b ACR mechanism on camshaft gear

*A Flyweights*
*B Tab that raises exhaust valve off its seat*

8.26 Remove the snap-ring (A), shim (B) and balance gear - (C) indicates lower balance gear removed from its shaft - then the spacer (D) from each balance gear shaft

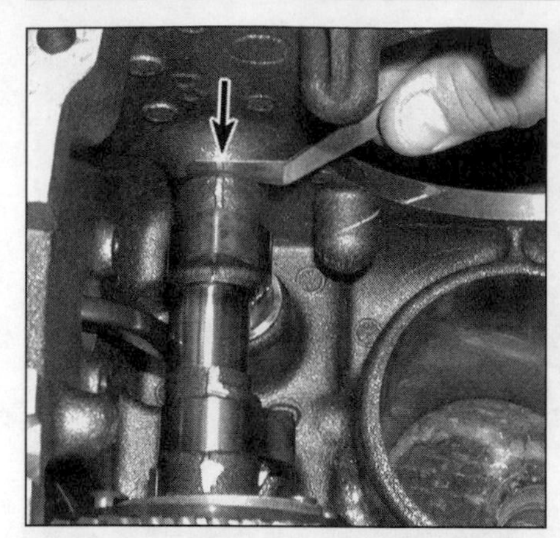

8.33 With the camshaft pin inserted through the case, through the shims and part way into the cam, measure the endplay with feeler gauges between the case and spacer (arrow)

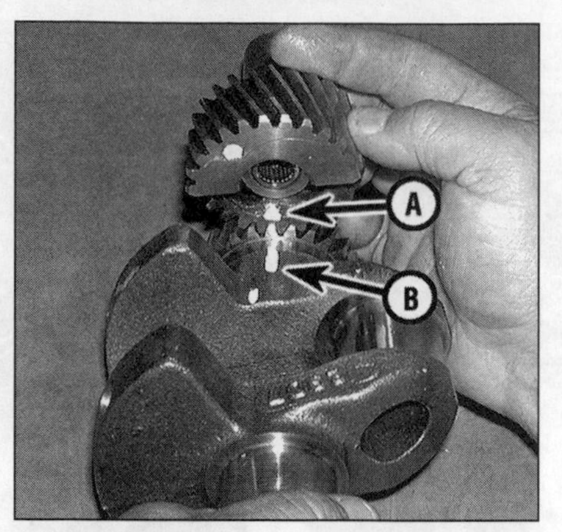


8.36a Align the two timing marks as the crankshaft is installed - (A) is the upper balance gear, (B) is the crankshaft gear (shown removed from the engine for clarity)

8.36b The crankshaft must also align with the camshaft gear (arrows) - turn the camshaft to align the gears, not the crankshaft

## Inspection

27 After the engine has been completely disassembled, refer to Chapter 5 for the cleaning, component inspection and valve lapping procedures.

28 After measuring both the cylinder bore and the piston (see Chapter 5), compare the results to Specifications. If the piston-to-bore clearance is within Specifications, the piston can be reused with new rings, and the bore honed for proper ring seating. If the cylinder is worn, it must be rebored at a machine shop and a new oversize piston used.

29 Inspect the valves, valve guides and valve seats (see Chapter 5). The valves and seats are generally ground during any overhaul, and the work should be done by a competent automotive or small-engine machine shop.

30 Inspect all of the bearings and measure the inside diameter of the connecting rod and compare it to the crankshaft journal size. If the clearance is beyond Specifications, a new rod must be purchased. If crankshaft bearings are worn, they must be pressed off, either from the crankshaft or the bearing plate. In either case, these bearings should be replaced at a shop with a hydraulic press. Never hammer on the bearings, either to install or remove.

31 Once you've inspected and serviced everything and purchased any necessary new parts, which should always include new gaskets and seals, reassembly may begin.

## Reassembly

32 Drive the old oil seal from the bearing plate with a punch and small hammer, being careful not to nick the seal bore in the plate (see Chapter 5). For removal, the plate should be supported with two pieces of 2x4 wood under the plate, close to the seal bore. New oil seals can be installed using a seal driver, or a large socket of the appropriate diameter (see Chapter 5). Drive the seal in squarely and only to the depth listed in the Specifications. The oil seal at the PTO side of the engine can be removed with a screwdriver or seal-removal tool.

33 Turn the case so that the bottom is up. Lubricate the tappets with engine oil and install them in the case **(see illustration 8.24a)**. Lubricate and install the camshaft in the case. Insert the long camshaft pin through the case (from the bearing plate side) and into the camshaft. The original shim stack should be in place between the case and the end of the camshaft. Measure the clearance (endplay) between the camshaft and the shims with feeler gauges **(see illustration 8.33)**. Add new shims as required until the endplay is within Specifications.

34 With the clearance set, drive the camshaft pin in until the end of the pin is 0.275 to 0.285-inch (K161, K181 and M8) or 0.300 to 0.330-inch (all except K161, K181 and M8) from the face of the case (bearing plate side). Using a hammer and punch, drive a new cup plug in over the pin, until the plug is flush to 0.030-inch from the edge of the case. **Note:** *Some older K-series models may be equipped with a longer camshaft pin and no cup plug, if so equipped, drive the pin in until flush with the case.*

35 Install the top (furthest from the oil pan) balance gear onto its stub shaft (with any spacer in place), then the shim and snap-ring. If they are not caged, use heavy grease to hold the needle bearings in place inside the balance gear during installation.

36 Place the crankshaft into the case and start it into the rear bearing. As it is going in, mesh the crankshaft and upper balance gears such that the two timing marks just start to align **(see illustrations 8.36a and 8.36b)**. At the same time, the crankshaft gear must begin alignment with the camshaft gear.

37 Push the crankshaft all the way into the rear bearing. Place the spacer and shim on the stubshaft for the lower balance gear (the one closest to the oil pan). Rotate the crankshaft until it is past Bottom Dead Center (BDC) by about 15 degrees. Now slip the lower balance gear onto its

8.37 When the lower balance gear is installed, align its secondary mark (A) with the primary mark on the crankshaft (B)

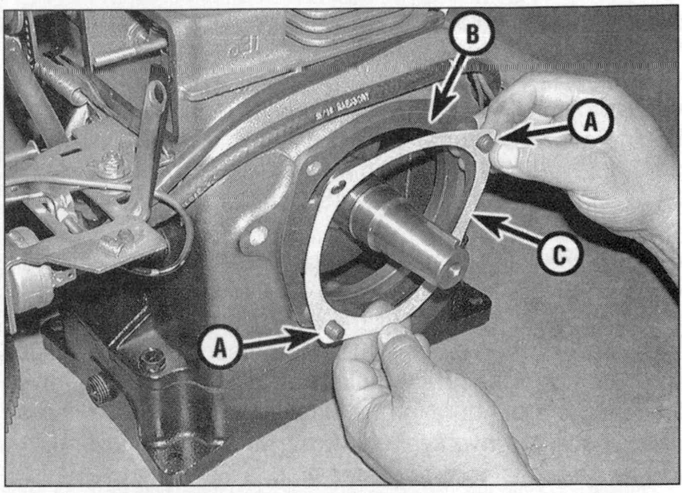

8.38 With two studs (A) used for alignment, install the crankshaft endplay shims (B) and a new gasket (C)

stubshaft, while aligning its secondary timing mark with the primary mark on the crankshaft **(see illustration 8.37)**.

38 Install two long studs in two of the bearing plate mounting holes on the case. Install the original crankshaft endplay shims over the studs, with a new gasket **(see illustration 8.38)**. Install the bearing plate with two of the bolts, then remove the studs and install the other bolts, tightening them to Specifications.

39 The gasket and shims control the endplay of the crankshaft, which can only be measured after the bearing plate is tightened in place. Use feeler gauges to measure the endplay and compare the results to Specifications **(see illustration 8.39)**. Add or remove shims (using a new gasket each time) to correct the crankshaft endplay. Lube the crankshaft and bearings liberally with new engine oil.

40 Drive the new crankshaft oil seal in on the PTO end of the engine, using an appropriate-size socket and hammer (see Chapter 5). Drive the seal in squarely and to the depth listed in the Specifications.

41 Install the piston and connecting rod (see Chapter 5), gently tapping the assembly down until the connecting rod cap can be bolted to the rod at the crankshaft journal. Make sure the match marks made earlier align. Tighten the rod cap nuts/bolts to the torque listed in this Chapter's Specifications. **Caution:** *Make sure the oil dipper is oriented as it was originally* **(see illustration 8.18)**. **Note:** *On engines with Posi-Lock rods (rod caps that have sleeves that extend into the rod), tighten as per Specification* **(see illustration 8.41)**. *On all other rods, tighten the rod cap fasteners to Specifications, then loosen them, then retighten to Specifications.*

42 Several piston designs have been used on Kohler engines. The top of the piston should have either an arrow or the letters "fly" stamped in, indicating that the piston/rod should be installed with this side of the piston toward the flywheel. If the piston has no markings on top, it can be installed either way.

8.39 Measure the crankshaft endplay with feeler gauges (arrow) between the crankshaft and the endplate bearing

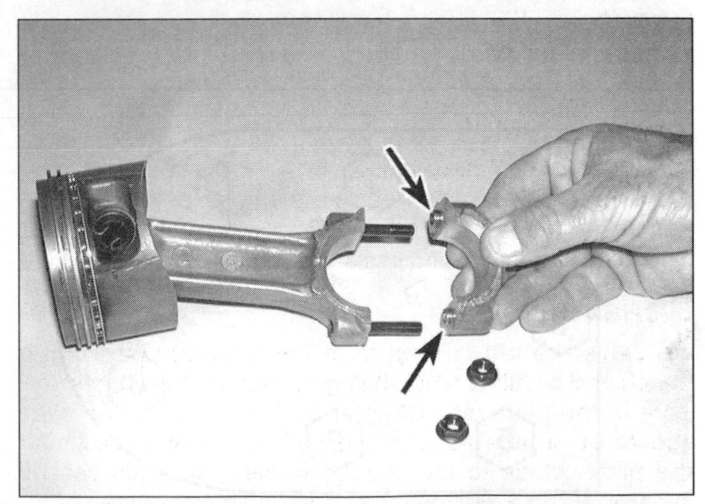
8.41 Posi-Lock connecting rods have sleeves (arrows) on the cap

43 Rotate the engine to TDC and install the oil pan with a new gasket. Tighten the fasteners to the torque listed in this Chapter's Specifications.
44 Coat the intake valve stem with clean engine oil or engine assembly lube, then reinstall it in the block. Make sure it's returned to its original location. **Note:** *If a seal is used on the intake valve, always install a new one when the engine is reassembled.*
45 Slip the valve spring and upper and lower retainers (if used) over the valve and attach a valve spring compressor (see Chapter 5). Compress the spring until the keepers can be installed, then release the compressor and make sure the retainer is securely locked on the end of the valve. Repeat the procedure for the exhaust valve. **Note:** *It is easier to install the two keepers with a small screwdriver if the keepers are coated with grease first.*
46 Install a new head gasket on top of the cylinder and install the cylinder head and its bolts. Tighten the bolts to the torque listed in this Chapter's Specifications following the recommended tightening sequence **(see illustrations 8.46a and 8.46b).**
47 To install the remaining components, refer to the text and illustrations for the disassembly procedure. Assembly is basically the reverse of the disassembly, provided that you use new gaskets wherever required. For adjustments to ignition, starting and charging systems, refer to Sections 3 through 7. Consult Chapters 4 and 5 as necessary. **Caution:** *Be sure to fill the crankcase to the correct level with the specified oil before attempting to start the engine.*

# Twin-cylinder engines

The models covered by this Section include the following:

**K-series and Magnum L-head twin-cylinder engines**

| | |
|---|---|
| MV16 | 16 HP |
| KT-17 | 17 HP |
| K482, M18, MV18 | 18 HP |
| KT-19 | 19 HP |
| K532, M20, MV20 | 20 HP |

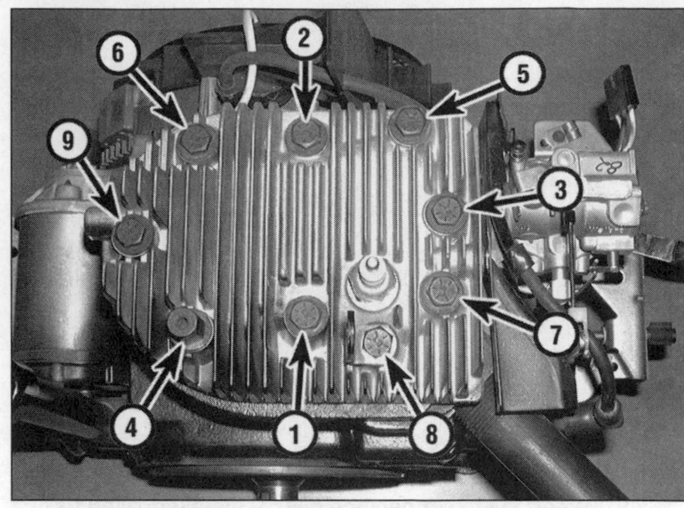

**8.46a Cylinder head bolt tightening sequence - models M10, M12, M14, K241, K301 and K321**

## Disassembly

48 All Kohler L-head engines share basic characteristics of design. The disassembly text and illustrations below cover the differences when working on a twin-cylinder version of the K-series or Magnum-series single-cylinder engine. Consequently, not every step of the disassembly or assembly procedures is described or illustrated here.
49 Begin disassembly by removing the engine from the equipment and draining the engine oil **(see illustration 8.49)**. Remove the exterior items from the engine, including the air filter assembly and both mufflers.
50 Remove the screws and take the air filter base from the top of the carburetor. Look at the arrangement of throttle, choke and governor linkages on the carburetor before disconnecting anything **(see illustration 8.50)**. If necessary, make a sketch to aid in reassembly. Disconnect the linkages and fuel and crankcase hoses, then unbolt the intake manifold from the engine with the carburetor still mounted. Refer to Sections 5 and 6 for carburetor overhaul and adjustment.

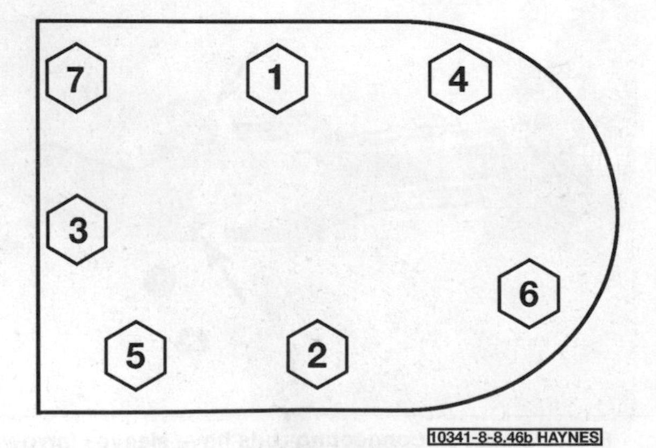

**8.46b Cylinder head bolt tightening sequence - models M8, K161 and K181**

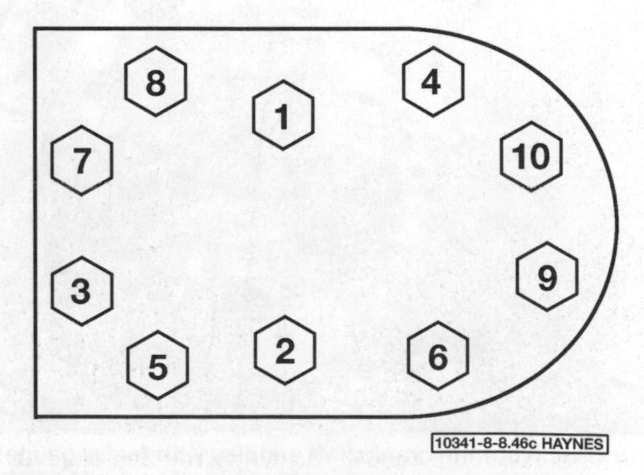

**8.46c Cylinder head bolt tightening sequence - models M16 and K341**

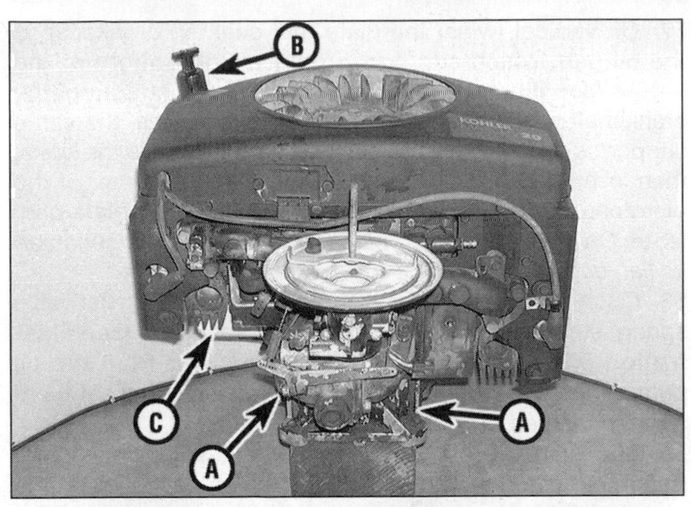

8.49 L-head twin-cylinder (vertical crankshaft model shown)

A Oil drain plugs
B Oil dipstick
C Oil filter

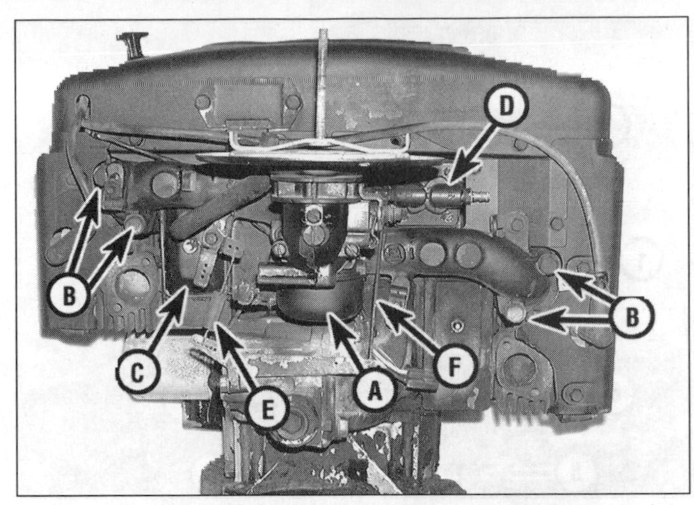

8.50 Fuel system components - Magnum twin-cylinder

A Carburetor
B Intake manifold bolts
C Crankcase breather
D Fuel pump
E Governor spring
F Governor linkage

51 Remove the mounting screws around the blower housing and remove the housing. On Magnum models, the regulator/rectifier is mounted to the housing, just behind the carburetor. Disconnect the wiring connector and remove the rectifier along with the housing.

52 Refer to the procedures in the disassembly of the Magnum single-cylinder engine for removal of the flywheel, stator, cylinder heads and ignition (Steps 8 through 15).

53 Because there are two cylinders, and each one has been manufactured with slight differences, make paint marks on the cylinder heads, cylinder barrels and case halves so all go back together on their original sides **(see illustration 8.53)**. **Note:** *Cylinder number 1 is the closest to the flywheel end*. On some models, the cylinder barrels are factory marked with a stamped number near the breather area, but the other components are not marked.

54 With the cylinder heads off, remove the barrel-to-case nuts, then tap the barrels carefully off with a rubber or soft-faced hammer to break the barrel-to-case seal **(see illustration 8.54)**. Once loose from the case, the barrels will come off slowly, because you are dragging the barrel off over the piston and rings, which cause some drag. **Caution:** *When the barrel is just about to come off the piston, have some rags or padding near the bottom of the case opening so that the piston and rod don't hit the case when released. This could damage the connecting rod.*

55 The valves and valve springs are contained within the cylinder barrel assembly. Put the barrels on the bench and remove the valves and springs for measurement and reconditioning (see Chapter 5).

56 Use paint marks to number each piston and its rod, then refer to Chapter 5 and remove the pistons from the rods by removing the retaining ring on one side with snap-ring pliers, then pushing the full-floating piston pin out.

8.53 Number the heads, cylinder barrels and case halves

8.54 Remove the four cylinder barrel nuts (arrows indicate three), then tap the barrel carefully to breaks its seal to the case

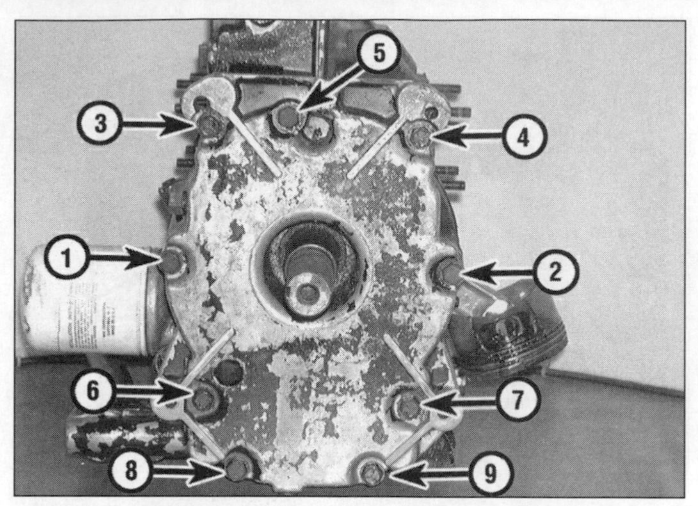

8.57 Remove the oil pan(shown on vertical engine) or closure plate bolts (numbers indicate the reassembly tightening sequence)

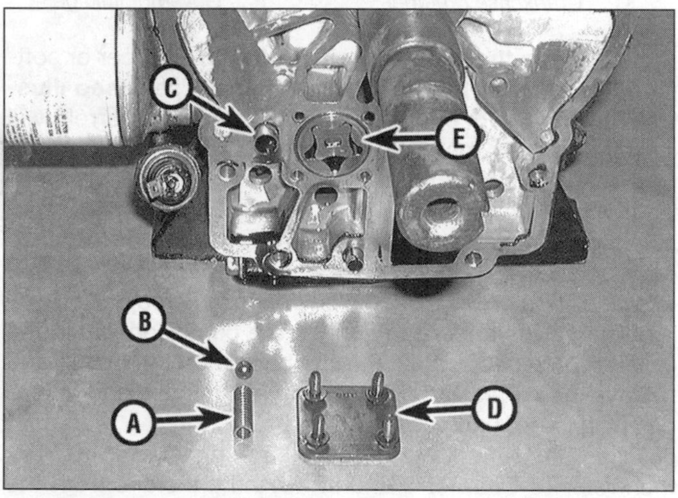

8.58 Oil pressure relief valve and oil pump components

| | | | |
|---|---|---|---|
| *A* | *Relief valve spring* | *D* | *Oil pump cover* |
| *B* | *Steel ball* | *E* | *Oil pump rotors* |
| *C* | *Sleeve* | | |

57 On vertical twins, the plate that seal the crankcase is the oil pan, which sits horizontally at the bottom of the engine **(see illustration 8.57)**. It support the bottom of the crankshaft on the PTO end. On horizontal twins, the same support is provided by the closure plate. Remove the bolts, then insert a large, dull screwdriver behind a corner of the plate/pan and break the gasket seal. Remove the plate/pan. **Note:** *On vertical engines, place the engine on the bench on its flat, engine-mount side to work on the oil pan.*

58 Once the pan is off, remove the oil pressure relief valve spring, steel ball and sleeve from the crankcase **(see illustration 8.58)**. Remove the bolts and cover from the oil pump. Clean all of the oil system components and set them aside in marked plastic bags.

59 The twin-cylinder engines have a split crankcase. Before removing the bolts to split the two halves, clean the camshaft bore plug (cylinder no. 1 side) and scribe an alignment mark across it and the case so the plug can be reinstalled the same way **(see illustration 8.59)**.

60 Place the engine with the number 1 side down, and wrap masking tape around the protruding tops of the tappets in cylinder no. 2 **(see illustration 8.60)**.

61 With the number 2 side still up, remove the crankcase fasteners and insert a large screwdriver in the splitting notch between them to separate the crankcase halves **(see illustration 8.61)**.

62 With the number 2 half removed, remove the camshaft and camshaft bore plug from the number 1 side **(see illustration 8.23)**.

63 Mark and remove the tappets from both cylinders **(see illustration 8.24)**.

64 Carefully lift the crankshaft, with both rods attached, from the number 1 case.

65 Mark both rods and rod caps as to cylinder number, then remove the rod cap nuts **(see illustration 8.65)**. Put the matching caps and nuts back on the rods and set them aside for reassembly. Also remove the front and rear sleeve bearings from the crankshaft and mark them.

66 Use a feeler gauge to measure the endplay of the oil pump drive gear and compare the measurement to this Chapter's Specifications **(see illustration 8.66)**. If endplay

8.59 Scribe an alignment mark across the camshaft bore plug and the block

8.60 Tape the number 2 cylinder's tappets (arrows) to hold them in place during case-splitting

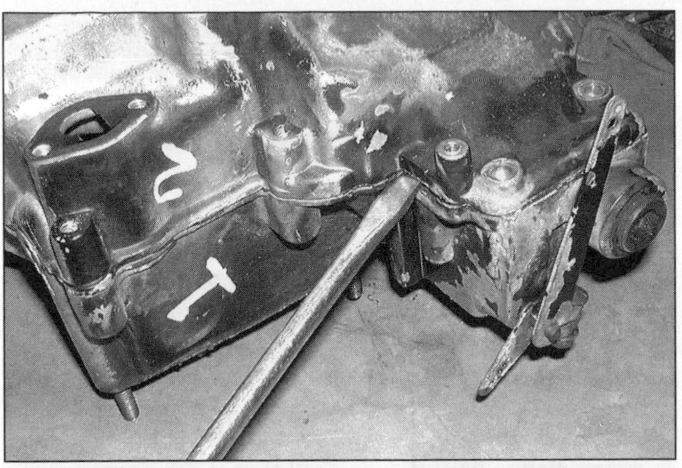

8.61 Remove all of the crankcase nuts and bolts, then pry the halves apart at the splitting notch

8.66 Measure the oil pump drive gear endplay with a feeler gauge (arrow)

is excessive add an additional shim to correct it.

67 To replace the oil pump drive gear (if necessary), use a long drift punch to drive the roll pin out of the shaft, then remove the shaft and gear **(see illustration 8.67)**.

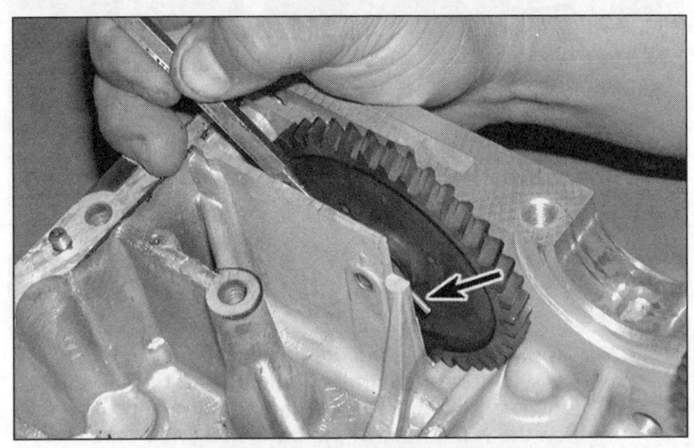

8.67 Drive the roll pin (arrow) out of the oil pump drive gear shaft

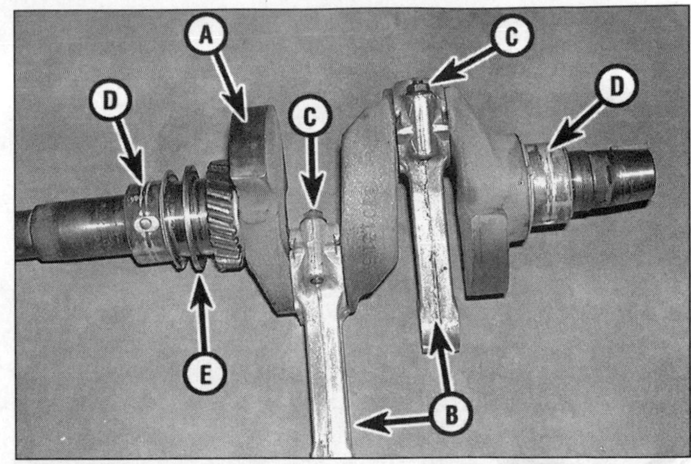


8.65 Crankshaft assembly

| | | | |
|---|---|---|---|
| *A* | *Crankshaft* | *D* | *Sleeve bearings* |
| *B* | *Connecting rods* | *E* | *Thrust washer* |
| *C* | *Rod cap nuts* | | |

## Inspection

68 Refer to the Inspection Section for the single-cylinder L-head engines above. All components should be cleaned, inspected, and replaced or renewed. See Chapter 5 for procedures for inspection and repair of valves, cylinders and pistons.

70 Remove the cover for the oil pump pickup and clean the pickup and tube. On vertical engines, the pickup is in the oil pan, and on horizontal engines in the number 1 case half. Measure the length of the oil pressure relief valve spring, and compare it to this Chapter's Specifications. Replace the spring if it does not meet Specifications.

## Reassembly

71 Clean the crankshaft and connecting rods thoroughly, then lubricate them with new engine oil or assembly lube. Install the rods on the crankshaft and tighten the rod cap nuts to the torque listed in this Chapter's Specifications. On some models of horizontal twins, the connecting rods are angled **(see illustration 8.71)**. When reinstalling these rods,

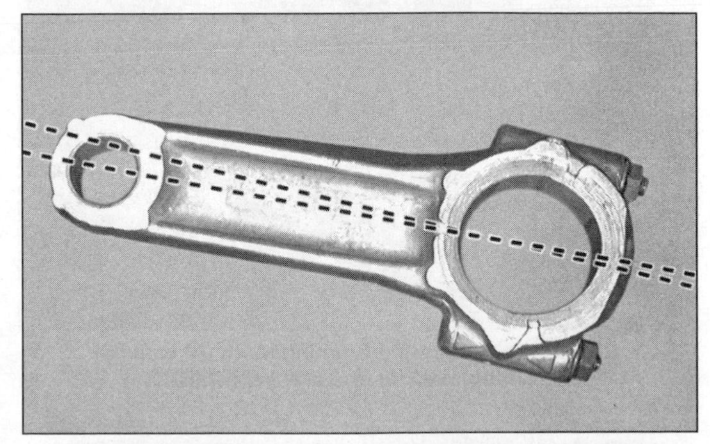

8.71 Horizontal twins with angled rods like this must be assembled with the angle down, away from the camshaft

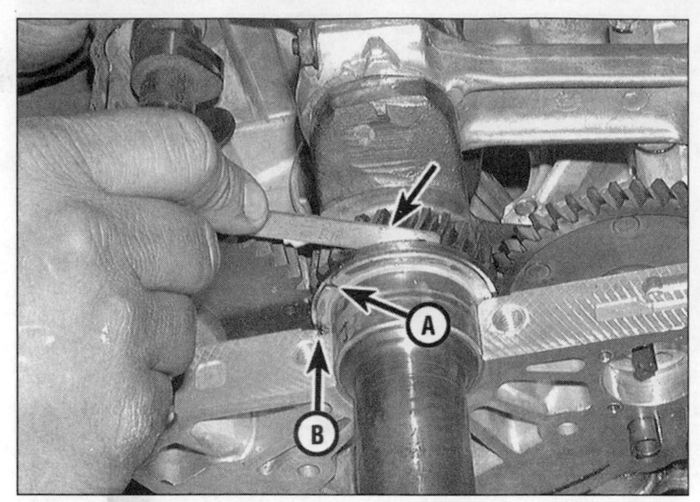

8.72a At the PTO end of the crankshaft, the sleeve bearing's tab (A) must be aligned with the notch in the case (B) - measure crankshaft endplay (arrow)

8.72b At the flywheel end of the crankshaft, align the hole in the bearing (A) with the oil passage (B) on the case

the angle must be down, away from the camshaft. **Note:** *On engines with Posi-Lock rods (rod caps that have sleeves that extend into the rod), tighten as per Specification. On all other rods, tighten the rod cap fasteners to Specifications, then loosen them and retighten to Specifications*.

72 Fit the thrust washer and new front and rear sleeve bearings to the crankshaft and install the assembly into the number 1 case half **(see illustration 8.65)**. Both front and rear bearings must have their oil holes align with the oil passages in the case **(see illustrations 8.72a and 8.72b)**.

73 Measure the crankshaft endplay with feeler gauges between the thrust washer and the sleeve bearing at the PTO end of the crankshaft **(see illustration 8.72a)**. Use a different thickness thrust washer to bring the clearance within Specifications. Various thickness are available from Kohler parts distributors.

74 Lubricate the tappets and install them in their original holes (new tappets can be installed initially in any hole). The tappets in the number 2 case need to be taped **(see illustration 8.60)**.

75 Install the camshaft in case number 1, aligning the timing marks on the crankshaft and camshaft gears **(see illustration 8.75)**.

76 Apply clean engine oil to the O-ring and install the camshaft bore plug in place on the number 1 case, making sure to align the scribe marks made during disassembly **(see illustration 8.59)**.

77 Clean the mating surfaces of each case half with

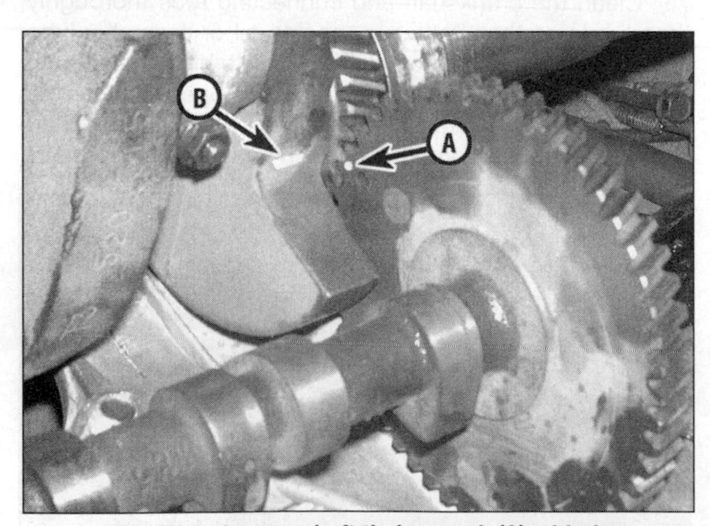

8.75 Align the camshaft timing mark (A) with the timing mark on the crankshaft (B)

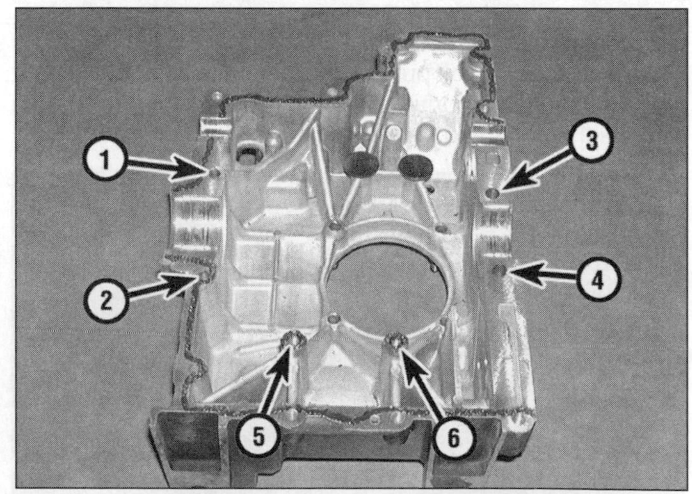

8.77a Apply sealant as indicated to case number 2 (on case flange and around bolts 5 and 6), and tighten the fasteners in sequence; after tightening bolts 1 through 6 tighten the remainder of the bolts in a criss-cross pattern - KT17 with serial no. 9755085 and later, KT19 and all Magnum series

lacquer thinner, then apply a bead of anaerobic sealer (Loctite 518 or equivalent) around the perimeter of the mating surface on case number 2 **(see illustrations 8.77a and 8.77b)**. Carefully lower case number 2 onto the studs in case number 1, then install all case bolts and nuts and following the recommended tightening sequence, tighten them to the torque listed in this Chapter's Specifications. **Caution:** *While the case fasteners are being tightened, keep constant finger pressure in on the camshaft bore plug to keep it seated (have an assistant help you if necessary).* After completing the case tightening procedure, use a centerpunch to stake the camshaft bore plug in place.

78 Drive a new oil seal into the case at the flywheel end, using a socket and hammer. Lube the seal with multi-purpose grease.

79 Install the oil pressure relief valve and oil pump components.

80 Refer to Chapter 5 and install the reconditioned valves in the cylinder barrels with their springs, spring seats and seals (where used).

81 Refer to Chapter 5 for piston ring installation and install the pistons in the cylinder barrels, orienting the piston top with the arrow or marking toward the flywheel end of the engine.

82 Push the piston through the cylinder barrel so that the pin area is exposed below the barrel, but keep the compressed rings within the bore. Apply anaerobic sealant (Loctite 518 or equivalent) to the bottom of the cylinder barrel. Have an assistant help you by holding the cylinder barrel/piston assembly near the connecting rod (at TDC), while you push the piston pin through the piston and rod and install the retaining clips **(see illustrations 8.82a and 8.82b)**.

83 Now push the barrel down over the piston until the barrel contacts the case, and install the barrel-to-case nuts and tighten them in sequence to the torque listed in this Chapter's Specifications **(see illustration 8.82a)**.

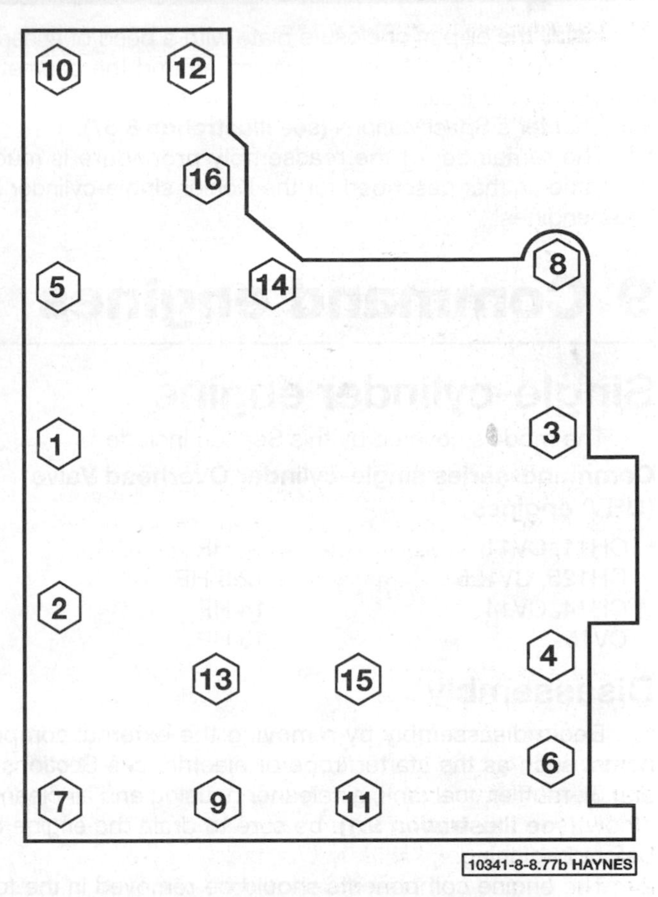


**8.77b Crankcase bolt tightening sequence - KT17 prior to serial no. 9755085**

84 Refer to Section 7 and check the valve clearance at TDC for each side.

85 Install the cylinder heads with new gaskets and tighten the bolts in sequence to the torque listed in this Chapter's Specifications **(see illustration 8.85)**.

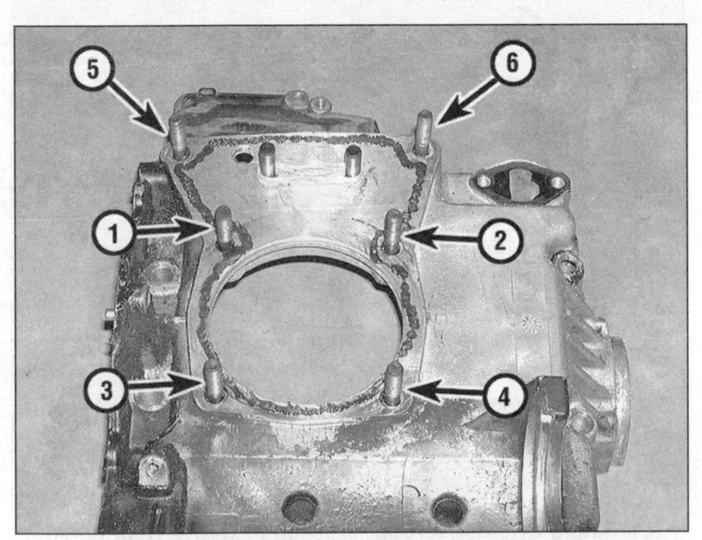


**8.82a Cylinder barrel tightening sequence and sealant application**

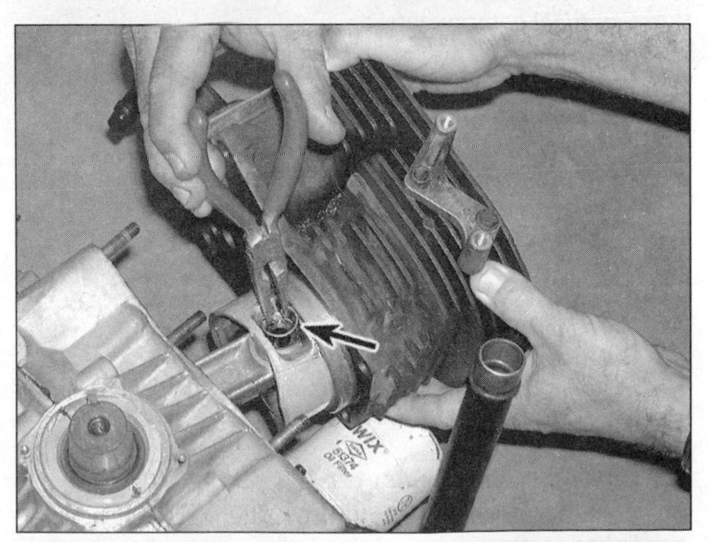

**8.82b Have an assistant hold the barrel while you insert the piston pin to connect the rod to the piston, then install the retaining clip (arrow)**

86 Install the oil pan or closure plate with a bead of anaerobic sealant (Loctite 518 or equivalent) around the perimeter and tighten the fasteners in sequence to the torque listed in this Chapter's Specifications **(see illustration 8.57)**.
87 The remainder of the reassembly procedure is much the same as that described for the Kohler single-cylinder L-head engines.

# 9 Command engines

## Single-cylinder engine

The models covered by this Section include:

**Command-series single-cylinder Overhead Valve (OHV) engines**

| | |
|---|---|
| CH11, CV11 | 11 HP |
| CH125, CV125 | 12.5 HP |
| CH14, CV14 | 14 HP |
| CV15 | 15 HP |

### Disassembly

1 Begin disassembly by removing the external components, such as the starter (rope or electric, see Sections 1 and 3), muffler, fuel tank, air cleaner housing and air cleaner elbow **(see illustration 9.1)**. be sure to drain the engine oil before beginning.
2 The engine components should be removed in the following general order:

*Oil filter*
*Oil-Sentry switch (if used)*
*Muffler*
*Air filter*
*Fuel tank*
*Cooling shroud/recoil starter*
*Carburetor/intake manifold*
*Electrical components*
*Fuel pump*
*Valve cover*
*Cylinder head and pushrods*
*Flywheel*
*Flywheel brake components (if equipped)*
*Stator*
*Ignition components*
*Crankcase breather assembly*
*Oil pan*
*Camshaft and hydraulic lifters*
*Balance shaft*
*Piston/connecting rod assembly*
*Crankshaft*
*Valves*
*Governor components*

3 Look at the arrangement of throttle, choke and governor linkages on the carburetor before disconnecting anything **(see illustration 9.3)**. If necessary, make a sketch to aid in reassembly.

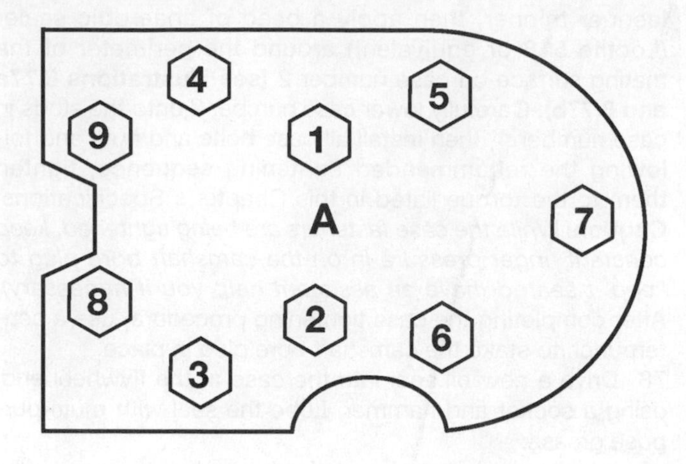


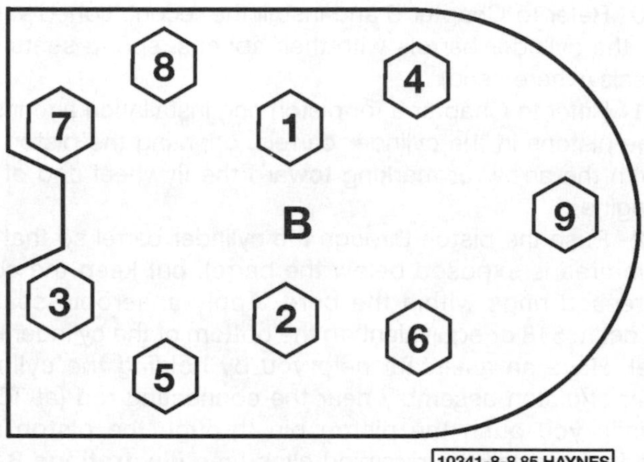


**8.85 Cylinder head tightening sequences**

*A K482, K532*
*B KT17, KT19 and Magnum twins*

**9.1 Single-cylinder vertical-crankshaft OHV Kohler engine**

*A Air cleaner*
*B Muffler*
*C Oil dipstick*
*D Oil filter*

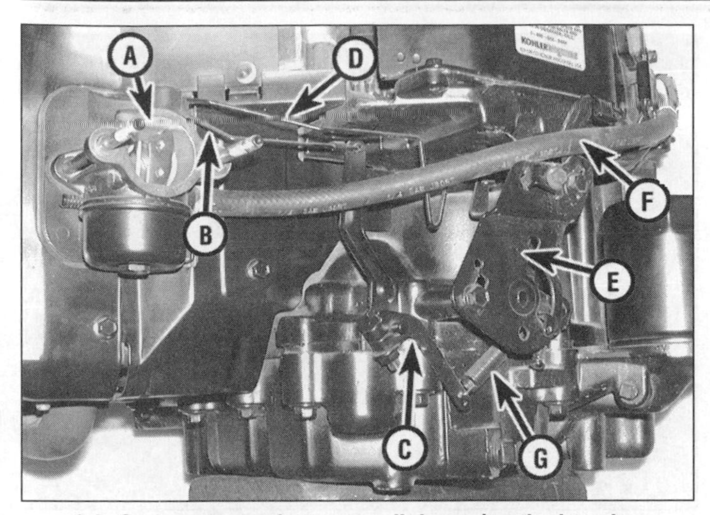


**9.3 Carburetor and governor linkage (vertical engine shown, horizontal similar)**

*A Carburetor*
*B Choke linkage*
*C Governor lever*
*D Throttle linkage*
*E Speed control bracket*
*F Carburetor fuel line*
*G Governor spring*

4 Disconnect the fuel line to the carburetor and remove the carburetor. **Note:** *On some models there may be a large heat deflector in between the intake manifold and carburetor. It slides off the two long carburetor mounting studs.*

5 Remove the valve cover **(see illustration 9.5)**. **Note:** *On some models, the muffler support bracket is retained by two of the valve cover screws.*

6 Remove the bolts holding the cylinder head cooling shrouds and blower housing. Disconnect the electrical connector from the regulator/rectifier on the blower housing. The regulator may be left mounted on the blower housing, but take care when setting the housing aside that the regulator isn't damaged.

7 Unbolt the ignition module and its air baffle from the engine, mark and disconnect any electrical connectors, and remove the module with the secondary spark plug lead **(see illustration 8.9)**.

8 Remove the cylinder head bolts, keeping note of the length of the bolts, any spacers and where they go **(see illustration 9.8)**. Remove the cylinder head with the rocker arm assembly in place. Remove the pushrods.

9 With the cylinder head off, use a valve spring compressor to remove the valves springs, retainers and keepers (see Chapter 5). Note that there are spring seats between each valve spring and the aluminum head, and that the intake valve has a seal **(see illustration 9.9)**.

10 Remove the plastic or metal debris screen from the fan. Hold the flywheel using a flywheel-strap tool and remove the center bolt with a breaker bar and socket **(see illustration 8.11)**. On models with a rope starter, remove the bolt, then take off the washer and starter drive cup. After the nut is removed, the starter cup and debris screen can be detached. **Caution:** *The bolt is under considerable torque. Make sure your socket is squarely on the bolt head - slipping off under pressure could cause some injury.*

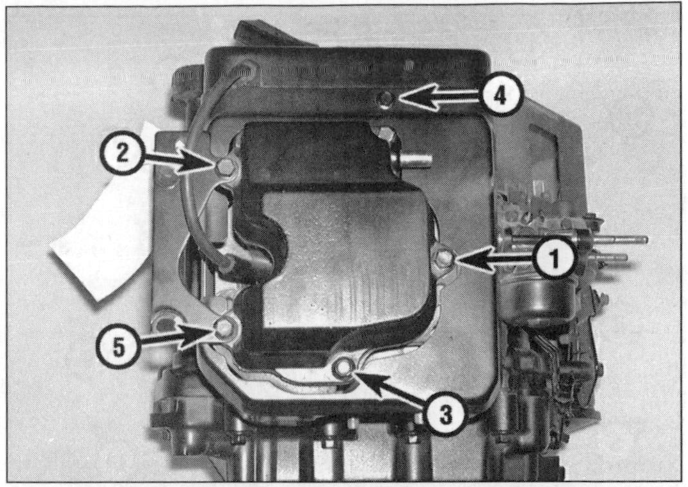


**9.5 Remove the valve cover mounting screws (arrows) - numbers indicate the tightening sequence for reassembly**

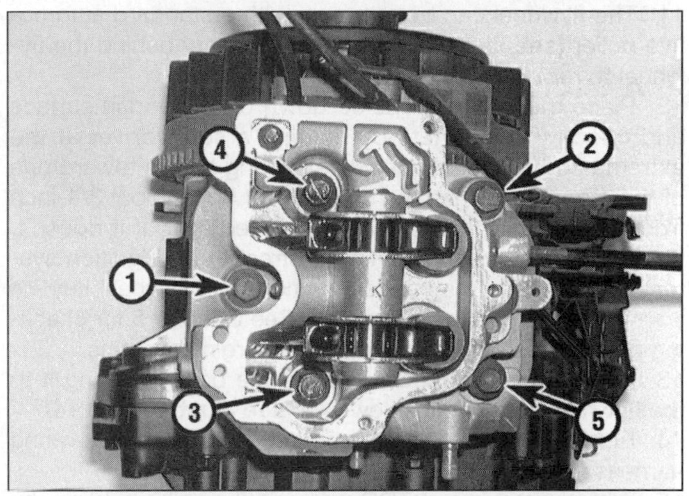


**9.8 Cylinder head bolts (arrows) - numbers indicate tightening sequence for reassembly**

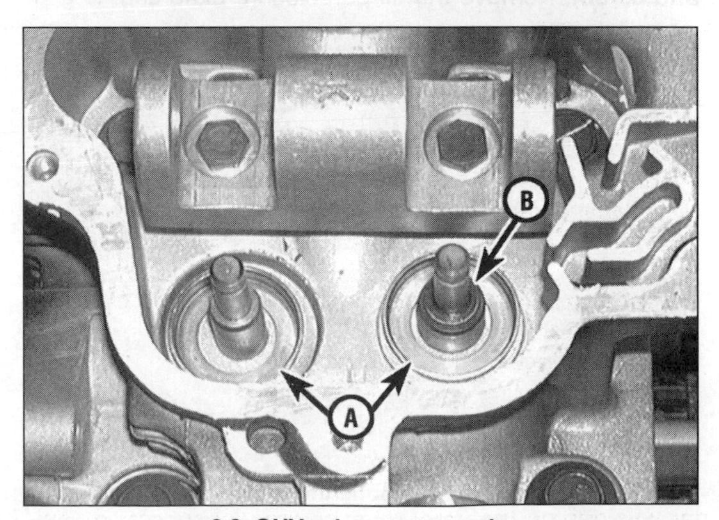


**9.9 OHV valve components**

*A Spring seats*
*B Intake valve seal*

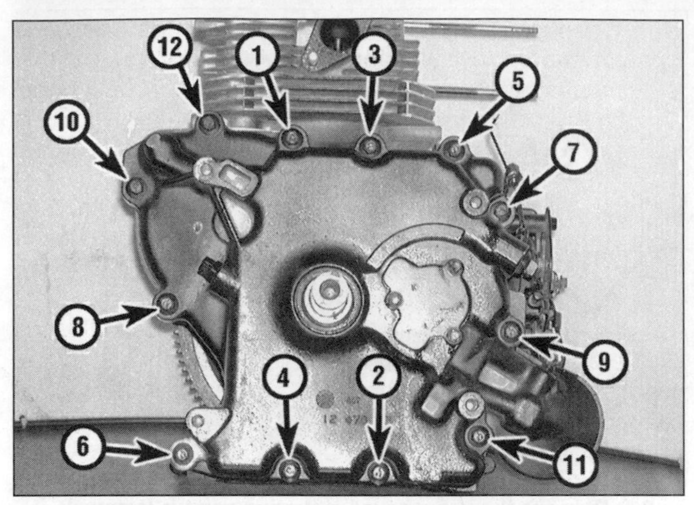


9.15a Oil pan bolts (arrows) - numbers indicate tightening sequence for reassembly

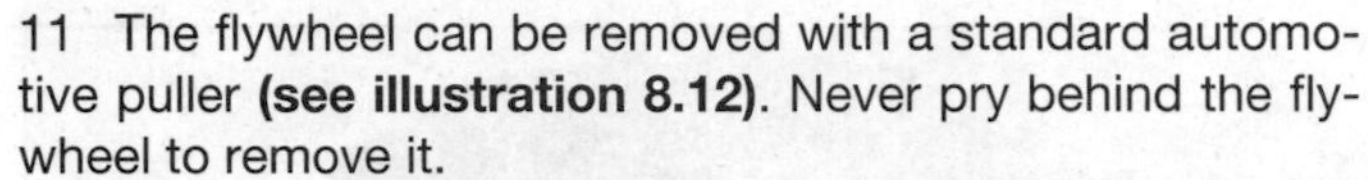

9.15b To separate the oil pan/closure plate, use a large screwdriver only in the "splitting notch" provided (arrow)

11 The flywheel can be removed with a standard automotive puller **(see illustration 8.12)**. Never pry behind the flywheel to remove it.

12 Place the flywheel upside-down on a wooden surface and check the magnets by holding a screwdriver at the extreme end of the handle while moving the tip toward one of the magnets - when the screwdriver tip is about 3/4-inch from the magnet, it should be attracted to it. If it doesn't, the magnets may have lost their strength and ignition system performance may suffer. Remove the flywheel key - if it's sheared off, install a new one. See Chapter 5 for checking procedures on the ignition and charging systems.

13 If the fan is damaged, remove the bolts securing it to the flywheel and install a new one **(see illustration 8.14)**.

14 Refer to Section 3 and remove the stator and its wiring harness from the engine's bearing plate.

15 With the engine at TDC, remove the bolts holding the oil pan/closure plate to the case **(see illustrations 9.15a and 9.15b)**. Remove the oil pan/closure plate and its gasket. **Note:** *On vertical engines, remove the oil pan, while on horizontal engines remove the closure plate.*

16 In the oil pan/closure plate, there are several oiling system components to be removed, cleaned and inspected, including the oil pickup, relief valve assembly and the oil pump **(see illustrations 9.16a and 9.16b)**.

17 Pull the camshaft and its shims from the case **(see illustration 9.17)**. Mark the hydraulic lifters (exhaust is closest to the pan/block gasket surface) and remove them from the case. Most OHV engines are equipped with ACR (Automatic Compression Release), which is a set of flyweights and a spring attached to the camshaft gear. Examine these components for wear.

18 Remove the balance shaft from the case **(see illustration 9.18)**.

19 With the case clear of the cam and balance shaft, remove the two connecting rod bolts and the rod cap **(see illustration 9.19)**.

20 Check the top of the cylinder with your finger for a

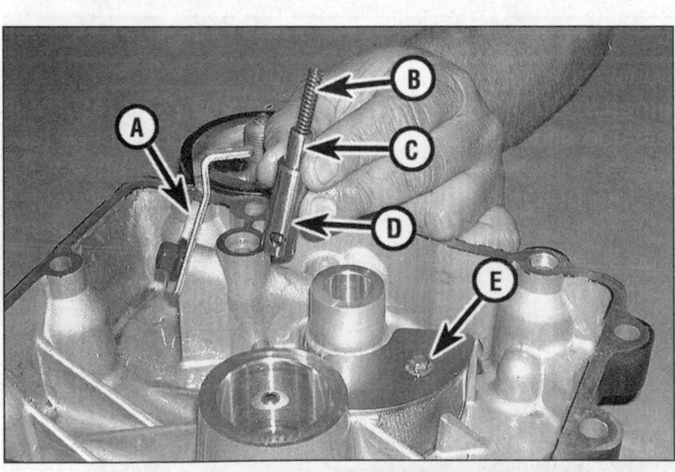


9.16a Oil system components (engine side of pan)

*A Relief valve bracket*
*B Spring*
*C Piston*
*D Relief valve body*
*E Oil pickup*

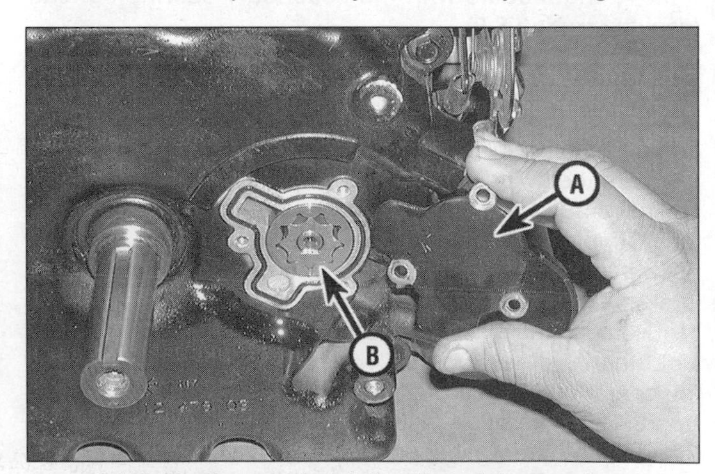


9.16b On the exterior of the oil pan/closure plate, remove the oil pump cover (A) and the rotors (B)

9.17 Remove the camshaft (A) with its shims (B) (not on all models are equipped with shims)

ridge, and use fine emery paper to remove it if present. Using a wooden hammer handle, push the rod and piston up and out of the engine from the bottom.

21 Refer to Chapter 5 for piston/rod inspection and measuring procedures. If it is determined that the piston is worn beyond Specifications or otherwise damaged, use snap-ring pliers to remove the piston pin snap-rings, then slide the piston pin out and separate the piston from the rod (see Chapter 5). A new piston generally comes with a new pin and snap-rings.

22 Slide the crankshaft out of the crankcase and set it aside.

23 Do not separate the lever from the governor shaft unless new parts are needed.

## Inspection

24 After the engine has been completely disassembled, refer to Chapter 5 for the cleaning, component inspection and valve lapping procedures.

25 After measuring both the cylinder bore and the piston (see Chapter 5), compare the results to this Chapter's Specifications. If the piston-to-bore clearance is within Specifications, the piston can be reused with new rings, and the bore honed for proper ring seating. If the cylinder is worn, it must be rebored at a machine shop and a new oversize piston used.

26 Inspect the valves, valve guides and valve seats (see Chapter 5). The valves and seats are generally ground during any overhaul, and the work should be done by a competent automotive or small-engine machine shop.

27 Inspect all of the bearings and measure the inside diameter of the connecting rod and compare it to the crankshaft journal size. If the clearance is beyond Specifications, a new rod must be purchased. If crankshaft bearings are worn, they must be pressed off, either from the crankshaft or the bearing plate. In either case, these bearings should be replaced at a shop with a hydraulic press. Never hammer on the bearings, either to install or remove.

28 Once you've inspected and serviced everything and purchased any necessary new parts, which should always include new gaskets and seals, reassembly can begin.

## Reassembly

29 Drive the old oil seal from the closure plate/oil pan with a punch and small hammer, being careful not to nick the seal bore in the plate. For removal, the plate should be supported with two wood blocks under the plate, close to the seal bore. New oil seals can be installed using a seal driver, or a large socket of the appropriate diameter (see Chapter 5). Drive the seal in squarely and only to the depth as originally installed. The oil seal at the other side of the engine can be removed with a screwdriver or seal-removal tool, and a new seal driven in.

30 If there is a crankshaft sleeve bearing in the case, it should be installed by a small engine repair shop with a press or the proper driver. Make sure the oil hole in the bearing aligns with the oil passage hole in the case **(see illustration 9.30)**. Install the crankshaft. **Note:** *Some engines do not have a sleeve bearing.*

9.18 Pull the balance shaft (A) from the case - (B) are the balance shaft-to-crankshaft timing marks

9.19 Remove the two connecting rod cap bolts (arrows) (be sure to mark the connecting rod and cap so they can be reinstalled in the correct position)

9.30 The oil passage hole (arrow) in the plate must align with the bearing's hole, if a sleeve bearing is used

9.33 Align the camshaft timing mark (A) with the matching mark on the crankshaft gear (B)

9.42 With the valve spring compressed by a tool over the rocker arm, slip the top of the pushrod (arrow) under its rocker arm

31 Install the connecting rod and piston/rings (see Chapter 5) through the bore until the connecting rod cap bolts can be installed over the crankshaft journal. **Warning:** *If an arrow or other mark was on top of the piston, install the piston with that mark toward the flywheel end of the engine*). **Note:** *On engines with Posi-Lock rods (rod caps that have sleeves that extend into the rod), tighten them to Specifications. On all other rods, tighten the rod cap fasteners to Specifications, then loosen them and retighten them to Specifications.*

32 Lubricate the balance shaft with clean engine oil and install it into the case, with its timing mark aligned with the mark on the (larger) gear on the crankshaft **(see illustration 9.18)**.

33 Install the camshaft, aligning its timing mark with the timing mark on the second (smaller) gear on the crankshaft **(see illustration 9.33)**.

34 Place a flat, thick section of metal plate across the case in the area of the camshaft snout and thrust shim. Clamp the plate to the case with C-clamps and measure the camshaft endplay between the plate and the shim. Compare your reading with the value listed in this Chapter's specifications. Various-thickness shims are available to bring the endplay within Specifications, if necessary.

35 Rotate the crankshaft so that the piston is at TDC. Lubricate the hydraulic lifters with engine oil and install them in the case.

36 With the oiling system components **(see illustrations 9.16a and 9.16b)** cleaned, inspected and reassembled into the oil pan or closure plate, apply a 1/8-inch wide bead of RTV sealant around the pan (going to the inside of all the bolts holes) and install the pan/plate. Tighten the fasteners to the torque listed in this Chapter's Specifications **(see illustration 9.15a for the tightening sequence)**.

37 Lubricate a new crankshaft seal with multi-purpose grease and install it (see Chapter 5).

38 Install the stator, flywheel/fan, wiring harness and ignition components in the reverse procedure of the disassembly process.

39 Coat the intake valve stem with clean engine oil or engine assembly lube, then reinstall it in the cylinder head. Make sure it's returned to its original location. **Note:** *If a seal is used on the intake valve, always install a new one when the engine is reassembled.*

40 Slip the valve spring and upper and lower retainers (if used) over the valve and attach a valve spring compressor (see Chapter 5). Compress the spring until the keepers can be installed, then release the compressor and make sure the retainer is securely locked on the end of the valve. Repeat the procedure for the exhaust valve, making sure the rotator (if used) is in place **(see illustration 9.9)**. **Note:** *It is easier to install the two keepers with a small magnetic screwdriver if the keepers are coated with grease first.*

41 Slip the pushrods in place, making sure their bottom ends fit into the sockets on top of the hydraulic lifters. Install a new head gasket on top of the cylinder and install the cylinder head and its bolts. Tighten the bolts to the torque listed in this Chapter's Specifications in the recommended sequence **(see illustration 9.8)**.

42 Using a lever-type valve spring compressor tool or a pair of locking pliers on the valve end of the rocker arms, compress the valve spring with the rocker arm and slip the upper ends of the pushrods under the pushrod end of the rocker arms, then release the tool **(see illustration 9.42)**. **Note:** *Because of the self-adjusting action of the hydraulic lifters, no manual valve adjustment is necessary on Kohler OHV engines.*

43 To install the remaining components, refer to the text and illustrations for the disassembly procedure. Assembly is basically the reverse of the disassembly, provided that you use new gaskets wherever required.

44 For adjustments to ignition, starting and charging systems, refer to Sections 3 through 7. Consult Chapters 4 and 5 as necessary. **Caution:** *Be sure to fill the crankcase to the correct level with the specified oil before attempting to start the engine.*

9.47 Disconnect the fuel and vacuum lines, then remove the pulse fuel pump (arrow)

## Command-series V-twin engines

The models covered by this Section include:
**Command-series V-twin Overhead Valve (OHV) engines**

| | |
|---|---|
| CH18, CV18 | 18 HP |
| CH20, CV20 | 20 HP |

### Disassembly

45 Begin disassembly by draining the oil and removing external components such as mufflers, air filter, oil filter, control panel (if used), and fuel tank.
46 After the spin-on oil filter is removed, use a hex wrench to remove the oil filter adapter, then remove the bolts and the oil cooler (not on all models) from the case.
47 The OHV twins use a vacuum-operated pulse-type fuel pump instead of the camshaft-driven mechanical pump used on other models. Disconnect the crankcase vacuum line (on the lower case on vertical models, near the intake manifold on the upper end of horizontal models), remove the pulse pump mounting screws from the case, and disconnect the fuel line to the carburetor and remove the pump with the lines **(see illustration 9.47)**.
48 Look at the arrangement of throttle, choke and governor linkages on the carburetor before disconnecting anything. If necessary, make a sketch to aid in reassembly. Remove the carburetor mounting bolts and the carburetor.
49 Remove the Oil-Sentry switch from the valve cover (or breather on some models).
50 Disconnect the electrical connectors at the starter and remove the starter motor (see Section 3).
50 Disconnect the electrical connector at the regulator/rectifier, then remove the screws holding the blower housing and side baffles.
51 Remove the inner baffles **(see illustration 9.51)**. On some models, the inner baffles simply pull out, while on others some of the crankcase breather assembly mounting screws hold the baffles in place.

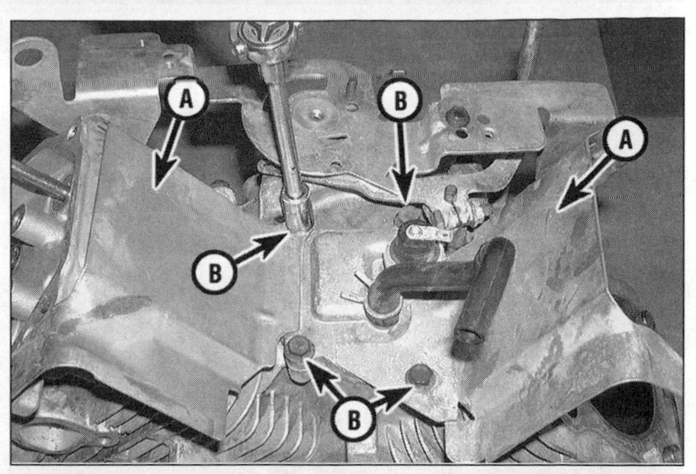


9.51 Remove the inner baffles (A) held by the crankcase breather cover bolts (B)

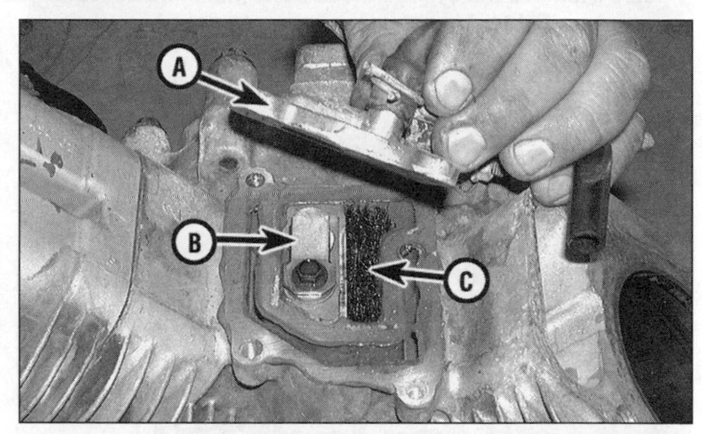


9.52 Crankcase breather components

*A Cover* *B Reed* *C Filter*

52 Remove the screws (if not removed in Step 51) and remove the crankcase breather cover **(see illustration 9.52)**. Pry the cover off with a thin putty knife or screwdriver, then remove and clean the breather components.
53 Remove the valve cover from each side **(see illustration 9.53)**.

9.53 Remove the valve cover bolts (numbers indicate tightening sequence for reassembly)

9.56 Remove the cylinder head bolts (numbers indicate tightening sequence for reassembly)

9.58 Closure plate/oil pan bolts (numbers indicate tightening sequence for reassembly) - a 1/2-drive breaker bar fits into the splitting tabs (A)

9.60 Remove the connecting rod cap bolts inside the case

54 Remove the flywheel, stator and ignition components much as described in Section 8 for the L-head engines.

55 Remove the intake manifold bolts and remove the intake manifold.

56 Remove the cylinder head mounting bolts, gently tap the head with a soft-faced hammer to break the gasket seal and remove the cylinder heads **(see illustration 9.56)**.

57 Unlike the L-head twin-cylinder engines, the OHV twins do not have separate cylinder barrels. They are part of the case. After the heads are removed, remove the pushrods and hydraulic lifters, labeling them with tape as to which side and which valve they are for.

58 Remove the mounting bolts and the closure plate (on horizontal twins) or oil pan on vertical twins **(see illustration 9.58)**. **Note:** *Use only the factory "splitting tabs" to pry the cover/pan off. Do not use a screwdriver anywhere else or the gasket surface could be damaged.*

59 Remove the camshaft and its endplay shim.

60 Working inside the case, remove the connecting rod cap bolts and cap from the closest cylinder, marking it as to which cylinder it is for **(see illustration 9.60)**.

61 Push the rod and piston assembly up through the cylinder and out, then repeat for the second rod and piston. **Note:** *Each rod, cap and piston should be marked as to their cylinder number and installed direction.*

62 Pull the crankshaft out of the case, being careful to hold it straight as it comes out, so as not to cock the rear bearing.

## Inspection

63 After the engine has been completely disassembled, refer to Chapter 5 for the cleaning, component inspection and valve lapping procedures.

64 After measuring both the cylinder bores and the pistons (see Chapter 5), compare the results to Specifications. If the piston-to-bore clearance is within Specifications, the pistons can be reused with new rings, and the bores honed for proper ring seating. If the cylinders are worn, they must be rebored at a machine shop and a new oversize piston used.

65 Inspect the valves, valve guides and valve seats (see Chapter 5). The valves and seats are generally ground during any overhaul, and the work should be done by a competent automotive or small-engine machine shop.

66 Inspect all of the bearing surfaces and measure the inside diameter of the connecting rod and compare it to the crankshaft journal size. If the clearance is beyond Specifications, a new rod must be purchased. If crankshaft bearing surfaces are worn, in the case or at the oil pan/closure plate end, either the case or closure plate will have to be replaced. Usually, the plate or case will wear before the hardened crankshaft itself will wear. **Note:** *On some horizontal OHV twins, there is a replaceable crankshaft sleeve bearing at the flywheel end* **(see illustration 8.72b)**.

67 The crankshafts used in the OHV twin engines have a drilled oil passage through the connecting rod journal, which is capped with a plug on the end. If the crankshaft must be reground for proper rod/journal fit, this cap should

9.68a The OHV twin oil pump (arrow) is located inside the closure plate

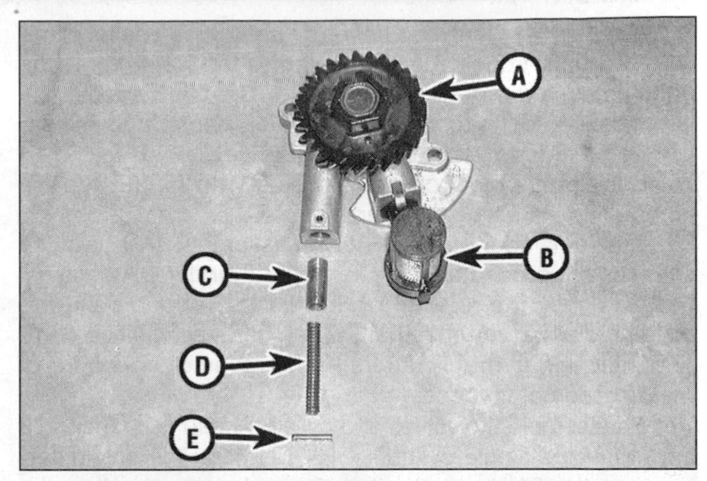

9.68b Oiling system components - OHV twins

*A Oil pump*
*B Oil pickup*
*C Oil pressure relief valve*
*D Relief valve spring*
*E Relief valve retaining pin*

be pulled out with a slide-hammer. After grinding, thoroughly clean the oil passages of any grinding grit, flush the passage with solvent, then drive a new cap in place. **Caution:** *If this is not done, grit can remain in the oil passage and damage the rebuilt engine.* Once you've inspected and serviced everything and purchased any necessary new parts, which should always include new gaskets and seals, reassembly can begin.

## Reassembly

68 Refer to Step 29 in the single-cylinder OHV section, and inspect or renew the oiling system components in the oil pan/closure plate, including the oil pump, pressure relief valve assembly and oil pickup **(see illustrations 9.68a and 9.68b)**.

69 Lubricate and install the crankshaft in the case.

70 Install the connecting rods and pistons/rings (see Chapter 5) through the bores until the connecting rod cap bolts can be installed over the crankshaft journal. **Warning:** *If an arrow or other mark was on top of the piston, install the piston with that mark toward the flywheel end of the engine.* The rod/piston furthest from the open end of the case should be installed first). **Note:** *On engines with Posi-Lock rods (rod caps that have sleeves that extend into the rod,* **(see illustration 8.41)**, *tighten them to Specifications. On all other rods, tighten the rod cap fasteners to Specifications, then loosen them and retighten them to Specifications.*

71 Install the camshaft, aligning its timing mark with the timing mark on the gear on the crankshaft **(see illustration 9.71)**.

72 Refer to Step 34 for the single-cylinder Command engine to measure the camshaft endplay. If its isn't within Specifications, change the shim size to correct the endplay.

73 Install the closure plate/oil pan with just a few bolts, and use a dial indicator to check the crankshaft endplay **(see illustration 9.73)**. If necessary, change the shim size to bring the endplay within Specifications, then remove the plate/pan.

9.71 Align the camshaft timing mark (A) with the mark on the crankshaft (B)

9.73 Check the crankshaft endplay with a dial indicator - push/pull the crank back and forth and measure the play

74 Apply a 1/8-inch wide bead of Loctite 518 gasket eliminator sealant around the pan (going to the inside of all the bolts holes) and install the pan/plate. Tighten the fasteners in sequence, to the torque listed in this Chapter's Specifications **(see illustration 9.58 for sequence)**.

75 Lubricate a new crankshaft seal with multi-purpose grease and install it (see Chapter 5).

76 The remainder of the OHV twin assembly is virtually the same as the procedures for the single-cylinder OHV engine, and basically the reverse of the disassembly procedure. Install the stator, flywheel/fan, wiring harness, ignition components, and hydraulic lifters in the reverse procedure of the disassembly process.

77 Reassemble the valves and cylinder head as in Steps 39 and 40 of the single-cylinder OHV Section. The only difference on the OHV twins is that the rocker arms bolt to the head individually, rather than as part of a shaft assembly. Install the heads with the rocker arms off, then install the pushrods, followed by the rocker arms, tightening all the fasteners to the torque listed in this Chapter's Specifications. **Note:** *Because of the self-adjusting action of the hydraulic lifters, no manual valve adjustment is necessary on Kohler OHV engines*.

78 To install the remaining components, refer to the text and illustrations for the disassembly procedure. Assembly is basically the reverse of the disassembly, provided that you use new gaskets wherever required. For adjustments to ignition, starting and charging systems, refer to Sections 3 through 7. Consult Chapters 4 and 5 as necessary. **Caution:** *Be sure to fill the crankcase to the correct level with the specified oil before attempting to start the engine.*

# 10 Specifications

## General

| | |
|---|---|
| Engine oil | |
| Type | API grade SH or better high quality detergent oil |
| Viscosity* | |
| K-series and Magnum-series | |
| Above 32-degrees F | SAE 30 |
| Zero to 32-degrees F | SAE 10W-30 or 10W-40 |
| Below 32-degrees F | SAE 5W-20 or 5W-30 |
| Command-series | |
| Above zero-degrees F | SAE 10W-30 or 10W-40 |
| Below 32-degrees F | SAE 5W-20 or 5W-30 |

**Use 10W-30 for the first five hours on new or overhauled engines*

| | |
|---|---|
| Engine oil capacity | |
| K-series, Magnum-series | |
| K141, K161, K181, M8 | 1.0 quart |
| K241, K301, K321, K341, M10 to M16 | 2.0 quarts |
| KT17, KT19 | 3.0 pints |
| K482, K532 | 3.0 quarts |
| MV16, MV18, MV20 | 1.75 quarts |
| M18, M20 | 1.5 quarts |
| Command-series | 2.0 quarts |

## Fuel system

| | |
|---|---|
| Idle speed | |
| K-series, Magnum-series | 1200 rpm |
| Command-series (except horizontal singles) | 1200 rpm |
| Command-series horizontal singles | 1500 rpm |
| Float level | |
| K-series, Magnum-series singles, Magnum horizontal twins | 11/64 inch |
| Magnum vertical twins | 45/64 inch |
| Command OHV vertical twin | 21/32 inch |
| Command OHV horizontal twin | 55/64 inch |
| Float drop | |
| K-series, Magnum-series singles | 1-1/32 inch |

## Ignition system

Spark plug type (Champion)
- K161, M8 ........ RCJ-8
- K241, K301, K321, K341, K482, K532, M10 to M16 ........ RH-10
- KT17, KT19 ........ RBL-15Y
- M18, M20 ........ RV-17YC
- MV16, MV18, MV20 ........ RV-15YC
- All Command-series ........ RC-12YC

Spark plug gap
- K161, Magnum singles, Magnum vertical twins ........ 0.025 inch
- K241 to K341, horizontal Magnum twins ........ 0.035 inch
- All Command-series ........ 0.040 inch

Ignition point gap (models with points) ........ 0.020 inch

Ignition module-to-magnet air gap
- Magnum-series singles ........ 0.012 to 0.016 inch
- Magnum-series twins, all Command-series ........ 0.008 to 0.012 inch

## Engine

Intake valve clearance (cold)
- K161, M8 ........ 0.006 to 0.008 inch
- K241, K301, K321, K341, K482, K532, M10 to M16 ........ 0.008 to 0.010 inch
- KT17, KT19, M18, M20, Magnum twins ........ 0.003 to 0.006 inch

Exhaust valve clearance (cold)
- All K-series and Magnum-series singles ........ 0.017 to 0.019 inch
- K482, K532 ........ 0.017 to 0.020 inch
- KT17, KT19 ........ 0.011 to 0.014 inch
- M18, M20 (before #1816500646) ........ 0.016 to 0.019 inch
- M18, M20 (after #1816500656) ........ 0.011 to 0.014 inch
- MV16 to MV20 (before #1816500656) ........ 0.016 to 0.019 inch
- MV16 to MV20 (between #1816500646 and 1917809296) ........ 0.011 to 0.014 inch
- MV16 to MV20 (after #1917809286) ........ 0.013 to 0.016 inch

Valve stem-to-guide clearance (minimum)
- Intake
  - K161 ........ 0.005 inch
  - K241, K301, K321, K341, M8 to M16, K482, K532 ........ 0.006 inch
  - KT17, KT19 ........ 0.0045 inch
  - Magnum-series twins ........ 0.005 inch
  - OHV Command-series ........ 0.0015 to 0.0030 inch
- Exhaust
  - K161 ........ 0.007 inch
  - K241, K301, K321, K341, M8 to M16, K482, K532 ........ 0.008 inch
  - KT17, KT19 ........ 0.0065 inch
  - Magnum-series twins ........ 0.007 inch
  - Command-series ........ 0.0020 to 0.0035 inch

Hydraulic lifter-to-bore clearance (Command-series) ........ 0.0005 to 0.0020 inch

Cylinder bore diameter (standard)
- K161, M8 ........ 2.938 inches
- K241, M10 ........ 3.250 inches
- K301, K532, M12 ........ 3.375 inches
- K321, M14 ........ 3.500 inches
- K341, M16 ........ 3.750 inches
- KT17, KT19 ........ 3.125 inches
- K482 ........ 3.251 inches

## Engine (continued)

Cylinder bore diameter (standard)

| | |
|---|---|
| Magnum-series twins | 3.120 inches |
| CV11, CV12.5, CV14, CH11, CH12.5, CH14 | 3.430 inches |
| CV15 | 3.600 inches |
| CH18, CH20, CV18, CV20 | 3.030 inches |

Piston diameter (minimum)

| | |
|---|---|
| K161, M8 | 2.925 inches |
| K241, M10 | 3.238 inches |
| K301, K532, M12 | 3.363 inches |
| K321, M14 | 3.491 inches |
| K341, M16 | 3.738 inches |
| KT17, KT19 | 3.1165 inches |
| K482 | 3.238 inches |
| Magnum-series twins | 3.120 inches |
| CV11, CV12.5, CV14, CH11, CH12.5, CH14 | 3.4179 inches |
| CV15 | 3.5363 inches |
| CH18, CH20, CV18, CV20 | 3.0252 inches |

Piston-to-cylinder bore clearance

| | |
|---|---|
| K-series, except horizontal twins | 0.007 to 0.010 inch |
| K-series horizontal twins | 0.006 to 0.008 inch |
| Magnum-series | 0.003 to 0.005 inch |
| Command-series singles, except CV15 | 0.0016 to 0.0017 inch |
| CV15 | 0.0012 to 0.0016 inch |
| Command-series twins | 0.0006 to 0.0023 inch |

Piston-to-pin clearance

| | |
|---|---|
| L-head engines | 0.0003 inch |
| OHV engines | 0.0002 to 0.0007 inch |

Piston ring side clearance

| | |
|---|---|
| K-series singles, K482, K532, all Magnum engines | 0.006 inch maximum |
| KT17, KT19 | |
| Top ring | 0.002 to 0.004 inch |
| Second ring, oil ring | 0.001 to 0.003 inch |
| Command-series | |
| Top ring | 0.0016 to 0.0041 inch |
| Second ring | 0.0016 to 0.0028 inch |
| Oil ring | 0.0217 to 0.0266 inch |

Piston ring end gap

| | |
|---|---|
| New bore | 0.010 to 0.020 inch |
| Used | 0.030 inch (maximum) |

Connecting rod bearing oil clearance

| | |
|---|---|
| New | 0.001 to 0.002 inch |
| Maximum | 0.003 inch |

Connecting rod side play

| | |
|---|---|
| K-series, Magnum-series | 0.005 to 0.016 inch |
| Command-series | 0.007 to 0.016 inch |

Crankshaft connecting rod journal diameter (minimum)

| | |
|---|---|
| K161, M8 | 1.1850 inches |
| K241, K301, K321, K341, M10 to M16 | 1.4990 inches |
| KT17, KT19 | 1.3738 inches |
| K482, K532 | 1.6240 inches |
| M18, MV16, MV18, MV20 | 1.3728 inches |
| M20 | 1.4988 inches |
| Command-series singles | 1.5328 inches |
| Command-series twins | 1.4150 inches |

| | |
|---|---|
| Crankshaft main bearing journal diameter | |
| K161, M8 | 1.1811 to 1.1814 inches |
| K241, K301, K321, K341, M10 to M16 | 1.5745 to 1.5749 inches |
| KT17, KT19 | 1.7412 to 1.7422 inches |
| K482, K532 | 1.7712 to 1.7721 inches |
| M18, MV16, MV18, MV20, M20 | 1.7407 inches (wear limit)) |
| Command-series singles | 1.6480 to 1.6510 inches |
| Command-series twins | 1.6080 to 1.6116 inches |
| Camshaft endplay | |
| K-series, Magnum singles | 0.005 to 0.010 inch |
| Magnum-series twins | 0.003 to 0.013 inch |
| Command-series | 0.003 to 0.005 inch |
| Camshaft bearing clearance | 0.0010 to 0.0025 inch |
| Camshaft-to-camshaft pin clearance, Magnum-series | 0.0010 to 0.0035 inch |
| Oil pump drive gear endplay | 0.010 to 0.029 inch |
| Oil pump relief valve spring free length | |
| KT17, KT19 | 0.940 inch |
| Magnum-series twins | 0.992 inch |
| CH11, CH125, CH14, CV11, CV125, CV14, CV15 | 0.992 inch |
| CH18, CH20 | 1.8 inch |

**Torque specifications**

| | |
|---|---|
| Connecting rod cap bolts/nuts | |
| K161, KT17, KT19, M8, Magnum twins | |
| Capscrews | 200 in-lbs |
| Posi-lock rods, new | 140 in-lbs |
| Posi-lock rods, used | 100 in-lbs |
| K241, K301, K321, K341, M10 through M16 | |
| Capscrews | 285 in-lbs |
| Posi-lock rods, new | 260 in-lbs |
| Posi-lock rods, used | 200 in-lbs |
| K482, K532 | |
| 5/16-inch capscrews | 200 in-lbs |
| 3/8-inch capscrews | 300 in-lbs |
| Command-series singles | 200 in-lbs |
| Command-series CV11 to CV15 (with step-down bolts) | 130 in-lbs |
| Command-series twins | 130 in-lbs |
| Cylinder head bolts | |
| K161, KT17, KT19, Magnum-series twins | 15 to 20 ft-lbs |
| K241, K301, K321, K341, M8 to M16, | 25 to 30 ft-lbs |
| K482, K532 | 35 ft-lbs |
| All Command-series | 30 ft-lbs |
| Rocker arm bolts (Command-series twins) | 124 in-lbs |
| Oil pan bolts | |
| K161 | |
| Grade 5 bolts | 250 in-lbs |
| Grade 8 bolts | 350 in-lbs |
| K241, K301, K321, K341, M8 to M16, K482, K532 | |
| Cast-iron pan | 35 ft-lbs |
| Sheetmetal pan | 200 in-lbs |
| Aluminum pan | 30 ft-lbs |
| Magnum-series vertical twins | 150 in-lbs |
| Command-series-series, vertical | 216 in-lbs |

## Torque specifications (continued)

Closure plate bolts
- KT17, KT19, Magnum-series horizontal twins .......... 150 in-lbs
- K482, K532, .......... 30 ft-lbs
- Magnum-series singles .......... 35 ft-lbs
- Command-series, horizontal .......... 216 in-lbs

Crankcase fasteners **(see illustrations 8.77a and 8.77b)**
- KT17 (prior to serial no. 9755085)
  - Fasteners 1 through 4, 12, 14 .......... 260 in-lbs
  - Fasteners 5 through 11, 13, 15 .......... 150 in-lbs
  - Fastener 16 .......... 35 in-lbs
- KT17 (serial no. 9755085 and later), and KT19 and Magnum twins
  - Fasteners 1, 2, 3, and 4 .......... 260 in-lbs
  - Fasteners 5 and 6 .......... 200 in-lbs
  - All remaining fasteners .......... 200 in-lbs

Cylinder barrel nuts
- KT17, KT19 .......... 360 in-lbs
- Magnum-series twins
  - Step 1 .......... 100 in-lbs
  - Step 2 .......... 200 in-lbs

Flywheel fan-to-flywheel .......... 115 in-lbs

Flywheel mounting fastener
- K-series singles .......... 50 to 60 ft-lbs
- K482, K532 .......... 115 ft-lbs
- M8 .......... 80 to 90 ft-lbs
- M10 to M16 .......... 40 to 45 ft-lbs
- Magnum-series twins .......... 40 ft-lbs
- Command-series .......... 49 ft-lbs

Fuel pump mounting screws .......... 40-45 in-lbs

Intake manifold mounting bolts
- KT17, KT19, Magnum-series twins .......... 150 in-lbs
- K482, K532 .......... 210 in-lbs
- Command-series twins .......... 88 in-lbs

Oil drain plug
- 1/4-inch, iron pan .......... 150 in-lbs
- 1/4-inch, aluminum pan .......... 100 in-lbs
- 3/8-inch, iron pan .......... 180 in-lbs
- 3/8-inch, aluminum pan .......... 120 in-lbs
- 1/2-inch, iron pan .......... 20 ft-lbs
- 1/2-inch, aluminum pan .......... 13 ft-lbs
- 3/4-inch, iron pan .......... 25 ft-lbs
- 3/4-inch, aluminum pan .......... 16 ft-lbs
- X-708-1 hex-head plug with gasket .......... 20 to 25 ft.-lbs

Spark plugs
- K-series, Magnum-series, Command-series twins .......... 18 to 22 ft-lbs
- Command-series singles .......... 28 to 32 ft-lbs

Starter mounting screws
- Recoil starters .......... 60 in-lbs
- Electric starters .......... 135 in-lbs

Valve cover screws* .......... 65 to 95 in-lbs

**Where screws are self-tapping design, lower figure is for used assembly, higher figure for new assembly.*

# 9 Honda engines

## Contents



## 1 Engine identification numbers/models covered

Honda G-series engines from 5.5 to 20 horsepower are covered in this Chapter. Included are the:

**GX160, GX240, GX270, GX340** and **GX390** horizontal crankshaft single cylinder engines

**GXV160, GXV270, GXV340** and **GXV390** vertical crankshaft single cylinder engines

**GX610** and **GX620** horizontal crankshaft V-twin cylinder engines

**GXV610** and **GXV620** vertical crankshaft V-twin cylinder engines

All models are overhead valve engines, which means the valves are mounted in the cylinder head. The "V" indicates the crankshaft is mounted vertically.

On vertical crankshaft engines, the model number, type and serial number are stamped into the crankcase near the spark plug **(see illustration 1.1)**. On horizontal crankshaft engines, both numbers are on the side between the carburetor and oil filler neck **(see illustration 1.2)**. Always have the model and serial numbers available when purchasing parts.

**1.1 Engine model number location (vertical shaft engine)**

**1.2 Engine model number location (horizontal shaft engine)**

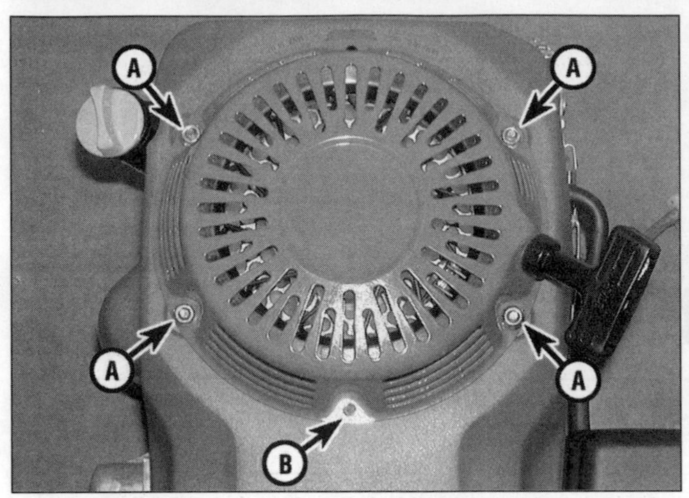


2.2 Remove the bolts (arrows to detach the cover from the recoil starter; on some vertical models, the crankshaft TDC mark is visible through a hole in the cover (B)

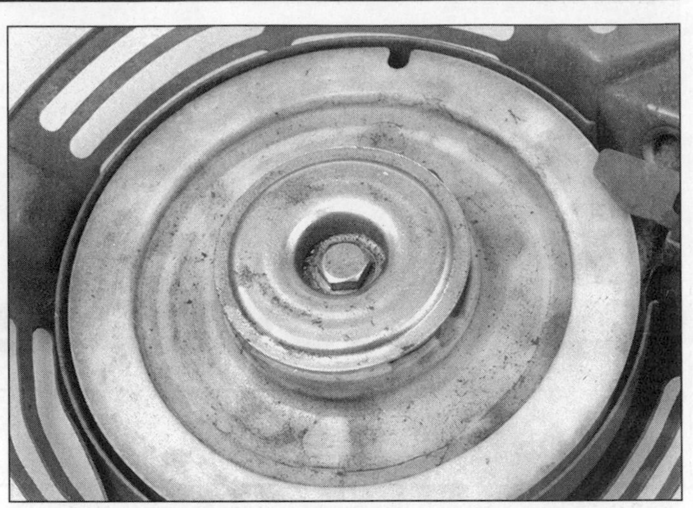

2.4 If necessary for access to the knot in the rope, remove the bolt to detach the friction plate or reel cover from the recoil starter

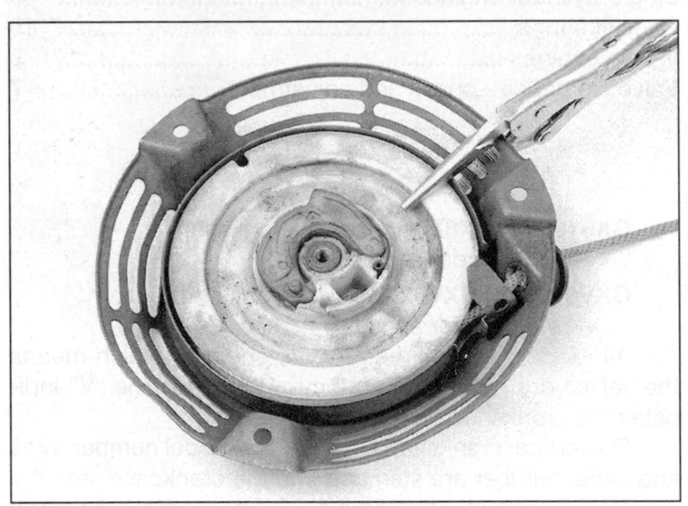

2.5a Restrain the recoil starter pulley with a C-clamp or locking pliers so the spring doesn't rewind

2.5b Cut the knot at the pulley to release the rope

# 2 Recoil starter service

## Rope replacement

1 If the rope breaks, the starter doesn't have to be disassembled to replace it, but it may be a good idea to take the opportunity to do a thorough cleaning job and check the spring and ratchet mechanism.

2 Unbolt the recoil starter housing from the engine and lift the starter off **(see illustration 2.2)**.

3 Hold the recoil starter housing in a vise or clamp it to the workbench so it doesn't move around as you're working on it. If necessary, use soft jaws in the vise to prevent damage to the housing. **Note:** *If you can't see the knot in the pulley end of the rope because it's under the reel cover, the starter will have to be partially disassembled to install the new one. If you can see the knot, disregard Step 4 and proceed to Step 5.*

4 If you're working on a starter with the rope knot located under the reel cover, remove the bolt (and the washer if equipped) and detach the reel cover from the center of the reel **(see illustration 2.4)**.

5 If the rope isn't broken, pull it all the way out. Hold the pulley with locking pliers or a C-clamp so the spring won't rewind and the pulley is held in position for installing the new rope **(see illustrations 2.5a and 2.5b)**.

6 Pull the knot out of the cavity with needle-nose pliers, then cut the knot off and pull the rope out **(see illustration 2.6)**. Note the type of knot used, then detach the handle - it can be used on the new rope.

7 Cut a piece of new rope the same length and diameter as the original.

8 Melt the ends of the nylon rope with a match to prevent fraying.

9 If the rope was broken (which means the starter spring

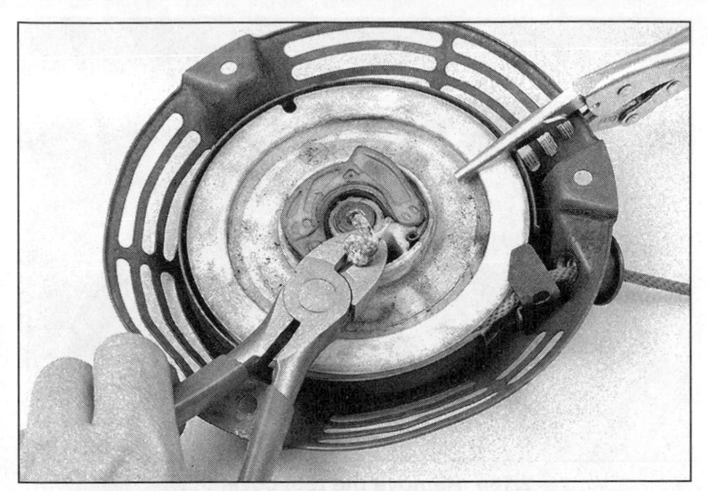

2.6 On some models, the knot is under the reel cover

2.10 If the rope is difficult to thread into the pulley, attach it to a piece of wire and use the wire to pull it into place

2.11 Insert the rope through the housing hole and into the pulley

2.16 Remove the bolt, spring, reel cover and pulley

unwound), you'll have to wind up the spring before installing the new rope. Turn the pulley about three turns counterclockwise (GXV270, GXV240, GXV340 and GXV390) or five turns (all others). Position the pulley so the opening for the rope is as close to the opening in the housing as possible. Refer to Step 5 above to restrain the pulley after the spring is tensioned.

10 If you're working on a starter with the rope knot located under the reel cover, insert the rope into the housing opening and out through the pulley opening. This can be tricky - if the rope won't cooperate, hook a piece of wire through the end of the rope and bend it over with a pliers, then thread the wire through the holes and use it to pull the rope into place **(see illustration 2.10)**. Tie a knot in the rope and manipulate it down into the cavity in the pulley. Install the friction plate or reel cover and nut or bolt.

11 If you're working on a starter with the rope knot on the pulley rim, insert the rope through the housing, then through the pulley **(see illustration 2.11)**. Tie a knot in the end of the rope and make sure it's secure.

12 Attach the handle to the rope (make sure it's secure or the rope will disappear into the starter and you'll have to start over).

13 Release the locking pliers or C-clamp while holding the rope, then allow it to rewind slowly onto the pulley.

14 Pull the handle several times and check the starter for proper operation.

# Recoil assembly service

## Disassembly

15 If the rope won't rewind and it isn't due to binding in the recoil starter, the spring may be broken. **Warning:** *Be sure to release the spring tension, if necessary, before disassembling the starter. This is done by cutting off or untying the knot in the handle and allowing the rope to slowly rewind into the pulley.*

### Dual-pawl starters

16 Remove the bolt (and upper friction plate if equipped) and detach the reel cover **(see illustration 2.16)**.

2.17 Remove the ratchet pawls and spring

2.18a Remove the reel cover . . .

2.18b . . . for access to the ratchet pawl

2.20a Typical dual-pawl starter spring

17 Remove the friction spring, ratchets and ratchet springs **(see illustration 2.17)**.

## Single-pawl starters

18 Remove the friction spring and the lower friction plate, then lift out the ratchet, guide plate and retaining pin **(see illustrations 2.18a and 2.18b)**.

## All models

19 Lift out the pulley and rope.

20 Remove the spring from the starter housing (or the spring base cover if equipped) **(see illustrations 2.20a and 2.20b). Warning:** *Wear gloves when handling the spring to avoid hand injuries.*

21 Remove the rope from the pulley.

# Inspection

22 Clean the parts with solvent and dry them with compressed air, if available, or a clean cloth, then check them for wear and damage. **Warning:** *Wear eye protection when using compressed air!* If the spring is distorted, bent or broken, install a new one.

# Assembly

23 Assembly is the reverse of the disassembly procedure, with the following additions:

a) *Coil the return spring tight enough that it will fit inside the base cover or pulley, then install it and engage its outer end with the slot in the base cover or pulley* **(see illustration 2.20a or 2.20b).**

b) *Install the return spring (and base cover, if equipped) on the pulley so the outer end of the spring is aligned with the tab on the pulley.*

c) *Apply a dab of grease to the spring notch in the starter case, then install the pulley and spring so the inner end of the spring engages the notch in the starter case* **(see illustration 2.23). Warning:** *Once the pulley and spring are installed, don't rotate the pulley. This might release the spring, which could fly out and cause serious injury.*

d) *Install the pawls and rope as described above.*

e) *Pull and release the handle several times to check the action of the assembled recoil starter.*

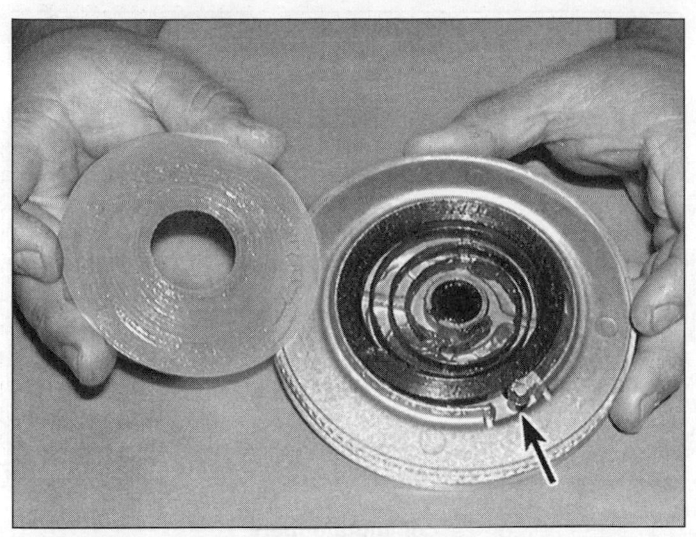

2.20b Typical single-pawl starter spring; the outer end (arrow) hooks the notch in the pulley

2.23 The inner end of the spring hooks the notch in the housing (arrow)

# 3 Charging and electric starting systems

## Charging system

1 The charging system on all models uses one or more charging coils positioned behind the flywheel. Some models also include a lamp coil. The current on some one-amp models passes through a diode. On others, it passes through a rectifier that converts the alternating current to direct current. Single cylinder engines with a 10-amp or 18-amp charging system, and V-twins with a 20-amp charging system, use a regulator-rectifier.

### Component removal and installation

2 Refer to Section 8 or Section 9 and remove the flywheel.

3 Follow the wiring harness from the charge coil(s) to the electrical connector and disconnect it. Remove the coil mounting bolts and take the coil(s) off.

4 If the charging system is equipped with a diode, unplug it.

5 If the charging system is equipped with a regulator/rectifier, disconnect its electrical connector. Note how the regulator/rectifier is mounted, then remove the mounting bolt or screw and lift it off the engine.

## Starting system

6 All the engines covered in this Chapter include an electric starter, either as an option or as standard equipment.

### Removal and installation

#### Starter motor

7 Disconnect the cable from the negative terminal of the battery.

8 If you're working on a horizontal-crankshaft V-twin, remove the control box.

9 Pull back the rubber cover, remove the nut retaining the starter cable to the starter and disconnect the cable.

10 Remove the starter mounting bolts. On some engines, one or more stay brackets are secured by these bolts; be sure to note the location of the brackets so they can be reinstalled correctly.

11 Pull the starter away from the engine. On vertical-crankshaft single cylinder engines and all V-twins, there's a locating dowel at each bolt hole. Find these and place them where they won't be lost.

12 Install the starter by reversing the removal procedure.

#### Starter solenoid

13 Disconnect the cable from the negative terminal of the battery.

14 Pull back the rubber covers, remove the nuts retaining the starter and battery cables to the solenoid and disconnect the cables.

15 Disconnect the switch (thin) wire from the solenoid. Remove the solenoid mounting bolts or nuts and detach it from the starter (single cylinder) or the engine (V-twin).

16 Refer to Chapter 3 to check the solenoid.

17 Installation is the reverse of removal. Make sure the cables and wire are in good condition. Also make sure the solenoid mounting points are clean and free of corrosion, and tighten the bolts or nuts securely.

### Component replacement

18 Mark the position of the housing to each end cover. Remove the through-bolts and detach both end covers. Pull the armature out of the housing and remove the brush holder **(see illustration 3.18)**. Separate the brushes from the holder.

19 The parts of the starter motor that will most likely require regular attention are the brushes. Measure the length of the brushes. Compare the results to the brush

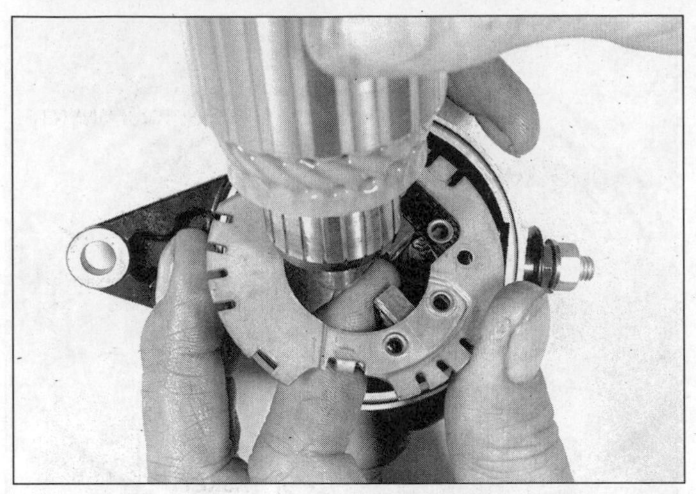

3.18 Pull the armature and commutator clear of the brushes

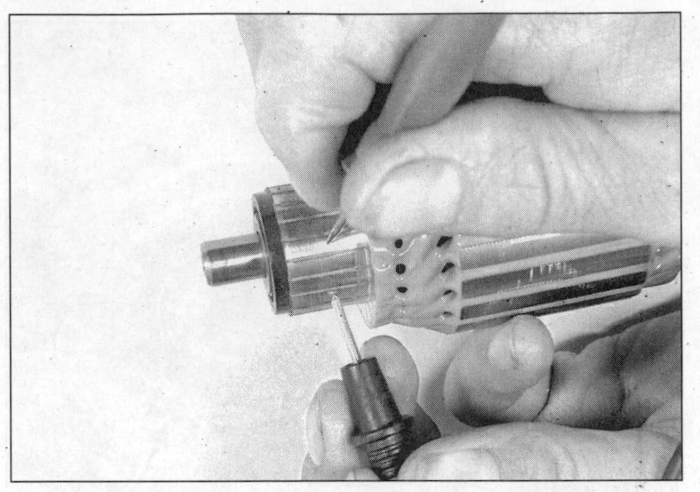

3.21 Continuity should exist between pairs of commutator bars

length listed in this Chapter's Specifications. If any of the brushes are worn beyond the specified limit, replace the brushes as a complete set. If the brushes are not worn excessively, cracked, chipped or otherwise damaged, they may be re-used. **Note:** *On horizontal-crankshaft single cylinder engines, the entire center bracket must be replaced if the positive brushes are worn.*

20 Inspect the commutator (the surface the brushes press against) for scoring, scratches and discoloration. The commutator can be cleaned and polished with crocus cloth, but do not use sandpaper or emery paper. After cleaning, wipe away any residue with a cloth soaked in electrical contact cleaner or denatured alcohol.

21 Using an ohmmeter or a continuity test light, check for continuity between the commutator bars **(see illustration 3.21)**. There should be no continuity between the commutator and the shaft. If the checks indicate otherwise, the armature is defective.

22 Measure the depth of the mica insulators between the commutator segments.

a) *On single-cylinder engines, if it's less than the value listed in this Chapter's Specifications, undercut the mica to the specified depth with a piece of hacksaw blade.*

b) *On V-twins, if it's less than the specified value, clean the grooves and measure again. If the depth is still less than specified, replace the armature.*

23 Check for continuity between the pair of negative brushes and the pair of positive brushes. There should be continuity in both cases.

24 Inspect the drive gears. If they're damaged or worn, press the spacer down off the stopper ring with an open-end wrench, then remove the stopper ring and slide the gears off the end of the shaft.

25 Assembly is the reverse of the disassembly procedure, with the following additions:

a) *If you're working on a horizontal-crank single, hold the brushes back with a piece of stiff wire so you can pass the commutator between them.*

b) *If you're working on a vertical-crank single, push the brushes back so the commutator can be passed between them. Once this is done, position the springs so they push the brushes against the commutator.*

# 4 Ignition system

1 The engines covered by this Chapter use a "electronic magneto" ignition system. The electronic ignition system, also called "breakerless" ignition, incorporates solid state electronic components. The electronic-magneto system operates similar to a points system but the method used to "break" the electronic signal differs. Flywheel rotation generates electricity in the primary circuit of the ignition coil assembly when the magnets rotate past the charge coil. As the flywheel rotates, it passes a trigger coil where a low voltage signal is produced. The low voltage signal allows the transistor to close allowing the energy stored in the coil to fire the secondary circuit or spark plug. The transistor acts as a primary circuit switching device in the same manner as the points in a mechanical (points) ignition system. Spark is produced whenever the engine is rotated. A terminal on the coil is connected to either a kill switch or the OFF side of an ignition switch, to ground the coil so the engine will stop.

## Ignition coil

### Removal and installation

2 Remove the recoil starter assembly (if equipped) and the fan cover.

3 Note how the coil wires are routed, then disconnect them. Unscrew the spark plug cap from the plug wire. Remove the coil mounting screws and take it off the engine. Installation is the reverse of removal. Adjust the ignition coil air gap.

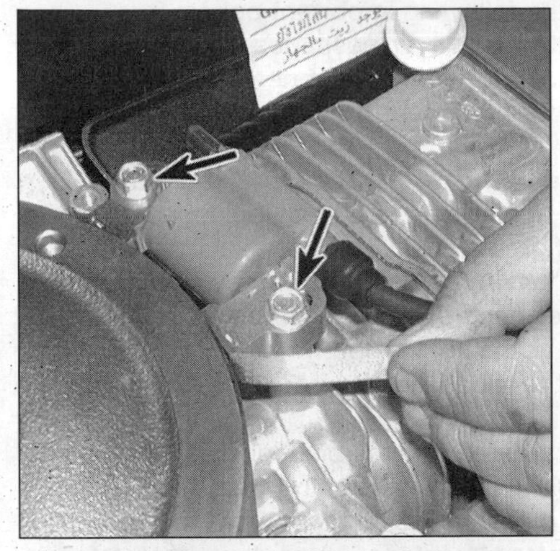

4.5 To adjust the ignition coil air gap, loosen the mounting bolts (arrows) and using the appropriate feeler gauge, adjust the air gap between the ignition coil armature segments and the flywheel

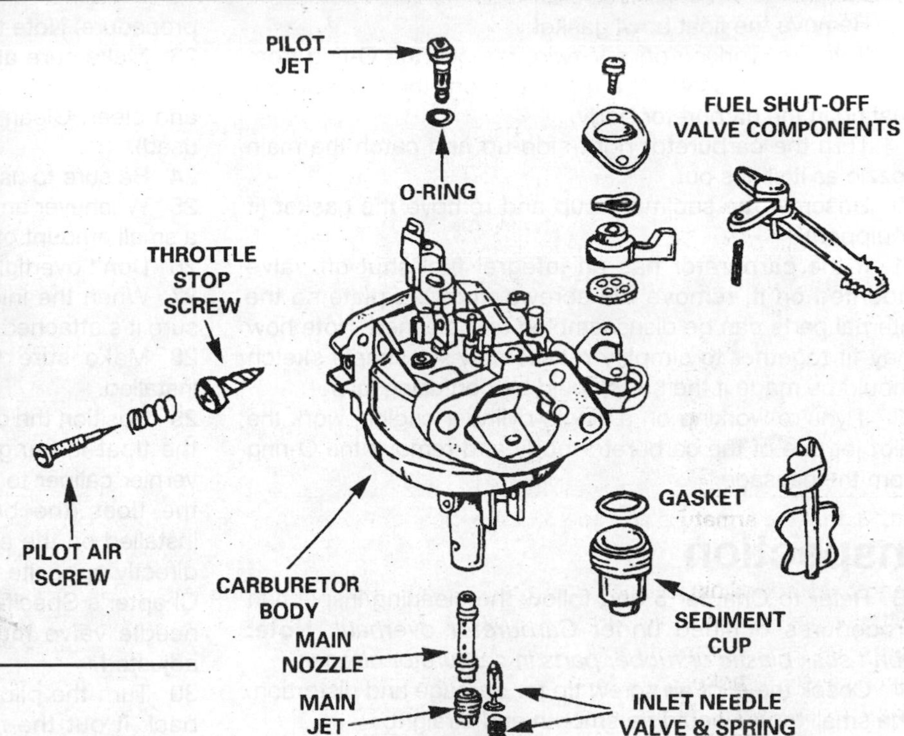


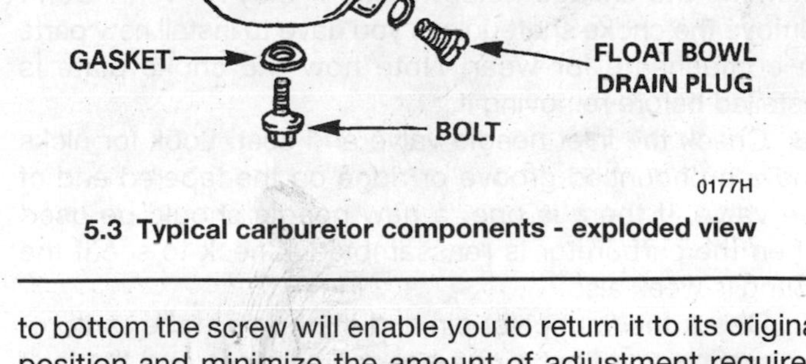


5.3 Typical carburetor components - exploded view

## Air gap adjustment

4 Rotate the flywheel so the magnet is not under the ignition coil.

5 Loosen the ignition coil mounting bolts and insert a 0.016-inch feeler gauge under the ignition coil armature segments **(see illustration 4.5)**. **Note:** *Use a feeler gauge long enough to reach under both segments.* Press the ignition coil against the feeler gauge, tighten the mounting bolts and remove the feeler gauge.

# 5 Carburetor disassembly and reassembly

## Disassembly

1 The following procedures describe how to disassemble and reassemble the carburetor so new parts can be installed. Read the sections in Chapter 5 on carburetor removal and overhaul before doing anything else.

2 In some cases it may be more economical (and much easier) to install a new carburetor rather than attempt to repair the original. Check with a dealer to see if parts are readily available and compare the cost of new parts to the price of a complete ready-to-install carburetor before deciding how to proceed.

3 If there's a limiter cap on the pilot air screw, break it off. The screw has a narrow section that's designed to break when the limiter cap is removed. While counting the number of turns, carefully screw the pilot air screw in until it bottoms, then remove it along with the spring **(see illustration 5.3)**. Counting and recording the number of turns required to bottom the screw will enable you to return it to its original position and minimize the amount of adjustment required after reassembly.

4 Detach the float bowl from the carburetor body. On single cylinder engines, it's held in place with a bolt. On V-twins, it's held on by screws. Be sure to note the locations of any gaskets/washers used. Some carburetors also have a drain plug in the float bowl - it doesn't have to be removed.

5 Push the float pivot pin out of the carburetor body (you may have to use a small punch to do this).

6 Remove the float assembly and the inlet needle valve (and spring, if used) by lifting the float straight up. Note how the inlet needle valve is attached to the float.

7 Remove the float bowl gasket.
8 If you're working on a V-twin, remove the O-ring from the main jet. Unscrew the main jet from the main nozzle casting in the carburetor body.
9 Turn the carburetor right-side-up and catch the main nozzle as it slides out.
10 Unscrew the sediment cup and remove the gasket (if equipped).
11 If the carburetor has an integral fuel shut-off valve mounted on it, remove the screws from the plate so the internal parts can be disassembled and cleaned. Note how they fit together to simplify reassembly - a simple sketch should be made if the parts could be confusing later.
12 If you're working on a single-cylinder engine, work the pilot jet free of the carburetor body and remove the O-ring from the passage.

## Inspection

13 Refer to Chapter 5 and follow the cleaning/inspection procedures outlined under *Carburetor overhaul.* **Note:** *Don't soak plastic or rubber parts in carburetor cleaner.*
14 Check the pilot air screw tip for damage and distortion. The small taper should be smooth and straight.
15 Check the throttle plate shaft for wear by moving it back-and-forth.
16 Check the throttle plate fit in the carburetor bore. If there's play in the shaft, the bore is worn excessively, which may mean a new carburetor is required (on some carburetors, replacing the throttle plate shaft or plate, or both, may cure the problem). The throttle plate and shaft don't have to be removed unless new parts are required. Note how the throttle plate is installed before removing it - if it is removed, make sure it's positioned exactly as it was originally.
17 Check the choke shaft for play in the same manner and examine the linkage holes to see if they're worn. Don't remove the choke shaft unless you have to install new parts to compensate for wear. Note how the choke plate is installed before removing it.
18 Check the inlet needle valve and seat. Look for nicks and a pronounced groove or ridge on the tapered end of the valve. If there is one, a new needle should be used when the carburetor is reassembled. Check to see if the spring is weak also.
19 Check the float pivot pin and the bores in the carburetor casting, the float hinge bearing surfaces and the inlet needle tab for wear - if wear has occurred, excessive amounts of fuel will enter the float bowl and flooding will result.
20 Shake the float to see if there's gasoline in it. If there is, install a new one.
21 Check the fuel inlet fitting to see if it's clean and unobstructed.

## Assembly

22 Once the carburetor parts have been cleaned thoroughly and inspected, reassemble it by reversing the above procedure. Note the following important points:
23 Make sure all fuel and air passages in the carburetor body, main jet, main nozzle and inlet needle seat are clean and clear. Clean the sediment bowl thoroughly as well (if used).
24 Be sure to use new gaskets, seals and O-rings.
25 Whenever an O-ring or seal is installed, lubricate it with a small amount of grease or oil.
26 Don't overtighten small fasteners or they may break off.
27 When the inlet needle valve assembly is installed, be sure it's attached to the float properly.
28 Make sure the float pivots freely after the pin is installed.
29 Position the carburetor so the pivot pin is at the top and the float is hanging down, vertically, then use a dial or vernier caliper to measure the distance from the bottom of the float (the bottom surface when the carburetor is installed on the engine) to the carburetor body on the side directly opposite the pivot pin. It should be as listed in this Chapter's Specifications. If it isn't, a new float and/or inlet needle valve must be installed - the height cannot be adjusted.
30 Turn the pilot air screw in until it bottoms lightly and back it out the number of turns listed in this Chapter's Specifications.
31 If the pilot air screw had a limiter cap, install a new one with its tab positioned against the stop on the carburetor body, so the pilot air screw cannot be turned counterclockwise. Secure the limiter cap to the screw with a fast-curing epoxy adhesive.

# 6 Carburetor and governor adjustment

## Carburetor

1 When making carburetor adjustments, the air cleaner must be in place and the fuel tank should be at least half full. The carburetor has a fixed main jet - no mixture screw is installed and no adjustment is required.
2 If there's a limiter cap on the pilot air screw, you'll have to break it off and install a new screw to make mixture adjustments (see Section 5).
3 If there's no limiter cap on the pilot air screw, turn it in until it bottoms lightly, then back it out the specified number of turns **(see illustrations 6.3a and 6.3b).**
4 To adjust the idle speed, run the engine until it reaches normal operating temperature, then turn the throttle stop screw until the idle speed is as listed in this Chapter's Specifications.
5 If there's no limiter cap on the pilot air screw, turn it in or out in small increments until the engine runs at the highest speed.
6 Readjust the idle speed if necessary.

6.3a Idle mixture screw (left arrow) and idle speed screw (right arrow) (horizontal crankshaft single cylinder engine shown; others similar)

## Governor

7 It is best not to change the relationship of any linkages or other components in the governor system, unless a worn or broken component is to be replaced. If the engine is being overhauled, do not change any adjustments. If the governor was working suitably before the overhaul, and none of the connections are altered, it will continue to work properly. While governor adjustment should be made with a small-engine tachometer to avoid setting the engine to a higher rpm than it is designed for, some basic adjustments can be made by the home mechanic.

8 On most engines, a governor shaft sticks through the side of the crankcase, while the governor mechanism is protected inside the crankcase. An arm attaches to this shaft with a clamp-bolt/nut. The basic initial adjustment on begins with loosening the nut/bolt where the arm attaches to the shaft. Pull the arm back away from the carburetor to its limit and hold it there. Grip the end of the shaft that sticks out beyond the arm with pliers and turn it counterclockwise to its limit. Tighten the bolt/nut on the arm while the arm and shaft are in this relationship.

9 Sensitivity is the other common adjustment for governors. If the engine seems like it is hunting for consistent speed as the load changes, the sensitivity is too sensitive and must be lowered. If the engine speed drops too much when the engine encounters a load, the governor adjustment isn't sensitive enough.

10 The governor sensitivity is adjusted by moving the position of the governor spring, where it hooks into the lower portion of the governor arm. A series of holes are provided. To increase the sensitivity, raise the spring into the next higher hole in the arm. To decrease sensitivity, hook the spring into a hole lower than its original position. The center of this arm is the factory position, but varying equipment and loads are accommodated by the extra adjustment holes.

| Engine model | |
|---|---|
| GX160 | |
| QXC type | 2-1/8 turns out |
| Except QXC type | 3 turns out |
| GX240 | 2 turns out |
| GX270 | |
| QXC type | 3 turns out |
| Except QXC type | 2-7/8 turns out |
| GX340 | |
| QXC and QXE types | 3 turns out |
| Except QXC and QXE types | 2-1/2 turns out |
| GX390 | 2-1/4 turns out |
| GX610 | 1-3/4 turns out |
| GX620 | 1-1/4 turns out |
| GXV160 | 2 turns out |
| GXV270 | 2-1/4 turns out |
| GXV340 | |
| 1995-on California models | 2-3/8 turns out |
| Except 1995-on California models | 2-1/2 turns out |
| GXV610 | 2 turns out |
| GXV620 | 2-1/4 turns out |

6.3b Carburetor initial setting

## 7 Valve clearance - check and adjustment

1 Valve clearance on all models is checked and adjusted with the engine cold. If possible, check the valve clearance before starting the engine for the day.

2 Remove the spark plug(s). This will make it easier to turn the engine by hand.

3 Remove the valve cover(s) from the engine (see Section 8 or Section 9).

4 Place the piston (the no. 1 piston on V-twin models) at Top Dead Center (TDC) on the compression stroke.

a) *On horizontal-crankshaft singles, align the triangular mark on the recoil starter pulley with the hole in the upper side of the fan cover* **(see illustration 7.4a).**

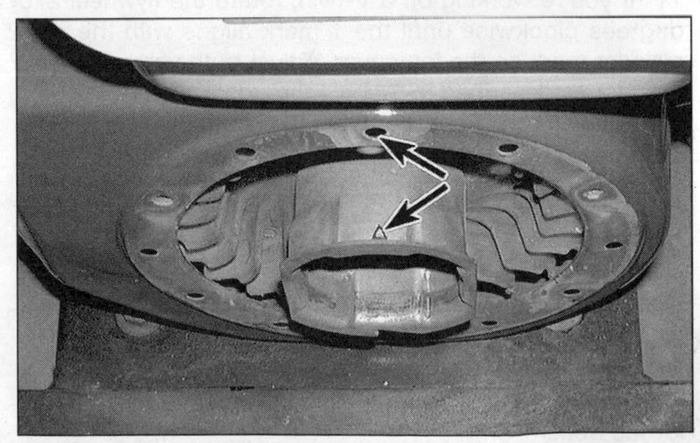
7.4a Crankshaft TDC marks - horizontal crankshaft single cylinder engines

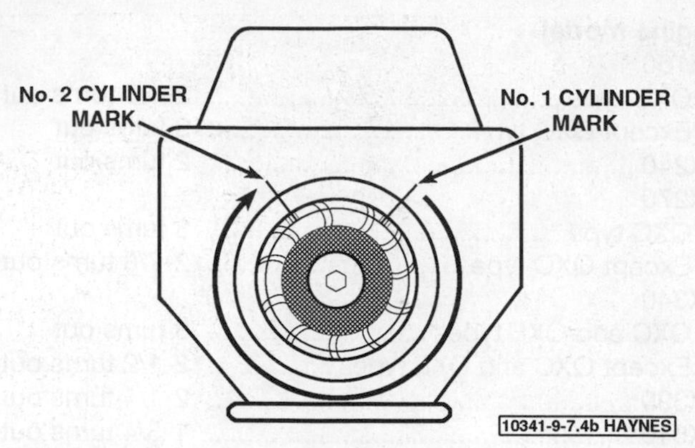


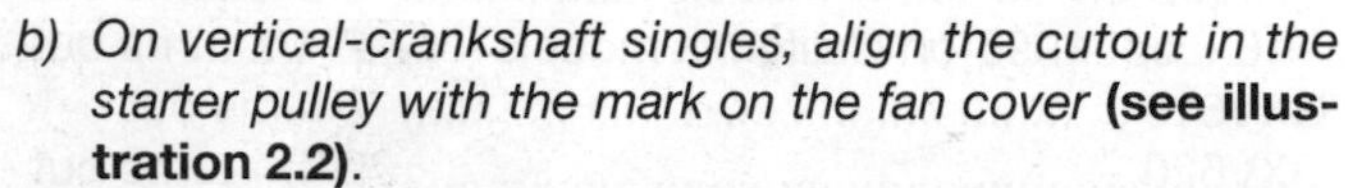
**7.4b On V-twin engines, align the mark on the flywheel with the no. 1 cylinder mark, adjust the no. 1 cylinder valves then rotate the flywheel 270-degrees clockwise, align the mark with the no. 2 cylinder mark and adjust the no. 2 cylinder valves**

*b) On vertical-crankshaft singles, align the cutout in the starter pulley with the mark on the fan cover* **(see illustration 2.2)**.

*c) On V-twins, align the line next to the T mark on the fan with the T mark on the fan cover* **(see illustration 7.4b)**.

5 Make sure the piston is on its compression stroke, not its exhaust stroke. To do this, wiggle the rocker arms. If they have some play, the cylinder is on the compression stroke. If they're tight, the cylinder is on its exhaust stroke. In this case, rotate the crankshaft one full turn and align the marks again.

6 Measure valve clearance with a feeler gauge **(see illustration 7.6)**. The feeler gauge should slip between the adjuster and valve stem with a light drag. If there's a heavy drag, or if the feeler gauge slides loosely, adjust the clearance as described below.

7 Loosen the adjuster locknut.

8 Turn the adjuster bolt (single-cylinder) or rocker arm pivot (V-twin) to obtain the correct clearance, then tighten the locknut.

9 Recheck the clearance to make sure it didn't change when you tightened the locknut.

10 Check the other valve and adjust it if necessary.

11 If you're working on a V-twin, rotate the flywheel 270-degrees clockwise until the T mark aligns with the no. 2 cylinder mark on the fan cover. Adjust both valves for no. 2 cylinder as described in Steps 4 through 8.

12 Install the valve cover(s) and the spark plug(s).

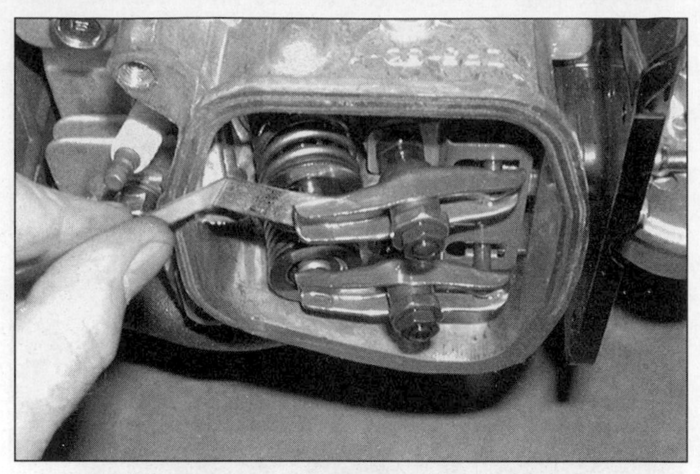
**7.6 Measure valve clearance with a feeler gauge**

# 8 Single-cylinder engines

## Disassembly

The engine components should be removed in the following general order:

*Engine cooling shroud/recoil starter*
*Fuel tank*
*Air cleaner*
*Carburetor/controls*
*Electric starter/control box (if used)*
*Muffler*
*Ignition components*
*Cooling fan*
*Flywheel*
*Charging coils*
*Flywheel brake components (if equipped)*
*Cylinder head/rocker arms/pushrods*
*Oil pan/crankcase cover*
*Camshaft*
*Tappets*
*Piston/connecting rod assembly(ies)*
*Balance shaft(s) (if equipped)*
*Crankshaft*
*Governor components*
*Valves*

1 For shroud/recoil starter, fuel tank, air cleaner, carburetor and muffler removal, refer to Chapters 4 and 5 as necessary. The remaining components can be removed to complete engine disassembly by following the sequence described below. **Note:** *Use a ridge reamer to remove the carbon/wear ridge (if present) from the top of the cylinder bore after the cylinder head is off. Follow the manufacturer's instructions included with the tool.*

2 Restrain the flywheel with a strap wrench and remove the large nut **(see illustration 8.2)**.

3 Remove the starter pulley and cooling fan.

4 The flywheel must be removed with a bolt-type puller (available at most automotive parts stores) **(see illustration 8.4). Caution:** *DO NOT strike the flywheel or crankshaft with a hammer!*

5 Remove the flywheel key and store it in a safe place so it doesn't get lost **(see illustration 8.5)**.

6 Refer to Section 4 and remove the charging coil(s).

7 If you're working on a vertical-crankshaft model, remove the bolts and detach the crankcase breather plate, then lift out the valve and wire mesh filter element **(see**

8.2 Remove the flywheel nut with a breaker bar or ratchet and socket

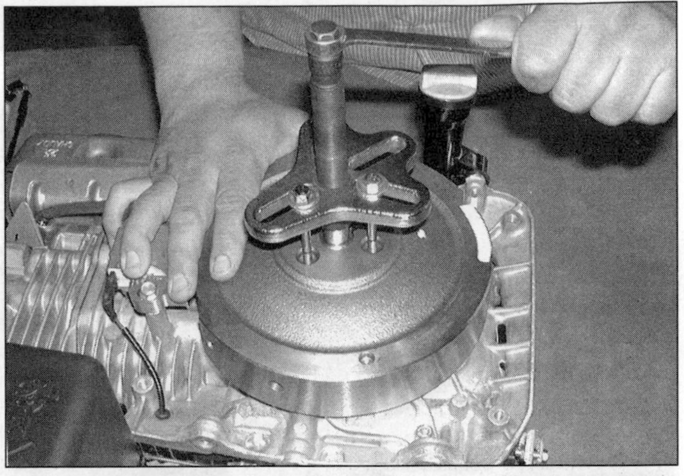

8.4 Use a bolt-type automotive puller to remove the flywheel

8.5 Remove the flywheel key and store it so it won't be lost

8.7a Remove the breather cover bolts (arrows) . . .

8.7b . . . and lift out the valve (and wire mesh screen if equipped)

**illustrations 8.7a and 8.7b)**.

8 Remove the bolts (arrows) and separate the under cover (if equipped) from the engine **(see illustration 8.8)**. Some engines have other brackets attached with bolts.

9 Remove the cover bolt, washer and rubber bushing, then detach the cover from the cylinder head - you may have to tap it with a soft-face hammer to break the gasket seal **(see illustration 8.9)**.

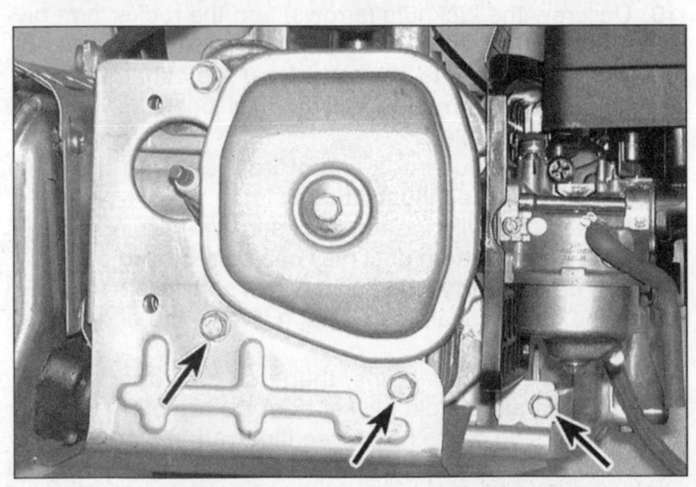

8.8 Remove the under cover bolts (arrows)

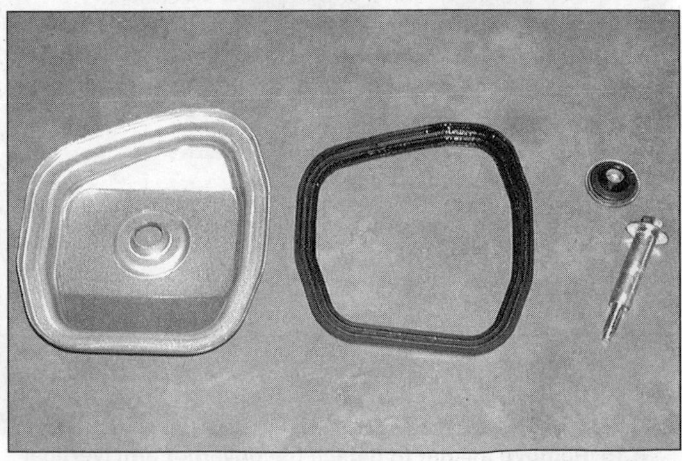

8.9 Valve cover details; note how the gasket fits into the cover groove

8.10 Unscrew the locknuts (arrows) to remove the rocker arms

8.11 Pull out the pushrods (arrows); label them so they can be installed in their original locations and directions

8.13 Tap the cylinder head with a plastic or rubber mallet to break the gasket seal

8.14 Pry the gasket loose; take care not to scratch the mating surface

8.15 Remove the nut to detach the governor arm (vertical crankshaft model shown; horizontal model similar)

10 Unscrew the locknuts (arrows) and the rocker arm pivots, then pull off the rocker arms **(see illustration 8.10)**. Store the intake parts separate from the exhaust parts - they should be returned to their original locations when the engine is reassembled.

11 Pull out the pushrods and store them with the other valve train parts **(see illustration 8.11)**.

12 Loosen the cylinder head bolts in 1/4-turn increments in a criss-cross pattern until they can be removed by hand.

13 Tap up on the cylinder head with a soft-face hammer to break the gasket seal, then detach it from the engine **(see illustration 8.13)**.

14 Pull out the dowel pins, then use a scraper or putty knife to separate the old gasket from the top of the cylinder or the underside of the head **(see illustration 8.14)**.

15 Note how it's positioned on the shaft (mark the shaft and arm if necessary), then remove the nut/bolt and detach

8.17 Loosen the crankcase cover or oil pan bolts evenly in stages to prevent warping the cover or pan

8.19 Remove the governor crank retaining clip and withdraw the crank from the engine

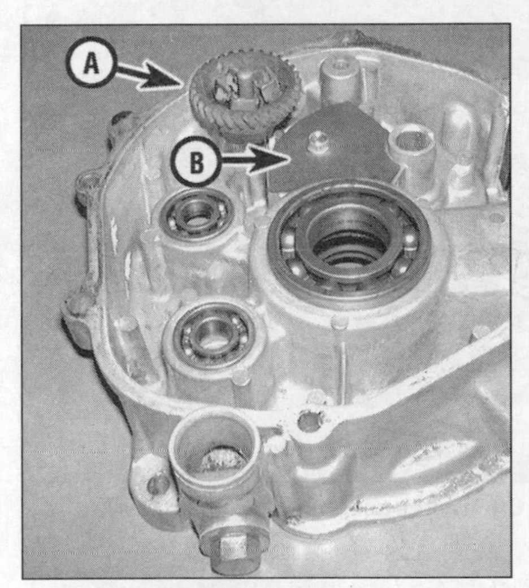


8.20 Oil pan (vertical crankshaft single cylinder engine)

*A Governor*
*B Oil screen cover*

8.22 Some single cylinder engines have two balancer shafts (shown); others have only one

*A Balancer-to-camshaft timing marks*
*B Balancer-to-balancer timing marks*

8.23 Typical camshaft-to-crankshaft timing marks; these must be aligned for the engine to run properly

the governor arm **(see illustration 8.15)**.

16 Use emery cloth to remove rust and burrs from the drive end of the crankshaft so the bearing in the oil pan or cover can slide over it.

17 Loosen the bolts that secure the crankcase cover (horizontal crankshaft) or oil pan (vertical crankshaft) evenly in a criss-cross pattern to avoid warping the cover, then remove them **(see illustration 8.17)**.

18 Tap the oil pan or crankcase cover with a soft-face hammer to break the gasket seal, then separate it from the engine block and crankshaft - if it hangs up on the crankshaft, continue to tap on it with the hammer, but be very careful not to crack or distort it. If a thrust washer is installed on the camshaft, slide it off and set it aside.

19 Remove the retaining clip and pull the governor shaft out of the crankcase bore **(see illustration 8.19)**.

## Vertical-crankshaft single cylinder engines

20 Remove the clip and separate the governor assembly from the oil pan **(see illustration 8.20)**. Remove the oil screen as well.

21 Vertical-crankshaft single cylinder engines have two balance shafts. They must be aligned with each other, as well as with the crankshaft, to prevent severe engine vibration.

22 Turn the crankshaft so the timing marks on the two balance shafts align with each other **(see illustration 8.22)**. Lift the balance shafts out of the crankcase.

23 Turn the crankshaft until the marks on the timing gears are aligned, then lift out the camshaft **(see illustration 8.23)**.

8.27 Remove the tappets from their bores

8.28 There should be a raised rib across the rod and cap (arrow); if you don't see one, make your own mark so the rod and cap can be reassembled correctly

8.29 Loosen the cap nuts evenly

8.32 Unscrew the rocker assembly studs (arrow)

## Horizontal-crankshaft single cylinder engines

24 Remove the clip and separate the governor assembly from the oil pan **(see illustration 8.20)**.

25 Turn the crankshaft until the marks on the timing gears are aligned, then lift out the camshaft **(see illustration 8.22)**.

26 Turn the crankshaft so the timing marks on the balance shaft gear and the crankshaft gear align with each other **(see illustration 8.23)**. Lift the balance shaft out of the crankcase.

## All models

27 After the camshaft is removed, pull out the tappets and store them in marked containers so they can be returned to their original locations **(see illustration 8.27)**.

28 Look for a raised rib (match mark) that extends across the rod and cap - if you can't see one, mark the side of the connecting rod and cap that faces out and note how the oil dipper (if used) is oriented **(see illustration 8.28)**. The parts must be reassembled in the exact same relationship to the crankshaft.

29 Loosen the connecting rod cap bolts in 1/4-turn increments until they can be removed by hand. Separate the cap from the connecting rod, move the end of the rod away from the crankshaft journal and push the piston/rod assembly out through the top of the bore **(see illustration 8.29)**. Place the connecting rod cap back on the connecting rod for safe keeping.

30 Remove the crankshaft.

# Inspection of components

31 After the engine has been completely disassembled, refer to Chapter 5 for the cleaning, component inspection and valve lapping procedures. Once you've inspected and serviced everything covered in Chapter 5 and purchased any necessary new parts, which should always include new gaskets and seals, follow the inspection procedures covered here before proceeding with engine reassembly.

32 Check the rocker arm pivot stud threads for wear and

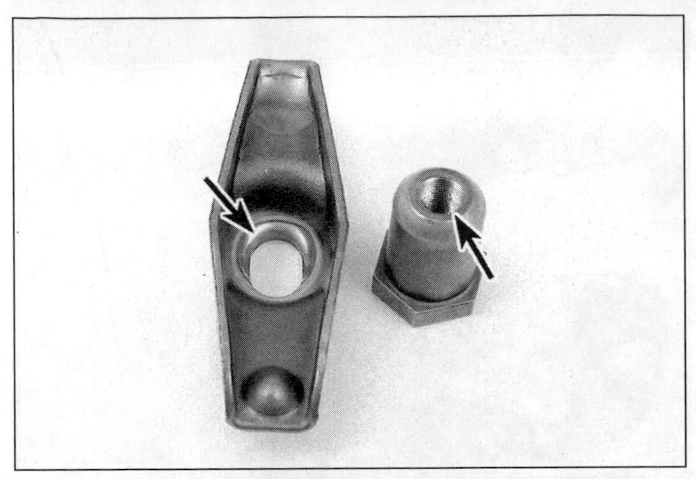

8.33 **Inspect the wear points of rocker arms and pivots (arrows) . . .**

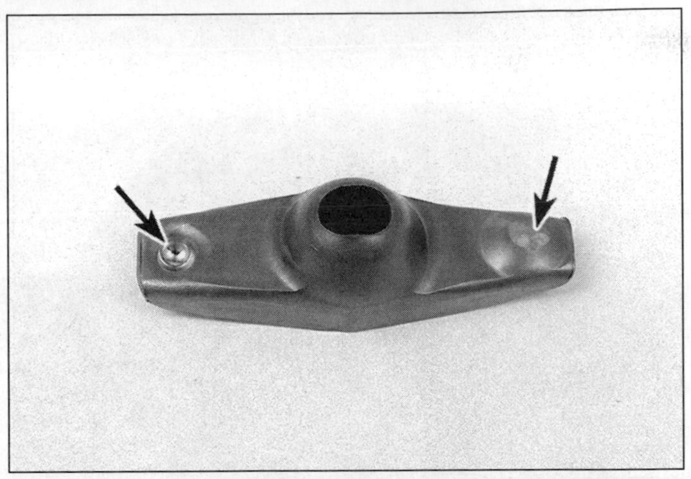

8.34 **. . . as well as the points where the rocker arms contact the pushrod and valve stem (arrows)**

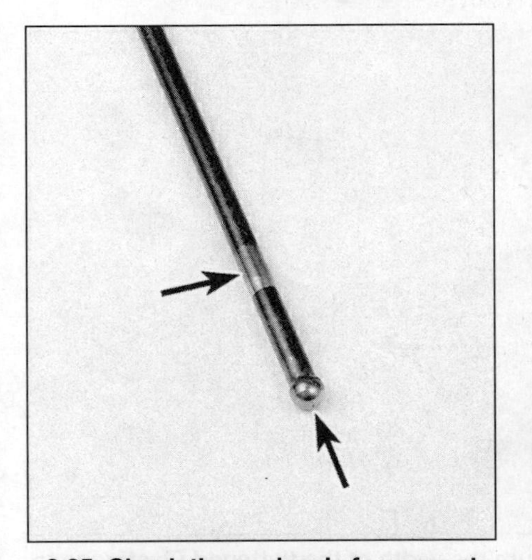

8.35 **Check the pushrods for wear at the ends and where they contact the guide plate (arrows)**

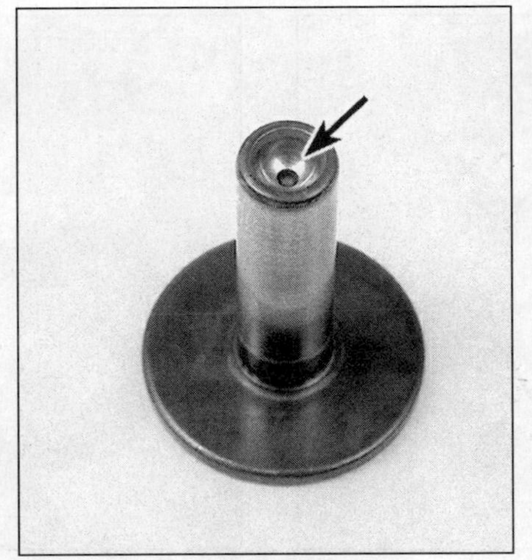

8.36 **Check the tappet sockets for wear**

8.37 **Spin the ball bearings to check for looseness, roughness or noise**

damage. If new ones must be installed, they can be unscrewed from the cylinder head with a socket on the hex **(see illustration 8.32)**.

33 Check the rocker arm sockets and the pivot balls for wear and galling **(see illustration 8.33)**.

34 Check the rocker arm surfaces that contact the valves and pushrods for wear, galling and pitting **(see illustration 8.34)**.

35 Check each pushrod for wear on the ends and where it rides in the guide **(see illustration 8.35)**. Look for any indication the pushrod is bent or otherwise distorted.

36 Check the tappet pushrod sockets for wear and galling **(see illustration 8.36)**.

37 Check the ball-bearing(s) for radial and side-to-side play and make sure they turn smoothly **(see illustration 8.37)**. They must be securely mounted in the engine bores.

38 Check the connecting rod bearing surface as described in Chapter 5 - if it's scored, it shouldn't be reinstalled **(see illustration 8.38)**.

39 If the corresponding journal on the crankshaft is also

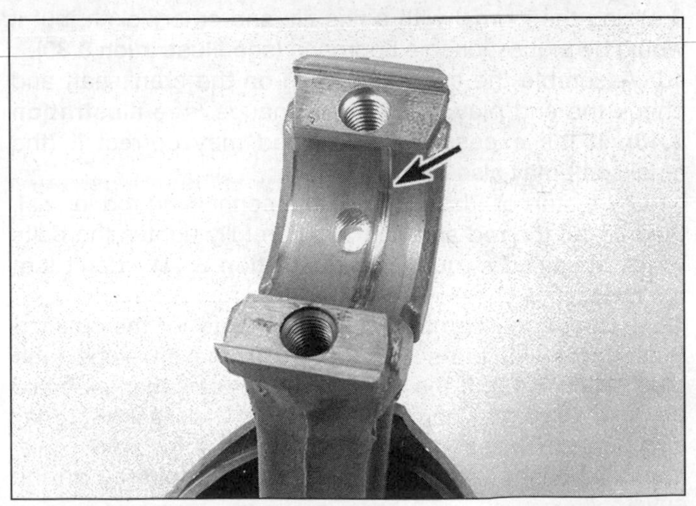

8.38 **If there's a groove like this on the connecting rod bearing surface, the rod should be replaced**

8.39 Replace the crankshaft if it's worn like this

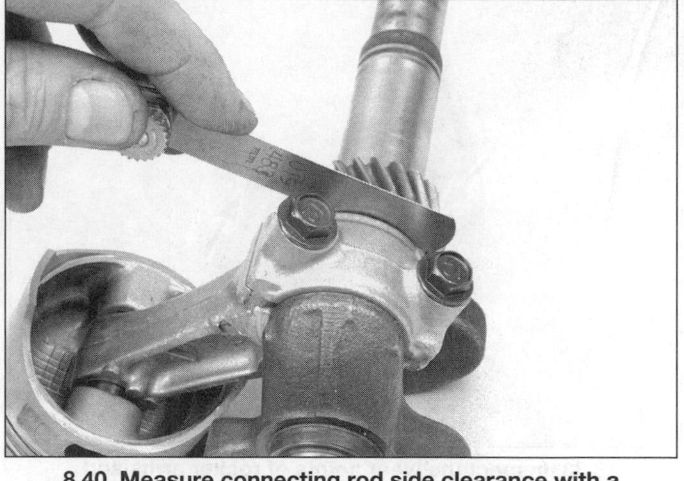

8.40 Measure connecting rod side clearance with a feeler gauge

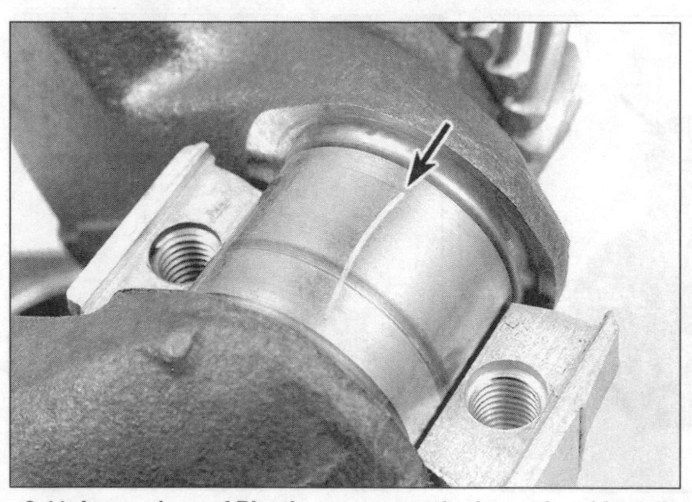

8.41 Lay a piece of Plastigage across the journal and install the rod cap . . .

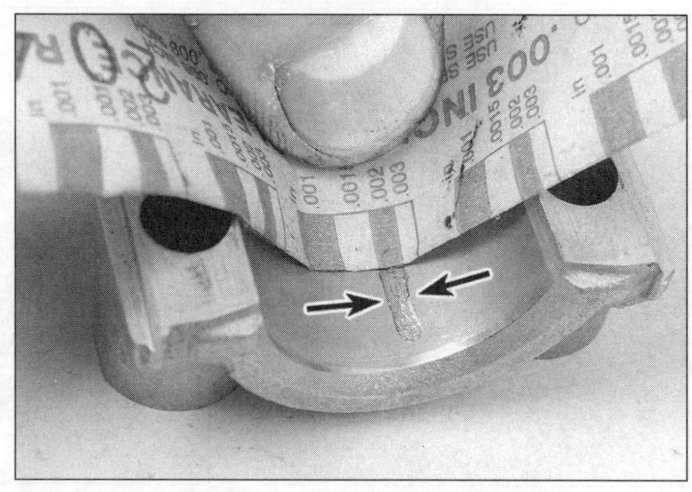

8.42 . . . then measure the width of the flattened Plastigage to determine bearing clearance

damaged, it may be possible to salvage the crankshaft by dressing the journal with a fine file and emery cloth, but it would be a questionable approach **(see illustration 8.39)**.

40 Assemble the connecting rod on the crankshaft and check the end play with a feeler gauge **(see illustration 8.40)**. If it's excessive, a new rod may correct it (the crankshaft may also be worn).

41 Lay a strip of Plastigage on the connecting rod journal, then install the rod and cap and carefully tighten the bolts to the specified torque **(see illustration 8.41)** - don't turn the rod as this check is done!

42 Remove the cap and check the width of the crushed Plastigage with the scale printed on the envelope **(see illustration 8.42)**. If the clearance is greater than it should be, a new rod may correct it (however, the crankshaft may also be worn excessively). Be sure to use the correct scale; standard (inch) and metric ones are both printed on the envelope.

43 Check the decompression mechanism on the camshaft to make sure it moves freely and the spring hasn't sagged **(see illustration 8.43)**.

44 Install a new governor shaft oil seal in the oil pan **(see illustration 8.44)**. Carefully pry the old one out and drive the new one in with a socket and hammer.

8.43 Check the decompressor on the camshaft (if equipped) for wear or damage

8.44 Replace the governor shaft seal (arrow)

8.50a Apply clean engine oil or multi-purpose grease the crankshaft seal before installing the crankshaft

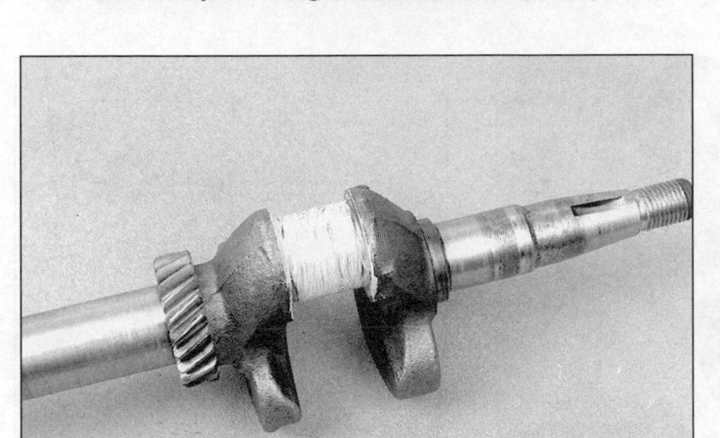

8.50b Coat the connecting rod journal with engine assembly lube

8.51 Install a piston ring compressor on the piston

45 If the engine has an oil pump, remove the cover. Measure the following dimensions:

*a) Outer rotor-to-body clearance*
*b) Rotor tip clearance*
*c) Rotor side clearance*

46 Replace the rotors if the tip clearance is not within the range listed in this Chapter's Specifications, or if the rotors are visibly scored or worn.
47 If the outer rotor-to-body clearance is excessive, remove the rotors and measure the diameter of the pump bore in the oil pan or crankcase cover. Replace the pan or cover if it's beyond the limit listed in this Chapter's Specifications.
48 If the rotor side clearance is excessive, measure the height of the outer rotor and the depth of the pump bore. Replace the rotors as a set or replace the oil pan/crankcase cover, whichever is worn.

# Engine reassembly

49 Install the governor assembly in the oil pan/crankcase cover **(see illustration 8.20)**. If you're working on a vertical crankshaft engine, install the oil filter screen in the cover, narrow edge first.
50 Apply clean engine oil or multi-purpose grease to the magneto side oil seal and engine assembly lube to the connecting rod journal on the crankshaft, then install the crankshaft in the engine **(see illustrations 8.50a and 8.50b)**.
51 Before installing the piston/connecting rod assembly, the cylinder must be perfectly clean and the top edge of the bore must be chamfered slightly so the rings don't catch on it. Refer to Chapter 5 and install the piston rings on the pistons. Stagger the piston ring end gaps approximately 120-degrees around the top of the piston and do not align any ring gap with the piston pin bore. Lubricate the piston and rings with clean engine oil, then attach a ring compressor to the piston **(see illustration 8.51)**. Leave the skirt protruding about 1/4-inch. Tighten the compressor until the piston cannot be turned, then loosen it until the piston turns in the compressor with resistance.
52 Rotate the crankshaft until the connecting rod journal is at TDC (Top Dead Center - top of the stroke) and apply a coat of engine oil to the cylinder walls. The triangular mark in the top of the piston must face the pushrod side of the engine. Make sure the match marks on the rod and cap will

8.52 Tap the piston into the bore, taking care that the rod doesn't damage the crankshaft

8.54 Note the installed direction of the mark on the piston

8.56 Lubricate the crankshaft ball bearing (arrow) with oil

8.57 Lubricate the tappets and install them in their bores

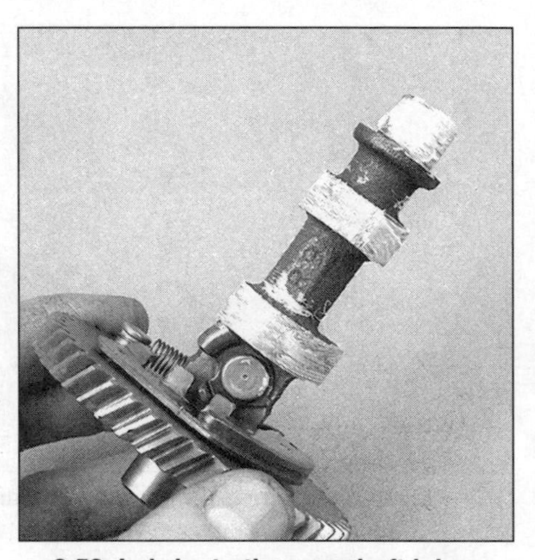

8.58 Lubricate the camshaft lobes and journal

be facing out when the rod/piston assembly is in place. Gently insert the piston/connecting rod assembly into the cylinder and rest the bottom edge of the ring compressor on the engine block **(see illustration 8.52)**. Tap the top edge of the ring compressor to make sure it's contacting the block around its entire circumference.

53 Carefully tap on the top of the piston with the end of a wooden or plastic hammer handle while guiding the end of the connecting rod into place on the crankshaft journal. The piston rings may try to pop out just before entering the bore, so keep some pressure on the ring compressor. Work slowly - if any resistance is felt as the piston enters the cylinder, stop immediately. Find out what's hanging up and fix it before proceeding. Do not, for any reason, force the piston into the cylinder - you'll break a ring and/or the piston.

54 When the piston/rod assembly is installed correctly, the triangular mark on the piston crown will be pointing at the pushrod side of the engine **(see illustration 8.54)**.

55 Install the connecting rod cap and the bolts. Make sure the marks you made on the rod and cap (or the manufacturer's rib) are aligned and facing out and the oil dipper (if used) is oriented correctly **(see illustration 8.28)**. Tighten the bolts to the torque listed in this Chapter's Specifications. Work up to the final torque in three steps. Temporarily install the camshaft and turn the crankshaft through two complete revolutions to make sure the rod doesn't hit the cylinder or camshaft. If it does, the piston/connecting rod is installed incorrectly.

56 Lubricate the magneto side ball-bearing with clean engine oil **(see illustration 8.56)**.

57 Apply clean engine oil or engine assembly lube to the tappets, then reinstall them - make sure they're returned to their original locations **(see illustration 8.57)**.

58 Apply clean engine oil or engine assembly lube to the camshaft lobes and the lower bearing journal **(see illustration 8.58)**.

59 Install the camshaft and balance shaft(s) by reversing the removal Steps. Be sure to align the index marks: camshaft to crankshaft, balancer to crankshaft, and balancer B (if equipped) to balancer A. Lubricate the main bearing journal on the crankshaft and the upper journal on the camshaft before proceeding.

60 Lubricate the governor shaft and install it in the

8.60 Lubricate and install the governor shaft; don't forget its washer

8.61 With the dowels in place and the governor shaft arm vertical (arrows), install the gasket

8.64 Make sure the cylinder head dowels (arrows) are in place

8.65 Cylinder head TIGHTENING sequence

crankcase, along with the washer (on the inside of the case) **(see illustration 8.60)**.

61 Make sure the dowel pins are in place and the governor shaft arm is vertical, then position a new gasket on the crankcase (the dowel pins will hold it in place) **(see illustration 8.61)**.

62 Lubricate the lip on the oil seal in the oil pan (or crankcase cover) with clean engine oil or engine assembly lube. If a ball bearing is used in the cover, lubricate it with clean engine oil. Carefully lower the oil pan (or crankcase cover) into place over the end of the crankshaft until it seats on the crankcase. **Caution:** *DO NOT damage the oil seal lip or leaks will result!*

63 Install and tighten the oil pan or crankcase cover mounting bolts to the torque listed in this Chapter's Specifications. Follow a criss-cross pattern and work up to the final torque in three equal steps to avoid warping the oil pan. Attach the governor lever to the outer end of the shaft and tighten the nut/bolt securely.

64 Make sure the dowel pins are in place, then position a new gasket on the head **(see illustration 8.64)**. DO NOT use sealant on the gasket.

65 Install the cylinder head and the bolts, then following the recommended tightening sequence, tighten the bolts to the torque listed in this Chapter's Specifications **(see illustration 8.65)**. Work up to the final torque in three equal steps.

66 Install the pushrods and make sure they're engaged in the tappet sockets, then lubricate the pushrod and valve

8.66 Lubricate the pushrod tips with engine assembly lube

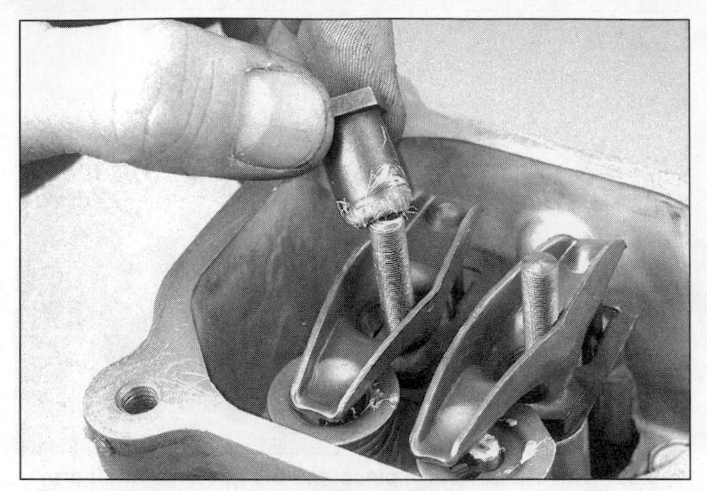

8.67 Lubricate the rocker pivots with engine assembly lube and install them

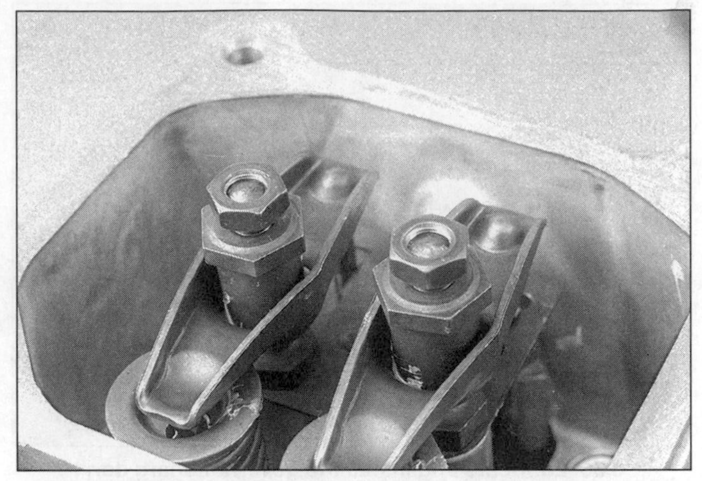

8.68 Install the locknuts after the pivots

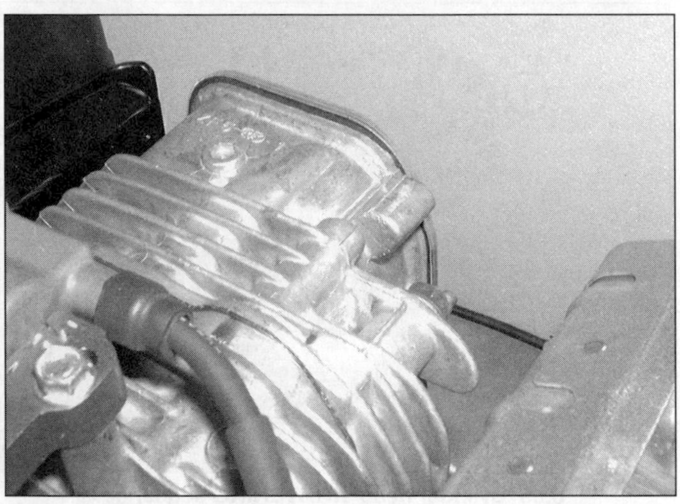

8.69 Make sure the valve cover gasket is installed right side up

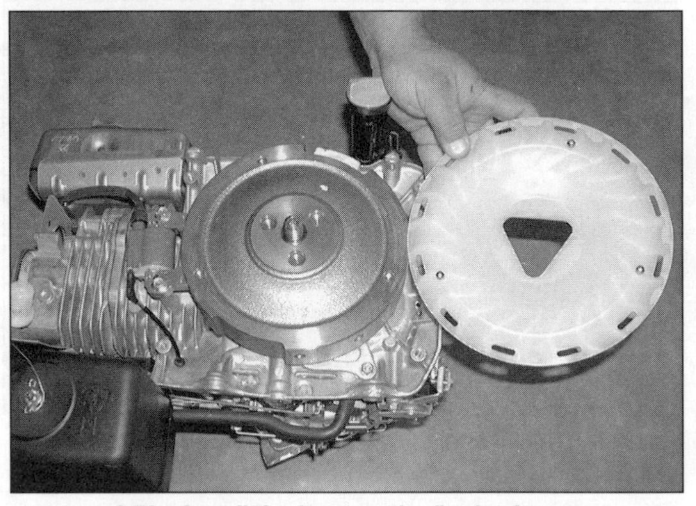

8.72a Install the fan over the flywheel pegs

stem ends with engine assembly lube **(see illustration 8.66)**.

67 Position the rocker arms, then lubricate and install the pivots **(see illustration 8.67)**.

68 Thread the locknuts onto the studs, then refer to Section 6 and adjust the valve clearance **(see illustration 8.68)**.

69 Install the valve cover and a new gasket **(see illustration 8.69)**.

70 Position the crankcase breather wire mesh filter and the valve in the recess, then install the cover and tighten the bolts securely **(see illustrations 8.7b and 8.7a)**.

71 Check the taper on the crankshaft to make sure it's clean, install the flywheel key then install the flywheel.

72 Install the plastic fan (make sure the pegs fit into the holes in the flywheel) and the starter pulley (make sure the pegs in the flywheel fit into the cup holes) **(see illustrations 8.72a and 8.72b)**.

73 Install the flywheel nut and tighten it to the torque listed in this Chapter's Specifications **(see illustration 8.2)**.

74 The remainder of installation is the reverse of the removal Steps. Refer to other Sections, and to Chapters 4 and 5, as necessary. **Caution:** *Be sure to fill the crankcase to the correct level with the specified oil before attempting to start the engine.*

8.72b Install the recoil starter pulley

# 9 V-twin engines

1 V-twin engines are similar to single cylinder versions, with the following differences.
2 The valve covers are secured by several bolts around the edge of the cover, rather than one bolt in the center.
3 The rocker arms pivot on shafts, which are a slip fit in the cylinder head. When removing the rocker shafts on a vertical crankshaft V-twin engine, thread a 6 mm bolt into the upper end of the rocker shaft to use as a handle so the shaft won't fall down out of its bore.
4 There's a matching cylinder number on each cylinder head and cylinder to prevent the heads from being installed on the wrong cylinders.
5 An oil pump and filter screen are mounted in the oil pan. To remove it, unbolt the pump cover and lift the pump drive gear, cover and inner rotor out of the outer rotor. The outer rotor can then be lifted out of the pump bore in the oil pan (vertical crankshaft) or crankcase cover (horizontal crankshaft). Inspection is the same as for single cylinder engines, described in Chapter 8.
6 The governor on horizontal crankshaft models is mounted on the outer end of the camshaft. It should move smoothly when the camshaft is spun, pushing the slider plate against the slider. Check the balls and their grooves in the camshaft for wear and replace the camshaft if any problems are found. It's a good idea to replace the snap-ring with a new one whenever the governor is disassembled.
7 The connecting rod uses replaceable plain bearings, similar to those used in automotive engines. The bearing inserts can be removed by pushing their centers to the side, then lifting them out. Note which insert came from the connecting rod and which from the cap. A color code painted on the edge of the bearing insert identifies its thickness **(see illustration 9.7a)**. To select a replacement bearing, refer to the letter stamped on the crankshaft throw at the timing gear end of the crankshaft, and the code number stamped across the joint of each connecting rod and cap. Use the letter and number together with the accompanying chart to select the bearing insert **(see illustration 9.7b)**.

**9.7a Bearing thickness is indicated by a color code painted on the side of the bearing**

8 When you install the pistons and connecting rods, make sure the FW mark on the top of each piston faces the flywheel side of the engine. There's a "1" mark on the no. 1 cylinder connecting rod (this is the cylinder on the left when the open end of the crankcase is toward you). The "1" mark should face toward you (toward the open side of the crankcase). The "2" mark on the no. 2 cylinder connecting rod should also face toward you.
9 When you install the connecting rod caps, make sure the number stamped across the rod and cap is legible. If the halves of the number on the rod and cap are misaligned, the wrong cap is on the rod.

| CONNECTING ROD MARK | CRANKPIN MARK A | B | C | D |
|---|---|---|---|---|
| 1 | RED | PINK | YELLOW | GREEN |
| 2 | PINK | YELLOW | GREEN | BROWN |
| 3 | YELLOW | GREEN | BROWN | BLACK |
| 4 | GREEN | BROWN | BLACK | BLUE |

**9.7b Use the crankpin letter mark and connecting rod number mark to select the connecting rod bearings**

# 10 Specifications

## Engine oil

Type ............ API grade SH or better high quality detergent oil

Viscosity

GX160, GX240, GX270, GX340, GX390, GX610, GX620, GXV610, GXV620*

Above 50 degrees F ............ SAE 30

Zero to 85 degrees F ............ SAE 10W-30

Below 32 degrees F ............ SAE 5W-30

GVX160, GXV270, GXV340, GXV390**

Above -5 degrees F ............ SAE 10W-40

-5 to 90 degrees F ............ SAE 10W-30

## Engine oil (continued)

Engine oil capacity

| | |
|---|---|
| GX160 | 0.63 quart |
| GXV160 | 0.69 quart |
| GX240, GX270, GX340, GX390 | 1.16 quarts |
| GXV270, GXV340, GXV390 | 1.2 quarts |
| GX610, GX620 | |
| without oil filter replacement | 1.27 quarts |
| with oil filter replacement | 1.59 quarts |
| GXV610, GXV620 | |
| without oil filter replacement | 1.90 quarts |
| with oil filter replacement | 2.23 quarts |

**SAE 10W-30 is recommended for general all temperature use.*
***SAE 10W-40 is recommended for general all temperature use.*

## Fuel system

| | |
|---|---|
| Idle speed | |
| GXV160 | 1700 rpm |
| All others | 1400 rpm |
| Float level | |
| Single cylinder engines | |
| GXV160 | 31/64 to 39/64 inch |
| All others | 33/64 inch |
| V-twin engines | 35/64 inch |

## Ignition system

| | |
|---|---|
| Spark plug type (NGK) | |
| Horizontal crankshaft engines | BPR6ES |
| Vertical crankshaft engines | BPR5ES |
| Spark plug gap | 0.028 to 0.031 inch |
| Ignition coil-to-flywheel gap | 0.016 inch |

## Starting system

| | |
|---|---|
| Brush length limit | |
| Horizontal crankshaft single-cylinder engines | |
| GX160 | 1/4 inch |
| All others | 9/64 inch |
| Vertical crankshaft single-cylinder engines | 11/32 inch |
| V-twin engines | 19/64 inch |
| Mica depth limit | |
| GX160 | 0.040 inch |
| All others | 0.010 inch |

## Engine

| | |
|---|---|
| Cylinder bore diameter limit | |
| Single-cylinder engines | |
| GX160, GXV160 | 2.684 inches |
| GX240 | 2.881 inches |
| GX270, GXV270 | 3.038 inches |
| GX340, GXV340 | 3.235 inches |
| GX390, GXV390 | 3.471 inches |
| V-twin engines | 3.038 inches |

| Specification | Limit |
|---|---|
| **Piston diameter limit** | |
| Single-cylinder engines | |
| GX160 | 2.671 inches |
| GXV160 | 2.675 inches |
| GX240 | 2.859 inches |
| GX270, GXV270 | 3.026 inches |
| GX340, GXV340 | 3.222 inches |
| GX390, GXV390 | 3.459 inches |
| V-twin engines | 3.026 inches |
| **Piston-to-cylinder bore clearance limit** | 0.005 inch |
| **Piston-to-pin clearance limit** | |
| GX160 | 0.002 inch |
| All others | 0.003 inch |
| **Piston ring side clearance limit** | 0.006 inch |
| **Piston ring end gap service limit** | 0.039 inch |
| **Connecting rod bearing clearance limit** | |
| Single-cylinder engines | 0.005 inch |
| Horizontal crankshaft V-twin engines | 0.003 inch |
| Vertical crankshaft V-twin engines | 0.002 inch |
| **Connecting rod side clearance limit** | |
| Horizontal crankshaft single-cylinder engines | |
| GX160 | 0.043 inch |
| All others | 0.039 inch |
| Vertical crankshaft single-cylinder engines | 0.043 inch |
| V-twin engines | 0.020 inch |
| **Connecting rod journal diameter limit** | |
| Single-cylinder engines | |
| GX160, GXV160 | 1.178 inch |
| GX240, GX270, GXV270 | 1.296 inch |
| GX340, GXV340, GX390, GXV390 | 1.415 inch |
| V-twin engines | 1.572 inch |
| **Camshaft journal diameter limit** | |
| Single-cylinder engines | |
| GX160, GXV160 | 0.548 inch |
| All others | 0.627 inch |
| V-twin engines | 0.670 inch |
| **Camshaft holder diameter limit (in crankcase cover)** | |
| Single-cylinder engines | |
| GX160 | 0.553 inch |
| All others | 0.632 inch |
| V-twin engines | 0.672 inch |
| **Cam lobe height limit** | |
| Single-cylinder engines | |
| GX160, GXV160 | |
| Intake | 1.081 inch |
| Exhaust | 1.083 inch |
| GX240 and GX270 | 1.234 inch |
| GX340 | |
| Intake | 1.224 inch |
| Exhaust | 1.252 inch |
| GX390 | |
| Intake | 1.270 inch |
| Exhaust | 1.250 inch |
| GXV270, GXV340 | |
| Intake | 1.289 inch |
| Exhaust | 1.274 inch |

## Engine (continued)

Cam lobe height limit
  Single-cylinder engines
    GXV390
      Intake ........ 1.274 inch
      Exhaust ........ 1.254 inch
  V-twin engines ........ 1.160 inch
Oil pump
  Vertical crankshaft single-cylinder engines
    Rotor tip clearance limit ........ 0.012 inch
    Outer rotor-to-body clearance limit ........ 0.010 inch
    Outer rotor height limit ........ 0.293 inch
    Pump body depth ........ 0.298 inch
    Rotor side clearance limit ........ 0.004 inch
  V-twin engines
    Rotor tip clearance limit ........ 0.012 inch
    Outer rotor-to-body clearance limit ........ 0.012 inch
    Rotor side clearance limit ........ 0.005 inch
Valve stem-to-guide clearance limit
  Intake ........ 0.004 inch
  Exhaust ........ 0.005 inch
Valve clearance (cold)
    Intake ........ 0.005 to 0.007 inch
    Exhaust ........ 0.007 to 0.009 inch

## Torque specifications

Connecting rod bolts
  Single-cylinder engines
    GX160, GXV160 ........ 9 ft-lbs
    All others ........ 10 ft-lbs
  V-twin engines ........ 12 ft-lbs
Crankcase cover bolts (horizontal crankshaft) ........ 17 ft-lbs
Cylinder head bolts
  Single-cylinder engines
    GX160, GXV160 ........ 18 ft-lbs
    All others ........ 25 ft-lbs
  V-twin engines ........ 22 ft-lbs
Flywheel nut
  Single-cylinder engines
    GX160, GXV160 ........ 55 ft-lbs
    All others ........ 83 ft-lbs
  V-twin engines ........ 145 ft-lbs
Oil drain plug
  Single-cylinder engines
    GX160, GXV160 ........ 13 ft-lbs
    GX240, GX270, GX340, GX390 ........ 17 ft-lbs
    GXV270, GXV340, GXV390 ........ 29 ft-lbs
  V-twin engines ........ 29 ft-lbs
Oil pan bolts (vertical crankshaft) ........ 17 ft-lbs
Rocker pivot locknuts (single-cylinder engines) ........ 84 inch-lbs
Rocker arm pivot bolts (single-cylinder engines) ........ 17 ft-lbs

# 10 Robin/Wisconsin Robin engines

## Contents



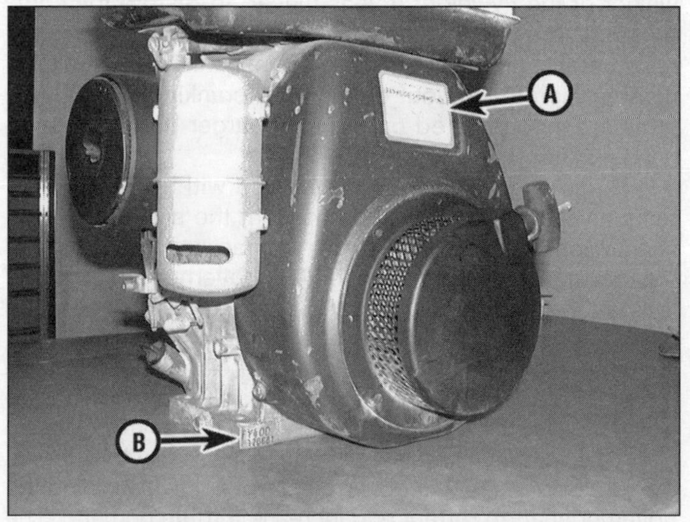


**1.1a Location of model identification decal (A) and serial number (B) on Robin/Wisconsin Robin L-head engines**

## 1 Engine identification numbers/models covered and general information

1 The engine designation system used by Robin consists of a model and serial number. The model number is stamped onto the recoil housing or shroud while the engine identification number is stamped onto the crankcase base. The number(s) can be used to determine the exact replacement parts number in the Robin parts catalogs **(see illustrations 1.1a and 1.1b)**. The letters in the model number can be explained generally as follows:

**1.1b Location of the serial number (arrow) on Robin/Wisconsin Robin OHV engines**

2 The first two letters in a model number indicate the basic engine type. The letter V at the end of the model number indicates a vertical shaft engine design:

*EH = Overhead valve design*
*EY = L-head valve design*
*V = Vertical shaft*

This manual covers the following L-head engines:

| Robin | Wisconsin Robin | Horsepower rating |
|---|---|---|
| EY23 | W1-230 | 6 HP |
| EY25 | EY25W | 6.5 HP |
| EY27 | EY27W | 7.5 HP |
| EY28 | W1-280 | 8 HP |
| EY35 | W1-340 | 9 HP |
| EY40 | W1-390 | 11 HP |
| EY45V | W1-450V | 12 HP |

OHV engines covered by this manual include:

| Robin | Wisconsin Robin | Horsepower rating |
|---|---|---|
| EH17 | WO1-170 | 6 HP |
| EH21 | WO1-210 | 7 HP |
| EH25 | WO1-250 | 8.5 HP |
| EH30, EH30V | WO1-300(V) | 9 HP |
| EH34, EH34V | WO1-340(V) | 11 HP |
| EH43V | WO1-430V | 14 HP |

3 These Robin engines are distinctly separated by the types of engine design and the metallurgy of the engine components. There are basically two types of engines, either the horizontal shaft design or the vertical shaft design. The horizontal engine positions the crankshaft horizontally (vertical piston) while the vertical crankshaft engine design positions the crankshaft vertically (horizontal piston). Some models use the L-head design which places the valves in the block allowing the air/fuel mixture to enter the combustion chamber in an L direction. Other models are equipped with overhead valves (OHV) which places the valves in the cylinder head directly over the combustion chamber. Vertical engines are commonly used in lawnmowers, while horizontal crankshaft designs are used in pumps, snowblowers, etc. Metallurgy changes with engine designs also. Small frame engines are equipped with aluminum blocks and cylinder bores. Medium frame engines are equipped with aluminum blocks with cast iron sleeves while the heavy frame engines are equipped with the cast iron blocks and cylinder heads.

4 There are several different types of oiling systems on these engines. Smaller horsepower horizontal engines are equipped with a "splash oiling system." A dipper is installed onto the end of the connecting rod to agitate the oil in the sump, causing oil to splash over the cylinder walls and the bearing assembly. Vertical crankshaft engines are equipped with a rotary oil pump (lobed gear type) located in the sump and driven from the camshaft. On smaller horsepower engines, this system is called the "forced splash oiling system" because it uses a pump to pressurize the oil and splashes the oil over the cylinder walls and bearings through a special oiling port. Larger horsepower engines are also equipped with a rotary oil pump that forces the pressurized oil through the oil jet on the crankshaft arm over the piston, cylinder and connecting rod. This is called the "pressurized oiling system."

5 Some of these models are equipped with an automatic decompression mechanism to allow for a much easier "pull" on the recoil starter. The automatic decompression assembly is mounted on the camshaft. It releases the compression of the engine by lifting the exhaust valve while the engine is cranking. There are basically two different type systems; a flyweight system and a camshaft lobe system. The flyweight system incorporates a release lever that activates by centrifugal motion, lifting the exhaust valve slightly. After the engine starts, the rpm levels overcome the weight of the flyweight (release lever) to restore full compression. The camshaft lobe system incorporates a slightly raised profile on the intake lobe of the camshaft that allows compression to leak out during slow cranking. These systems are incorporated onto mostly larger horsepower L-head engines with vertical crankshafts.

6 These engines are also equipped with governor systems to ensure constant operation at the selected speed against load variations. The governor assembly consists of a centrifugal flyweight design. The governor assembly is located in the crankcase cover or the oil pan depending upon the type of crankshaft arrangement (vertical or horizontal). As the engine speed increases, the weights on the governor assembly move outward. The shape of the weights force the governor spool to lift thereby forcing the governor linkage using lever action to close the throttle. As the rpm decreases, the governor spool will fall, allowing the governor shaft to retract and increase throttle demand. The governor acts as a throttle and rpm limiter.

2.3 Remove the recoil starter mounting bolts (arrows) - L-head engine with a horizontal crankshaft shown

2.5 Pull the starter rope out of the reel and lock it into the notch (arrow) to hold it in place

# 2 Recoil starter service

**Note:** *These recoil assemblies may differ in construction but not in the way they operate. Follow the procedure carefully and be sure to make notes or pictures of the location of the various components to insure correct reassembly.*

## Rope replacement

1 Rope replacement in these models will require disassembly of the recoil starter assembly in order to access the slot where the rope end is installed. Rope replacement is much easier if the original rope remains intact inside the recoil assembly but, in the event the old rope has broken off inside, the remaining piece of old rope can be removed from the recoil assembly with needle-nose pliers. Use caution when working with the spring assembly to prevent accidents and lost parts.

2 Recoil starter related problems such as broken ratchet mechanism, incomplete rope recoiling, etc. will require the recoil starter to be removed from the engine and visually inspected. Rope replacement should be done with the correct size and length rope to ensure easy starting and pulling.

3 Remove the starter assembly mounting bolts and separate the recoil starter assembly from the blower **(see illustration 2.3)**.

4 If the rope has not been broken, pull the starter knob and pull out the starter rope at least 14 inches until the notch on the reel (recoil assembly) aligns with the outlet hole for the starter rope. If the rope has been broken off inside the recoil assembly, continue with the disassembly process.

5 Prevent the recoil assembly from rotating by holding it firmly with your thumb, then pull the rope slack out using a screwdriver and jam the rope into the notch in the reel **(see illustration 2.5)**.

6 Allow the recoil assembly to rewind slowly by releasing thumb pressure slight while gripping the rope. It should rewind in a clockwise direction. The recoil assembly will eventually stop and there should be sufficient length rope (slack) to remove the recoil assembly from the housing.

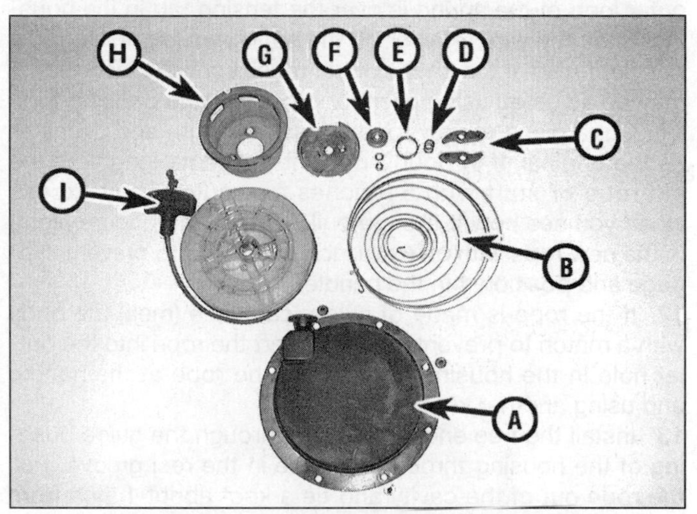


2.7 Exploded view of the recoil starter assembly on the L-head engine

*A Recoil assembly starter cover*
*B Recoil spring*
*C Ratchets*
*D Friction spring*
*E Return spring*
*F Thrust washer and snap-ring*
*G Friction spring cover*
*H Pulley*
*I Reel*

7 Remove the snap-ring or retaining nut, the thrust washer(s), the friction spring cover, spring and ratchet mechanism from the recoil assembly **(see illustration 2.7)**. Some recoil assemblies use a set screw to retain the recoil components. **Note:** *It is important to note the exact position of these components to insure correct reassembly. Make notes or pictures.*

8 Release the main recoil spring by moving the reel back and forth until it unlocks from the reel. Remove the reel from the housing. Lift carefully so as not to jolt the reel and release the main recoil spring from the reel. **Warning:** *Wear eye protection and gloves when handling the reel to prevent any accidents that may occur if the recoil spring accidentally disengages from the reel.*

9 Untie the knot from the old rope and remove the excess rope. **Note:** *The recoil spring is very difficult to install back into the reel. It is best if the spring is secured in a coiled position (wound) using a wire wrapped completely around the circumference of the spring. Tighten the spring by twisting the ends of the wire.* **Warning:** *Use caution when handling the recoil spring to avoid injury or damaged parts.*

10 Double-check the recoil spring to make sure it hooks into the reel securely. Place the new spring assembly over the recess in the housing and make sure the hook in the outer loop of the spring is over the tension tab in the housing. Slide the wire retainer off the spring while pressing into position. Also, if the shape of the recoil spring is incorrect, it may be adjusted using pliers or other suitable bending tool.

11 Cut a piece of new rope the same length and diameter as the original. If in doubt, make it the same length as the old rope or start with 54-inches and cut off any excess when you see how it fills the pulley. Tie a knot (figure eight) in the new rope with an extra inch at the end to prevent slippage and position it in the handle.

12 If the rope is made of nylon, cauterize (melt) the ends with a match to prevent fraying. Insert the rope into the outlet hole in the housing and secure the rope at the handle end using another knot.

13 Install the free end of the rope through the guide bushing of the housing through the hole in the reel groove. Pull the rope out of the cavity and tie a knot about 1 inch from the end (figure eight). Tuck the knot into the cavity opening. **Note:** *On some smaller horsepower L-head engines, place the slip knot around the center bushing and pull the knot tight. Tuck the end of the rope (figure eight) into the reel cavity.*

14 Before installing the reel into the housing, wind the rope around the reel 2 to 3 times in a counterclockwise direction. Bring up the slack in the rope and line up the reel hook with the inner end of the spring and install the reel into the housing. Rotate the reel counterclockwise until the tang on the reel engages the hook on the on the inner loop of the rewind spring.

15 Hook the loop of the rope into the reel notch and using your hand, preload the recoil spring by rotating the pulley counterclockwise four full turns and retaining the recoil in that position with the thumb. Carefully wind the starter by holding tension with the rope handle until the rope completely coils around the reel inside the housing. **Note:** *On some smaller horsepower L-head engines that lock the rope end over the center guide bushing, preload the recoil assembly by rotating the reel seven full turns counterclockwise.*

16 Install the ratchet, spring, friction plate and spring and the center screw into the reel. Make sure the ratchets are tensioned by the ratchet springs toward the center of the recoil. Install the friction plate with the bosses set inside the bent portions of the ratchets. Be sure to apply a small amount of thread sealing compound to the set screw before installation.

17 Lubricate the rotating parts with heat-temperature grease or light oil.

18 Test the recoil operation to make sure it ratchets properly and returns the handle to the stop position.

19 In case the main recoil spring jumps out of the locating notches in the reel, it will be necessary to devise a special tool to retain the spring inside the reel while assembling.

20 Installation is the reverse of removal.

# 3 Charging and electric starting systems

## Charging system

1 The electrical system consists of three main components - the battery, the starting circuit and charging circuit. The battery functions within both the starting circuit and the charging circuit. Be sure the battery is checked and replaced, if necessary, before attempting to diagnose any charging circuit or starting circuit malfunction.

2 The charging system consists of the alternator charge coil or stator, a voltage regulator/rectifier assembly (selenium or solid state), an ignition switch, flywheel magnets and the battery. The charging system works independently of the starting circuit and other control circuits that regulate lighting, accessories, etc. The engine must be rotating at speed to produce electrical current flow. Most small engine alternators produce alternating current (AC). From there, it is necessary to convert the current from AC to DC (direct current) to charge the battery. A rectifier or diode uses only one half of the AC signal thereby allowing the conversion process from AC to DC. **Note 1:** *The charge coil is located behind the flywheel, mounted on the front crankcase cover. Do not confuse the ignition exciter coil with the alternator charge coil. The ignition exciter coil has a harness lead that connects to the ignition coil circuit.* **Note 2:** *Vertical L-head engines (EY45V) are equipped with a stator assembly under the flywheel and an ignition module/coil assembly on the front cover above the flywheel.*

3 There are several variations of the charging system. Some heavy duty applications produce 12.5 amp power for the lighting system on tractors, lawnmowers, generators, etc. Some light duty applications produce a 3 or 5 amp charge for the battery. When buying replacement parts, the parts retailer will need the serial number and model identification number to make the correct selection.

## Component removal and installation

### Regulator

4 Disconnect the negative battery cable.

5 Remove the blower housing. **Note:** *Some regulator/rectifier assemblies are mounted under the blower housing while others are mounted on the top or the side of the blower housing. Refer to the appropriate engine section for blower housing removal. All others are mounted externally.*

6 To remove the regulator/rectifier, mark and disconnect the harness connector from the regulator.

7 Remove the bolts holding the regulator/rectifier to the engine shroud or equipment and remove the regulator.

### Charge coil or stator

**Note:** *The charge coil is located behind the flywheel, mounted on the front crankcase cover. Do not confuse the ignition exciter coil with the alternator charge coil. The ignition exciter coil has a secondary lead that connects to the spark plug.*

8 To remove the charge coil or stator, refer to the appropriate engine section and remove the flywheel.

9 Disconnect the leads from the charge coil.

10 Remove the screws and remove the charge coils from the engine's bearing plate.

11 Installation (regulator/rectifier and charge coil) is the reverse of the removal procedure. Refer to Section 4 for the charge coil air gap adjusting procedure.

# Starting system

12 The starting system consists of a 12 volt battery, battery cables, a starter and ignition switch, safety switches and a starter solenoid. There are two different types of starters; a starter assembly with the solenoid and starter mounted together (integral) or a remote mounted solenoid (external). The externally mounted solenoids (switches) are the most common. Brush assemblies can be accessed by removing the cover at the commutator end of the starter (opposite the drive end).

## Removal and installation

### Solenoid

13 Disconnect the negative battery cable, then remove the cable from the battery to the starter solenoid (internally mounted solenoids) or from the solenoid (switch) to the starter (externally mounted solenoids).

14 Remove any other connectors from the solenoid, then remove the mounting screw(s).

15 Separate the solenoid from the engine (externally mounted solenoids) or from the starter assembly (internally mounted solenoids).

### Starter motor

16 There are several different makes of starter motors used on the various Robin engines, depending upon engine size and application.

17 Disconnect the negative battery cable at the battery, and the positive cable from the starter motor (the cable from the solenoid).

18 Remove the mounting bolts.

19 Make matching marks on the case and both ends before disassembling the starter.

20 Installation is the reverse of the removal procedure.

## Brush replacement

21 There are two problem areas concerning starter motors. If the motor has been abused (cranked regularly for more than 15 seconds at a time without a cooling off period) or has been in service for many years, the armature or field coils may be damaged. Examination of the starter with the end cap off will indicate it the commutator is worn or damaged. These problems are generally not user-serviced, and a rebuilt or new replacement starter is in order.

22 More common is the general wear of the brushes that ride on the commutator, and this is a normal replacement procedure that can be handled by the home mechanic.

23 The commutator end is the opposite end from the starter drive. Remove the long through-bolts and remove the commutator end cap.

24 The brushes are contained within the end cap. A new brush kit will include the brushes and brush springs. **Note:** *Some brush assemblies are attached to the brush holder with set screws while others are attached by solder joints.*

25 When all the brushes and springs are back in place in the commutator end cap, they must be kept in position while the end cap is slipped over the starter shaft. Insert the through-bolts, align the case markings and tighten the bolts with nuts on the drive end.

26 Starter installation is the reverse of the removal procedure.

# 4 Ignition systems

## Points ignition systems

1 This mechanical ignition system relies on the opening and the closing of the points to trigger the secondary voltage to the spark plug. The breaker points and the condenser are located under the flywheel. As the flywheel rotates, the magnet mounted on the flywheel passes the coil mounted on the engine case. The magnetic field from the flywheel magnet (magneto) causes a current to generate in the ignition coil primary circuit (winding). As the flywheel continues to rotate, it passes the last pole in the lamination stack on the coil. This produces a large change in the magnetic field ultimately producing high current in the primary circuit. This current change in the primary circuit builds until the points open and the secondary circuit is induced to flow to the spark plug and fire, releasing a high voltage spark.

2 Most points ignition systems include the ignition points, the coil, the condenser, the flywheel magneto, the ignition cam and the breaker arm (attached to the ignition points).

## Removal and installation

### Points

3 Remove the nut and detach the primary wires from the point terminal. Remove the mounting screws for the ignition points and the condenser.

4 Remove the point assembly from the crankcase cover and then remove the breaker arm (stiff spring).

5 Install the new condenser in its place and route the wire over to the point terminal. **Note:** *A condenser should always be replaced when the points are replaced.*

6 Clean the ignition point cavity with contact cleaner and wipe it out with a rag. Use the contact cleaner to remove oil and dirt from the new ignition point contact faces as well.

7 Install the new set of points and leave the screw loose enough to allow movement of the plate.

8 Slip the primary and condenser wires onto the terminal and install the nut.

9 Turn the crankshaft very slowly until the cam opens the movable point as far as possible.

10 Insert a clean feeler gauge - 0.014-inch thick - between the contact points and move the fixed point very carefully with a screwdriver until the gap between the points is the same thickness as the feeler gauge. Be careful not to change the position of the movable point as this is done.

11 Turn the crankshaft and make sure the movable arm opens and closes.

## Ignition timing (TDC locating)

12 All points-type systems require that the engine be timed statically, either with an ohmmeter or a test light.

13 Align the "D", "B" or "P" timing mark on the crankcase cover with the "M" on the flywheel. The actual timing setting is determined by the point gap. If the point gap is too wide or too small, the timing marks will not align properly when the ignition points are tested statically with a test light. **Note:** *The timing setting with the timing marks aligned and the points gapped correctly is 23-degrees BTDC.*

14 Follow the primary wire from the ignition points and detach it. With a self-powered test light or ohmmeter connected to the wire and ground, rotate the crankshaft until the contact points separate. This will be indicated by an open circuit on the ohmmeter, or the test light will go out. Confirm that the timing marks are aligned. If the timing marks are not perfectly aligned, increase or decrease the gap on the points 0.001 inch and recheck the timing. Make the necessary adjustments and tighten the set screw. **Note:** *On "B" engines, turn the flywheel counterclockwise to "break" the ignition points and on all others, turn the flywheel clockwise.*

# Electronic ignition systems

15 The electronic ignition system also called the "solid state ignition" system (SSI) incorporates solid state electronic components with no mechanical features. These systems are equipped with flywheel magnets (magneto), exciter coil (internal), pulser coil (external), pulse transformer, trigger coil and spark plug. On early SSI systems, the exciter coil is located behind the flywheel, the pulser coil above the flywheel and the ignition coil is mounted on the blower housing. On later SSI systems, the exciter coil, trigger coil, pulse transformer and capacitor are all contained within the ignition coil/module assembly which is mounted on the crankcase cover above the flywheel.

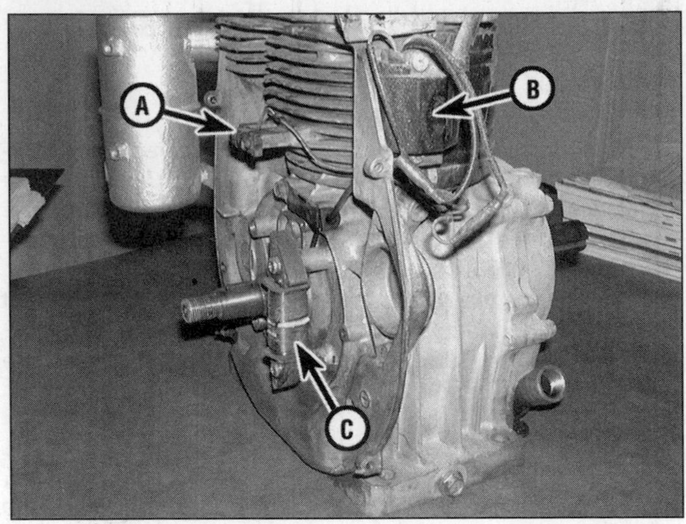


**4.19 Location of the ignition components on a Solid State Ignition (SSI) system on an L-head engine**

A *Pulser coil*
B *Ignition coil*
C *Exciter coil*

16 This electronic ignition system is more durable than the conventional points system that commonly fail due to burning, oxidation and mechanical wear. The SSI system operates in a similar fashion to a points system, but the method used to "break" the electronic signal differs. Flywheel rotation generates electricity in the primary circuit of the ignition coil assembly when the magnets rotate past the charge coil. As the flywheel rotates, it passes a trigger coil where a low voltage signal is produced. The low voltage signal allows the transistor to close allowing the energy stored in the capacitor to fire the secondary circuit or spark plug. The transistor acts as a primary circuit switching device in the same manner as the points in a mechanical (points) ignition system.

17 These systems do not require an ignition timing procedure.

## Removal and installation

### Exciter coil, pulser coil and ignition coil (early) or coil/module assembly (late)

18 Disconnect the spark plug lead from the plug. On models with battery systems, disconnect the negative cable from the battery. **Caution:** *Never connect battery voltage to any wire or component of the electronic-magneto ignition system, or the module will be damaged.*

4.20 Location of the coil/module assembly mounting bolts (arrows) on a OHV horizontal crankshaft engine (EH34)

19 On electronic ignition systems that mount the exciter coil behind the flywheel **(see illustration 4.19)**, remove the flywheel (refer to the appropriate engine section in this Chapter).

20 At the module, disconnect the electrical harness connectors from the module. Remove the two module mounting screws and remove the module **(see illustration 4.20)**.

21 Installation is the reverse of the removal procedure.

22 When installing the module, the air gap between the flywheel magnet and the module must be set.

23 Turn the flywheel so that the magnet is directly under the module. Loosen the mounting screws slightly and insert a clean feeler gauge of the correct thickness (0.012 to 0.020 inch) between the magnet and the module. The magnet will pull the module down to squeeze the feeler gauge between the module and magnet. Tighten the module mounting screws.

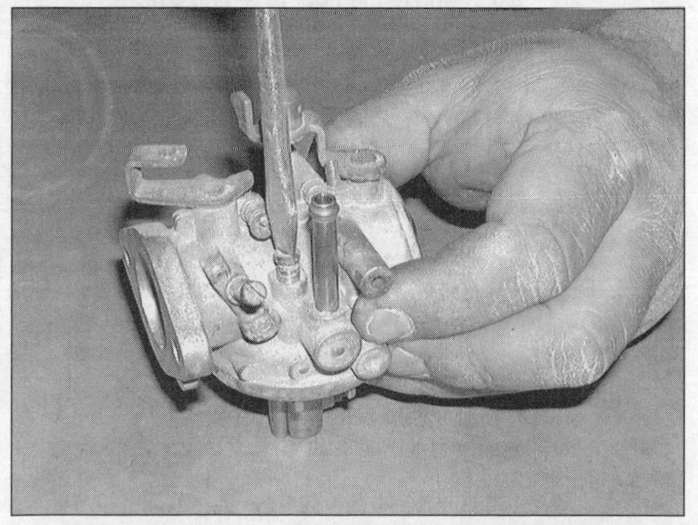

5.3 Remove the idle mixture jet (pilot jet) from the carburetor body

24 Rotate the flywheel a few revolutions to make sure that the magnet never touches the module's legs. Use feeler gauges to check that both legs of the module are the same distance from the magnet. Tighten the mounting bolts.

### Ignition timing (TDC locating)

25 Even though these electronic ignition systems do not require timing adjustment, there are some engine repair procedures that require the engine to be positioned at Top Dead Center (TDC). Align the "T" mark on the crankcase cover with the "T" on the flywheel. **Note:** *On engines that do NOT have any type of TDC locating marks, check the movement of the valves to verify the exact position of the piston* (see Section 7).

## 5 Carburetor disassembly and reassembly

### General information

1 The following procedures describe how to disassemble and reassemble the carburetor so new parts can be installed. Read the Sections in Chapter 5 on carburetor removal and overhaul before doing anything else. In some cases it may be more economical (and much easier) to install a new or rebuilt carburetor rather than attempt to repair the original. Check with a parts retailer or small engine repair shop to see if parts are readily available and compare the cost of new parts to the price of a complete ready-to-install (standard service) carburetor before deciding how to proceed. The carburetor model number and date code are stamped on the edge of the mounting flange.

### Carburetor disassembly

2 The following procedure is divided into two categories; the L-head carburetor and the OHV carburetor disassembly procedures. Although each carburetor is the same design, a float type carburetor, there are some differences in the construction and the adjustment procedures.

#### L-head engines

3 While counting the number of turns, carefully screw the idle mixture adjusting screw in until it bottoms, then remove it along with the spring **(see illustration 6.4)**. Counting and recording the number of turns required to bottom the screw(s) will enable you to return them to their original positions and minimize the amount of adjustment required after reassembly. Also, remove the idle mixture jet (pilot jet) **(see illustration 5.3)**.

4 Remove the idle speed adjustment screw and spring from the carburetor body in the same manner **(see illustration 5.4)**.

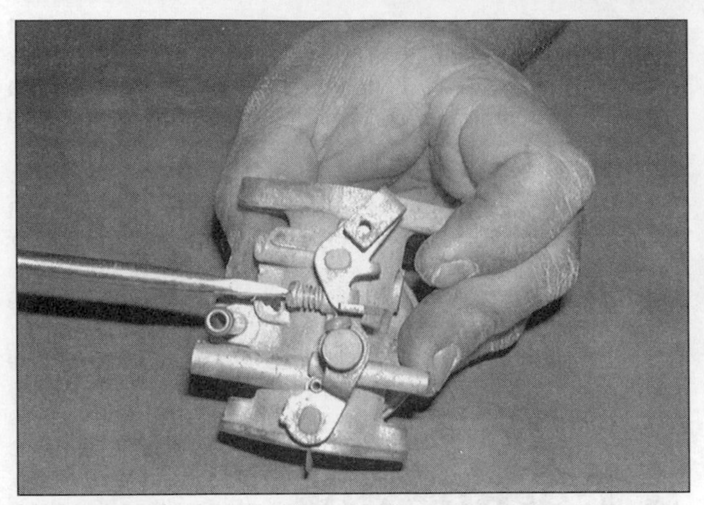
5.4 Remove the idle speed screw from the side of the carburetor body

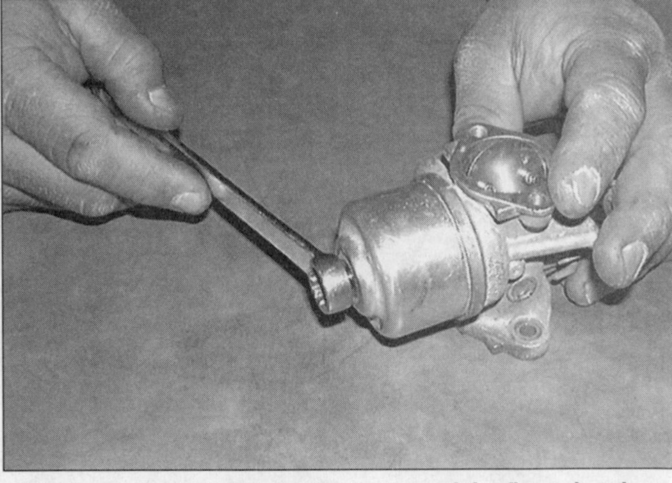
5.5 Remove the bolt from the bottom of the float chamber

5 Detach the float bowl from the carburetor body by removing the bolt from the float chamber **(see illustration 5.5)**. Be sure to note the locations of any gaskets and washers used.

6 Separate the float from the carburetor body, noting how the inlet needle valve is positioned before removing the float. It may be a good idea to draw a simple sketch to refer to during reassembly. Push the float pivot pin out of the carburetor body. You may have to use a small punch to do this **(see illustration 5.6)**.

7 Remove the float assembly and the inlet needle valve. Note how the inlet needle valve is attached to the float. The retainer clip, if used, must be positioned the same way during reassembly.

8 Remove the float bowl gasket.

9 Remove the main nozzle and the main jet **(see illustration 5.9)**.

10 Do a thorough cleaning job **(see illustration 5.10)**. **Warning:** *Wear protective gloves and goggles when using carburetor cleaner.* **Caution:** *Do not remove any brass cups or ball plugs (if equipped).*

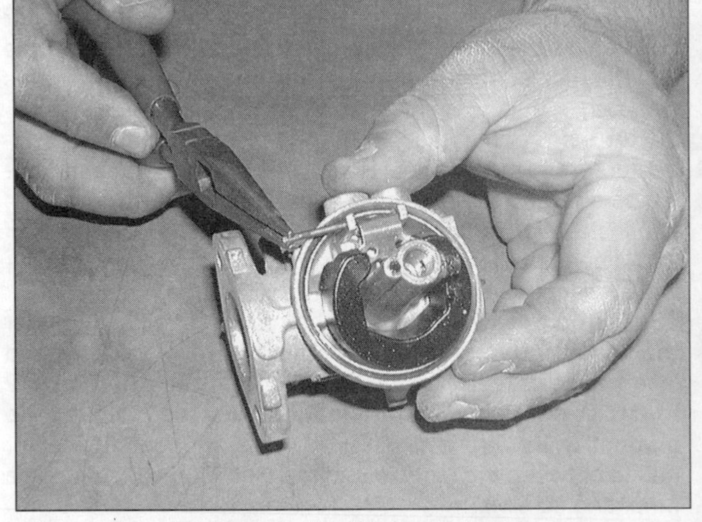
5.6 Remove the float retaining pin

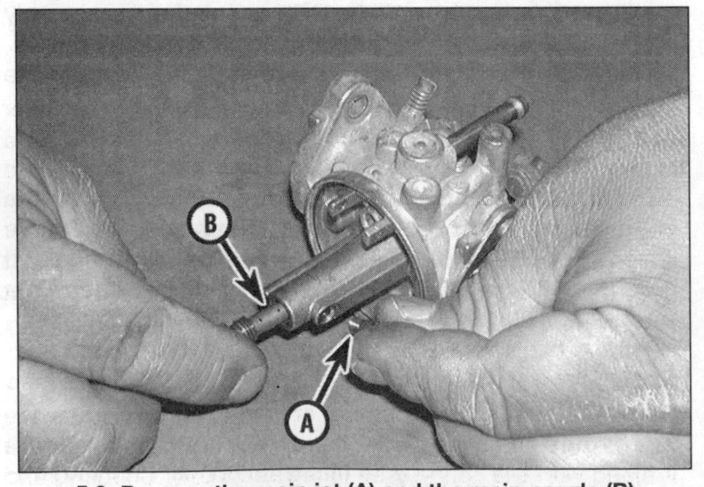

5.9 Remove the main jet (A) and the main nozzle (B)

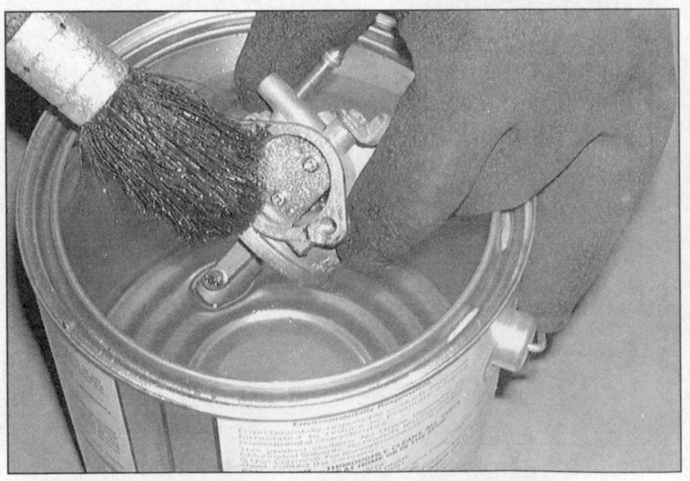
5.10 Using carburetor cleaner or acetone, thoroughly brush the carburetor clean

5.13 **With one hand firmly grasping the carburetor body, check the throttle shaft and choke shaft for excess play**

11 Refer to Chapter 5 and follow the cleaning/inspection procedures outlined under Carburetor Overhaul. **Note:** *Don't soak plastic or rubber parts in carburetor cleaner. If the float is made of cork, don't puncture it or soak it in carburetor cleaner.*

12 Check each mixture adjusting screw tip for damage and distortion **(see illustration 5.14 in Chapter 7)**. The small taper should be smooth and straight. Install the mixture adjusting screw and the idle speed screw back into the carburetor body after it is cleaned.

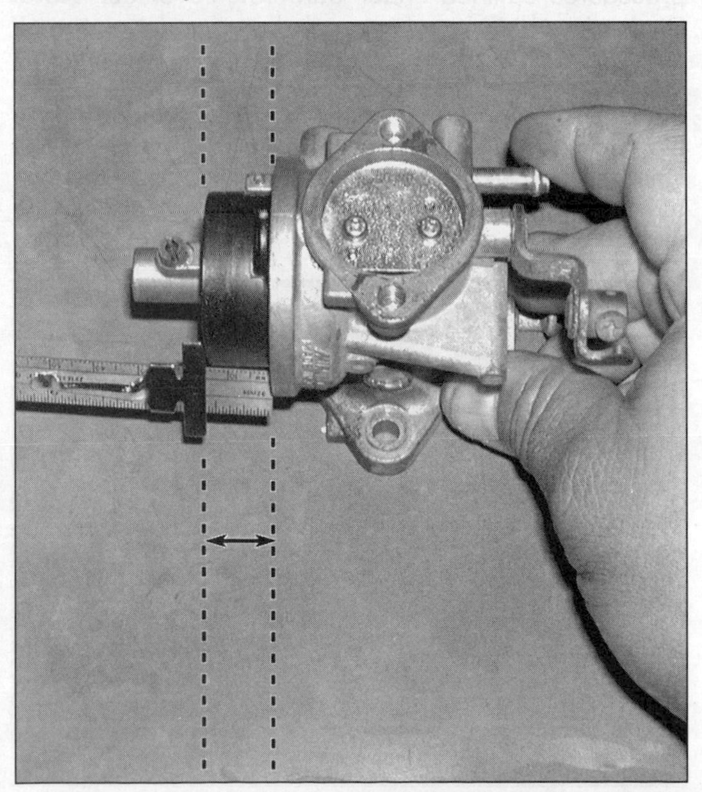

5.20 **Check the float height from the base of the body (float chamber) to the top of the float directly opposite the float hinge**

13 Check the throttle plate shaft for wear by moving it back-and-forth **(see illustration 5.13)**. If the linkage is excessively loose, have new bushings pressed into the body and reamed to the correct diameter by a qualified machine shop or carburetor specialty shop.

14 Check the throttle plate fit in the carburetor bore. If there's play in the shaft, the bore is worn excessively, which may mean a new carburetor is required. On some carburetors, replacing the throttle plate shaft or plate, or both, may cure the problem. The throttle plate and shaft don't have to be removed unless new parts are required.

15 Check the choke shaft for play in the same manner and examine the linkage holes to see if they're worn. Don't remove the choke shaft unless you have to install new parts to compensate for wear. Note how the choke plate is installed before removing it. The flat side must face down, toward the float bowl. They will operate in either direction, so make sure it's reassembled correctly. If dust seals are used, they should be positioned next to the carburetor body.

16 Check the inlet needle valve and seat. Look for nicks and a pronounced groove or ridge on the tapered end of the valve. If there is one, a new needle and seat should be used when the carburetor is reassembled. They are normally replaced as a matched set.

17 Check the float pivot pin and the bores in the carburetor casting, the float hinge bearing surfaces and the inlet needle tab for wear. If wear has occurred, excessive amounts of fuel will enter the float bowl and flooding will result.

18 Shake the float to see if there's gasoline in it. If there is, install a new one.

19 Once the carburetor parts have been cleaned thoroughly and inspected, reassemble it by reversing the above procedure. Note the following important points:

a) *Make sure all fuel and air passages in the carburetor body, main nozzle, inlet needle seat and float bowl mounting bolt are clean and clear.*
b) *Be sure to use new gaskets, seals and O-rings.*
c) *Whenever an O-ring or seal is installed, lubricate it with a small amount of grease or oil.*
d) *Don't overtighten small fasteners or they may break off.*

20 Set the float height. With the carburetor standing on the manifold flange, measure the float height using a small ruler from the base of the carburetor to the top of the float at a point opposite (180 degrees) from the float hinge pin and needle assembly **(see illustration 5.20)**. The tab on the float should just contact the needle valve. Adjust the float by gently bending the float lever tang (tab) until the correct float height is obtained (see the Specifications Section at the end of this Chapter.

## OHV engines

21 While counting the number of turns, carefully screw the idle mixture adjusting screw in until it bottoms, then remove it along with the spring **(see illustration 6.4)**. Counting and recording the number of turns required to bottom the

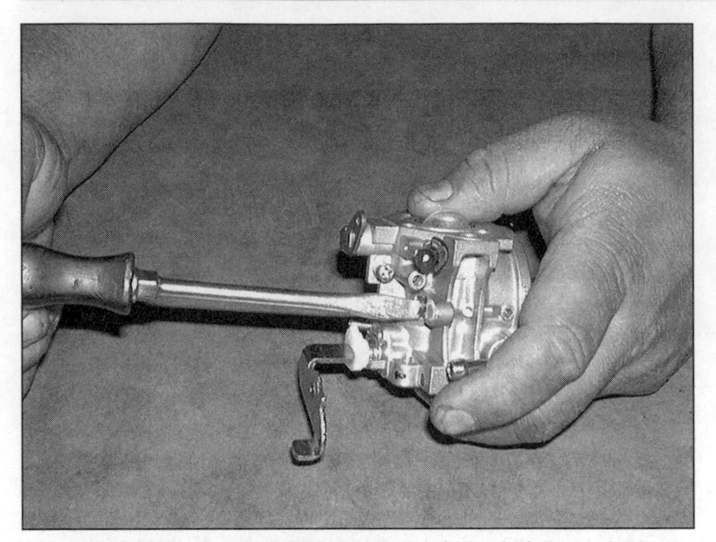

**5.21 Remove the idle mixture jet (pilot jet) from the carburetor body**

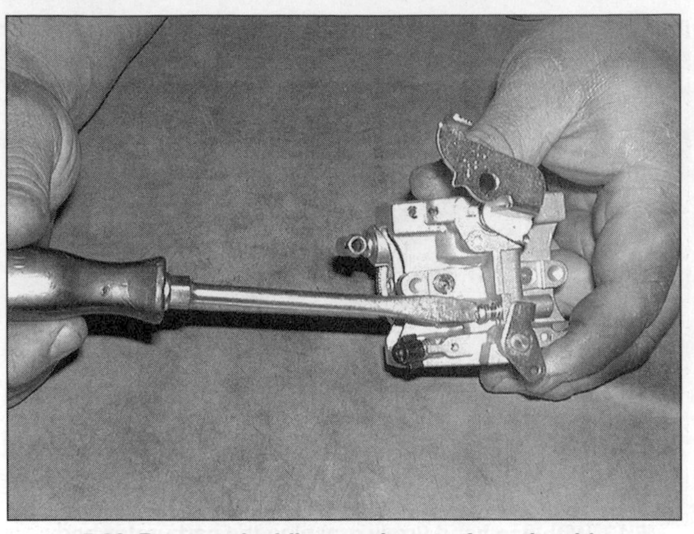

**5.22 Remove the idle speed screw from the side of the carburetor body**

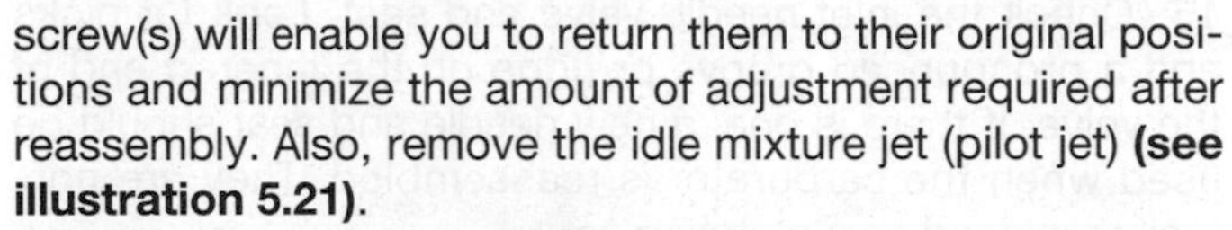

screw(s) will enable you to return them to their original positions and minimize the amount of adjustment required after reassembly. Also, remove the idle mixture jet (pilot jet) **(see illustration 5.21)**.

22 Remove the idle speed adjustment screw and spring from the carburetor body in the same manner **(see illustration 5.22)**.

23 Detach the float bowl from the carburetor body by removing the bolt from the float chamber **(see illustration 5.23)**. Be sure to note the locations of any gaskets and washers used.

24 Separate the float from the carburetor body, note how the inlet needle valve is positioned before removing the float. It may be a good idea to draw a simple sketch to refer to during reassembly. Push the float pivot pin out of the carburetor body. You may have to use a small punch to do this **(see illustration 5.24)**.

25 Remove the float assembly and the inlet needle valve. Note how the inlet needle valve is attached to the float. The retainer clip, if used, must be positioned the same way during reassembly.

26 Remove the float bowl gasket.

27 Remove the main nozzle and the main jet **(see illustration 5.27)**.

28 Do a thorough cleaning job. **Caution:** *Do not remove any brass cups or ball plugs (if used).*

29 Refer to Chapter 5 and follow the cleaning/inspection procedures outlined under Carburetor overhaul. **Note:** *Don't soak plastic or rubber parts in carburetor cleaner. If the float is made of cork, don't puncture it or soak it in carburetor cleaner.*

30 Check each mixture adjusting screw tip for damage and distortion **(see illustration 5.14 in Chapter 7)**. The small taper should be smooth and straight. Install the mixture adjusting screw and the idle speed screw back into the carburetor body after it is cleaned.

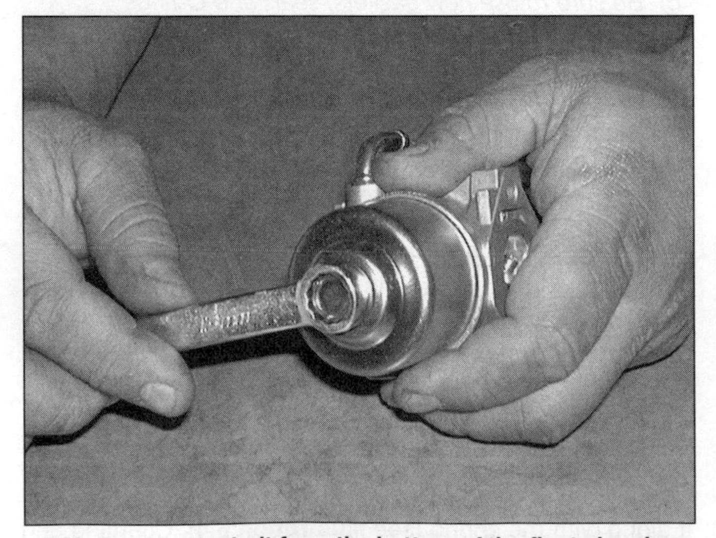

**5.23 Remove the bolt from the bottom of the float chamber**

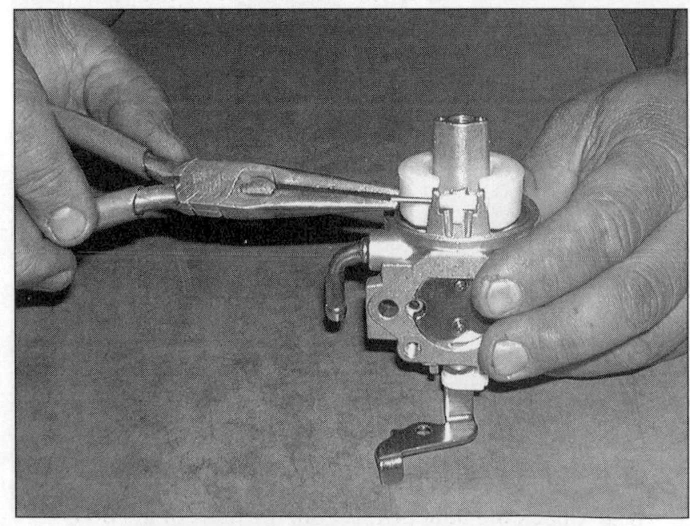

**5.24 Remove the float retaining pin using a pair of needle-nose pliers**

5.27 Remove the main nozzle the main jet

5.31 With one hand firmly grasping the carburetor body, check the throttle shaft and choke shaft for excess play

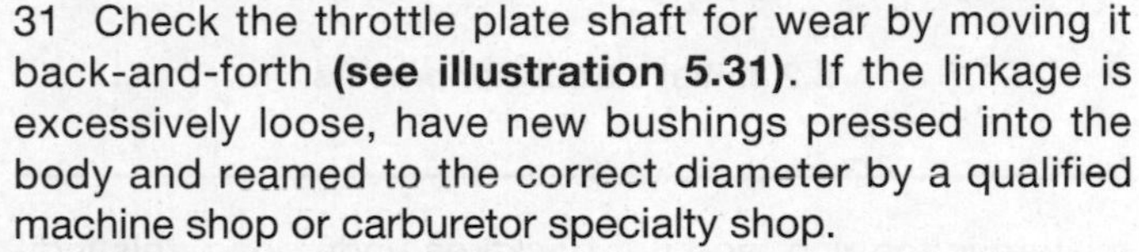

31 Check the throttle plate shaft for wear by moving it back-and-forth **(see illustration 5.31)**. If the linkage is excessively loose, have new bushings pressed into the body and reamed to the correct diameter by a qualified machine shop or carburetor specialty shop.

32 Check the throttle plate fit in the carburetor bore. If there's play in the shaft, the bore is worn excessively, which may mean a new carburetor is required. On some carburetors, replacing the throttle plate shaft or plate, or both, may cure the problem. The throttle plate and shaft don't have to be removed unless new parts are required.

33 Check the choke shaft for play in the same manner and examine the linkage holes to see if they're worn. Don't remove the choke shaft unless you have to install new parts to compensate for wear. Note how the choke plate is installed before removing it. The flat side must face down, toward the float bowl. They will operate in either direction, so make sure it's reassembled correctly. If dust seals are used, they should be positioned next to the carburetor body.

34 Check the inlet needle valve and seat. Look for nicks and a pronounced groove or ridge on the tapered end of the valve. If there is one, a new needle and seat should be used when the carburetor is reassembled. They are normally installed as a matched set.

35 Check the float pivot pin and the bores in the carburetor casting, the float hinge bearing surfaces and the inlet needle tab for wear. If wear has occurred, excessive amounts of fuel will enter the float bowl and flooding will result.

36 Shake the float to see if there's gasoline in it. If there is, install a new one.

37 Once the carburetor parts have been cleaned thoroughly and inspected, reassemble it by reversing the above procedure. Note the following important points:

- a) *Make sure all fuel and air passages in the carburetor body, main nozzle, inlet needle seat and float bowl mounting bolt are clean and clear.*
- b) *Be sure to use new gaskets, seals and O-rings.*
- c) *Whenever an O-ring or seal is installed, lubricate it with a small amount of grease or oil.*
- d) *Don't overtighten small fasteners or they may break off.*

38 Set the float height. Using a small ruler, measure the float height from the base of the carburetor to the top of the float at a point opposite (180 degrees) the float hinge pin and needle assembly. The tab on the float should just contact the needle valve. Adjust the float by gently bending the float lever tang (tab) until the correct measurement is attained **(see illustration 5.38)**. These models specify a parallel measurement from the base of the carburetor to the top of the float. Check the measurement at several different locations on the float and make sure the measurement is exactly the same (parallel) at each location.

5.38 Check the float height from the base of the body (float chamber) to the top of the float directly opposite the float hinge

# 6 Carburetor adjustment

## Carburetor

1 When making carburetor adjustments, the air cleaner must be in place and the fuel tank should be at least half full.

2 If not already done, turn the idle speed and idle mixture adjusting screws clockwise until they seat lightly, then back them out the same number of turns as you recorded during disassembly. The accompanying chart of suggested initial settings should be a good starting point before making running adjustments **(see illustration 6.2)**:

3 Start the engine and allow it to reach operating temperature before making the final adjustments.

4 With all carburetor mixture adjustments, the ideal setting point is halfway between the rich and lean positions **(see illustration 6.4)**.

5 Start with the idle speed screw. The engine should be completely warmed up to operating temperature. A tachometer must be used to monitor the engine speed. Follow the tool manufacturer's installation procedure and observe the engine rpm. Turn the idle speed screw to obtain 1200 rpm.

6 Next, adjust the idle mixture screw **(see illustration 6.6)**. With the engine running at 1200 rpm, observe the tachometer as the idle mixture screw is turned. In small increments, turn the idle mixture screw out until the engines runs slower - this is the rich position. Adjust the screw inward until the engine speed increases again (to where it started) and then begins to decrease. This is the lean position. Set the screw to halfway between the rich and lean positions.

7 If the engine smokes under load, this indicates the idle mixture is too rich, and if it backfires under load, this indicates the mixture is too lean. If the engine was adjusted in warm weather, but stops frequently in cold weather, try adjusting the idle mixture slightly richer.

| L-head model number | Idle mixture screw |
|---|---|
| EY23 (W1-230) | 1-3/8 turns |
| EY25 (EY25W) | 1-5/8 turns |
| EY27 (EY27W) | 1-1/2 turns |
| EY28 (W1-280) | 1-1/4 turns |
| EY35 (W1-340) | 1 turn |
| EY40 (W1-390) | 1-1/2 turns |
| EY45V (W1-450V) | 1-1/2 turns |
| **OHV model number** | **Idle mixture screw** |
| EH17 (WO1-170) | 1-1/2 turns |
| EH21 (WO1-210) | 1-1/2 turns |
| EH25 (WO1-250) | 1 turn |
| EH30 (WO1-300) | 2 turns |
| EH30V (WO1-300V) | 3/4 turn |
| EH34 (WO1-340) | 1-1/2 turns |
| EH34V (WO1-340V) | 2 turns |
| EH43V (WO1-430V) | 1-1/2 turns |

**6.2 Initial carburetor presets**

8 Move the speed control lever from SLOW to FAST - the engine should accelerate without hesitating or sputtering.

9 If the engine dies, it's too lean - turn the adjusting screws out in small increments.

10 If the engine sputters and runs rough before picking up the load, it's too rich - turn the adjusting screws in slightly.

11 If the adjustments are "touchy", check the float level and make sure the float isn't sticking.

**6.4 Location of the idle mixture screw (A) and the idle speed screw (B) on an L-head engine**

**6.6 Location of the idle mixture screw and the idle speed screw on an OHV engine**

## Governor

12 It is best not to change the relationship of any linkages or other components in the governor system, unless a worn or broken component is to be replaced. If the engine is being overhauled, do not change any adjustments. If the governor was working suitably before the overhaul, and none of the connections are altered, it will continue to work properly. While governor adjustment should be made with a small-engine tachometer to avoid setting the engine to a higher rpm than it is designed for, some basic adjustments can be made by the home mechanic.

13 On most engines, a governor shaft sticks through the side of the crankcase, while the governor mechanism is protected inside the crankcase. An arm attaches to this shaft with a clamp-bolt/nut. The basic initial adjustment on Robin governors begins with loosening the nut/bolt on the governor lever **(see illustration 8.3 and 9.3)**. Pull the arm back away from the carburetor to its limit (wide open throttle) and hold it there. Use a screwdriver and turn the governor shaft clockwise (left side mounted levers) or counterclockwise (right side mounted levers) as far as it will go (end) **(see illustration 6.13)**. Tighten the bolt/nut on the clamp in this position. **Note:** *Governor levers that are mounted to the left of the carburetor will require the governor shaft to be turned clockwise, while governor levers that are mounted to the right of the carburetor will require the governor shaft to be turned counterclockwise.*

14 Sensitivity is the other common adjustment for Robin governors. If the engine seems like it is hunting for a consistent speed as the load changes, the sensitivity is too sensitive and must be lowered. If the engine speed drops too much when the engine encounters a load, the governor adjustment isn't sensitive enough.

15 The governor sensitivity is adjusted by moving the position of the governor spring, where it hooks into the lower portion of the governor arm. A series of holes are provided. To increase the sensitivity, raise the spring into the next higher hole in the arm. To decrease sensitivity, hook the spring into a hole lower than its original position. The center of this arm is the factory position, but varying equipment and loads are accommodated by the extra adjustment holes.

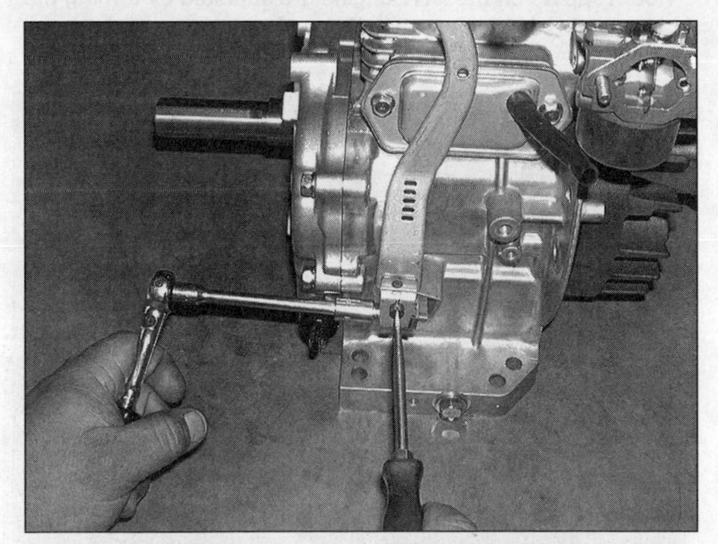

**6.13 Pull back the governor arm (A) until the throttle valve is at wide open throttle (WOT) while turning the governor shaft end (B) clockwise (OHV engine shown)**

# 7 Valve clearance - check and adjustment

1 Correct valve tappet clearance is essential for efficient fuel use, easy starting, maximum power output, prevention of overheating and smooth engine operation. It also ensures the valves will last as long as possible.

2 When the valve is closed, clearance should exist between the end of the stem and the tappet. The clearance is very small - measured in thousandths of an inch - but it's very important. The recommended clearances are listed in the Specifications at the end of this Chapter. Note that intake and exhaust valves often require different clearances. **Note:** *The engine must be cold when the clearances are checked.* A feeler gauge with a blade thickness equal to the valve clearance(s) will be needed for this procedure.

3 On L-head engines, if the clearances are too small, the valves will have to be removed and the stem ends ground down carefully and lapped to provide more clearance. This is a major job, covered in the overhaul and repair procedures. If the clearances are too great, new valves will have to be installed. On OHV engines, the tappets are adjustable, and with two wrenches, the clearance can be set to the Specifications.

4 Disconnect the wire from the spark plug and ground it on the engine.

5 On L-head engines, remove the bolts and detach the tappet cover plate or the crankcase breather assembly **(see illustration 8.10)**. On OHV engines, remove the bolts and remove the valve cover **(see illustration 9.9a)**.

6 Turn the crankshaft by hand and watch the valves to see if they stick in the guide(s).

7 Turn the crankshaft until the intake valve is wide open, then turn it an additional 360-degrees (one complete turn). This will ensure the valves are completely closed for the clearance check. L-head engines are designed so that the valves are easily viewed with the tappet cover off. On OHV engines, make sure the engine is positioned on TDC. Refer to the timing procedure in Section 4.

8 Select a feeler gauge thickness equal to the specified valve clearance and slip it between the valve stem end and the tappet **(see illustrations 7.8a and 7.8b)**.

9 If the feeler gauge can be moved back-and-forth with a slight drag, the clearance is correct. If it's loose, the clearance is excessive; if it's tight (watch the valve to see if it's forced open slightly), the clearance is inadequate. On OHV engines with adjustable tappets, loosen the locknut and turn the adjusting screw (tappet) with an Allen wrench to achieve the proper valve clearances, then tighten the locknut while holding the adjusting screw from turning.

10 If the clearance is incorrect on an L-head engine, refer to Chapter 5 for valve service procedures.

11 Reinstall the crankcase breather or tappet cover plate, using a new gasket if necessary.

7.8a Checking the valve adjustment on the L-head engine

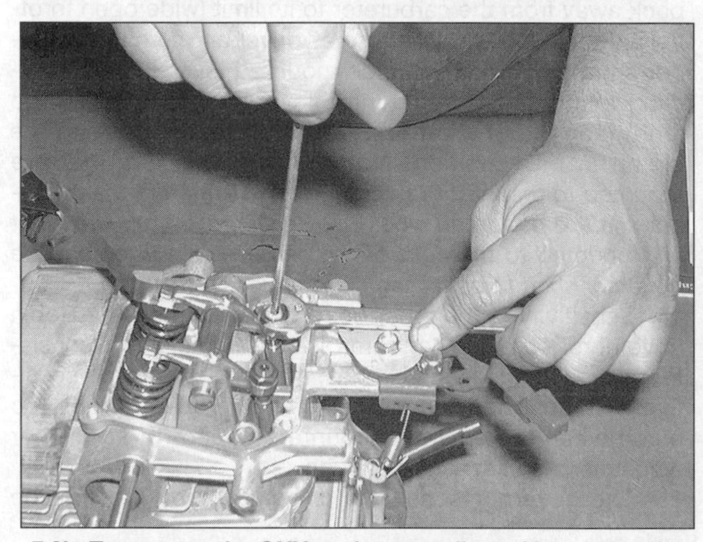

7.8b Tappets on the OHV engine are adjusted by turning the tappet with an Allen wrench, then using an open-end wrench to tighten the locknut

# 8 L-head engines

## Disassembly

The engine components should be removed in the following general order:

*Engine cover (if used)*
*Fuel tank*
*Cooling shroud/recoil starter*
*Carburetor/intake manifold*
*Muffler*
*Cylinder head*
*Flywheel*
*Flywheel brake components (if equipped)*
*Ignition components*
*Crankcase breather assembly*
*Oil sump/crankcase cover*
*Crankshaft*
*Camshaft*
*Tappets*
*Piston/connecting rod assembly*
*Valves*
*Governor components*

1 For fuel tank, carburetor/intake manifold, muffler and oil draining procedures, refer to Chapters 4 and 5 as necessary. For the shroud and recoil assembly removal, refer to the beginning of this Chapter.

2 Remove the fuel tank mount **(see illustration 8.2)**.

3 Disconnect the governor linkage from the carburetor. Be sure to make notes or a drawing of the linkage to insure correct reassembly **(see illustration 8.3)**.

4 First, remove the pulley mounting bolts **(see illustration 8.4)**, remove the pulley and hold the flywheel using a flywheel tool (strap wrench) and remove the large nut and washer.

5 Install a small puller onto the flywheel and remove the flywheel from the crankshaft **(see illustration 8.5)**. DO NOT hammer on the end of the crankshaft and DO NOT use a jaw-type puller that applies force to the outer edge of the flywheel. **Note:** *If a flywheel brake is used, remove the brake and related components.*

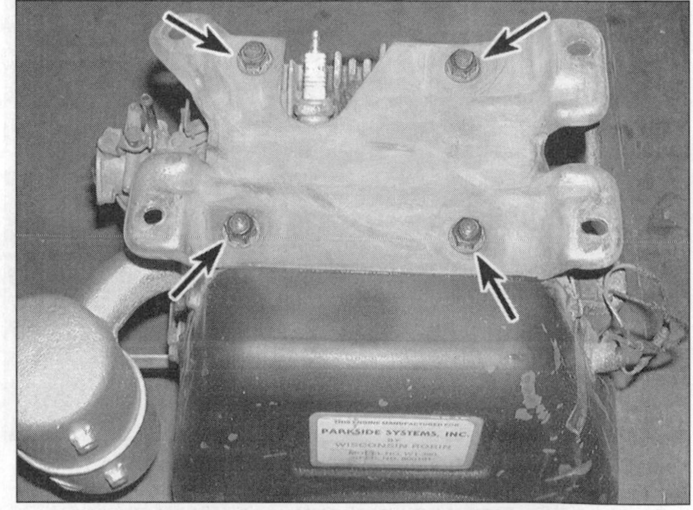


8.2 Remove the bolts (arrows) from the fuel tank mounting plate

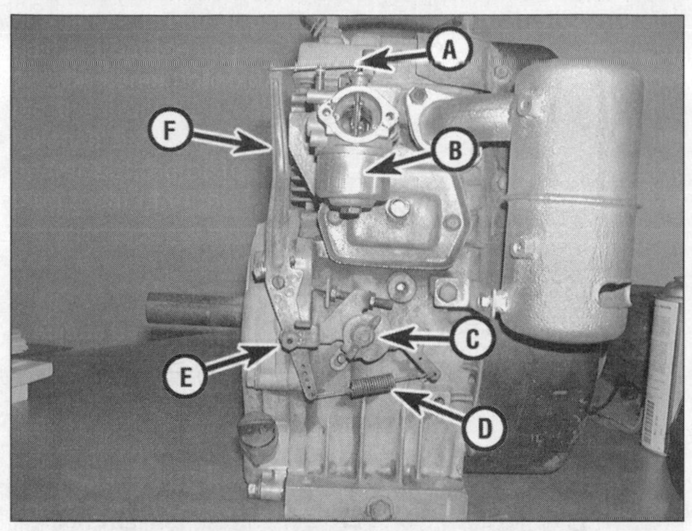


8.3 Location of the governor components

*A* *Governor rod*
*B* *Carburetor*
*C* *Speed control lever*
*D* *Governor spring*
*E* *Governor shaft*
*F* *Governor lever*

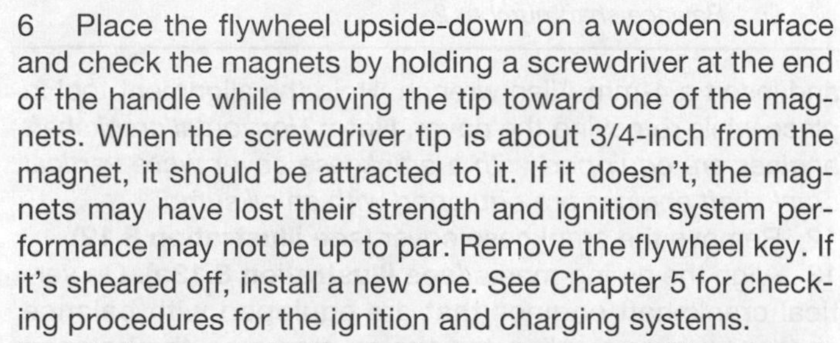

6 Place the flywheel upside-down on a wooden surface and check the magnets by holding a screwdriver at the end of the handle while moving the tip toward one of the magnets. When the screwdriver tip is about 3/4-inch from the magnet, it should be attracted to it. If it doesn't, the magnets may have lost their strength and ignition system performance may not be up to par. Remove the flywheel key. If it's sheared off, install a new one. See Chapter 5 for checking procedures for the ignition and charging systems.

7 First remove the flywheel key, charge coil (if equipped) and baffle plate, then refer to Chapter 4 and remove the ignition points and related parts (if equipped).

8 Remove the cylinder head bolts, working in an order opposite that of the tightening sequence **(see illustration 8.43b)**. Remove the cylinder head.

9 Note how the wires to the coil are routed. Use masking tape to write each wire designation. Mark the coil bracket and engine bosses with a scribe or center punch, then remove the bolts and detach the coil **(see illustration 4.20)**. Note that this engine has the exciter coil mounted under the flywheel. Some engines are equipped with an ignition coil/module assembly mounted outside the flywheel. The removal procedures are basically the same for both types.

10 If it's still in place, remove the bolts and separate the intake manifold from the engine - remove the gasket and discard it. Remove the mounting bolts and detach the crankcase breather assembly **(see illustration 8.10)** and gasket from the engine.

11 On some horizontal crankshaft engines, before removing the crankcase cover, it will be necessary to align the balance shaft (inside cover) to allow clearance between the gears. Set the piston to TDC (see Section 5), remove the plug located in the lower section of the crankcase cover

8.4 Remove the crankshaft pulley bolts

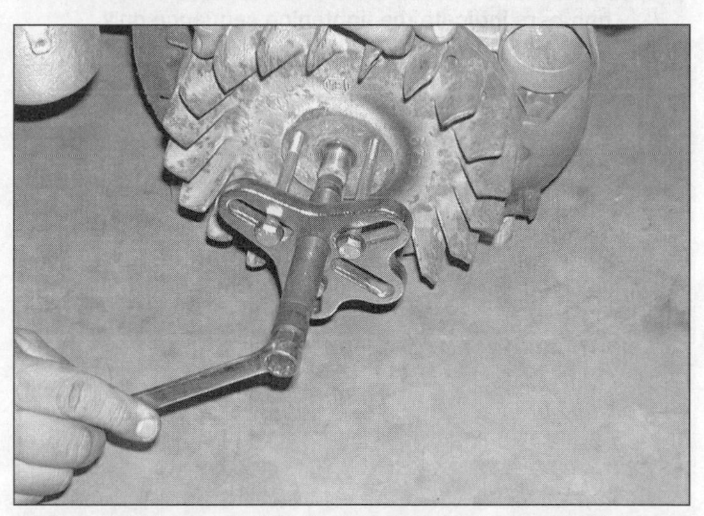

8.5 Use a small puller to separate the flywheel from the crankshaft

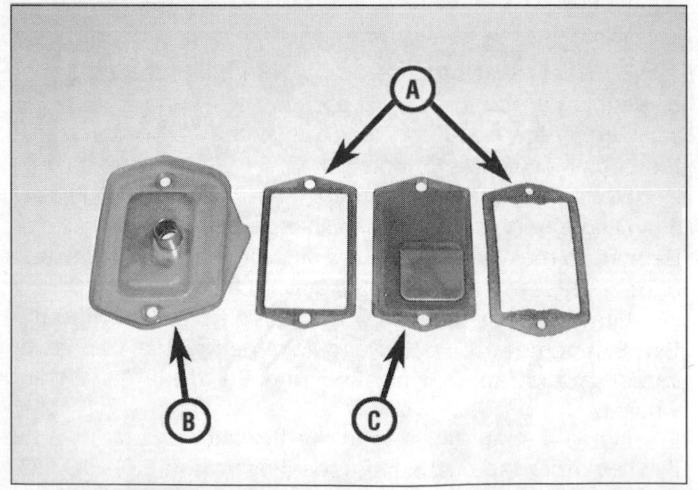


8.10 Crankcase breather components

*A* *Gasket*
*B* *Cover*
*C* *Filter*

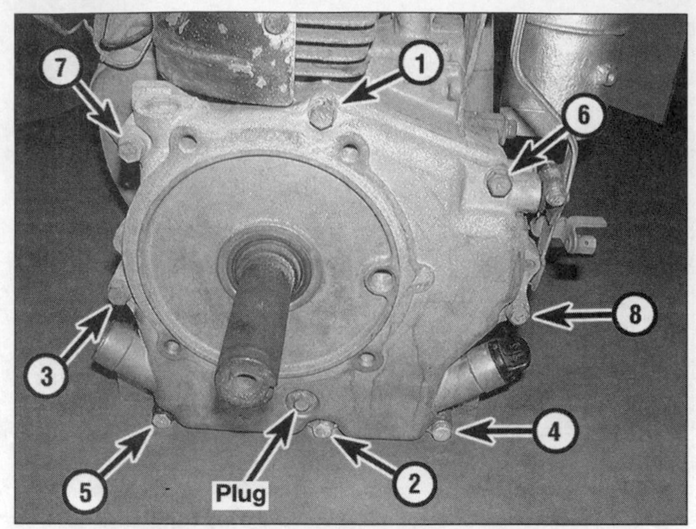


8.12 Remove the crankcase cover bolts (arrows) - the numbers indicate the tightening sequence only

8.13b Crankshaft (A) and camshaft (B) timing marks

8.13c Remove the camshaft but make sure the tappets do not interfere with the camshaft lobes

*A Camshaft* *B Tappets*

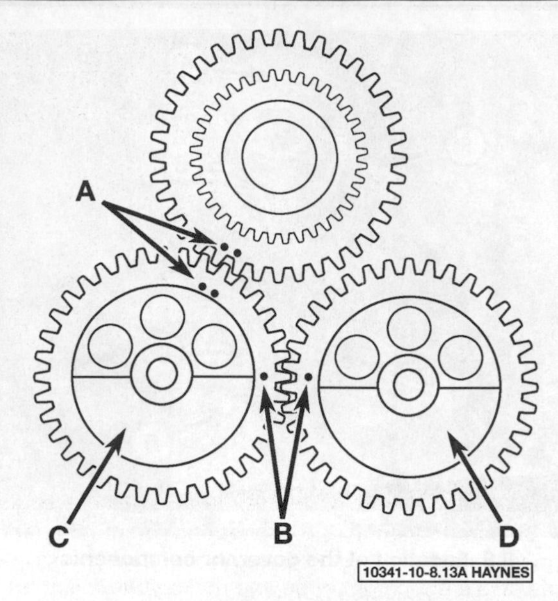


8.13a Balance shaft alignment marks on double balance shaft systems

*A Crankshaft-to-balance shaft number 1 timing marks*
*B Balance shaft number 1-to-balance shaft number 2 timing marks*
*C Balance shaft number 1*
*D Balance shaft number 2*

and insert a 4-mm Allen wrench. Hold the alignment tool in place while removing the cover. **Note:** *Horizontal crankshaft engines are equipped with a crankcase cover while vertical crankshaft engines are equipped with an oil sump.*

12 Remove the crankcase cover **(see illustration 8.12).**

13 Align the timing marks **(see illustration 8.13a).** On vertical crankshaft engines that are equipped with balance shafts, be sure to align the timing marks on the balance shafts before removing them and checking the alignment marks on the camshaft and the crankshaft **(see illustration 8.13b).** Remove the camshaft from the engine but make sure the tappets do not interfere with the camshaft when lifting. Sometimes they will fall out if the engine is tilted **(see illustration 8.13c).**

14 Remove the tappets and store them in marked containers so they can be returned to their original locations. Remove the connecting rod cap bolts and the oil dipper (horizontal crankshaft engines). Look for match marks on the connecting rod and cap **(see illustration 8.14).** If you can't see any, mark the side of the connecting rod and cap that faces OUT and note how the oil dipper (if used) is installed (the parts must be reassembled in the exact same relationship to the crankshaft).

15 Use a ridge reamer to remove the carbon/wear ridge from the top of the cylinder bore after the cylinder head is off. Follow the manufacturer's instructions included with the tool.

16 Turn the crankshaft so the rod journal is at the bottom of its stroke (Bottom Dead Center). Flatten the locking tabs (if used) on the connecting rod bolts with a punch and hammer.

17 Loosen the bolts or nuts in 1/4-turn increments until they can be removed by hand. Separate the cap (and

**8.14 Be sure to identify the connecting rod matchmarks (arrows) before removing the cap bolts. If no marks are visible, use a punch or scribe and make your own marks**

**8.17 Remove the crankshaft from the crankcase**

washers, if used) from the connecting rod, move the end of the rod away from the crankshaft journal and push the piston/rod assembly out through the top of the bore. The crankshaft can now be lifted out **(see illustration 8.17)**.

18 On vertical crankshaft engines, remove the oil pump assembly from the sump (lower cover), noting how the parts fit in relation to each other. **Note:** *There are several different types of oiling systems on these engines. Smaller horsepower horizontal engines are equipped with the "splash oiling system". A dipper is installed on the end of the connecting rod to agitate the oil in the sump causing oil to splash over the cylinder walls and bearings. Vertical crankshaft engines are equipped with a rotary oil pump located in the sump, driven from the camshaft. On smaller horsepower engines, this system is called the "forced splash oiling system" because it uses a pump to pressurize the oil and splashes the oil over the cylinder walls and bearings through a special oiling port. Larger horsepower engines are also equipped with a rotary oil pump that forces the pressurized oil through the oil jet on the crankshaft arm over the piston, cylinder and connecting rod. This is called the "pressurized oiling system".*

19 Compress the intake valve spring and remove the retainer, then withdraw the valve through the top of the engine. Pull out the spring, then repeat the procedure for the exhaust valve. Some engines have a retainer at the base of the spring as well.

20 Do not separate the lever from the governor shaft unless new parts are needed. The lever mount will be damaged during removal.

21 The governor assembly, located in the crankcase cover or on the camshaft, can be withdrawn from the gear shaft. Note how the parts fit together to simplify reassembly (a simple sketch would be helpful). Check the governor parts for wear and damage. If the gear shaft or yoke must be replaced, measure how far it protrudes before removing it and don't damage the crankcase boss. Clamp the shaft in a vise and tap the crankcase boss with a soft-face hammer to extract the shaft from the hole. **Caution:** *DO NOT twist the shaft with pliers or the mounting hole will be enlarged and the new shaft won't fit into it securely.*

22 When installing the new shaft, coat the serrated end with stud and bearing mount liquid after the shaft has been started in the hole with a soft-face hammer. Use a vise or press to finish installing the shaft and make sure it protrudes the same amount as the original (or the distance specified on the instruction sheet included with the new part). Wipe any excess stud and bearing mount liquid off the shaft and mounting boss flange.

## Inspection of components

23 After the engine has been completely disassembled, refer to Chapter 5 for the cleaning, component inspection, cylinder honing and valve lapping procedures.

24 Once you've inspected and serviced everything and purchased any necessary new parts, which should always include new gaskets and seals, reassembly can begin.

25 Begin by reinstalling the PTO components (if used) in the crankcase, then proceed as follows:

# Reassembly

26 Coat the intake valve stem with clean engine oil, moly-base grease or engine assembly lube, then reinstall it in the block. Make sure it's returned to its original location. **Note:** *If a seal is used on the intake valve, always install a new one when the engine is reassembled.*

27 Compress the valve spring with both retainers in place, then pull the valve out enough to position the spring and install the slotted retainer. Release the compressor and make sure the retainer is securely locked on the end of the valve. Repeat the procedure for the exhaust valve. **Note:** *Some models are equipped with dampening coils (tight wound coils) on the end of the valve spring. Position the dampening coils near the engine block away from the valve spring retainers.*

28 Install the governor assembly. Install the governor sleeve, plate and flyweights onto the cover or the camshaft assembly (see Step 35).

29 Lubricate the crankshaft oil seal lip, the plain bearing (if applicable) and the connecting rod journal with clean engine oil, moly-base grease or engine assembly lube, then reposition the crankshaft in the crankcase. If a ball-bearing is used on the magneto side, lubricate it with clean engine oil.

30 Before installing the piston/connecting rod assembly, the cylinder must be perfectly clean and the top edge of the bore must be chamfered slightly so the rings don't catch on it. Stagger the piston ring end gaps and make sure they're positioned opposite the valve seats when the piston is installed (see Chapter 5). Lubricate the piston and rings with clean engine oil, then attach a ring compressor to the piston. Leave the skirt protruding about 1/4-inch. Tighten the compressor until the piston cannot be turned, then loosen it until the piston turns in the compressor with resistance.

31 Rotate the crankshaft until the connecting rod journal is at TDC (Top Dead Center - top of the stroke) and apply a coat of engine oil to the cylinder walls. If the piston has an arrow in the top, it must face the flywheel side of the engine (if possible) **(see illustration 8.31)**. Make sure the match marks on the rod and cap will be facing out when the rod/piston assembly is in place. Gently insert the piston/connecting rod assembly into the cylinder and rest the bottom edge of the ring compressor on the engine block. Tap the top edge of the ring compressor to make sure it's contacting the block around its entire circumference. **Note:** *The piston(s) used on these models will have either an square or an arrow stamped on the top of the piston. The piston must be installed with these marks toward the flywheel (front).*

32 Carefully tap on the top of the piston with the end of a wooden or plastic hammer handle while guiding the end of the connecting rod into place on the crankshaft journal. The piston rings may try to pop out just before entering the bore, so keep some pressure on the ring compressor. Work slowly, if any resistance is felt as the piston enters the cylinder, stop immediately. Find out what's hanging up and fix it before proceeding. Do not, for any reason, force the piston into the cylinder. You'll break a ring and/or the piston.

33 Install the connecting rod cap, a NEW lock plate, the oil dipper (if used) and the bolts (or nuts) and washers (if used). Make sure the marks you made on the rod and cap (or the manufacturer's marks) are aligned and facing out and the oil dipper is oriented correctly.

34 Tighten the bolts or nuts to the specified torque. Refer to the Specifications listed at the end of this Chapter. Work up to the final torque in three steps. Temporarily install the camshaft and turn the crankshaft through two complete revolutions to make sure the rod doesn't hit the cylinder or camshaft. If it does, the piston/connecting rod is installed incorrectly.

35 Apply clean engine oil, moly-base grease or engine assembly lube to the tappets, then reinstall them. Make sure they're returned to their original locations.

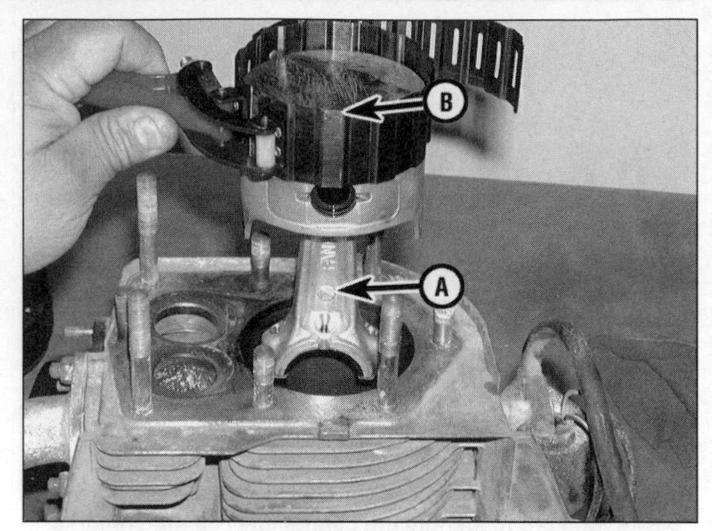


**8.31 Be sure the connecting rod MAG MARK (A) and the square (B) on the piston crown face toward the flywheel**

36 If it's not already in place, install the camshaft. Apply clean engine oil, moly-base grease or engine assembly lube to the lobes and bearing journals and align the timing marks on the gears **(see illustration 8.13a)** - this is very important! The marks are usually dimples near the outer edge of the gears. On vertical crankshaft engines, install the balance shafts and be sure the alignment marks are correct **(see illustration 8.13b)**.

37 Be sure the mechanical compression release is installed and in working order. **Note 1:** *Some models are equipped with the Mechanical Compression Release (MCR) camshafts that are equipped with a locking pin in the camshaft that extends over the exhaust cam lobe to lift the valve and relieve engine compression during cranking. After the engine has started, centrifugal force moves the weight outward, dropping the pin down and out of the way of the exhaust valve, bringing the engine compression back to normal. Be sure this pin is locked into the correct position after installation.* **Note 2:** *Some models are equipped with a Bump Compression Release (BCR) camshaft. This type of camshaft is equipped with a small bump on the exhaust lobe to allow slight compression release when the engine is cranking slow during start-up. Make sure the correct type of camshaft is installed in the particular engine by verifying the part numbers and camshaft types with a small engine parts distributor.*

38 Install the rotary type oil pump (vertical crankshaft engines) after lubricating it with clean engine oil.

39 Lubricate the crankshaft main bearing journal and the lip on the oil seal in the sump (or crankcase cover) with clean engine oil, moly-base grease or engine assembly lube.

40 Make sure the dowel pins are in place, then position a new gasket on the crankcase (the dowel pins will hold it in place). Carefully lower the oil sump or crankcase cover into place over the end of the crankshaft until it seats on the crankcase. DO NOT damage the oil seal lip or leaks will

8.41a Install a dial indicator mounted to the crankcase cover and check the endplay by moving the crankshaft along its axis

8.41b Install a larger shim to compensate for the excess play or a smaller shim to increase the endplay

result! The governor shaft must match up with the spool end and the oil pump shaft ball end must be engaged in the recess on vertical crankshaft engines. Torque the bolts to the Specifications listed at the end of this Chapter. **Note:** *On engines equipped with a balance shaft system, be sure to have the engine set at TDC and a 4 mm Allen wrench installed though the access hole (see Step 11).*

41 The crankshaft endplay must be checked with a dial indicator. Refer to the Specifications listed at the end of this Chapter **(see illustration 8.41a)**. If the endplay is not correct, remove the crankcase cover or sump and install the correct size shim **(see illustration 8.41b)**.

42 Once the crankshaft endplay is correct, install the cover, apply non-hardening thread locking compound to the bolt threads, then install and tighten them to the specified torque. Follow a criss-cross pattern and work up to the final torque in three equal steps to avoid warping the oil sump/cover **(see illustration 8.12)**.

43 Install a new cylinder head gasket with the mark "TOP" facing up **(see illustration 8.43a)** Install the cylinder head. Tighten the bolts in several passes, using the correct tightening sequence **(see illustration 8.43b)**, to the torque listed in the Specifications at the end of this Chapter.

44 Install the crankcase breather **(see illustration 8.10)** and intake manifold. Use new gaskets and tighten the bolts securely.

45 Install the ignition coil/spark plug wire assembly and align the marks on the coil bracket and crankcase bosses, then tighten the coil mounting bolts securely. **Note:** *If the coil is mounted above the flywheel, don't tighten the bolts completely - just snug them up. The bolt holes are slotted; move the coil as far away from the flywheel as possible before snugging up the bolts. Follow the air gap procedure in Section 5 to complete the adjustment.*

8.43a Install the cylinder head gasket with the TOP mark (arrow) facing up

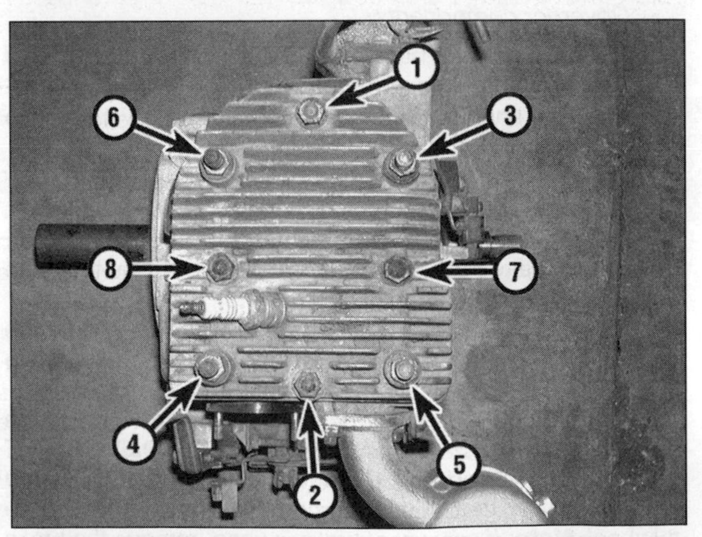

8.43b Cylinder head bolt tightening sequence

46 Refer to Section 5 and install the ignition points, if equipped, with the TOP side of the point cam facing up, then make sure the tapered portion of the crankshaft and the inside of the flywheel hub are clean and free of burrs. Position the Woodruff key in the crankshaft keyway and install the flywheel. Install the starter cup, washer and nut. Tighten the nut to the specified torque.
47 Install the remaining components, refer to Chapters 4 and 5 as necessary. **Caution:** *Be sure to fill the crankcase to the correct level with the specified oil before attempting to start the engine.*

# 9 Overhead Valve (OHV) engines

## Disassembly

The engine components should be removed in the following general order:

*Engine cover (if used)*
*Fuel tank*
*Cooling shroud/recoil starter*
*Carburetor/intake manifold*
*Muffler*
*Flywheel*
*Flywheel brake components (if equipped)*
*Ignition components*
*Cylinder head*
*Crankcase breather assembly*
*Oil sump/crankcase cover*
*Balance shaft(s) (if equipped)*
*Crankshaft*
*Camshaft*
*Tappets*
*Piston/connecting rod assembly*
*Valves*
*Governor components*

1 For fuel tank, carburetor/intake manifold, muffler and oil draining procedures, refer to Chapters 4 and 5 as necessary. For the shroud and recoil assembly removal, refer to the beginning of this Chapter.
2 Remove the cylinder head baffle plate **(see illustration 9.2)**.
3 Disconnect the governor linkage from the carburetor. Be sure to make notes or a drawing of the linkage to insure correct reassembly **(see illustration 9.3)**.
4 Remove the pulley mounting bolts and pulley. Hold the flywheel using a flywheel tool (strap wrench) and remove the large nut and washer **(see illustration 9.4)**.
5 Install a hub-type puller on the flywheel and remove the flywheel from the crankshaft **(see illustration 9.5)**. DO NOT hammer on the end of the crankshaft and DO NOT use a jaw-type puller that applies force to the outer edge of the flywheel. **Note:** *If a flywheel brake is used, remove the brake and related components.*

**9.2 Remove the cylinder head baffle plate mounting bolt (arrow)**

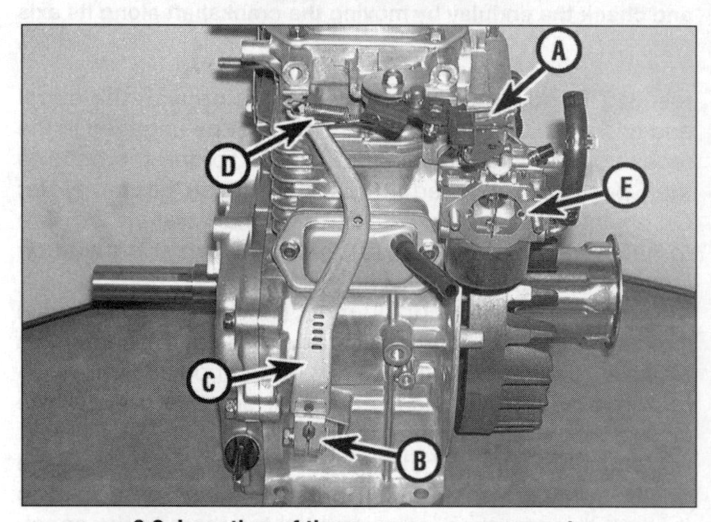


**9.3 Location of the governor components**

| | | | |
|---|---|---|---|
| *A* | *Governor rod* | *D* | *Governor spring* |
| *B* | *Governor shaft* | *E* | *Carburetor* |
| *C* | *Governor lever* | | |

**9.4 Remove the flywheel nut while locking the flywheel in position using a strap wrench**

9.5 Use a hub-type puller to separate the flywheel from the crankshaft

9.7 Remove the oil level sensor from the lower section of the crankcase

9.8 Remove the breather cover mounting bolts (arrows)

6 Place the flywheel upside-down on a wooden surface and check the magnets by holding a screwdriver at the end of the handle while moving the tip toward one of the magnets. When the screwdriver tip is about 3/4-inch from the magnet, it should be attracted to it. If it doesn't, the magnets may have lost their strength and ignition system performance may not be up to par. Remove the flywheel key. If it's sheared off, install a new one. See Chapter 5 for checking procedures on the ignition and charging systems.

7 Remove the flywheel key, charge coil (if equipped) and baffle plate, then refer to Chapter 4 and remove the ignition points and related parts (if equipped). If equipped, remove the oil level sensor **(see illustration 9.7)**.

8 Remove the crankcase breather plate and gaskets from the side of the engine **(see illustration 9.8)**.

9 Remove the rocker arm assembly and cylinder head with the valve assemblies intact. First, remove the valve cover bolts **(see illustration 9.9a)**. Remove the rocker arm assembly from the cylinder head **(see illustration 9.9b)**. Remove the rocker arms, bearings (if equipped), nuts,

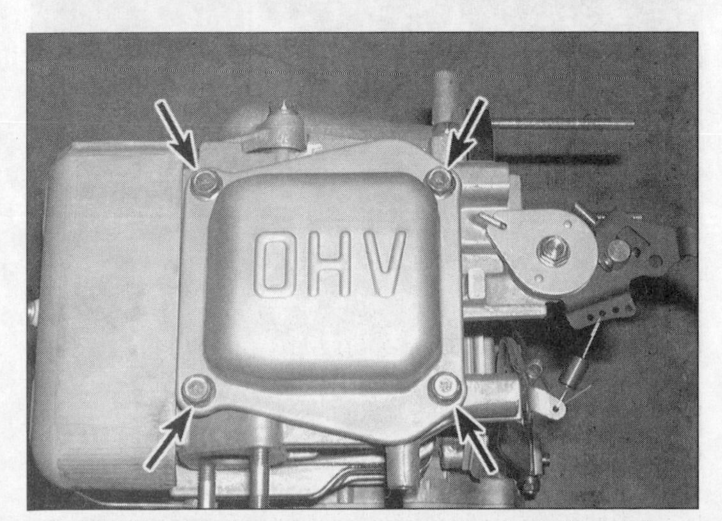


9.9a Remove the valve cover mounting bolts (arrows)

9.9b Loosen the rocker arm adjusters and drive the shaft toward the flywheel side

9.9c Remove the rocker arms and pushrods from the cylinder head

9.12 Use a mallet or brass hammer to break the cylinder head loose from the crankcase

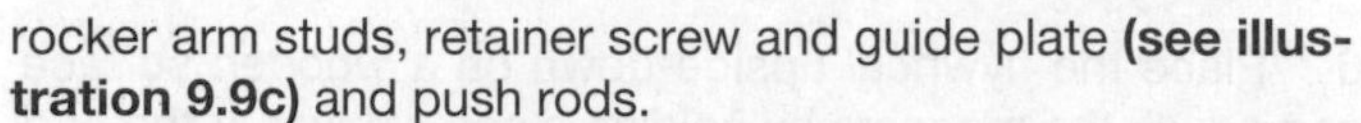

rocker arm studs, retainer screw and guide plate **(see illustration 9.9c)** and push rods.

10 If it's still in place, remove the bolts and separate the intake manifold from the engine. Remove the gasket and discard it. Remove the mounting bolts and detach the crankcase breather assembly and gasket from the engine.

11 Remove the cylinder head bolts, working in a criss-cross pattern. **Note:** *Some smaller horsepower engines are equipped with 7/16 or 1/2 inch head bolts while larger horsepower engines will require an Allen wrench to remove the head bolts.* Mark the locations of the bolts to ensure that they are returned to their original locations upon reassembly.

12 Break the cylinder head loose from the crankcase using a mallet or brass hammer **(see illustration 9.12)**.

13 On horizontal shaft engines, remove the crankcase cover **(see illustration 9.13)**.

14 On vertical shaft engines, loosen the oil sump-to-engine block bolts in 1/4-turn increments to avoid warping the sump, then remove them. Tap the sump with a soft-face hammer to break the gasket seal, then separate it from the engine block and crankshaft. If it hangs up on the crankshaft, continue to tap on it with the hammer, but be very careful not to crack or distort it (especially if it's made of aluminum). If thrust washers are installed on the crankshaft or camshaft, slide them off and set them aside.

15 Align the timing marks **(see illustrations 9.15a and 9.15b)**. On engines equipped with a balance shaft system, it will be necessary to make two sets of timing marks; one for the counterbalance shaft gears and the other set for the camshaft and crankshaft gears. Remove the counterbalance shaft gears to access the camshaft and crankshaft gears. **Note:** *Some engines may be equipped with an Ultra-Balance Counterbalance system. This system uses a single weighted shaft that drives off the crankshaft. This shaft functions as a counterweight to the weights on the crankshaft that cause excessive engine vibration. Be sure to paint the timing marks before disassembly and make notes or pictures to help the reassembly procedure.*

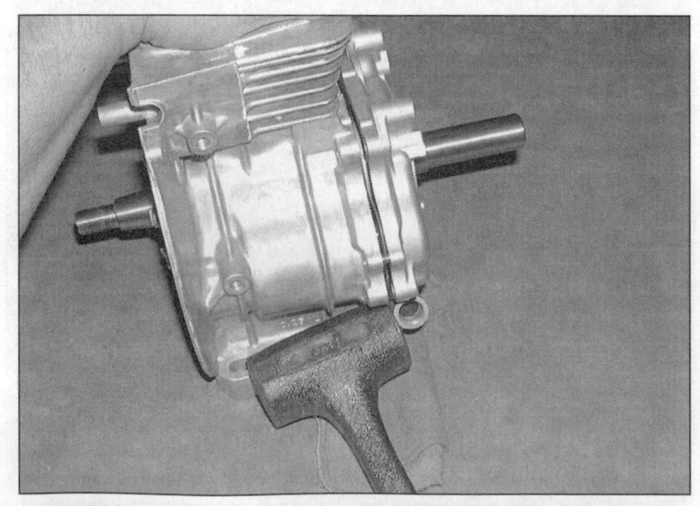

9.13 Use a mallet to separate the crankcase cover from the crankcase (horizontal crankshaft engine shown)

9.15a Location of the timing marks (arrows)

*A Crankshaft-to-camshaft timing marks*
*B Balance shaft-to-crankshaft timing marks*

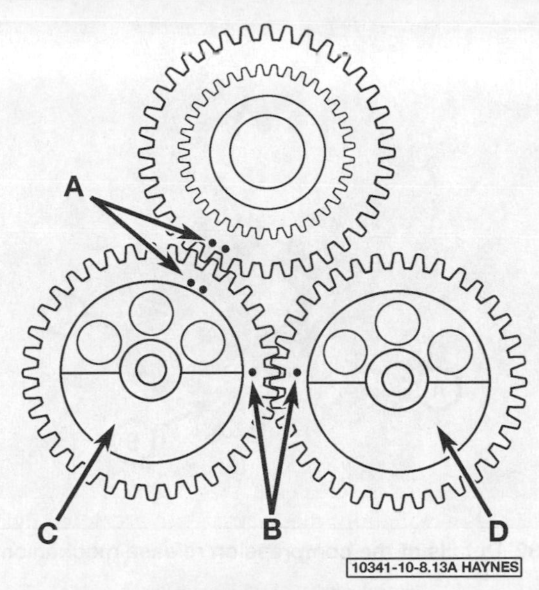


**9.15b Balance shaft alignment marks on double balance shaft systems**

*A Crankshaft-to-balance shaft number 1 timing marks*
*B Balance shaft number 1-to-balance shaft number 2 timing marks*
*C Balance shaft number 1*
*D Balance shaft number 2*

16 Remove the balance shaft(s), if the engine is equipped with the balance shaft system **(see illustration 9.16).** Remove the camshaft.

17 Use a ridge reamer to remove the carbon/wear ridge from the top of the cylinder bore after the cylinder head is off. Follow the manufacturer's instructions included with the tool (see Chapter 5).

18 Vertical crankshaft engines are equipped with a rotary-type oil pump that's driven by a slot on the camshaft end. Remove the bolts and lift out the oil pump assembly. **Note:** *There are several different types of oiling systems on these engines. Smaller horsepower horizontal engines are equipped with the "splash oiling system". A dipper is installed on the end of the connecting rod to agitate the oil in the sump, causing oil to splash over the cylinder walls and bearings. Vertical crankshaft engines are equipped with a rotary oil pump located in the sump driven from the camshaft. On smaller horsepower engines, this system is called the "forced splash oiling system" because it uses a pump to pressurize the oil and splashes the oil over the cylinder walls and bearings through a special oiling port. Larger horsepower engines are also equipped with a rotary oil pump that forces the pressurized oil through the oil jet on the crankshaft arm over the piston, cylinder and connecting rod. This is called the "pressurized oiling system".*

**9.16 Remove the balance shaft (single balance shaft system)**

19 After the camshaft is removed pull out the tappets **(see illustration 9.19)** and store them in marked containers so they can be returned to their original locations.

20 Look for match marks on the connecting rod and cap. If you can't see any, mark the side of the connecting rod and cap **(see illustration 9.20)** that faces OUT and note how the oil dipper (if used) is installed (the parts must be

**9.19 Remove the tappets**

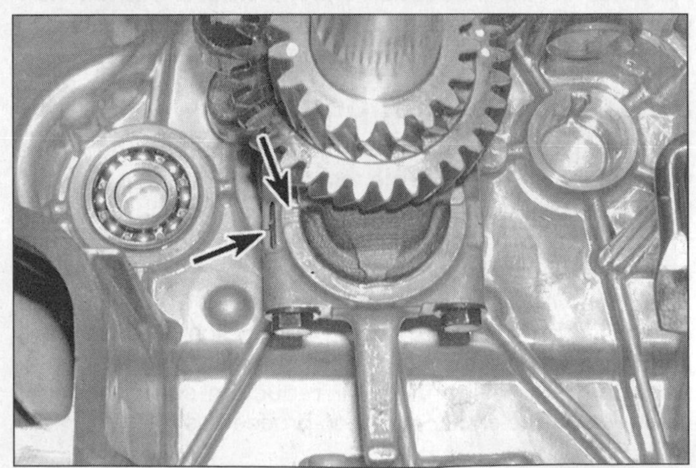

**9.20 Be sure to identify the connecting rod matchmarks (arrows) before removing the cap bolts and piston/connecting rod assembly from the engine. If no marks are visible, use a punch or scribe and make your own marks**

reassembled in the exact same relationship to the crankshaft). **Note:** *Some models use offset piston and connecting rods. Be sure the match marks on the connecting rods are facing OUT when installing the connecting rods on all types of engines.*

21 Turn the crankshaft so the rod journal is at the bottom of its stroke (Bottom Dead Center).

22 Flatten the locking tabs (if used) on the connecting rod bolts with a punch and hammer.

23 Loosen the bolts in 1/4-turn increments until they can be removed by hand. Separate the cap (and washers, if used) from the connecting rod, move the end of the rod away from the crankshaft journal and push the piston/rod assembly out through the top of the bore. The crankshaft can now be lifted out.

24 Do not separate the lever from the governor shaft unless new parts are needed. The lever mount will be damaged during removal.

25 The governor assembly can be withdrawn from the gear shaft but it must first be removed from the crankcase cover. The governor assembly is pressed into the case and can be removed by prying under the assembly. In most cases, the governor assembly should be left intact to ensure correct operation after the engine is rebuilt. Note how the parts fit together to simplify reassembly (a simple sketch would be helpful). Check the governor parts for wear and damage. If the gear shaft must be replaced, measure how far it protrudes before removing it and don't damage the crankcase boss. Clamp the shaft in a vise and tap the crankcase boss with a soft-face hammer to extract the shaft from the hole. **Caution:** *DO NOT twist the shaft with pliers or the mounting hole will be enlarged and the new shaft won't fit into it securely.*

26 When installing the new shaft, coat the serrated end with stud and bearing mount liquid after the shaft has been started in the hole with a soft-face hammer. Use a vise or press to finish installing the shaft and make sure it protrudes the same amount as the original (or the distance specified on the instruction sheet included with the new part). Wipe any excess stud and bearing mount liquid off the shaft and mounting boss flange.

## Inspection of components

27 Working on the cylinder head, compress the intake valve spring and remove the retainer (see Chapter 5).

28 After the engine has been completely disassembled, refer to Chapter 5 for the cleaning, component inspection and valve lapping procedures.

29 Once you've inspected and serviced everything and purchased any necessary new parts, which should always include new gaskets and seals, reassembly can begin. Begin by reinstalling the gear reduction components (if equipped) in the crankcase, then proceed as follows:

# Reassembly

30 Working on the cylinder head, coat the intake valve stem with clean engine oil, moly-base grease or engine assembly lube, then reinstall it in the block. Make sure it's returned to its original location.

**9.32 Details of the compression release mechanism**

A *Hinge pin*
B *Locking pin*
C *Lever*

31 Compress the valve spring with both retainers in place, then pull the valve out enough to position the spring and install the slotted retainer. Release the compressor and make sure the retainer is securely locked on the end of the valve. Repeat the procedure for the exhaust valve. **Note:** *Some models are equipped with dampening coils (tight wound coils) on the end of the valve spring. Position the dampening coils near the engine block away from the valve spring retainers.*

32 If equipped, assemble the compression release components onto the camshaft **(see illustration 9.32). Note 1:** *Some models are equipped with the Mechanical Compression Release (MCR) camshafts that are equipped with a locking pin in the camshaft that extends over the exhaust cam lobe to lift the valve to relieve engine compression during cranking. After the engine has started, centrifugal force moves the weight outward, dropping the pin down and out of the way of the exhaust valve, thereby bringing the engine compression back to normal. Be sure this pin is locked into the correct position after installation.* **Note 2:** *Some models are equipped with a Bump Compression Release (BCR) camshaft. This type of camshaft is equipped with a small bump on the exhaust lobe to allow slight compression release when the engine is cranking slow during start-up. Make sure the correct type of camshaft is installed in the particular engine by verifying the part numbers and camshaft types with a small engine parts technician.*

33 Lubricate the crankshaft magneto side oil seal lip, the plain bearing (if applicable) and the connecting rod journal with clean engine oil, moly-base grease or engine assembly lube, then reposition the crankshaft in the crankcase **(see illustration 9.33)**. If a ball-bearing is used on the magneto side, lubricate it with clean engine oil.

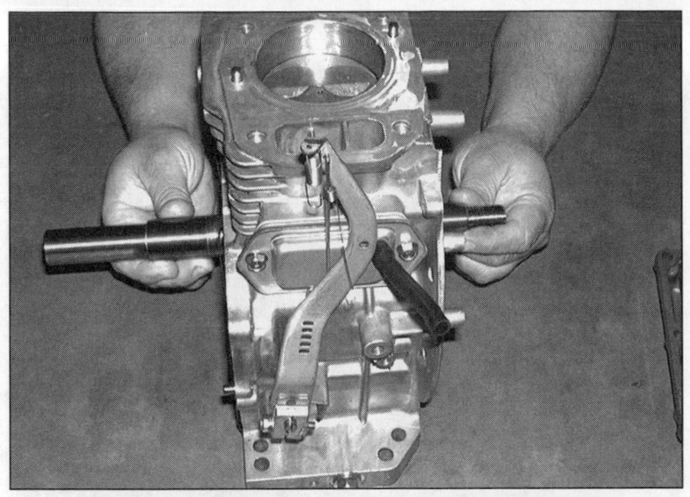
9.33 Install the crankshaft into the crankcase

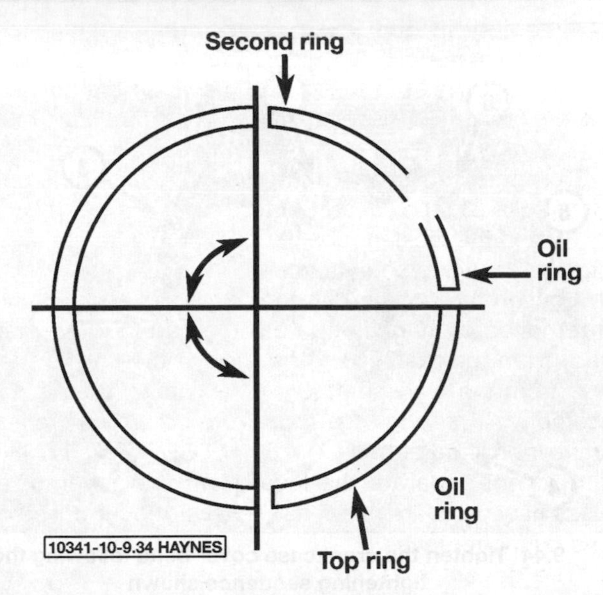

9.34 Be sure the rings are staggered at 90 degree angles to each other

34 Before installing the piston/connecting rod assembly, the cylinder must be perfectly clean and the top edge of the bore must be chamfered slightly so the rings don't catch on it. Stagger the piston ring end gaps and make sure they're positioned opposite the valve seats when the piston is installed (see Chapter 5). Lubricate the piston and rings with clean engine oil, then attach a ring compressor to the piston. Leave the skirt protruding about 1/4-inch. Tighten the compressor until the piston cannot be turned, then loosen it until the piston turns in the compressor with resistance. **Note:** *These models require the rings to be staggered at 90 degree angles* **(see illustration 9.34)**. *Make sure the ring marks face UP.*

35 Rotate the crankshaft until the connecting rod journal is at TDC (Top Dead Center - top of the stroke) and apply a coat of engine oil to the cylinder walls. If the piston has an arrow in the top, it must face OUT of the engine (if possible) or toward the carburetor when looking at the engine with the connecting rod pointing down. Make sure the match marks on the rod and cap will be facing out when the rod/piston assembly is in place. Gently insert the piston/connecting rod assembly into the cylinder and rest the bottom edge of the ring compressor on the engine block. Tap the top edge of the ring compressor to make sure it's contacting the block around its entire circumference. **Note:** *The piston(s) used on these models will have either a square or an arrow stamped on the top of the piston. The piston must be installed with these marks pointing toward the flywheel* **(see illustration 8.31)**.

36 Carefully tap on the top of the piston with the end of a wooden or plastic hammer handle while guiding the end of the connecting rod into place on the crankshaft journal. The piston rings may try to pop out just before entering the bore, so keep some pressure on the ring compressor. Work slowly. If any resistance is felt as the piston enters the cylinder, stop immediately. Find out what's hanging up and fix it before proceeding. Do not, for any reason, force the piston into the cylinder, as you will break a ring and/or the piston.

37 Install the connecting rod cap, a NEW lock plate, the oil dipper (if used) and the bolts. Make sure the marks you made on the rod and cap (or the manufacturer's marks) are aligned and facing out and the oil dipper is oriented correctly.

38 Tighten the bolts to the torque listed in the Specifications at the end of this Chapter. Work up to the final torque in three steps. Temporarily install the camshaft and turn the crankshaft through two complete revolutions to make sure the rod doesn't hit the cylinder or camshaft. If it does, the piston/connecting rod is installed incorrectly.

39 Apply clean engine oil, moly-base grease or engine assembly lube to the tappets, then reinstall them. Make sure they're returned to their original locations.

40 If it's not already in place, install the camshaft. Apply clean engine oil, moly-base grease or engine assembly lube to the lobes and bearing journals and align the timing marks on the gears - this is very important! The marks are usually dimples/lines or beveled teeth (or a combination of them) near the outer edge of the gears **(see illustrations 9.15a and 9.15b)**.

41 Install the rotary-type oil pump (vertical crankshaft engines) after lubricating it with clean engine oil. **Note 1:** *On rotary type oil pumps, make sure the alignment slot in the camshaft end is positioned to engage the oil pump drive when the crankcase cover/sump is installed onto the engine.* **Note 2:** *On vertical crankshaft engines, install a new oil filter onto the crankcase cover sump. The oil filter is mounted near the oil pump.*

42 Lubricate the crankshaft main bearing journal and the lip on the oil seal in the sump (or crankcase cover) with clean engine oil, moly-base grease or engine assembly lube.

43 Install the counterbalance gears. Set the engine to TDC (see Section 4). Make sure the alignment marks are correct **(see illustrations 9.15a and 9.15b)**.

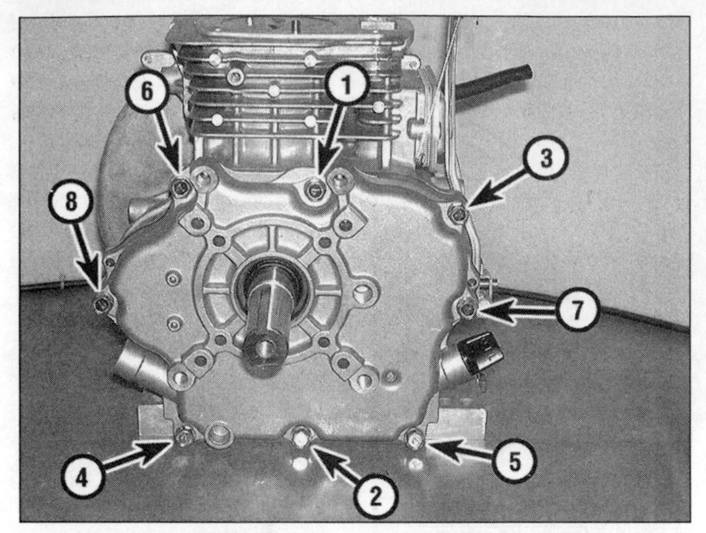


9.44 **Tighten the crankcase cover bolts following the tightening sequence shown**

9.45 **Install a larger shim to compensate for excess play or a smaller shim to increase the endplay**

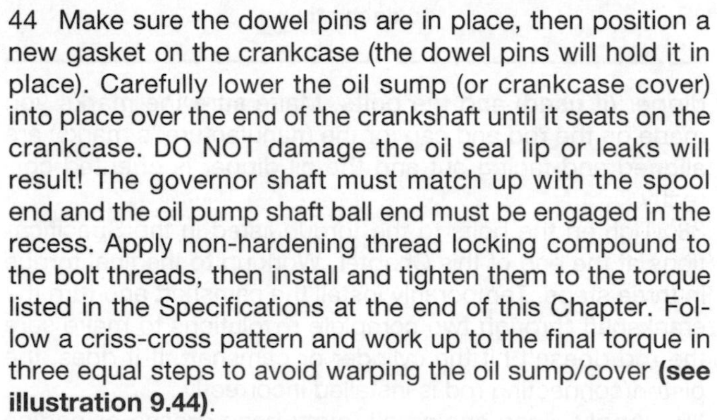

44 Make sure the dowel pins are in place, then position a new gasket on the crankcase (the dowel pins will hold it in place). Carefully lower the oil sump (or crankcase cover) into place over the end of the crankshaft until it seats on the crankcase. DO NOT damage the oil seal lip or leaks will result! The governor shaft must match up with the spool end and the oil pump shaft ball end must be engaged in the recess. Apply non-hardening thread locking compound to the bolt threads, then install and tighten them to the torque listed in the Specifications at the end of this Chapter. Follow a criss-cross pattern and work up to the final torque in three equal steps to avoid warping the oil sump/cover **(see illustration 9.44)**.

45 The crankshaft endplay must be checked with a dial indicator **(see illustration 8.41a)**. If it's excessive, install the correct size shims **(see illustration 9.45)**.

46 Install the cylinder head gasket **(see illustration 9.46a)** and cylinder head, then tighten the bolts in a criss-cross pattern **(see illustration 9.46b)** to the torque listed in the Specifications at the end of this Chapter. Work up to the final torque in three passes.

47 Install the crankcase breather (with the small hole facing down) and intake manifold (unless the carburetor and manifold were removed as an assembly). Use new gaskets and tighten the bolts securely.

48 Adjust the valves with the engine cold (see Section 7).

49 Install the valve cover with a new gasket. Be sure to install new O-rings onto the cover bolts, if equipped.

50 Install the governor assembly **(see illustration 9.3)**. Refer to the carburetor and governor adjustment procedure in Section 6 and make sure the governor linkage and carburetor is adjusted properly.

51 Install the ignition coil/spark plug wire assembly and align the marks on the coil bracket and crankcase bosses, then tighten the coil mounting bolts securely. **Note:** *If the coil is mounted on the outside of the flywheel, don't tighten the bolts completely - just snug them up. The bolt holes are slotted; move the coil as far away from the flywheel as*

9.46a **Install a new cylinder head gasket**

9.46b **Cylinder head bolt tightening sequence**

*possible before snugging up the bolts. Check the coil/module air gap adjusting procedure in Section 4. Be sure to reroute the wires from the ignition coil properly.*

52 Position the Woodruff key in the crankshaft keyway and install the flywheel. Install the starter cup, washer and nut. Tighten the nut to the specified torque. **Note:** *If the ignition coil is mounted outside the flywheel, turn the flywheel so the magnets are facing away from the coil assembly, then insert a feeler gauge equal to the thickness of the air gap listed in the specifications at the end of this Chapter, between the flywheel and the legs of the coil armature. Turn the flywheel until the magnets are aligned with the armature legs, then loosen the coil mounting bolts so the magnets will draw the armature against the flywheel. Tighten the coil mounting bolts securely, then turn the flywheel to release the feeler gauge.*

53 If a flywheel brake is used, install it now. To install the remaining components, refer to Chapters 4 and 5 as necessary. **Caution:** *Be sure to fill the crankcase to the correct level with the specified oil before attempting to start the engine.*

# 10 Specifications

### General

| | |
|---|---|
| Engine oil | |
| Type | API grade SH or better high quality detergent oil |
| Viscosity* | |
| Below 5 degrees F | SAE 5W |
| 5 to 32 degrees F | SAE 10W |
| 23 to 60 degrees F | SAE 20W |
| 50 to 95 degrees F | SAE 30W |
| Above 77 degrees F | SAE 40W |
| 5 to 95 degrees F | SAE 10W-30 |

**Use SAE 10W-30 for the first five hours on a new or overhauled engine.*

| | |
|---|---|
| Engine oil capacity | |
| L-head engines | |
| EY23 | 1.5 pints |
| EY25 and EY27 | 2.4 pints |
| EY28 | 1.5 pints |
| EY35 and EY40 | 2.5 pints |
| EY45V | 2.7 pints |
| OHV engines | |
| EH17 | 1.4 pints |
| EH21 | 1.4 pints |
| EH25 | 1.4 pints |
| EH30 and EH34 | 2.5 pints |
| EH30V and EH34V | 2.3 pints |
| EH43V | 2.7 pints |

### Fuel system

| | |
|---|---|
| Idle speed | |
| L-head engines | |
| EY23, EY25 and EY27 | 1250 rpm |
| EY28 | 1200 rpm |
| EY35 and EY40 | 1150 rpm |
| Y45V | 1200 rpm |
| OHV engines | |
| EH17, EH21, EH25, EH30, EH34, 0V and EH34V | 1200 rpm |
| EH43V | 1250 rpm |

**Fuel system (continued)**

| | |
|---|---|
| Float height | |
| EY23 (W1-230) | not available |
| EY25 (EY25W) | 29/32 to 63/64 inch |
| EY27 (EY27W) | 29/32 to 63/64 inch |
| EY28 (W1-280) | 33/64 to 19/32 inch |
| EY35 (W1-340) | not available |
| EY40 (W1-390) | not available |
| EY45, EY45V (W1-450(V)) | not available |

**Ignition system**

| | |
|---|---|
| Spark plug type (Champion) | |
| L-head engines | |
| EY23 | RL95YC |
| EY25 | L86C |
| EY27 | L90C |
| EY28 | L86C |
| EY35 and EY40 | RL95YC |
| Y45V | CJ8Y |
| OHV engines | |
| EH17, EH21 and EH25 | L87YC |
| EH30, EH34, EH30V, EH34V and EH43V | N9YC |
| Spark plug gap (all) | 0.025 inch |
| Ignition point gap (models with points) | 0.014 inch |
| Ignition coil/module-to-flywheel air gap | |
| EY45V | 0.020 inch |
| All others | 0.012 to 0.020 inch |

**Engine**

| | |
|---|---|
| Cylinder bore diameter (standard) | |
| L-head engines | |
| EY23 | 2.68 inches |
| EY25 | 2.83 inches |
| EY27 | 2.91 inches |
| EY28 | 2.95 inches |
| EY35 | 3.07 inches |
| EY40 | 3.30 inches |
| Y45V | 3.54 inches |
| OHV engines | |
| EH17 | 2.64 inches |
| EH21 | 2.83 inches |
| EH25 | 2.95 inches |
| EH30 and EH30V | 3.07 inches |
| EH34 and EH34V | 3.31 inches |
| EH43V | 3.50 inches |
| Piston diameter (minimum) | |
| L-head engines | |
| EY23 | 2.672 inches |
| EY25 | 2.822 inches |
| EY27 | 2.908 inches |
| EY28 | 2.942 inches |
| EY35 | 3.062 inches |
| EY40 | 3.292 inches |
| Y45V | 3.492 inches |

OHV engines
EH17 .......... 2.632 inches
EH21 .......... 2.822 inches
EH25 .......... 2.941 inches
EH30 and EH30V .......... 3.066 inches
EH34 and EH34V .......... 3.302 inches
EH43V .......... 3.492 inches

Piston-to-cylinder bore clearance
L-head engines
EY23, EY28 and EY45V .......... 0.001 to 0.003 inch
EY25 and EY40 .......... 0.002 to 0.004 inch
EY27 .......... 0.003 to 0.005 inch
EY35 .......... 0.002 to 0.005 inch
OHV engines .......... 0.001 to 0.003 inch

Piston-to-pin clearance
L-head engines .......... 0.0003 to 0.0004 inch
OHV engines .......... 0.0002 to 0.0005 inch

Piston ring side clearance
L-head engines
EY23
Top ring .......... 0.0020 to 0.0037 inch
Second ring .......... 0.0016 to 0.0033 inch
Oil ring .......... 0.0004 to 0.0025 inch
EY25 and EY27
Top ring .......... 0.0020 to 0.0037 inch
Second ring .......... 0.0016 to 0.0033 inch
Oil ring .......... 0.0004 to 0.0020 inch
EY28
Top ring .......... 0.0020 to 0.0035 inch
Second ring .......... 0.0012 to 0.0028 inch
Oil ring .......... 0.0004 to 0.0035 inch
EY35 and EY40
Top ring .......... 0.0020 to 0.0035 inch
Second ring .......... 0.0020 to 0.0035 inch
Y45V
Top ring .......... 0.0043 to 0.0059 inch
Second ring .......... 0.0024 to 0.0038 inch
Oil ring .......... 0.0004 to 0.0021 inch
OHV engines
EH17
Top ring .......... 0.0014 to 0.0031 inch
Second ring .......... 0.0010 to 0.0030 inch
Oil ring .......... 0.0004 to 0.0026 inch
EH21
Top ring .......... 0.0020 to 0.0037 inch
Second ring .......... 0.0016 to 0.0033 inch
Oil ring .......... 0.0004 to 0.0025 inch
EH25, EH30 and EH34
Top ring .......... 0.0020 to 0.0035 inch
Second ring .......... 0.0012 to 0.0028 inch
Oil ring .......... 0.0004 to 0.0026 inch
EH30V, EH34V and EH43V
Top ring .......... 0.0020 to 0.0035 inch
Second ring .......... 0.0012 to 0.0028 inch
Oil ring .......... 0.0004 to 0.0026 inch

**Engine (continued)**

Piston ring end gap
- New bore ........ 0.010 to 0.020 inch
- Used ........ 0.030 inch (maximum)

Connecting rod bearing oil clearance
- New ........ 0.001 to 0.002 inch
- Maximum ........ 0.003 inch

Connecting rod side play
- L-head engines ........ 0.005 to 0.016 inch
- OHV engines ........ 0.007 to 0.016 inch

Crankshaft connecting rod journal diameter (wear limit)
- L-head engines
  - EY23 ........ 1.0216 to 1.0222 inches
  - EY25 and EY27 ........ 1.1003 to 1.1008 inches
  - EY28 ........ 1.1011 to 1.1016 inches
  - EY35 and EY40 ........ 1.3352 to 1.3358 inches
  - Y45V ........ 1.4935 to 1.4941 inches
- OHV engines
  - EH17 ........ 1.1798 to 1.1803 inches
  - EH21 and EH25 ........ 1.3370 to 1.3376 inches
  - EH30, EH34, EH30V, EH34V and EH43V ........ 1.4943 to 1.4949 inches

Crankshaft main bearing journal diameter
- L-head engines
  - EY23, EY35 and EY40 ........ 0.9843 inches
  - EY25, EY27 and EY28 ........ 1.1811 inches
  - EY45V
    - Flywheel end ........ 1.3774 to 1.3780 inches
    - PTO end ........ 1.3778 to 1.3782 inches
- OHV engines
  - EH17 ........ 0.9838 to 0.9841 inches
  - EH21 ........ 1.1807 to 1.1811 inches
  - EH25
    - Flywheel end ........ 1.1807 to 1.1811 inches
    - PTO end
      - D type ........ 1.1806 to 1.1810 inches
      - B type ........ 1.1019 to 1.1022 inches
  - EH30, EH34, EH30V and EH34V ........ 1.3774 to 1.3778 inches
  - EH43V ........ 1.3773 to 1.3780 inches

Crankshaft endplay (limit)
- L-head engines
  - EY23, EY35 and EY40 ........ 0.008 inch
  - EY25, EY27, EY28 and EY45V ........ 0.009 inch
- OHV engines
  - EH17, EH21, EH25 and EH43V ........ 0.008 inch
  - EH30, EH34, EH30V and EH34V ........ 0.009 inch

Camshaft bearing clearance ........ 0.0010 to 0.0025 inch

Valve stem-to-guide clearance (minimum) start here
- L-head engines
  - EY23 ........ 0.0022 to 0.0038 inches
  - EY25 and EY27 ........ 0.0015 to 0.0042 inches
  - EY28
    - Intake valve ........ 0.0010 to 0.0024 inches
    - Exhaust valve ........ 0.0022 to 0.0038 inches
  - EY35, EY40 and EY45V
    - Intake valve ........ 0.0012 to 0.0036 inches
    - Exhaust valve ........ 0.0028 to 0.0048 inches

OHV engines
EH17
Intake valve ........................................................ 0.0018 to 0.0031 inch
Exhaust valve ...................................................... 0.0022 to 0.0036 inch
EH21 and EH25
Intake valve ........................................................ 0.0020 to 0.0034 inch
Exhaust valve ...................................................... 0.0022 to 0.0039 inch
EH30, EH34, EH30V, EH34V and EH43V
Intake valve ........................................................ 0.0022 to 0.0034 inch
Exhaust valve ...................................................... 0.0022 to 0.0039 inch
Valve clearance (cold)
L-head engines
EY23, EY25, EY27and EY28................................. 0.006 to 0.008 inches
EY35 and EY40 ..................................................... 0.005 to 0.007 inches
EY45V ................................................................... 0.003 to 0.005 inches
OHV engines.............................................................. 0.003 to 0.005 inches

**Torque specifications**

Connecting rod cap bolts/nuts
L-head engines
EY23................................................................... 168 in-lbs
EY25 and EY27 ..................................................... 18 ft-lbs
EY28................................................................... 168 in-lbs
EY35, EY40 and EY45V ......................................... 21 ft-lbs
OHV engines
EH17 ................................................................... 168 in-lbs
EH21 ................................................................... 16 ft-lbs
EH25 ................................................................... 19 ft-lbs
EH30, EH34, EH30V, EH34V and EH43V ............. 20 ft-lbs
Cylinder head bolts
L-head engines
EY23................................................................... 25 ft-lbs
EY25 and EY27 ..................................................... 26 ft-lbs
EY28................................................................... 18 ft-lbs
EY35, EY40 and EY45V ......................................... 28 ft-lbs
OHV engines
EH17 ................................................................... 20 ft-lbs
EH21 ................................................................... 17 ft-lbs
EH25, EH30, EH34, EH30V, EH34V and EH43V.... 30 ft-lbs
Oil sump (vertical engines) or crankcase cover bolts (horizontal engines)
L-head engines.......................................................... 156 in-lbs
OHV engines
EH17 ................................................................... 84 in-lbs
EH21 ................................................................... 120 in-lbs
EH25, EH30, EH34, EH30V, EH34V and EH43V.... 156 in-lbs
Flywheel fan-to-flywheel ............................................ 108 in-lbs
Flywheel mounting bolt
L-head engines
EY23, EY25 and EY27 .......................................... 47 ft-lbs
EY28................................................................... 48 ft-lbs
EY35, EY40 and EY45V ......................................... 72 ft-lbs
OHV engines
EH17 ................................................................... 47 ft-lbs
EH21 ................................................................... 68 ft-lbs
EH25 ................................................................... 48 ft-lbs
EH30, EH34, EH30V, EH34V and EH43V ............. 68 ft-lbs

**Torque specifications (continued)**

| | |
|---|---|
| Spark plugs | |
| L-head engines | |
| EY23 and EY28 | 18 ft-lbs |
| EY25, EY27, EY35 and EY40 | 20 ft-lbs |
| EY45V | 132 in-lbs |
| OHV engines | 120 in-lbs |
| Starter mounting screws | |
| Recoil starters | 60 in-lbs |
| Electric starters | 135 in-lbs |
| Valve cover screws* | 65 to 95 in-lbs |

**Where screws are self-tapping design, lower figure is for used assembly, higher figure for new assembly*

# Index

## L



## M



## O



## P



## R



## S

# Haynes Automotive Manuals

*NOTE: New manuals are added to this list on a periodic basis. If you do not see a listing for your vehicle, consult your local Haynes dealer for the latest product information.*

**ACURA**
*12020 **Integra** '86 thru '89 **& Legend** '86 thru '90

**AMC**
**Jeep CJ** - *see JEEP (50020)*
14020 **Mid-size models,** Concord, Hornet, Gremlin & Spirit '70 thru '83
14025 **(Renault) Alliance & Encore** '83 thru '87

**AUDI**
15020 **4000** all models '80 thru '87
15025 **5000** all models '77 thru '83
15026 **5000** all models '84 thru '88

**AUSTIN-HEALEY**
**Sprite** - *see MG Midget (66015)*

**BMW**
*18020 **3/5 Series** not including diesel or all-wheel drive models '82 thru '92
*18021 **3 Series** except 325iX models '92 thru '97
18025 **320i** all 4 cyl models '75 thru '83
18035 **528i** & 530i all models '75 thru '80
18050 **1500 thru 2002** except Turbo '59 thru '77

**BUICK**
**Century (front wheel drive)** - *see GM (829)*
*19020 **Buick, Oldsmobile & Pontiac Full-size (Front wheel drive)** all models '85 thru '98 **Buick** Electra, LeSabre and Park Avenue; **Oldsmobile** Delta 88 Royale, Ninety Eight and Regency; **Pontiac** Bonneville
19025 **Buick Oldsmobile & Pontiac Full-size (Rear wheel drive)** **Buick** Estate '70 thru '90, Electra'70 thru '84, LeSabre '70 thru '85, Limited '74 thru '79 **Oldsmobile** Custom Cruiser '70 thru '90, Delta 88 '70 thru '85,Ninety-eight '70 thru '84 **Pontiac** Bonneville '70 thru '81, Catalina '70 thru '81, Grandville '70 thru '75, Parisienne '83 thru '86
19030 **Mid-size Regal & Century** all rear-drive models with V6, V8 and Turbo '74 thru '87
**Regal** - *see GENERAL MOTORS (38010)*
**Riviera** - *see GENERAL MOTORS (38030)*
**Roadmaster** - *see CHEVROLET (24046)*
**Skyhawk** - *see GENERAL MOTORS (38015)*
**Skylark '80 thru '85** - *see GM (38020)*
**Skylark '86 on** - *see GM (38025)*
**Somerset** - *see GENERAL MOTORS (38025)*

**CADILLAC**
*21030 **Cadillac Rear Wheel Drive** all gasoline models '70 thru '93
**Cimarron** - *see GENERAL MOTORS (38015)*
**Eldorado** - *see GENERAL MOTORS (38030)*
**Seville '80 thru '85** - *see GM (38030)*

**CHEVROLET**
*24010 **Astro & GMC Safari Mini-vans** '85 thru '93
24015 **Camaro V8** all models '70 thru '81
24016 **Camaro** all models '82 thru '92
**Cavalier** - *see GENERAL MOTORS (38015)*
**Celebrity** - *see GENERAL MOTORS (38005)*
24017 **Camaro & Firebird** '93 thru '97
24020 **Chevelle, Malibu & El Camino** '69 thru '87
24024 **Chevette & Pontiac T1000** '76 thru '87
**Citation** - *see GENERAL MOTORS (38020)*
*24032 **Corsica/Beretta** all models '87 thru '96
24040 **Corvette** all V8 models '68 thru '82
*24041 **Corvette** all models '84 thru '96
10305 **Chevrolet Engine Overhaul Manual**
24045 **Full-size Sedans** Caprice, Impala, Biscayne, Bel Air & Wagons '69 thru '90
24046 **Impala SS & Caprice and Buick Roadmaster** '91 thru '96
**Lumina** - *see GENERAL MOTORS (38010)*
24048 **Lumina & Monte Carlo** '95 thru '98
**Lumina APV** - *see GM (38035)*
24050 **Luv Pick-up** all 2WD & 4WD '72 thru '82
*24055 **Monte Carlo** all models '70 thru '88
**Monte Carlo** '95 thru '98 - *see LUMINA (24048)*
24059 **Nova** all V8 models '69 thru '79
*24060 **Nova and Geo Prizm** '85 thru '92
24064 **Pick-ups '67 thru '87** - Chevrolet & GMC, all V8 & in-line 6 cyl, 2WD & 4WD '67 thru '87; Suburbans, Blazers & Jimmys '67 thru '91
*24065 **Pick-ups '88 thru '98** - Chevrolet & GMC, all full-size pick-ups, '88 thru '98; Blazer & Jimmy '92 thru '94; Suburban '92 thru '98; Tahoe & Yukon '98
24070 **S-10 & S-15 Pick-ups** '82 thru '93, **Blazer & Jimmy** '83 thru '94,
*24071 **S-10 & S-15 Pick-ups** '94 thru '96 **Blazer & Jimmy** '95 thru '96
*24075 **Sprint & Geo Metro** '85 thru '94
*24080 **Vans - Chevrolet & GMC,** V8 & in-line 6 cylinder models '68 thru '96

**CHRYSLER**
25015 **Chrysler Cirrus, Dodge Stratus, Plymouth Breeze** '95 thru '98
25025 **Chrysler Concorde, New Yorker & LHS, Dodge** Intrepid, **Eagle** Vision, '93 thru '97
10310 **Chrysler Engine Overhaul Manual**
*25020 **Full-size Front-Wheel Drive** '88 thru '93
**K-Cars** - *see DODGE Aries (30008)*
**Laser** - *see DODGE Daytona (30030)*
*25030 **Chrysler & Plymouth Mid-size** front wheel drive '82 thru '95
**Rear-wheel Drive** - *see Dodge (30050)*

**DATSUN**
28005 **200SX** all models '80 thru '83
28007 **B-210** all models '73 thru '78
28009 **210** all models '79 thru '82
28012 **240Z, 260Z & 280Z** Coupe '70 thru '78
28014 **280ZX** Coupe & 2+2 '79 thru '83
**300ZX** - *see NISSAN (72010)*
28016 **310** all models '78 thru '82
28018 **510 & PL521 Pick-up** '68 thru '73
28020 **510** all models '78 thru '81
28022 **620 Series Pick-up** all models '73 thru '79
**720 Series Pick-up** - *see NISSAN (72030)*
28025 **810/Maxima** all gasoline models, '77 thru '84

**DODGE**
**400 & 600** - *see CHRYSLER (25030)*
*30008 **Aries & Plymouth Reliant** '81 thru '89
30010 **Caravan & Plymouth Voyager Mini-Vans** all models '84 thru '95
*30011 **Caravan & Plymouth Voyager Mini-Vans** all models '96 thru '98
30012 **Challenger/Plymouth Saporro** '78 thru '83
30016 **Colt & Plymouth Champ (front wheel drive)** all models '78 thru '87
*30020 **Dakota Pick-ups** all models '87 thru '96
30025 **Dart, Demon, Plymouth Barracuda, Duster & Valiant** 6 cyl models '67 thru '76
*30030 **Daytona & Chrysler Laser** '84 thru '89
**Intrepid** - *see CHRYSLER (25025)*
*30034 **Neon** all models '95 thru '97
*30035 **Omni & Plymouth Horizon** '78 thru '90
*30040 **Pick-ups** all full-size models '74 thru '93
*30041 **Pick-ups** all full-size models '94 thru '96
*30045 **Ram 50/D50 Pick-ups & Raider and Plymouth Arrow Pick-ups** '79 thru '93
30050 **Dodge/Plymouth/Chrysler** rear wheel drive '71 thru '89
*30055 **Shadow & Plymouth Sundance** '87 thru '94
*30060 **Spirit & Plymouth Acclaim** '89 thru '95
*30065 **Vans - Dodge & Plymouth** '71 thru '96

**EAGLE**
**Talon** - *see Mitsubishi Eclipse (68030)*
**Vision** - *see CHRYSLER (25025)*

**FIAT**
34010 **124 Sport Coupe & Spider** '68 thru '78
34025 **X1/9** all models '74 thru '80

**FORD**
10355 **Ford Automatic Transmission Overhaul**
*36004 **Aerostar Mini-vans** all models '86 thru '96
*36006 **Contour & Mercury Mystique** '95 thru '98
36008 **Courier Pick-up** all models '72 thru '82
36012 **Crown Victoria & Mercury Grand Marquis** '88 thru '96
10320 **Ford Engine Overhaul Manual**
36016 **Escort/Mercury Lynx** all models '81 thru '90
*36020 **Escort/Mercury Tracer** '91 thru '96
*36024 **Explorer & Mazda Navajo** '91 thru '95
36028 **Fairmont & Mercury Zephyr** '78 thru '83
36030 **Festiva & Aspire** '88 thru '97
36032 **Fiesta** all models '77 thru '80
36036 **Ford & Mercury Full-size,** Ford LTD & Mercury Marquis ('75 thru '82); Ford Custom 500,Country Squire, Crown Victoria & Mercury Colony Park ('75 thru '87); Ford LTD Crown Victoria & Mercury Gran Marquis ('83 thru '87)
36040 **Granada & Mercury Monarch** '75 thru '80
36044 **Ford & Mercury Mid-size,** Ford Thunderbird & Mercury Cougar ('75 thru '82); Ford LTD & Mercury Marquis ('83 thru '86); Ford Torino,Gran Torino, Elite, Ranchero pick-up, LTD II, Mercury Montego, Comet, XR-7 & Lincoln Versailles ('75 thru '86)
36048 **Mustang V8** all models '64-1/2 thru '73
36049 **Mustang II** 4 cyl, V6 & V8 models '74 thru '78
36050 **Mustang & Mercury Capri** all models Mustang, '79 thru '93; Capri, '79 thru '86
*36051 **Mustang** all models '94 thru '97
36054 **Pick-ups & Bronco** '73 thru '79
36058 **Pick-ups & Bronco** '80 thru '96
36059 **Pick-ups, Expedition & Mercury Navigator** '97 thru '98
36062 **Pinto & Mercury Bobcat** '75 thru '80
36066 **Probe** all models '89 thru '92
36070 **Ranger/Bronco II** gasoline models '83 thru '92
*36071 **Ranger** '93 thru '97 & **Mazda Pick-ups** '94 thru '97
36074 **Taurus & Mercury Sable** '86 thru '95
*36075 **Taurus & Mercury Sable** '96 thru '98
*36078 **Tempo & Mercury Topaz** '84 thru '94
36082 **Thunderbird/Mercury Cougar** '83 thru '88
*36086 **Thunderbird/Mercury Cougar** '89 and '97
36090 **Vans** all V8 Econoline models '69 thru '91
*36094 **Vans** full size '92-'95
*36097 **Windstar Mini-van** '95-'98

**GENERAL MOTORS**
*10360 **GM Automatic Transmission Overhaul**
*38005 **Buick Century, Chevrolet Celebrity, Oldsmobile Cutlass Ciera & Pontiac 6000** all models '82 thru '96
*38010 **Buick Regal, Chevrolet Lumina, Oldsmobile Cutlass Supreme & Pontiac Grand Prix** front-wheel drive models '88 thru '95
*38015 **Buick Skyhawk, Cadillac Cimarron, Chevrolet Cavalier, Oldsmobile Firenza & Pontiac J-2000 & Sunbird** '82 thru '94
*38016 **Chevrolet Cavalier & Pontiac Sunfire** '95 thru '98
38020 **Buick Skylark, Chevrolet Citation, Olds Omega, Pontiac Phoenix** '80 thru '85
38025 **Buick Skylark & Somerset, Oldsmobile Achieva & Calais and Pontiac Grand Am** all models '85 thru '95
38030 **Cadillac Eldorado** '71 thru '85, **Seville** '80 thru '85, **Oldsmobile Toronado** '71 thru '85 **& Buick Riviera** '79 thru '85
*38035 **Chevrolet Lumina APV, Olds Silhouette & Pontiac Trans Sport** all models '90 thru '95
**General Motors Full-size Rear-wheel Drive** - *see BUICK (19025)*

***(Continued on other side)***

** Listings shown with an asterisk (*) indicate model coverage as of this printing. These titles will be periodically updated to include later model years - consult your Haynes dealer for more information.*

# Haynes Automotive Manuals (continued)

*NOTE: New manuals are added to this list on a periodic basis. If you do not see a listing for your vehicle, consult your local Haynes dealer for the latest product information.*

### GEO
**Metro** - *see CHEVROLET Sprint (24075)*
**Prizm** - *'85 thru '92 see CHEVY (24060), '93 thru '96 see TOYOTA Corolla (92036)*
*40030 **Storm** all models '90 thru '93
**Tracker** - *see SUZUKI Samurai (90010)*

### GMC
**Safari** - *see CHEVROLET ASTRO (24010)*
**Vans & Pick-ups** - *see CHEVROLET*

### HONDA
42010 **Accord CVCC** all models '76 thru '83
42011 **Accord** all models '84 thru '89
42012 **Accord** all models '90 thru '93
42013 **Accord** all models '94 thru '95
42020 **Civic 1200** all models '73 thru '79
42021 **Civic 1300 & 1500 CVCC** '80 thru '83
42022 **Civic 1500 CVCC** all models '75 thru '79
42023 **Civic** all models '84 thru '91
*42024 **Civic & del Sol** '92 thru '95
*42040 **Prelude CVCC** all models '79 thru '89

### HYUNDAI
*43015 **Excel** all models '86 thru '94

### ISUZU
**Hombre** - *see CHEVROLET S-10 (24071)*
*47017 **Rodeo** '91 thru '97; **Amigo** '89 thru '94; **Honda Passport** '95 thru '97
*47020 **Trooper & Pick-up**, all gasoline models Pick-up, '81 thru '93; Trooper, '84 thru '91

### JAGUAR
*49010 **XJ6** all 6 cyl models '68 thru '86
*49011 **XJ6** all models '88 thru '94
*49015 **XJ12 & XJS** all 12 cyl models '72 thru '85

### JEEP
*50010 **Cherokee, Comanche & Wagoneer Limited** all models '84 thru '96
50020 **CJ** all models '49 thru '86
*50025 **Grand Cherokee** all models '93 thru '98
50029 **Grand Wagoneer & Pick-up** '72 thru '91 Grand Wagoneer '84 thru '91, Cherokee & Wagoneer '72 thru '83, Pick-up '72 thru '88
*50030 **Wrangler** all models '87 thru '95

### LINCOLN
**Navigator** - *see FORD Pick-up (36059)*
59010 **Rear Wheel Drive** all models '70 thru '96

### MAZDA
61010 **GLC Hatchback (rear wheel drive)** '77 thru '83
61011 **GLC (front wheel drive)** '81 thru '85
*61015 **323 & Protogé** '90 thru '97
*61016 **MX-5 Miata** '90 thru '97
*61020 **MPV** all models '89 thru '94
**Navajo** - *see Ford Explorer (36024)*
61030 **Pick-ups** '72 thru '93
**Pick-ups** '94 thru '96 - *see Ford Ranger (36071)*
61035 **RX-7** all models '79 thru '85
*61036 **RX-7** all models '86 thru '91
61040 **626** (rear wheel drive) all models '79 thru '82
*61041 **626/MX-6 (front wheel drive)** '83 thru '91

### MERCEDES-BENZ
63012 **123 Series Diesel** '76 thru '85
*63015 **190 Series** four-cyl gas models, '84 thru '88
63020 **230/250/280** 6 cyl sohc models '68 thru '72
63025 **280 123 Series** gasoline models '77 thru '81
63030 **350 & 450** all models '71 thru '80

### MERCURY
***See FORD Listing.***

### MG
66010 **MGB** Roadster & GT Coupe '62 thru '80
66015 **MG Midget, Austin Healey Sprite** '58 thru '80

### MITSUBISHI
*68020 **Cordia, Tredia, Galant, Precis & Mirage** '83 thru '93
*68030 **Eclipse, Eagle Talon & Ply. Laser** '90 thru '94
*68040 **Pick-up** '83 thru '96 & **Montero** '83 thru '93

### NISSAN
72010 **300ZX** all models including Turbo '84 thru '89
*72015 **Altima** all models '93 thru '97
*72020 **Maxima** all models '85 thru '91
*72030 **Pick-ups** '80 thru '96 **Pathfinder** '87 thru '95
72040 **Pulsar** all models '83 thru '86
*72050 **Sentra** all models '82 thru '94
*72051 **Sentra & 200SX** all models '95 thru '98
*72060 **Stanza** all models '82 thru '90

### OLDSMOBILE
*73015 **Cutlass** V6 & V8 gas models '74 thru '88
***For other OLDSMOBILE titles, see BUICK, CHEVROLET or GENERAL MOTORS listing.***

### PLYMOUTH
***For PLYMOUTH titles, see DODGE listing.***

### PONTIAC
79008 **Fiero** all models '84 thru '88
79018 **Firebird** V8 models except Turbo '70 thru '81
79019 **Firebird** all models '82 thru '92
***For other PONTIAC titles, see BUICK, CHEVROLET or GENERAL MOTORS listing.***

### PORSCHE
*80020 **911** except Turbo & Carrera 4 '65 thru '89
80025 **914** all 4 cyl models '69 thru '76
80030 **924** all models including Turbo '76 thru '82
*80035 **944** all models including Turbo '83 thru '89

### RENAULT
**Alliance & Encore** - *see AMC (14020)*

### SAAB
*84010 **900** all models including Turbo '79 thru '88

### SATURN
87010 **Saturn** all models '91 thru '96

### SUBARU
89002 **1100, 1300, 1400 & 1600** '71 thru '79
*89003 **1600 & 1800** 2WD & 4WD '80 thru '94

### SUZUKI
*90010 **Samurai/Sidekick & Geo Tracker** '86 thru '96

### TOYOTA
92005 **Camry** all models '83 thru '91
92006 **Camry** all models '92 thru '96
92015 **Celica Rear Wheel Drive** '71 thru '85
*92020 **Celica Front Wheel Drive** '86 thru '93
92025 **Celica Supra** all models '79 thru '92
92030 **Corolla** all models '75 thru '79
92032 **Corolla** all rear wheel drive models '80 thru '87
92035 **Corolla** all front wheel drive models '84 thru '92
*92036 **Corolla & Geo Prizm** '93 thru '97
92040 **Corolla Tercel** all models '80 thru '82
92045 **Corona** all models '74 thru '82
92050 **Cressida** all models '78 thru '82
92055 **Land Cruiser** FJ40, 43, 45, 55 '68 thru '82
92056 **Land Cruiser** FJ60, 62, 80, FZJ80 '80 thru '96
*92065 **MR2** all models '85 thru '87
92070 **Pick-up** all models '69 thru '78
*92075 **Pick-up** all models '79 thru '95
*92076 **Tacoma** '95 thru '98, **4Runner** '96 thru '98, **& T100** '93 thru '98
*92080 **Previa** all models '91 thru '95
92085 **Tercel** all models '87 thru '94

### TRIUMPH
94007 **Spitfire** all models '62 thru '81
94010 **TR7** all models '75 thru '81

### VW
96008 **Beetle & Karmann Ghia** '54 thru '79
96012 **Dasher** all gasoline models '74 thru '81
*96016 **Rabbit, Jetta, Scirocco, & Pick-up** gas models '74 thru '91 & Convertible '80 thru '92
96017 **Golf & Jetta** all models '93 thru '97
96020 **Rabbit, Jetta & Pick-up** diesel '77 thru '84
96030 **Transporter 1600** all models '68 thru '79
96035 **Transporter 1700, 1800 & 2000** '72 thru '79
96040 **Type 3 1500 & 1600** all models '63 thru '73
96045 **Vanagon** all air-cooled models '80 thru '83

### VOLVO
97010 **120, 130 Series & 1800 Sports** '61 thru '73
97015 **140 Series** all models '66 thru '74
*97020 **240 Series** all models '76 thru '93
97025 **260 Series** all models '75 thru '82
*97040 **740 & 760 Series** all models '82 thru '88

### TECHBOOK MANUALS
10205 **Automotive Computer Codes**
10210 **Automotive Emissions Control Manual**
10215 **Fuel Injection Manual, 1978 thru 1985**
10220 **Fuel Injection Manual, 1986 thru 1996**
10225 **Holley Carburetor Manual**
10230 **Rochester Carburetor Manual**
10240 **Weber/Zenith/Stromberg/SU Carburetors**
10305 **Chevrolet Engine Overhaul Manual**
10310 **Chrysler Engine Overhaul Manual**
10320 **Ford Engine Overhaul Manual**
10330 **GM and Ford Diesel Engine Repair Manual**
10340 **Small Engine Repair Manual**
10345 **Suspension, Steering & Driveline Manual**
10355 **Ford Automatic Transmission Overhaul**
10360 **GM Automatic Transmission Overhaul**
10405 **Automotive Body Repair & Painting**
10410 **Automotive Brake Manual**
10415 **Automotive Detaiing Manual**
10420 **Automotive Eelectrical Manual**
10425 **Automotive Heating & Air Conditioning**
10430 **Automotive Reference Manual & Dictionary**
10435 **Automotive Tools Manual**
10440 **Used Car Buying Guide**
10445 **Welding Manual**
10450 **ATV Basics**

### SPANISH MANUALS
98903 **Reparación de Carrocería & Pintura**
98905 **Códigos Automotrices de la Computadora**
98910 **Frenos Automotriz**
98915 **Inyección de Combustible 1986 al 1994**
99040 **Chevrolet & GMC Camionetas** '67 al '87 Incluye Suburban, Blazer & Jimmy '67 al '91
99041 **Chevrolet & GMC Camionetas** '88 al '95 Incluye Suburban '92 al '95, Blazer & Jimmy '92 al '94, Tahoe y Yukon '95
99042 **Chevrolet & GMC Camionetas Cerradas** '68 al '95
99055 **Dodge Caravan & Plymouth Voyager** '84 al '95
99075 **Ford Camionetas y Bronco** '80 al '94
99077 **Ford Camionetas Cerradas** '69 al '91
99083 **Ford Modelos de Tamaño Grande** '75 al '87
99088 **Ford Modelos de Tamaño Mediano** '75 al '86
99091 **Ford Taurus & Mercury Sable** '86 al '95
99095 **GM Modelos de Tamaño Grande** '70 al '90
99100 **GM Modelos de Tamaño Mediano** '70 al '88
99110 **Nissan Camionetas** '80 al '96, **Pathfinder** '87 al '95
99118 **Nissan Sentra** '82 al '94
99125 **Toyota Camionetas y 4Runner** '79 al '95

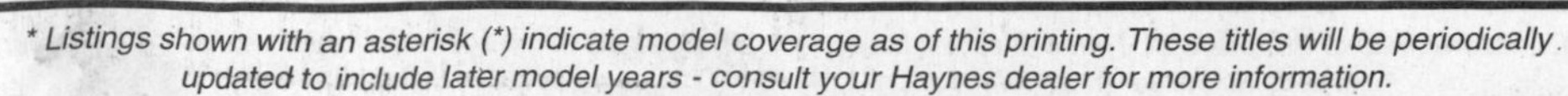
** Listings shown with an asterisk (*) indicate model coverage as of this printing. These titles will be periodically updated to include later model years - consult your Haynes dealer for more information.*

Over 100 Haynes motorcycle manuals also available

5-98